THE PICTURE OF THE TAOIST GENII PRINTED ON THE COVER
of this book is part of a painted temple scroll, recent but traditional, given to
Mr Brian Harland in Szechuan province (1946). Concerning these four divini-
ties, of respectable rank in the Taoist bureaucracy, the following particulars
have been handed down. The title of the first of the four signifies 'Heavenly
Prince', that of the other three 'Mysterious Commander'.

At the top, on the left, is Liu *Thien Chün*, Comptroller-General of Crops and
Weather. Before his deification (so it was said) he was a rain-making magician and weather forecaster named Liu Chün, born in the Chin dynasty about
+340. Among his attributes may be seen the sun and moon, and a measuring-
rod or carpenter's square. The two great luminaries imply the making of the
calendar, so important for a primarily agricultural society, the efforts, ever
renewed, to reconcile celestial periodicities. The carpenter's square is no ordinary tool, but the gnomon for measuring the lengths of the sun's solstitial
shadows. The Comptroller-General also carries a bell because in ancient and
medieval times there was thought to be a close connection between calendri-
cal calculations and the arithmetical acoustics of bells and pitch-pipes.

At the top, on the right, is Wên *Yuan Shuai*, Intendant of the Spiritual Officials
of the Sacred Mountain, Thai Shan. He was taken to be an incarnation of one
of the Hour-Presidents (*Chia Shen*), i.e., tutelary deities of the twelve cycli-
cal characters (see Vol. 4, pt. 2, p. 440). During his earthly pilgrimage his name
was Huan Tzu-Yü and he was a scholar and astronomer in the Later Han (b.
+142). He is seen holding an armillary ring.

Below, on the left, is Kou *Yuan Shuai*, Assistant Secretary of State in the
Ministry of Thunder. He is therefore a late emanation of a very ancient god,
Lei Kung. Before he became deified he was Hsin Hsing, a poor woodcutter,
but no doubt an incarnation of the spirit of the constellation Kou-Chhen (the
Angular Arranger), part of the group of stars which we know as Ursa Minor.
He is equipped with hammer and chisel.

Below, on the right, is Pi *Yuan Shuai*, Commander of the Lightning, with his
flashing sword, a deity with distinct alchemical and cosmological interests.
According to tradition, in his early life he was a countryman whose name was
Thien Hua. Together with the colleague on his right, he controlled the Spirits
of the Five Directions.

Such is the legendary folklore of common men canonised by popular acclama-
tion. An interesting scroll, of no great artistic merit, destined to decorate a
temple wall, to be looked upon by humble people, it symbolises something
which this book has to say. Chinese art and literature have been so profuse,
Chinese mythological imagery so fertile, that the West has often missed other
aspects, perhaps more important, of Chinese civilisation. Here the graduated scale of Liu Chün, at first sight unexpected in this setting, reminds us of
the ever-present theme of quantitative measurement in Chinese culture; there
were rain-gauges already in the Sung (+12th century) and sliding calipers in
the Han (+1st). The armillary ring of Huan Tzu-Yü bears witness that
Naburiannu and Hipparchus, al-Naqqash and Tycho, had worthy counterparts in China. The tools of Hsin Hsing symbolise that great empirical tradition which informed the work of Chinese artisans and technicians all through
the ages.

SCIENCE AND CIVILISATION
IN CHINA

Wang Shou was travelling along with written materials on his back. At a big crossroads he caught sight of Hsü Fêng. Said Hsü Fêng: 'Conduct consists of actions. Actions arise from circumstance. The person who knows has no constant pattern of conduct. Books consist of sayings. Sayings arise from knowing. Therefore the knowing person does not keep written materials. Why are you travelling along with these things on your back?' At this point Wang Shou burnt his written materials and danced round the bonfire.

HAN FEI, 21–15–1

中國科學技術史

李約瑟著

冀朝鼎

JOSEPH NEEDHAM

SCIENCE AND CIVILISATION IN CHINA

VOLUME 7

PART I: LANGUAGE AND LOGIC

BY

CHRISTOPH HARBSMEIER

PROFESSOR, DEPARTMENT OF EAST EUROPEAN AND
ORIENTAL STUDIES, UNIVERSITY OF OSLO

EDITED BY
KENNETH ROBINSON

CAMBRIDGE
UNIVERSITY PRESS

PUBLISHED BY THE PRESS SYNDICATE OF THE UNIVERSITY OF CAMBRIDGE
The Pitt Building, Trumpington Street, Cambridge CB2 1RP, United Kingdom

CAMBRIDGE UNIVERSITY PRESS
The Edinburgh Building, Cambridge CB2 2RU, United Kingdom
40 West 20th Street, New York, NY 10011-4211, USA
10 Stamford Road, Oakleigh, Melbourne 3166, Australia

First published 1998

Printed in the United Kingdom at the University Press, Cambridge

Typeset in Baskerville MT 11.25/13pt, in Quark XPress™   [GR]

A catalogue record for this book is available from the British Library

ISBN 0 521 57143 X hardback

Joseph Needham
(1900–1995)

SCIENCE AND CIVILISATION IN CHINA

"Certain it is that no people or group of peoples has had a monopoly in contributing to the development of Science. Their achievements should be mutually recognised and freely celebrated with the joined hands of universal brotherhood"

Science and Civilisation in China VOLUME I, PREFACE.

*

Joseph Needham directly supervised the publication of 17 books in the *Science and Civilisation in China* series, from the first volume, which appeared in 1954, through to volume 6.3, which was in press at the time of his death in March 1995.

The planning and preparation of further volumes will continue. Responsibility for the commissioning and approval of work for publication in the series is now taken by the Publications Board of the Needham Research Institute in Cambridge, under the chairmanship of Dr. Christopher Cullen, who acts as general editor of the series.

To

ANGUS CHARLES GRAHAM

8 July 1919–26 March 1991

philosopher, poet, teacher and friend
this volume is respectfully
dedicated

CONTENTS

LIST OF ABBREVIATIONS

The following abbreviations are used in the text. For abbreviations used for journals and similar publications in the bibliographies, see pp. 422ff.

LSCC	*Lü Shih Chhun Chhiu*
NP	*Nyayapraveśa*
SKCS	*Ssu Khu Chhüan Shu*
SPPY	*Ssu-pu-pei-yao*
SPTK	*Ssu-pu-tshung-khan*
WT	*Wen-ti* (Yale University)
WYWK	*Wan-you-wen-khu*

ACKNOWLEDGEMENTS

This book owes its inception to Janusz Chmielewski of Warsaw University, who suggested, conceived, planned and began it many years ago. I only wish I could have written a section more worthy of its originator.

Kenneth Robinson has edited this volume, and has given me invaluable help and encouragement at every stage, far beyond the call of duty. His loyal friendship and hospitality over the years has been crucial and wonderful.

Geoffrey Lloyd, Cambridge, has provided always incisive and often crucial comments on every single sub-section in this book.

R. E. Allinson, Hong Kong; Chad Hansen, Virginia; D. Kehren, Bonn; Gregor Paul, Osaka, have all given generous comments on early drafts. Robert Wardy has kindly re-read, and commented on, the final draft.

Yu Min of the Normal University in Peking has given detailed advice on the philological sub-sections.

I also wish to thank the following for their help on specific points: M. Broido, Oxford; D. D. Daye, Boulder Green; G. Dudbridge, Oxford; S. Egerod, Copenhagen; H. Eifring, Oslo; D. Føllesdal, Oslo; Nathan Sivin, Pennsylvania.

I owe the Institute of East Asian Philosophies, Singapore, a profound debt of gratitude for their generous support during a six-month fellowship there in 1986, and particularly to the superb librarians Lilian Phua and Li Kin-seng of that institution where I wrote the first draft for this book, the writing of which was finished in 1988. During the post-final stages I enjoyed the most congenial environment of the Institute for Advanced Study, Berlin, to see the book through its final stages of preparation.

Above all, I owe a profound debt of gratitude to the late A. C. Graham, from whom I have constantly learnt for almost a quarter of a century, and who has given his very detailed advice on several complete drafts of this book. In 1986 I spent six crucial months on this book as a fellow of the Institute of East Asian Philosophies, Singapore, in close collaboration with him. Angus Graham died as this book neared completion. In deep sorrow I dedicate this volume to the memory of this great philosopher and sinologist whose friendship had sustained me for more than twenty years.

Material from the present book has been used in several articles I have published over the years:

'The mass noun hypothesis and the part–whole analysis of the White Horse Dialogue', in H. Rosemont, ed., *Chinese Texts and Philosophical Contexts. Essays dedicated to A. C. Graham* (La Salle: Open Court, 1991), pp. 49–66.
'Marginalia sino-logica', in R. Allinson, ed., *Understanding the Chinese Mind: The Philosophical Roots* (London: Oxford University Press, 1989), pp. 59–83.

'Conceptions of knowledge in ancient China', in H. Lenk and G. Paul, eds., *Epistemological Issues in Classical Chinese Philosophy* (New York: SUNY Press, 1993), pp. 11–31.

'John Webb and the early history of the study of the classical Chinese language in the West', Proceedings of the International Conference on the History of Sinology (Taipei, London: Han Shan Tang, 1995), pp. 297–338.

Considering the extraordinarily generous instruction and help I have received from Angus Graham, and from so many other distinguished people, I sombrely reflect that even twelve years ago I should have been able to write a much better survey than the one I am able to present here. The responsibility for this failure remains entirely my own.

FOREWORD

Since the winter of 1986 we have been installed in our new and permanent building, of which the East Wing was built primarily by contributions from Hongkong (through our Trust there, of which the chairman is Dr Mao Wên-Chhi), while the Centre Block is owing to the generosity of Tan Sri Tan Chin Tuan of the Overseas Chinese Banking Corporation of Singapore. The architect was Christophe Grillet of Caius and the builder was Roger Bailey of Messrs Johnson & Bailey, Ltd. The building has vermilion pillars (like a Chinese Temple) and there is white Chinese lattice-work in the veranda railings. The building has won several awards, notably one from the Royal Institute of British Architects; others for the excellence of its brickwork and its woodwork. It stands at the corner of Herschel Road and Sylvester Road, Cambridge, on land belonging to Robinson College. It is in this new building that we come at last to Volume 7.

More than forty years have passed since Volume 1 in this series was published. We imagined then that Volume 7 would be the seventh in a series of only seven slim volumes, and not the last in a series of thirty. But so great was the wealth which we discovered inside the Chinese Treasure Mountain that our original plans had to be adapted. From the main shafts which have been sunk, subsidiary tunnels now run in all directions. Inevitably some lines of exploration have had to be abandoned; with others work is postponed, and sometimes the order in which work had been planned for publication has had to be altered. Volume 7 more than any other has been subjected to the pressures of time and change, but it is with great pleasure that I now present Volume 7, part 1.

For the support of our research and writing, we must mention the National Science Foundation (USA), the Mellon and the Luce Foundations, and the National Institute for Research Advancement (Japan). Without their continued support, our work would have been impossible. Most of their help was mediated through our New York Trust (Chairman Mr John Diebold); and this is also deserving of our warmest thanks for obtaining from the Kresge Foundation a grant of US$150,000 towards the building of the South Wing of our Institute.

In Volume 2 we wrote in large measure about the impact of Chinese people, societies and philosophies on the growth of science. Thereafter we wrote about specific fields of Chinese scientific thought – in mathematics, astronomy, meteorology, geography and cartography, geology, seismology, mineralogy, and then, turning to physics, on heat, light, sound, magnetism and electricity, and, in a new set of volumes, to engineering – mechanical, civil and nautical. This was to be followed by things chemical in several volumes, some of which are not yet completed, on military and textile technology, and on paper and printing, Chinese inventions which were of decisive importance in the growth of modern science. Then came the great effort to disentangle Chinese alchemy and the beginnings of modern chemistry, which required four volumes. These were followed by specific applications of chemical discoveries to technology, such as the making and use of gunpowder, and the use of steroids in medicine.

In what had originally been thought of as a slim Volume 6 we now have or shall soon have several volumes on such subjects as botany and agriculture with their related studies and technologies, to be followed by contributions on things medical and pharmaceutical.

All these were reasonably specific and objective. But now, in Volume 7, we return to that most proper study of mankind, namely human speech and thought processes, as they occurred in China.

I had originally arranged with the distinguished logician, Janusz Chmielewski of Warsaw University, to write a part of Section 49 in the seventh volume of *Science and Civilisation in China*, covering Chinese language and Chinese logic, as we then expressed it. But by September 1983 he had made it clear to us that his own failing eyesight, the illness of his wife, and the sheer physical difficulties of living in Poland at that time, made it impossible for him to go beyond the first two chapters which he had by then written. This was but the first of many setbacks we received in the writing of Volume 7. Janusz Chmielewski did, however, recommend that the work should be entrusted to Christoph Harbsmeier. We gratefully followed up this suggestion. Christoph went to see Janusz, and a smooth and cordial handover was arranged, which the author of this volume refers to on p. 1. We were very sorry that Janusz was unable to bring to a successful conclusion the work to which he had devoted much precious time, but are most grateful to him for ensuring its continuation in the hands of his gifted pupil.

Christoph Harbsmeier, in writing on Language and Logic in Traditional China, has given us a contribution which is not only erudite in its ability to draw together East and West, but also stimulating and entertaining. He feels that what logic amounts to is the history of the philosophy of science, and that this should be recognised as central to the intellectual scheme in *Science and Civilisation in China*.

In the pages which follow the reader will find many common preconceptions challenged. Early Chinese, for example, was an isolating, but not an isolated language. Literary Chinese was no vague and poetic language unsuitable for science, provided it was used by a competent scientific thinker. Nor were the Chinese uninterested in logic, explicit or implicit. Twice the Chinese appear to have been interested in explicit logic for its own sake, once in the Mohist School, and once again with the Chinese Buddhist commentators, in whose logical minds the ambiguities of Sanskrit were resolved when translated into Chinese. But the Chinese were always more interested in the truth on which assumptions were based than on the verbal machinery for developing these assumptions. Explicit logic did not therefore have that continuously sustained interest which it has received in the West.

Many readers will be anxious to know how Chinese compares for clarity with, say, classical Greek, and will turn to Section (*c*,6) on 'Complexity', where translations from Plato's works into Literary Chinese by skilled translators are compared. It would be wrong of me here, however, to anticipate the author's findings.

Granted that Literary Chinese was capable of expressing scientific ideas, what actually happened when it was so used? This, as the Americans would say, is a whole new ball-game for which the reader will have to await a subsequent volume.

JOSEPH NEEDHAM

Cambridge
9 May 1994

AUTHOR'S NOTE

The Chinese were superficial – *out of profundity.*[1]

There is only one culture in the world which has developed systematic logical definitions and reflections on its own and on the basis of a non-Indo-European language. This is the Chinese culture. The history of logical reflection in China is therefore of extraordinary interest for any global history of logic and hence for any global history of the foundations of science.

It has become an unquestioned assumption that there is one golden age of rational logical inquiry in China: the −5th to −3rd centuries, the later part of the Warring States period. The present volume will correct this assumption. It will demonstrate that there are two such periods of intense logical activity in China, by far the most productive being the +7th and early +8th centuries, where Buddhist logic made tremendous headway in China and produced a significant literature both in terms of sheer size and of philosophical quality. Buddhist logic was Indian in origin, and the Chinese language turned out to be no obstacle for a remarkable intellectual efflorescence of Buddhist logic in China. For an empirical global account of the natural relations between logical theorising and the language in which this theorising is going on, again, the case of China is of unique importance.

Chinese Buddhist logic remained limited to a small subculture. This is why it could be so largely overlooked by students of Chinese intellectual history. But then, even the Later Mohist achievements of the Warring States period, for all their sparkling intellectualism on many points represented a highly marginal subculture. The Later Mohist texts have been transmitted to us in a shape that bears witness to the fact that the Chinese transmitters did not understand much of what they were transmitting.

The Later Mohists and sophists like Kungsun Lung did have a certain influence on the philosophy of their times, but on the whole their achievements were not absorbed into the mainstream of Chinese intellectual culture. Mohist ethics maintained at least some, albeit highly marginal, presence, Mohist logic was simply forgotten. Buddhist logic, though not forgotten in Buddhist circles, was disregarded. It never became a central discipline for large numbers of Chinese Buddhists as it did among Tibetan and Indian Buddhists.

One cannot emphasise enough: logic, no matter whether Buddhist or Mohist, remained marginal in Chinese culture until modern intellectuals needed to demonstrate that China had its own logical traditions, just like the West. Chinese logic was rediscovered in an attempt to prove that China was the intellectual equal of the West. The result was a large modern indigenous literature on indigenous Chinese logic.

---

[1] Nietzsche, *Die fröhliche Wissenschaft*, end of the *Vorrede: Die Griechen waren oberflächlich – aus Tiefe*!

At a very elementary level, logical thought originates as a theoretical concern with certain logically crucial words such as negations like 'not' and various logical conjunctions and quantifiers like 'or' or 'all'. The Chinese clearly developed such specific logical concerns. From a global perspective, the Chinese general comments on their own language (*b*,3) are of concern because there are very few cultures indeed whose comments on their own languages are not directly or indirectly inspired by the Greek and Latin or the Sanskrit tradition. The Chinese are also the only people operating with a non-Indo-European language who developed an entirely indigenous interest in some grammatical features of their own language: a global history of grammar would have to accord a central place to Chinese traditional grammar as the only non-Indo-European pre-modern tradition. Chinese civilisation is the only non-Indo-European civilisation in the world which has developed independently of outside influences an indigenous and powerful lexicographic tradition and a sustained systematic interest in the definition of terms. Chinese interest in the lexicon of their own language is paralleled in Europe by that of the French in more modern times. I trace the Chinese development in Sections (*b*,4) and (*b*,5).

The significance of the lexicographic tradition for the study of the history of science is evident: the dictionaries and the definitions are a main source of our knowledge of the Chinese conceptual world.

Many exotic logical features have been claimed for the Chinese language throughout the ages. For example, it has been claimed that negation somehow does not have its full logical force in Chinese. In fact it turns out that such illogicalities as cumulative negation and such phrases as 'all that glitters is not gold' are in fact quite alien to Chinese grammar, and if we are to compare logicality with respect to negation, then it turns out that Chinese is very considerably more logically transparent and rigid than ancient Greek (*c*,1). It has been claimed that Chinese is very strange in lacking proper word classes, but Shakespearean English turns out to be very close indeed in its treatment of word classes to Classical Chinese (compare 'but me no buts'), and in any case there is a sound basis for distinguishing the functional properties of ancient Chinese words, not to speak of modern ones. They often even have morphological word class characteristics (*c*,4). The stylistic differences in explicitness in Greek versus implicitness in ancient Chinese are real enough, but they are only matters of degree, and it must be pointed out that Greek can be elliptic in many places where Classical Chinese cannot (*c*,5). I find that there are indeed many semantic/logical configurations that are perfectly possible but cannot be represented in Classical Chinese. But it turns out that these do not generally seem to be essential for the articulation of scientific thought. The syntax of Classical Chinese turns out to be rich enough to express the thought of Plato, though it is poor enough to necessitate considerable syntactic and logical revamping (*c*,6).

The history of logical concepts in China has, in recent years, received much more attention than the grammatical and linguistic problems introduced so far, and this area remains controversial. A. C. Graham remarked that 'it is perhaps inevitable that in the no-man's-land on the common borders of linguistics, philosophy, and sinology, among those from whom one looks for stimulating new approaches (Rosemont, Hansen, Hall) most generalisations about the Chinese language start from totally obsolete assumptions'. Graham goes on to give an example: 'To speak of Chinese sentences as "strings of names" is to revert to the grammatical knowledge of the ancient Chinese themselves (and not quite catching up with *Names and Objects*).' Graham argues: 'But a claim that, for example, there are no sentences in English, offered without even an alternative account of the grammatical differences between the so-called "sentence" and the nominalised clause, would be meaningless; why is the claim supposed to be meaningful if the language is Chinese?'[1]

Section (*d*) in the present work is a systematic study of some logically fundamental concepts like that of a sentence, of meaning, truth, and necessity, and some problems of their application to Chinese. I argue that there is as much – or if you prefer: as little – of a concept of a sentence in Chinese as there is in English or Greek (*d*,1). I show that the concept of truth is manifestly present, although it is indeed not regarded as all-importantly central to philosophy which in China has always been predominantly pragmatic in orientation (*d*,3). There is even some truly surprising evidence of the development of an abstract notion of a class, of subsumption, and of an abstract property (*d*,6)–(*d*,8). Moreover, there is no reason to attribute to the ancient Chinese an exotic absence of a notion of propositional knowledge. On the other hand it is quite correct to point out that the ancient Chinese words within the semantic field of knowledge – like their ancient Greek counterparts – also mean knowledge by acquaintance and 'knowing how to' (*d*,9). On close examination the Chinese language turns out to be less exotic than might appear from the current literature.

It turns out that the Chinese language is reasonably well equipped to express rational argumentation, essential to science, and the ancient Chinese have many current forms of argument in common with their contemporary Greeks (*e*,1)–(*e*,2).

My survey of the Chinese tradition of disputation and logic shows that – like its Greek counterpart – the Chinese tradition owed much to the legal tradition on the one hand and to intellectual entertainment on the other. Creative frivolity is shown to have been a crucial factor in the thought of Kungsun Lung (*f*,2). The intellectual response to logic by the culture at large is crucial for our appreciation of the role of logic in Chinese culture. In China, as in Greece, the response was predominantly negative, but unlike in Greece, even the argumentative philosophers in ancient China did not systematically deploy the insights of the logicians and their techniques in other areas than those of formal logic (*f*,5).

Finally, I present a critical survey of the system of Chinese Buddhist logic and of its terminology (*g*,2). Discussing the problems of Hsüan-Tsang's translation from the Sanskrit original into Chinese I argue that these translations make crucial

[1] Graham (1989), pp. 392 and 394.

additions to the Sanskrit original and thus contribute significantly to the progress of Buddhist logic (*g*,4). Finally, I have found it useful to highlight the contrast between the formal character of Aristotle's syllogistic and the more rhetoric-based system of Indian logic which should be viewed as a formal logic of dialogue and disputation rather than of formal logical relations as such.

It will be noticed that this volume is almost entirely concerned not with the history of science, but with the comparative history of the foundations of science. The scientific texts investigated are those concerned with logic and linguistic analysis. The detailed impact of Chinese approaches to and experiences of logic on the practice of concrete sciences originally, for good reasons, was intended to form part of this volume, which will be treated in a separate volume by another author.

The highly abstract and complex issues raised in this volume are, to use Winston Churchill's famous phrase from 1939, 'a riddle wrapped in a mystery inside an enigma'. I have looked for a key in a combination of philology and logical analysis. Now, many years after the writing of this book, the result of my efforts leaves me diffident in more ways than one. I must now quote from one of Churchill's speeches:

Writing a book is an adventure: to begin with it is a toy. And an amusement. And then it becomes a mistress. And then it becomes a master. And then it becomes a tyrant. And the last phase is that just about as you are reconciled to your servitude, you, you – kill the monster. Hm. And fling him about to the public.

In my present case I have found the last clause, after the hesitation, is the most problematic part. The fate of the present book was not so simple, after the killing of the monster: the thing – un-flung to any public wider than that of the Needham Research Institute and a few friends and colleagues – became a ghost, waiting for a companion to form a volume. And then a forgotten skeleton in a drawer. And just as I was reconciled to my indifference, then I had to – revive the skeleton. Now, in the end I fling him about to the public, diffidently. Hoping that the gentle readers will not see as many juvenile weaknesses in it as its author does.

The main part of this book was written in 1985, in the most congenial company of Angus Graham who had invited me to join him at the Institute of Far Eastern Philosophies, Singapore. Throughout its incubation it has benefited tremendously from the editorial help of Kenneth Robinson and, during the final stages, Christine Salazar. The manuscript was finished in 1988, after which date I have only been able to make very minor changes.

49. LANGUAGE AND LOGIC IN TRADITIONAL CHINA

(a) METHOD

(1) METHODOLOGICAL REMARKS[1]

The theory and practice of science and technology are inextricably bound up with language and logic. Scientific insights become transmittable cultural heritage to the extent that they are articulated in language. The insights add up to a scientific explanation to the extent that they are organised into a coherent argument. The explanations add up to a scientific system to the extent that they are organised into a general logical scheme.

In this Section I shall address three main questions:

1. What are the basic logical and linguistic features of the Literary Chinese language that enabled it to serve as a medium for scientific discourse? I shall try to answer this question in terms of grammar and conceptual history.
2. What (if any) were the strategies of argumentation and proof employed by the ancient Chinese?[2] I shall try to answer this question by a logical examination of ancient Chinese arguments.
3. What (if any) were the main logical theories advocated by the ancient Chinese before the impact of Western philosophy on science and logic? I shall summarise the history of logic in China.

To some extent there will be a temporal progression from this Sub-section to the next. Here I shall be mainly concerned with the period from Confucius (−551 to −479) to Wang Chung 王充 (+27 to +100).[3] As far as indigenous Chinese logic is concerned this turns out to be the only really productive period. Moreover, insofar as Classical Chinese remained the main focus of intellectual attention among all later Chinese thinkers (except for some Buddhists) and insofar as the Classical Chinese literature defined a large part of the basic linguistic and logical conceptual framework for later Chinese thought, it deserves our special attention.[4]

---

[1] This Section owes its inception to Janusz Chmielewski of Warsaw University, who suggested, conceived, planned and began it many years ago. In taking over from him, I have tried to live up to his high standards of philological and logical rigour, and I have occasionally drawn on his notes. I only wish I could have written a section more worthy of its distinguished originator.

[2] It will be found that the emphasis in this section is largely on the period from Confucius's time to the −2nd century. To some extent this simply reflects the specialisation of the present writer and the limitations of his knowledge. On the other hand it also seems right to concentrate on the most formative and creative period in Chinese intellectual history. In any case, I have done my best to cover important later developments as well, such as Buddhist logic of the +7th century.

[3] The following Section will deal mainly with later developments.

[4] Our enterprise is not uncontroversial. Marcel Granet, in his 'Quelques particularités de la langue et de la pensée chinoise' argues forcefully for the unique and special character of the Chinese language and of Chinese thought which make Chinese unsuitable for an analysis in Western logical terms.

Just as we should try to learn something about the relevant parts of optics or biology in general before we expound the history of optics or biology in China, so in the case of logic and the philosophy of language we should feel obliged to familiarise ourselves with some basic relevant parts of logic before we embark on the study of the history of logic in China.

Basic modern logic and analytical philosophy provide us with the best theoretical account to date of the complex relations between language and logic. This is essential in order to study the particular case of language and logic in ancient China.[1]

I shall aspire to treat ancient Chinese ways of thinking about language and logic entirely on the basis of ancient Chinese texts which I present in English translations.[2] I shall try to avoid saying anything whatsoever about ancient Chinese language and logic which I cannot or do not either document through quotation or demonstrate by the absence of certain phenomena in the texts we have. I am not abstractly philosophising about Chinese texts. I am practising the kind of logical *philology* which interprets, translates and explains relevant Chinese texts in the light of what is known about language and logic in general and Classical Chinese grammar and philology in particular.[3]

Every translation, however well founded it may be on the necessary detailed grammatical and philological research, will inevitably have to beg a host of questions of interpretation. Translations are never evidence. They embody interpretations that are open to questioning. They are part of the argument and must be treated as such, with the appropriate critical awareness that there may always be alternative translations.

In translating Chinese texts I apply modern principles of Western linguistics, logic and analytical philosophy. In this sense I am always begging questions concerning the 'prelogicality' of the Chinese. I am assuming that the Chinese, when properly understood, did make logical sense. I am in sympathy with W. V. O. Quine

---

Both in China and in the West there have been those who, like Lionel Giles in 1945, would want to claim 'Chinese cannot really be said to have any grammar at all.' (BSOAS 11 (1943–6), p. 236). In 1986 Carl B. Becker published an article 'Reasons for the Lack of Argumentation and Debate in the Far East'.

In 1939 Chang Tung-Sun shrank from denying that the Chinese had logic. Instead, he set up a curiously opaque distinction between four types of logic: mathematical logic (applicable to mathematics only), formal logic (applicable to Indo-European languages only), metaphysical logic (typical of Indian thought), and socio-political logic (characteristic of the Chinese mind). Chang argues that Chinese thought cannot be placed in the Western logical framework.

[1] In teaching basic formal logic I have found most useful E. J. Lemmon (1965) *Beginning Logic* and the more advanced Benson Mates (1968) *Elementary Logic*, which includes a concise survey of the history of logic in the West. For the more philosophical aspects of the matter P. F. Strawson (1952) *Introduction to Logical Theory* and the more advanced W. V. O. Quine (1970) *Philosophy of Logic* provide useful introductions.

[2] In so doing I am aware that I will still be able to put forward in each case only *one* of several possible interpretations of the texts I present. Moreover, there is a sense in which all modern interpretations of ancient texts are inevitably anachronistic.

[3] If, to some, the word 'philology' sounds old-fashioned and out of date like 'classical philology', I can only plead guilty on all charges and admit that the high standards of Latin and Greek philology have indeed been an ideal I have aspired to live up to. As an exemplification of the standards I have set myself I refer to the book *Die Aristotelische Syllogistik, Logisch-philologische Untersuchungen über das Buch A der 'Ersten Analytiken'* (3rd. ed. 1969, English translation *Aristotle's Theory of the Syllogism*. (1969) by my philosophical mentor Günther Patzig, and to the article 'Syllogistics' in *Encyclopedia Britannica*, 15th ed., 1982 vol. 17, pp. 890–8, by the same author.

when he remarks: 'Wanton translation can make natives sound as queer as one pleases. Better translation imposes our logic upon them, and would beg the question of prelogicality if there were a question to beg.'[1]

However, it turns out that one can make good logical sense in radically different ways, and we need to find out exactly how different the Chinese ways were from ours. Whatever the needs of idiomatic translation into English, one may still wonder whether widely different cultures may not each carry their own logical notions (as they may carry their own notions of gods), or whether logic is something trans-cultural (like arithmetic, where cultures only differ in their strategies for representing the same arithmetic truths). One may wonder whether the apparent 'logicality' of the translations from the Chinese is indeed only the result of the sort of interpretive imposition Quine speaks of.[2]

I do rather expect that people who have learnt to count, in all cultures and at all times, will agree that two apples added to two apples make four apples, no matter how they may express this thought in their language. But do people of all cultures and times have to agree that a statement cannot be at the same time true and untrue? Do the Chinese, for example, have a notion of a statement and of truth in the first place?[3]

By introducing Coca-Cola in all the markets of the world we impose a culture-bound American taste and replace indigenous tradition. By teaching logic in all the universities of the world, are we imposing a culture-specific logical scheme on peoples who have independent, different, but equally sound, logical schemes of their own? By interpreting ancient Chinese texts in terms of modern formal and philosophical logic, are we imposing on Chinese thought an intellectual scheme which is so alien to the Chinese that they might not even know whether to disagree with it?

Marcel Granet has been the foremost modern proponent of the relativist position, deploring the application of even elementary 'Western' logic to Chinese. Jacques Gernet (1983) has argued eloquently that Chinese categories of thinking are fundamentally different from Western ones, thus making effectively impossible the spread of Western Christianity to China.

On the other hand Chinese historians of philosophy like Hu Shih (as well as his critics like Fêng Yu-lan) take the opposite view by not only attributing 'Western' logic to the ancient Chinese, but by even interpreting most of Chinese intellectual

---

[1] Quine (1969), p. 58.

[2] In one sense one can say that even when members of our own culture seem to make perfectly good and ordinary logical sense, appearances may be deceptive. Such members of our own society may theoretically just have developed an uncannily pervasive habit of behaving linguistically *as if* they had my basic notions of logic when in fact they live in their own totally different logical universe. Since their behaviour would be so pervasively deceptive, we have no way of ever finding out. Similarly, we have no way of *proving* that the ancient Chinese did not really live in a totally different logical universe. But I propose to demonstrate inductively that they certainly often wrote as if they didn't. Indeed, as for example in the case of multiple negation, we shall find that the ancient Chinese adhered more rigidly to logical hygiene than their Greek contemporaries, or their English-speaking successors.

[3] We shall take up these questions in detail in Sections (d,1) and (d,3) below.

history through predominantly Western categories of thought. Hu Shih calls his history of Chinese philosophy *The Development of Logical Method in Ancient China*, and by 'logical method' he understands something close to Western logical methods.

These approaches to the problem of logic and culture may seem irreconcilable, but they each make an important point. Jacques Gernet and his predecessors are right that Chinese ways of articulating and structuring thought are profoundly different from Greek and European ones. However, in using their very different linguistic and intellectual strategies, Aristotelians as well as Confucians will still use essentially the same kinds of logical tools, such as negations, conditional statements and quantifiers like 'some' and 'all'. There is nothing Western about negation, conditionals or quantifiers any more than there is anything Western about the numbers 'one' and 'two'. I shall not argue this philosophical point in detail here, but a comparison of logic with arithmetic may help to clarify the approach I shall follow in this Section.

Notice first that notions of the Holy Trinity and the like do not belong to the realm of arithmetic. We cannot point to the notion of the Trinity and argue that the notion of 'three' is culture-specific. Certainly, different cultures do different things with the notion 'three', but they do their different things with fundamentally the same notion of 'three' which is systematically treated for example in 'Western' number theory. Plato, we are told, regarded the ability to count as one of the defining general characteristics which give man his dignity.[1]

Some peoples might, I suppose, not be able to count as far as three, counting 'one', 'two', 'many'. (Indeed, as Yü Min from Peking has pointed out to me, the early meaning 'many' for *san* 三 may possibly be a fossil of an earlier stage of Chinese when indeed the Chinese counted this way.) If there really were such people I should simply say that they are at a certain arithmetical disadvantage.[2]

Let us now turn to the more controversial subject of logic. Modern formal logic[3] can be construed as the study of the foundations of mathematics (including number theory among many other things). Bertrand Russell and Alfred North Whitehead's classic *Principia Mathematica* (1910) attempts to deduce the truths of mathematics from some principles of logic. They did not quite succeed, but on the other hand they showed in detail how the propositions of logic may usefully be considered as propositions belonging to the field of the foundations of mathematics, and that is the point that matters for us.

Just as one can study not only number theory but also the anthropology of counting, so one can study the anthropology of logical articulation that has to do with the logic of sentences (propositional logic), and the logic of properties and relations

---

[1] Aristotle, *Problemata* 30.6, 956a11ff.

[2] Whether a people express 'three' in a binary or a decimal system is interesting, but of little consequence in our context. What cultural significance a population attaches to the notion of 'three' is again historically very interesting, but it is mathematically irrelevant. It makes sense to detach the abstract arithmetic notion of 'three' as universal from its culture-specific cultural contexts such as that of numerology or the Holy Trinity.

[3] Modern formal logic must, of course, be distinguished from traditional Aristotelian logic, which raises entirely different problems.

(first order predicate logic).[1] In propositional logic we study logical sentence connectives, connectives naturally (but only approximately) expressed in English by such words as 'and', 'if', 'if and only if', 'either . . . or', and propositional logic makes these concepts precise by interrelating them through formal definitions of logical relations between propositions p and q which we could translate roughly as:

Definition: If p then q = either not $-p$ or q.

In first order predicate logic we study logical relations between predicates (like 'be a number') symbolised by letters like 'F', and relations (like 'be larger than') symbolised by letters like 'R', and such more complicated logical concepts as quantifiers, naturally expressed in English by such words as 'all', 'some' and the like. We can try to interrelate these concepts by definitions which we can translate informally as:

Definition: For all x: Fx = Not (for some x: not Fx)

(I.e., 'all things are F' is defined as 'it is not the case that there are some things which are not F'.)

As anthropologists of logic we can then go on to study the ways in which various peoples articulate or have articulated logical relations between sentences and between predicates (just as anthropologists of mathematics may study the way different peoples have articulated notions like that of addition, division, etc.). We then need to investigate such questions as whether these peoples employ sentences (and can distinguish these from other strings of morphemes), whether they apply truth predicates like the English 'is true' to sentences (or whether they manage without such truth predicates), whether they conceive of sentences as things that one can believe or know to be true (or whether they think along entirely different lines).

A people might, I suppose, turn out to have no word for 'not' and not to be able to make explicit any logical relations between propositions in any way. Indeed, they might be unable to distinguish between sentences and other strings of morphemes. They might have no words for 'all' and 'some'. In that case I should simply find them to be at a certain logical disadvantage. I have yet to hear of any such languages. I propose to demonstrate that *Classical Chinese is in any case certainly not such a language.*

I read in the *Corpus Hermeticum* (+1st to +3rd centuries) that thought and language are sisters, or rather instruments for each other.[2] Language cannot be used in separation from thought, nor can thought manifest itself in separation from language. I shall show that Classical Chinese is indeed an efficient instrument for the articulation of logically structured thought. Languages generally do have expressions or constructions which we are tempted to translate as 'not', 'all', 'is true' and the like, but these words will often have widely different non-logical semantic connotations

---

[1] The most interesting way of proving that this can be done is to go ahead and do it as best one can. In what follows I shall attempt this form of proof, taking Classical Chinese as an example.

[2] Nock and Festugière (1960), vol. 1, p. 97.

and are subject to non-logical grammatical restrictions.[1] I shall not be outraged by the fact that the ancient Chinese had half a dozen common negations. Neither shall I be outraged by the fact that they happen to express the thought 'either p or q' by saying something like 'if not–p then q (*fei* 非 p *tsê* 則 q)'.

It may be tempting to attribute to the ancient Chinese a logical insight or a definition to the effect that

$$\text{Definition: } p \text{ or } q = \text{if not–}p \text{ then } q$$

But this would be misleading unless we found such a definition or equivalence explicitly stated in the ancient texts. It so happens that the ancient Chinese never stated such a theorem. All we can say is that they most commonly expressed what we would call the notion of 'or' by using a certain combination of negation and implication. In so doing they were adhering to well-known principles of logic, but they were not necessarily discussing these principles. One can follow logical rules without ever discussing them or even reflecting on them.

In what follows I shall carefully distinguish between saying that the Chinese adhere to a logical rule on the one hand, and saying that they stated or focussed on a logical rule on the other. I shall try to distinguish between Chinese logical practice and Chinese logical theory. Similarly, I shall have to distinguish between saying that the Chinese followed a certain linguistic pattern, and saying that they focussed on or discussed this linguistic pattern. I shall try to distinguish between their linguistic practice and their language theory.[2]

All peoples, I believe, can count in one way or another, but few have developed a mathematical number theory. All peoples I know of perform basic logical operations such as negation and conjunction, but few indeed have developed theories of propositional logic or of predicate logic. I intend to investigate to what extent different peoples, and the ancient Chinese in particular, have found it useful to articulate logical structures, and to what extent they have focussed or reflected on logical notions such as 'all', 'some', 'if' and 'not' and have thus become theoreticians of logic.

Among the users of Indo-European languages it was the Greeks and the Indians who discussed such problems profusely. Aristotle's achievements in this area were unsurpassed for many centuries.[3] The Indian tradition of *Nyāya* was preoccupied with issues closely related to logic. It started out with the *Nyāya Sutras* attributed to Gotama (*c.* +3rd century) and achieved considerable theoretical rigour and logical preciseness with Dignāga (*c.* +6th century). All other logical reflection within the Indo-European area derives from these two sources. Outside the Indo-European area there is only one people for whom it has been claimed that they developed an

---

[1] Such, of course, is the case also for English, as P. F. Strawson has emphasised in his still useful book *Introduction to Logical Theory* (1952).

[2] In Sections (*c*), Logical features, to (*e*), Logical practice, we are exclusively concerned with Chinese linguistic and logical practice. In (*f*), Explicit logic, we shall study Chinese theories concerning language and logic.

[3] His achievements remind me of the astonishing technological achievements in the casting of bronze during the Shang dynasty, a feat which was also not be repeated or reproduced for centuries, if ever. In the intellectual sphere the grammarian Pânini (−5th or −6th century) is another case in point.

indigenous independent tradition of logical reflection, namely the Chinese. For example, the Later Mohists (*c.* −3rd century) defined:

> (Definition:) 'Some' = 'not all'.
> (Definition:) 'All' = 'none not'.[1]

Whoever wrote these definitions was doing logic.

The mathematician and linguist Y. R. Chao complains: 'Thus while aiming at finding out how Chinese logic operates, we shall probably end up with finding out how logic operates in Chinese.'[2] I am afraid I have ended up as Y. R. Chao feared we would. However, I have constantly tried to avoid looking in Chinese logical texts only for exemplification of what we are familiar with from our Western logical tradition. One has to try to expect to be logically surprised, and trying to be surprised is not at all an easy thing to do.

Logic, like chemistry, is basically the same subject in China and in Greece. But that certainly does not mean logic and chemistry are conducted in the same sorts of ways in traditional China and in ancient Greece. As we shall see, they evidently were not. The differences are profound. They need to be emphasised and studied in detail.

When it comes to grammar and the study of the linguistic strategies the Chinese used for communicating with each other, I shall certainly not want to find out 'how Latin grammar operates in Chinese'. On the contrary, I shall try to explore the deep differences between the traditionally most important Western languages and ways of thinking on the one hand and Chinese on the other.

One basic contrast is this: In China the thinkers who were pursuing logical matters were part of a small subculture, whereas in India and in the West the logicians belonged to the mainstream of intellectual endeavour. In Europe logic belonged to the standard obligatory curriculum of higher education. This explains why Matteo Ricci, writing some time after +1599, could be tempted to claim categorically that 'they (i.e., the Chinese) have no logic'.[3] In +1701 Father Jean-François Fouquet (+1665 to +1741) maintained that the Chinese have 'little aptitude for the subtleties of dialectic', although he did feel that 'for the most part they have good minds.'[4] G. B. Bilfinger wrote in +1724 about Chinese (particularly Confucian) philosophy: *Est practica tota, a subtilitatibus perfecte omnibus pura.*[5]

In 1979 the analytical philosopher Anthony G. N. Flew (1979) still writes in a similar vein in a thoroughly entertaining paper illustrating the effects of a philosophical culture clash. Basing himself on Western sources Flew not only finds no logic in China, but also a very remarkable lack of the analytical spirit of philosophy. I submit that if Ricci had known more than he did about Later Mohist logic he might have qualified his judgement. And I insist that if Fouquet had read +7th-century Chinese

---

[1] *Mo Tzu*, ed. Graham (1978) A 43 and NO 5.
[2] Chao (1976), p. 250. [3] Pasquale M. D'Elia (1949) vol. 2, p. 77.
[4] Letter from Nanchhang, quoted according to J. Gernet (1985), p. 242. Cf. Anon (1717), vol. 5, p. 165.
[5] 'It is totally practical and completely devoid of all subtleties.' G. B. Bülffingerius (=G. B. Bilfinger) (1724), p. 14.

Buddhist logic he, too, might have expressed himself somewhat differently. Moreover, I think that the present work may help analytical philosophers to develop a more balanced and well-informed view of Chinese intellectual achievements. And yet, when one considers the dominant main currents of Confucian thought of Ricci's, Fouquet's and Bilfinger's times, as well as the authoritative English sources on which Flew quite reasonably thought he might base his judgement, one readily understands their negative reactions.

Janusz Chmielewski, one of the pioneers of the study of Chinese logic in the West, observes:

It has long been a common opinion among sinologists that the linguistic structure of Chinese largely accounts for what might be termed the 'logical underdevelopment' of Chinese philosophy, and for the lack of logic as a distinct philosophical discipline in particular.[1]

His own considered conclusion is this:

Such characteristic features of early Chinese as monosyllabism of lexical units, lack of inflections and lack of clearly delimited grammatical word-classes (especially the lack of a clear morphological distinction between nominal and verbal forms) could hardly have any negative bearing on Chinese implicit logic; in fact they are beneficial rather than detrimental to this logic, since they make the Chinese language more similar to the symbolic language of modern logic than any tongue of the Indo-European type can claim to be.[2]

Janusz Chmielewski's intuition here is fundamental. The point is not that Classical Chinese is like symbolic logic. Classical Chinese is most certainly not like symbolic logic. No natural language is. But inspired by Janusz Chmielewski's remarks we can indeed inquire whether it is not part of what Wilhelm von Humboldt called *le génie de la langue chinoise*, that being an isolating language it is in some quite fundamental ways more logically transparent than the Indo-European languages we are familiar with. There is no reason why some languages might not be logically more transparent than others. Neither is there any reason why all languages should be equally powerful with regard to the representation of logical complexity. These are among the philological and philosophical issues which I propose empirically to investigate in some detail below.

(2) The History of the Study of Classical Chinese Language and Logic in the West

The pre-history of Chinese linguistics in the West

William of Ruysbroek (*c.* +1215 to +1270) noted in the diary of his travels to Mongolia in +1253 to +1255:

The inhabitants of Cathay are slight in size, and as they speak they breathe heavily through the nose.[3]

---

[1] Chmielewski (1969), Part IV, p. 103.　　[2] *Ibid.*　　[3] Risch (1934), p. 171.

The inhabitants of Cathay write with a brush, like that painters use, and they make several letters in a single character to make up one expression (*faciunt in una figura plures literas comprehendentes unam dictionem*). The Tibetans write as we do and have letters very similar to ours. The Tanguts write from right to left like the Arabs, but they put the rows of letters vertically next to each other, beginning at the bottom with each line and going upwards. The Uighur writing, as mentioned before, reads from top to bottom.[1]

After the remarkable William of Ruysbroek, who even pointed out the connection between Uighur and Turkish,[2] contact with and detailed knowledge of Far Eastern languages and writing systems declined considerably. Marco Polo, for all his perceptiveness, tells us disappointingly little of the languages current in the China of his day, although he was famous for knowing no less than four of them. (None of these, incidentally, would seem to have been Chinese.) Language, to Marco Polo, was a tool, not the focus of his attention.

In his *The Advancement of Learning* (1605) Francis Bacon wrote the now-famous lines:

For the organ of tradition, it is either speech or writing: for Aristotle saith well *Words are the images of cogitations, and letters are the images of words*; but yet it is not of necessity that cogitations be expressed by the medium of words. For *whatsoever is capable of sufficient differences and those perceptible by the sense is in nature competent to express cogitations*. And therefore we see in the commerce of barbarous people, that understand not one another's language, and in the practice of divers that are dumb and deaf, that men's minds are expressed in gestures, though not exactly, yet to serve the turn. And we understand further, that it is the use of China, and the kingdoms of the high Levant, to write in characters real, which express neither letters nor words in gross, but things or notions; insomuch as countries and provinces, which understand not one another's language, can nevertheless read one another's writings, because the characters are accepted more generally than the languages do extend; and therefore they have a vast multitude of characters, as many, I suppose, as radical words.[3]

More detailed Western knowledge about the Chinese language is almost entirely due to reports from the Jesuits.[4] The first of the Jesuits in China to have made a serious effort at learning Chinese was Michele Ruggieri (+1543 to +1607). He translated the Four Books (*Lun Yü* 論語, *Mêng Tzu* 孟子, *Ta Hsüeh* 大學, and *Chung Yung* 中庸) into Latin. However, only the first lines of the *Ta Hsüeh* (Great Learning) were published under the title *Liber sinensium* in Antonio Possevino's huge *Bibliotheca Selecta qui Agitur de Ratione Studiorum* (first published in +1593, also in +1603, and +1608).[5] The general standard of Chinese studies among the Jesuits has remained very high ever

---

[1] *Ibid.*, ch. 39, pp. 231–2. The earlier reference to Uighur writing will be found in ch. 27, p. 158. For early general accounts of China see the classic Henry Yule (1866).

[2] Risch (1934), p. 163.   [3] Bacon (1605), pp. 136ff.

[4] The standard reference works on the early history of sinology remain Cordier (1895a) and Cordier (1895b). For the general cultural background see Reichwein (1923), and from a Chinese point of view Chu Chhien-Chih (*1983*). More specifically see Duyvendak (1950) and Kraft (1976). In addition we now have Mungello (1985), Widmaier (1983) and Knud Lundbæk (1986), which give a wealth of biographical as well as bibliographical references for early studies of the Chinese language in Europe.

[5] See Lundbæk (1979).

since.[1] But as orientalists in Europe were quick to complain, the Jesuits remained remarkably possessive about their knowledge for a long time. Given that there were a considerable number of Europeans who were superb connoisseurs of Chinese language and culture, it is surprising how little of this was transmitted competently to the European public.[2]

Matteo Ricci (+1552 to +1610) had a superb command of Classical Chinese which was the envy even of some Chinese of his time. His report on the language of the Chinese in his *De Christiana Expeditione apud Sinas* (1615) as well as Juan Gonzales de Mendoza's remarkable bestseller *The Historie of the Great and Mightie Kingdome of China, and the Situation thereof: Togither with the Great Riches, Huge Cities, politike Government and rare Inventions in the same, translated out of Spanish by R. Parke* (London, 1588, first Spanish ed. 1586[3]), included some remarks on the Chinese language without going into any detail about the matter. Ricci's enthusiasm for Chinese culture in general and for the Chinese language in particular, on the other hand, had a considerable intellectual impact in Europe. In a letter to his rhetoric teacher Martino de Fornari, written in +1582 when he had just arrived in China and still knew little of the language, Ricci wrote:

I have recently given myself to the study of the Chinese language and I can promise you that it's something quite different from either Greek or German. In speaking it, there is so much ambiguity that there are many words that can signify more than a thousand things, and at many times the only difference between one word and another is the way you pitch them high or low in four different tones. Thus when (the Chinese) are speaking to each other they write out the words they wish to say so that they can be sure to understand – for all the written letters are different from each other. As for these written letters you would not be able to believe them had you not both seen and used them, as I have done. They have as many letters as there are words and things, so that there are more than 70,000 of them, every one quite different and complex. If you would like to see examples I can send you one of their books with an explanation appended.[4]

The excellence of the Chinese language demanded an explanation, and Athanasius Kircher (+1602 to +1680)[5] provided such an explanation in his book *Oedipus aegyptiacus* (1652). Kircher claimed that the Chinese received their language from wise Egyptian priests. In +1667 Athanasius Kircher published his important *China Monumentis, qua sacris qua profanis . . . illustrata*, which contained the Nestorian Stele text in translation and also reproduced the Chinese text. (The French

---

[1] Witness the high sinological standards maintained even today by such Jesuit publications as the *Bibliotheca Instituti Historici Societatis Jesu Roma*.

[2] This is not exclusively the Jesuits' fault. For example, Father Joseph Henri-Marie de Prémare (+1666 to +1736) submitted a very remarkable grammar of Chinese to Etienne Fourmont (+1683 to +1745) of the French Académie des Inscriptions et Belles-Lettres in +1730, but Fourmont perhaps failed to realise, and certainly failed properly to acknowledge, the excellent quality of this work.

[3] There were forty-six editions and reprints of this book in seven languages within the years +1585 and +1600 alone.

[4] I quote the English translation in Spence (1986), pp. 136ff.

[5] For a bibliography of this early orientalist see Really (1974) and Godwin (1979).

translation of this book (+1670) contained in addition a Chinese-French vocabulary, the *Dictionnaire Chinois et Français* on pp. 324–67.) Kircher claims that although the characters were originally pictographic they had partly lost this character since. Already Kircher considered Chinese as a suitable model for an artificial universal language built up according to rational principles.

The origins of the project of a universal artificial language go back at least to the mystic abbess Hildegard of Bingen (+1098 to +1179).[1] John Wilkins (+1614 to +1672), a leading exponent of the project, explained in his important theoretical essay *Mercury, or the Secret and Swift Messenger* (1641) how the Chinese characters fitted into this scheme of things:

After the Fall of Adam, there were two General Curses inflicted on Mankind: The one upon their Labours, the other upon their Language.

Against the first of these we do naturally endeavour to provide, by all those common Arts and Professions about which the World is busied; seeking thereby to abate the Sweat of their Brows in the Earning of their Bread.

Against the other, the best Help that we can yet boast of, is the Latin Tongue, and the other learned Languages, which by Reason of their Generality, do somewhat restore us from the first Confusion. But now if there were such an Universal Character to express Things and Notions, as might be legible to all People and Countries, so that Men of several Nations might with the same ease both write and read it, this Invention would be a far greater Advantage to this Particular, and mightily conduce to the spreading and promoting of all Arts and Sciences: Because that great part of our Time which is now required to the Learning of Words, might then be employed in the Study of Things. Nay, the Confusion at Babel might this Way have been remedied, if every one could have expressed his own meaning by the same kind of Character. But perhaps the Art of Letters was not invented.

That such manner of Writing is already used in some Parts of the World, the Kingdoms of the high Levant, may evidently appear from divers credible Relations. Trigaultius affirms, that though those of China and Japan, do as much differ in their Language as the Hebrew and the Dutch; yet either of them can, by this Help of a common Character, as well understand the Books and Letters of the others, as if they were only their own.

Unfortunately, the Chinese characters proved difficult. This extraordinary difficulty demanded and found an explanation: it was said that this was because Chinese was an invention of the Evil One to prevent the spread of Christianity in Eastern Asia: the protestant theologian Elias Grebniz maintained some time before +1682 that the Chinese characters '*durch Gottes Verhängniss vom Teuffel eingeführet/ damit er die elende Leute in der Finsterniss der Abgötterei destomehr verstricket halte* (introduced through God's fate by the Devil, so that he could better keep the miserable people in the darkness of superstition)'.[2]

Rumours about the Chinese tones inspired F. Godwin in his imaginary travelogue *The Man in the Moone: or a Discourse of a Voyage Thither by Domingo Gonsales, the Speedy*

---

[1] See Schrader and Fürkötter (1956) and Wilkins (1641), pp. 57f.

[2] This account is contained in Andreas Müller's book *Andreae Mülleri Besser Unterricht von der Sinenser Schrifft und Druck, als etwa in Hrn. D. Eliae Grebenitzen Unterricht von der Reformirten und Lutherischen Kirchen enthalten ist*, Berlin 1682, p. 5.

Messenger (London, 1638) where he explains that the strange language of the 'moone'

hath no affinitie with any other that ever I heard. (. . .) it consisteth not so much of words and Letters, as of tunes and uncouth sounds (. . .) you have few words but they signifie divers and severall things, and they are distinguished onely by their tunes.[1]

The difficulties of learning the Chinese language were reduced by dictionaries and word lists. Already Matteo Ricci was engaged in such a project, and one of the earliest important such lists was handwritten by the Dominican Francesco Diaz (died +1646) in the +1640s. The copy of it kept in the Royal Library of Berlin in the +18th century had 598 pages with three columns on each page. This dictionary comprised 7,169 characters. A copy of it is preserved in the Vatican library.[2]

The first large-scale publication on the Chinese language to be published in the West is John Webb's (+1611 to +1672).[3] *An Historical Essay Endeavouring a Probability That the Language of the Empire of China is the Primitive Language* (London, 1669). The title of the second edition of +1678 is even more specific: *The Antiquity of China, or an Historical Essay, Endeavouring a Probability That the Language of the Empire of China is the Primitive Language Spoken Through the Whole World Before the Confusion of Babel.* Webb's book is of considerable historical interest. It is facile to treat it as an entertaining curiosity. In reality it represented a very serious effort to sort out the character and importance of the Chinese language before much was known about the language.

And as if all things conspired to prove this the Primitive Tongue, we may observe how forceably Nature struggles to demonstrate so much. The very first expression we make of life, at the very instant minute of our Births, is, as was touched on before, by uttering the *Chinique* word *Ya*. Which is not only the first but indeed the sole and only expression that Mankind from Nature can justly lay claim unto.[4]

Webb is in no doubt of the moral excellence of the language he describes:

And what is more, they have not any Character whereby to write the privy parts.[5]

On the other hand he finds

. . . devout Ejaculations, such as cannot (oh the shame!) among Christians without difficulty be found.[6]

There is also entertaining detail on Chinese phonology:

The Chinois have not the letter R, nor can ever by any possible means be brought to express or pronounce the same, whatever labour or diligence is used by them. And when our Children attain to riper age; as if Nature abhorred the *Confusion*, what care and pains do we take, what opportunity not lay hold of, by practising and repeating to make them pronounce this letter, till education after long contest prevailing they arrive thereat? Thus from

---

[1] Cf. Wilkins (1641), p. xvii.
[2] *Borgia Cinese*, 412. For the earliest German word list by Florian Bahr see Fuchs (1937), pp. 68–72.
[3] See Ch'en Shou-Yi (1935).   [4] Webb (1669), p. 196.
[5] *Ibid.*, p. 203.   [6] *Ibid.*, p. 206.

our Births to our Infancy, and from our Infancy to Riper Age till Nature is compelled to yeeld by the enforced power of instruction, unto corrupt speech, we generally throughout the Universe appear in our Language direct *Chinois*.[1]

There is also a lyrical account of the exquisite naturalness of the Chinese system of morphology and syntax:

Furthermore the Chinois are never put to that irksome vexation of searching out a radix for the derivation of any of their words, as generally all other Nations are, but the radix is the word and the word is the radix. . . . Besides they are not troubled with variety of Declensions, Conjugations, Numbers, Genders, Moods, Tenses and the like grammatical niceties, but are absolutely free from all such perplexing accidents, having no other Rules in use than what the light of nature has dictated unto them; whereby their language is plain, easie and simple as NATURAL speech ought to be.[2]

Webb was aware of a special relation between written and spoken Chinese:

Hence it is, that the style they write is far different from that they speak, although sayth Semedo (and mark him, I pray) the words are the same, so that when one goeth about to write, he had need to collect his wits; for he that will write as commonly they speak, may worthily be laughed at.[3]

We do need to mark Alvarez Semedo (+1585 to +1658) and his famous *Imperio de la China*, published +1642 in Madrid, and all Webb's other sources such as Golius and Spizelius as well.[4] For in fact Webb's contribution to sinology was that he summarised what could be gleaned from the published Western literature. His originality was limited to constructing out of these reports a case that Chinese was the original language of mankind before the building of the tower of Babel.[5]

It appears that most early Jesuits believed that the Chinese were descendants of Shem, whose children were said to have been shrewd and wise – as the Chinese remained, according to the Jesuit view. The identification, of course, of the first emperor Yao Thang with Joktan, the great grandson of Shem, was perhaps problematic, though plausible, but the general picture was clear enough: the Chinese, according to many Jesuits, preserved the speech of one branch of Noah's family.[6]

The Dutch mathematician and linguist Golius (Jakob Gohl, +1596 to +1667) placed the Chinese language in a systematic and philosophical as well as a religious context:

---

[1] *Ibid.*, p. 197.   [2] *Ibid.*, p. 192.   [3] *Ibid.*, p. 186.

[4] We note particularly Theophilus Spizelius's *De Re litteraria Sinensium Commentarius*, published +1660 in Leiden, a city which was to become one of the leading modern centres of Chinese studies.

[5] For a survey of the doxography on the Chinese language during the +17th century see Mungello (1985).

[6] Joseph de Guignes (+1721 to +1800) became famous for a *Mémoire dans lequel on prouve, que les chinois sont une colonie égyptienne* (published in +1759). The case is made in glorious graphic detail in a book published anonymously by Pierre Martial Cibot (+1727 to +1780) which I have at hand: *Lettre de Pékin, sur le génie de la langue chinoise, et la nature de leur écriture symbolique, comparée avec celle des anciens égyptiens* (published 'avec approbation et permission', Brussells, +1773). In a *certificat* printed on p. v of this book no less than ten distinguished citizens of Rome are listed by name and said to have confirmed the authenticity of part of the material presented. The book continues with a detailed anonymous *Lettre sur les caractères chinois, par le reverend père ***, de la Compagnie de Jesus* (in fact by a certain John Turberville Needham) followed by twenty-seven exquisite plates demonstrating the close connection between Egyptian and Chinese characters.

The artificiality of their language means that it was invented at one point in time by a skillful person in order to establish verbal communication between the number of different nations who live in that large country which we call China, although it has to be said that this language might be changed now through long usage.[1]

Leibniz was thoroughly fascinated by accounts such as these. His feelings are summarised by his dictum 'If God had taught man a language, that language would have been like Chinese.'[2] But what particularly aroused his philosophical interest was the nature of the Chinese characters. Leibniz's enthusiasm is clear, as when he writes to La Croze in +1707:

This enquiry seems to me to be all the more important since I imagine that if we were able to discover the key to the Chinese characters, we would have found something which could serve for the analysis of thought.

Leibniz believes he would have something to contribute:

It does appear that if we Europeans were well enough informed about Chinese literature, then the aid of logic, critical thinking, mathematics, and our way of expressing ourselves which is more explicit than theirs, would make us discover in these Chinese monuments of such remote antiquity many things unknown to the modern Chinese and even to their later interpreters no matter how classical one takes them to be.[3]

Leibniz expresses here a feeling of European *Besserwisserei* or analytical superiority, which has remained important in Western attitudes to the Chinese language.

Leibniz's intensive analytical efforts did lead to a clear conclusion. The Chinese characters, for all their intrinsic interest, did not after all supply a suitable model for his philosophical alphabet of human thought, *alphabetum cogitationum humanarum* or *characteristica universalis*:

If we understood the characters of the Chinese I think we would find some more connections (with a *characteristica universalis*), but at bottom these characters are undoubtedly far removed from such an analysis of thought which is the essence of my plan.[4]

What he missed in the Chinese characters was a unified underlying rational principle of their construction or what he called a '*filum Ariadnae*' in the labyrinth of Chinese writing.[5] John Wilkins (+1614 to +1672) had arrived at similar conclusions already in +1668 when he published his ambitious *Essay towards a real character and a philosophical language*.[6]

---

[1] Letter by Leibniz to Christian Mentzel, dated 15 October 1698.
[2] Cf. Lundbæk (1986), pp. 97, 83, and 103.
[3] 'Lettre sur la philosophie chinoise à Nicolas de Remond' dated +1715/16, ed. Loosen and Vonessen (1968), pp. 126f.
[4] Letter to Herzog Johann Friedrich dated April (?) +1679; see Widmaier (1983), p. 36.
[5] Letter to Johan Christian Mentzel dated 21 January +1699; see Widmaier (1983), p. 36.
[6] The nature of the Chinese language is discussed on pp. 450–2 of that work.

Early grammars

Seventeenth-century concern with the Chinese language was dominated by speculation and hampered by lack of specific information.[1] But in +1703 the missionary
Francisco Varo's (+1627 to +1687) *Arte de la lengua mandarina* was published in
Canton. This pioneering grammar avoided the use of characters and introduced
the Chinese language entirely on the basis of transliterations. Unfortunately, it
remained inaccessible to most scholars in Europe and is excessively rare. S.
Fourmont, to whom we shall turn in a moment, was one of the few to have used it.[2]

On 18 June +1700 Peter the Great of Russia issued an *ukaz* which recommended
the finding of 'two or three good and learned men, not too old, that would be able to
teach Chinese and Mongolian language and grammar'.[3] T. S. Bayer (+1694 to
+1738) was called to St. Petersburg to fill one of these posts, and he apparently had
no access to Varo's grammar when he published his *Museum sinicum, in quo sinicae linguae et literaturae ratio explicatur* (+1730), which was the first grammatical account of
the Chinese language published in Europe.[4] This grammar was the beginning of a
great tradition of Chinese linguistics in Russia, a tradition which was to have a very
profound influence on modern linguistic developments in the People's Republic of
China, and to which we shall return in due course.[5] Bayer's introduction to his
Museum sinicum contains a most remarkable document: a detailed history of sinology
from its beginnings to +1730.[6]

T. S. Bayer died in profound distress over a negative review from S. Fourmont
(+1683 to +1745) of the Académie des Inscriptions et Belles-Lettres in Paris. S.
Fourmont's *Meditationes Sinicae* (+1737) and his *Linguae Sinarum mandarinicae hieroglyphicae grammatica duplex* (+1742) began an important French tradition in Chinese
linguistics which thrives to this day.[7]

Fourmont's *Grammatica duplex* has been considered by Abel-Rémusat as plagia-
rism on Varo's earlier work. There is no doubt that Fourmont freely drew on
Francisco Varo's *Arte de la lengua mandarina*. But what is worse: Fourmont also had

---

[1] See particularly Cornelius (1965) which includes a fine bibliography on pp. 159–72 and David (1965).

[2] Abel-Rémusat (1826), vol. 2, p. 107.   [3] Bartold (1925), p. 197.

[4] The sinological tradition in St. Petersburg did not die with Bayer: in 1832 N. J. Bichurin (+1770 to 1853) (alias
monachus Hyacinthus (lakinf)) published his *Kitajskaja grammatika, sochinennaja monakhom lakinfom*. In his preface,
N. J. Bichurin surveys a number of earlier grammars by Varo, Bayer, Fourmont, Prémare, Marshman, Morrison,
Rémusat and Gonçalves. Bichurin's effort was continued by an exceptional line of Russian Chinese linguists
including such scholars as A. A. Dragunov, and S. E. Jakhontov in our century.

[5] One special strength of the Russian linguists was lexicography. In 1867 the first dictionary of colloquial
Chinese was published in Peking, the *Russko-kitajskij slovar' razgovornogo jazyka, Pekinskogo narechija* by Isaija (Polikin)
with a supplement (*Pribavlenie* I) published in Tientsin in 1868. This began a distinguished tradition of Russian
lexicography culminating with P. P. Popovs *Kitajsko-russkij slovar'* (2 vols., Peking, 1888). For an exhaustive account
of early Chinese lexicography in Russia see Petrov (1961). In modern times we have V. M. Oshanin's important
Kitajsko-Russkij slovar' (Moskow, 1952) and the best Chinese dictionary in any Western language, the four-volume
Bol'shoj Kitajsko-Russkij slovar' completed in 1984. This dictionary attempts for the first time to give a detailed
description of stress phenomena within Chinese words.

[6] This history has now been presented in a masterly and memorable annotated translation in Lundbæk
(1986), pp. 39–101.

[7] See the work of the French linguists Alexis Rygaloff, Marie-Claude Paris and Viviane Alleton on modern
Chinese grammar, and of Alain Peyraube on historical grammar.

available to him another most remarkable piece of scholarship: the *Notitia Linguae sinicae* by Joseph Henri-Marie de Prémare (+1666 to +1736), which he mentions in his preface and dismisses as distinctly inferior to his own work. It so happens that this *Notitia* was a mature work of scholarship containing no less than 50,000 characters and – as the Chinese proverb has it – Fourmont's effort 'cannot be mentioned on the same day' as the *Notitia*. In good faith de Prémare had sent a copy of his manuscript to Fourmont in +1729 but the *Notitia* was first published more than a hundred years later, in 1831 in Malacca.[1]

De Prémare was undoubtedly the most outstanding grammarian of Chinese in the 18th century. He was an unusually thoughtful man. Like most Jesuits he did not believe that the grammar of Chinese could be reduced to mechanical rules. Instead of reducing Chinese language to an inevitably latinising grammatical system he decided to demonstrate its regularities through copious examples. He observes: *Absit ut ad nostras linguas sinicam revocare velim.*[2] This is a memorable remark the full significance of which first began to be appreciated centuries later. Indeed, even the modern sinologist will still find de Prémare's *Notitia* fascinating reading.[3] For its time, it is a simply astonishing scholarly achievement vastly superior to what preceded it, and quite arguably also superior to much more celebrated works such as Abel-Rémusat's *Elémens de la grammaire chinoise* of 1822, which – as we shall see – was in turn accused of being a selective plagiarisation of the *Notitia*. De Prémare deserves a very special place of honour in the history of Chinese linguistics in the West.

De Prémare's grammar was published a hundred years late, but at least it did get published in the end and did become very influential indeed for later developments. The same can unfortunately not be said of the efforts by Juan Rodriguez (+1724 to +1785), another Spanish grammarian of the Chinese language. We have an English translation, *A Grammar of the Chinese language expressed by the letters that are commonly used in Europe. From the Latin of F. John-Anthony Rodriguez. With a dedicatory letter from the translator, John Geddes, to the R. Honorable Mr. Dundas, dated May 29th 1792,*[4] of an early manuscript by Rodriguez. But neither this translation nor the original or its much improved later versions ever got published.[5]

Generally, the books on Chinese published during the 18th century were still found by most people to be pitifully inadequate for practical purposes. When

---

[1] I only have the unchanged reprint (Hong Kong, 1898) available.

[2] 'Far be it from me to wish to reduce the Chinese language to our kinds of languages.' Lundbæk (1980), p. 269.

[3] For those unfamiliar with Latin, it may be tempting to consult the translation of the *Notitia* by the American missionary J. G. Bridgman, Canton, 1847. But for some reason this translation pervasively perverts the meaning of the Latin original in most surprising ways. It often manages to say the plain opposite of what Prémare's original means, and it almost consistently cuts out the more interesting asides that makes Prémare such exquisite reading in the Latin. J. G. Bridgman, who died at the age of about thirty in 1850, is one of the more puzzling figures in the history of sinology.

[4] China Factory Records, vol. 20, India Office Records, Commonwealth Office, London, 55 pages, manuscript, Cf. Lundbæk (1980), p. 265.

[5] A 198-page careful manuscript of the latest version of Rodriguez' grammar, revised by José Villanueva, may be found in Bibl. Nacional, Madrid, Sigla: 2511 (H. 303). For a detailed account of Rodriguez's studies see Lundbæk (1980).

Rodriguez asked some officials at the factory offices in Canton what they had learned from Fourmont's *Linguae Sinarum grammatica* which was on their shelves, they all replied that they had learnt nothing. They complained that the more they worked with the book, the less they knew about Chinese. Looking at Fourmont's work today one can certainly understand this reaction.

Nineteenth-century grammatical scholarship

During the 19th century the basic issues concerning the nature of Chinese characters remained topical and controversial. In 1801 Joseph Hager wrote a little book entitled *An Explanation of the Elementary Characters of the Chinese; with an Analysis of their ancient Symbols and Hieroglyphics*, which prompted a splendid reply with the gruesome title *Leichenstein auf dem Grabe der chinesischen Gelehrsamkeit des Herrn Joseph Hager, Doctors auf der Hohen Schule zu Pavia* (Tombstone on the Grave of the Chinese learning of Mr. J. Hager . . .) published by Julius Klaproth apparently in or just before 1811.[1] Five years later, unperturbed, a certain J. C. F. Meister published his treatise *Ganz neuer Versuch auch freien Denkern aus der Chinesischen Schriftsprache eine symbolische Ansicht zu eröffnen unter welcher das Gemüth empfänglicher wird für das Geheimnis der christlichen Dreieinigkeit* (An entirely new attempt, on the basis of the written Chinese language, to open up a symbolic perspective under which the mind becomes more receptive for the mysteries of the Christian holy trinity . . .) (Leipzig, 1816). The study of such controversial material, which may strike us as idle today, would provide a historically more realistic picture of Western intellectual concerns with Chinese than an exclusive concentration on 'serious' philological scholarship. But unfortunately there is not the space here to enter into any detail about it. We shall have to concentrate on those intellectual efforts concerning the Chinese language which proved fruitful and instrumental in future sinological research.

The crucial new feature of the 19th century was the rapidly increasing flow of public information on the Chinese language. In the Danish colony of Serampore, India, the Baptist missionary J. Marshman produced a remarkably well-organised and thoughtful survey, his *Elements of Chinese Grammar, Clavis Sinica* (1814) which may be read with profit to this day. The contrast with Andreas Müller's abortive – or as we now know, non-existent – *Clavis Sinica* is striking indeed. R. Morrison's *A Grammar of the Chinese Language* (1815), also published in Serampore, was another useful detailed description of the language.[2]

The 'Seramporean' grammars were, however, eclipsed by Abel-Rémusat's elegant and influential *Elémens de la grammaire chinoise* published in 1822 in Paris. Abel-Rémusat was a most remarkable orientalist and comparative linguist. His

---

[1] For another eloquent pamphlet attacking British sinology, dating from 1918, see Schindler and Erkes (1918), pp. 105–15, which includes a wonderfully biased overview of the history of European sinology from a thoroughly German point of view.

[2] Robert Morrison (+1782 to 1834) also published a dictionary of Chinese in six volumes (1815 to 1823) which is remarkable not only for its high quality, but also for the exquisite printing.

grammar, which owes more than it acknowledges, and more than is generally recognised, to de Prémare's *Notitia*[1] established France as the European centre of Chinese linguistics throughout most of the 19th century.[2]

Abel-Rémusat's *Elémens* inspired one of the greatest general linguists of his time, Wilhelm von Humboldt, to write his long *Lettre à M. Abel-Rémusat sur la nature des formes grammaticales en général, et sur le génie de la langue chinoise en particulier* (1827) which remains to this day perhaps the finest introduction to the philosophical and general linguistic questions raised by the Chinese language.[3] Humboldt recognised Chinese as structurally the diametrical opposite of languages like Sanskrit and Greek, but as a perfect language in its own way. In the wake of Humboldt, a number of traditional German philosophers of language such as H. Steinthal (1823 to 1899) continued to speculate on the significance of Chinese for a general philosophy of language in the latter part of the 19th century. Steinthal finds 'the contrast between the means the Chinese language employs and the effects it achieves a phenomenon quite unique in the history of language.'[4]

One careful reader of Abel-Rémusat was the German idealist philosopher Friedrich Schelling (+1775 to 1854) who summarised an emotional as well as philosophical response to the strangeness of the Chinese language which was to remain prevalent for a long time to come when he wrote:

> The Chinese language is for us like a language from another world. And if one were to give a definition of language according to which all the other idioms are called languages, then one would have to admit that the Chinese language is not a language at all, just as the Chinese people is not a people.

Only ten years after Abel-Rémusat's *Elémens* the Russian missionary Nikita Jakojalevich Bichurin published his *Kitajskaja grammatika* (Peking, 1832), in which he develops the characteristic idea of *umstvennoe slovoizmenenie* or 'mental inflection'. This work deserves careful comparison with its Western European contemporaries and has been sadly neglected by Western sinologists.

Abel-Rémusat's *Elémens* (as well as Prémare's *Notitia*) included a survey not only of literary Chinese but also of the colloquial language. In our present account we disregard the study of colloquial Chinese.

In France Stanislas Julien provided an immensely practical handbook in his *Syntaxe nouvelle de la langue chinoise* (1869/70). Classical Chinese was now a language which, given patience, one could go ahead and learn on the basis of published Western books. 19th-century French scholarship in Classical Chinese was the inspiration and the essential background to Georg von der Gabelentz's *Chinesische Grammatik* (1881) which openly acknowledged its debts to de Prémare and to

---

[1] Cf. Carl Friederich Neumann (+1793 to 1870) in Neumann (1834), 2. Hälfte, pp. 1042–52 and pp. 1061–3. Neumann does seem to go too far when he refers to Abel-Rémusat's grammar as an excerpt from Prémare's *Notitia* throughout his detailed account, but he does have his point.

[2] Curiously, Varo and Rodriguez had no significant successors in their own country: Spain never again became a centre for Chinese linguistic studies.

[3] See Harbsmeier (1979).   [4] Steinthal (1860), p. 137.

Stanislas Julien. Georg von der Gabelentz was a distinguished general linguist. His *Grammatik* remains recognised until today as probably the finest overall grammatical survey of the language. Gabelentz's *Anfangsgründe der chinesischen Grammatik* (1883) is a useful supplement to the *Grammatik* itself.

In spite of the achievements of 19th-century grammarians there remained a conviction, even among the finest scholars of the Chinese language, that Classical Chinese does not have a grammar to speak of. H. A. Giles, in the introduction to the first edition of his quite outstanding work *A Chinese-English Dictionary*[1], advocated this view. Gustave Schlegel (1840 to 1903) of Leiden University became famous for his slogan: *Lisez, lisez; jetez la grammaire*, and Achilles Fang sympathises with him when he writes in 1953: 'Gustave Schlegel was not the best of Sinologists; yet he had a modicum of sound sense when he advised his students to forget their grammar. . . . The sooner we forget grammar, the speedier will we recover our sanity.'[2] Attitudes of this kind continue to have their profound indirect effect on Classical Chinese studies, particularly translations, to our own day.[3]

An authoritative treatment of general issues in the study of the Chinese language will be found in P. Demiéville (1943), pp. 33–70.

Historical phonology

There may be those who doubt that Chinese has a grammar, but no one has been able to doubt that Chinese has a phonology. When it comes to historical phonology the picture is totally different from the study of syntax: progress since the 19th century has been systematic and simply enormous. Bernhard Karlgren's *Etudes sur la phonologie chinoise* (1919) was the pioneering work in the reconstruction of earlier Chinese pronunciation, and his work inspired an important linguistic tradition in Scandinavia.[4] Henri Maspero's *Le Dialecte de Tch'ang-ngan sous les T'ang* (1920) was another crucial contribution – although it turns out that the dialect Maspero described was in fact not that of Chhang-an but mainly that of the area around Nanking.[5] Karlgren's classical work *Grammata Serica Recensa* (1952), Mantaro Hashimoto's *The Phonology of Ancient Chinese* (Tokyo, 1978), Chou Fa-Kao's *Pronouncing Dictionary of Chinese* (Hong Kong, 1973) and most recently E. G. Pulleyblank's *Middle Chinese: A Study in Historical Phonology* (1984) summarise a wealth of insight that was almost totally inaccessible in the West (and to a large extent in China) during the 19th century.[6]

---

[1] Giles (1892), pp. x–xii.   [2] A. Fang (1953), p. 282.

[3] It is as if one were to translate ancient Greek texts without taking note of the results in handbooks like J. D. Denniston's standard compendium *The Greek Particles* (1st ed. Oxford, 1934). Moreover, one needs specialised grammars for special texts: for what kind of a philologist or theologian would interpret the Bible without constant reference to Blass, Debrunner and Rehkopf (1896) to ensure consistency of interpretation? Chinese philology, it seems, has much to learn from Greek philology. Harbsmeier (1986) illustrates the point by showing how the particle *i* has been simply neglected even by the best of translators.

[4] See particularly the work of G. Malmqvist and of S. Egerod.   [5] See Malmqvist (1968).

[6] As Roy Andrew Miller (1973) shows, the efforts towards a reconstruction of ancient Chinese phonology are still controversial. But there can be no doubt of the tremendous overall progress in the art of historical phonology.

Historical syntax

A great deal of attention has been given to pre-Chhin Classical Chinese. Joseph Mullie's Dutch *Grondbeginselen van de Chinese letterkundige taal* (1948) provides a detailed traditional survey based on the grammatical categories of Latin applied to Chinese. S. E. Jakhontov's *Drevnekitajskij jazyk* (1965) provides a very fine but unfortunately short account of the Classical Chinese language. More recent grammars including those by W. A. C. H. Dobson and R. Shadick (1968) are theoretically more ambitious and terminologically much more prolific. The most substantial progress in the understanding of Classical Chinese grammar has mostly been presented in detailed papers on individual grammatical particles. F. M. Uhle's fascinating book *Die Partikel wei ilm Shu-king und Schi-king. Ein Beitrag zur Grammatik des vorklassischen Chinesisch* (1880), A. Conrady's *Über einige altchinesische Hilfswörter* (1933), J. Misch's fine dissertation *Der Konditionalsatz im klassischen Chinesisch* (1935), J. Mullie's book-length article 'Le mot-particule TCHE' (1942) and Walter Simon's long series of articles on the particle *êrh* 而 (1952 to 1954) set significant precedents. R. Gassmann's *Das grammatische Morphem ye* 也 (1980) and *Das Problem der Einbettungsstrukturen* (1984) attempt to apply transformational generative grammar to Chinese. Significant detailed contributions have been made in articles by such scholars as Ting Sheng-Shu, Lü Shu-Hsiang, Chou Fa-Kao, A. Conrady, W. Simon, N. C. Bodman, G. A. Kennedy, Hugh S. Stimson, E. G. Pulleyblank, A. C. Graham, G. Malmqvist, and J. Cikoski.[1] My own book *Aspects of Classical Chinese Syntax* (1981) is an attempt to continue this line of detailed research by applying it to some logical particles in Classical Chinese.

On the early history of Chinese syntax we have W. A. C. H. Dobson's *Late Archaic Chinese* (1959), *Early Archaic Chinese* (1962), *Late Han Chinese* (1964), and the best of the series, his *The Language of the Book of Songs* (1968). These provide a starting point for future research.

For the post-Han period, remarkably many significant contributions are in Russian, written in the tradition of A. A. Dragunov and under the general guidance of the great linguist S. E. Yakhontov of Leningrad. We have the following important books: M. V. Krjukov *The Language of the Shang Inscriptions* (1973), S. E. Yakhontov *The Ancient Chinese Language* (1965), I. S. Gurevich *Outline of the Grammar of the Chinese Language of the +3rd to +5th centuries* (1974), I. T. Zograf '*Outline of the Grammar of Medieval Chinese*' (1972), I. T. Zograf *The Popular Tale of the Rewards of Kindness, part 2: Grammar and Vocabulary* (1972) I. T. Zograf *The Medieval Chinese Language, Origins and Tendencies of Development* (1979).[2] In Western European languages the most important contributions to the field are Bernhard Karlgren (1952a), Gerty Kallgren's *Studies in Sung Time Colloquial Chinese as Revealed in Chu Hi's Ts'üanshu* (1958), M. A. K. Halliday's book *The Language of 'The Secret History of the Mongols'* (1959) and now Alain Peyraube's *Syntaxe*

---

[1] See also Gassmann (1980) and (1982).
[2] Unfortunately, these books have been largely ignored in the West.

diachronique du chinois: évolution des constructions datives du XIVe siècle av. J.C. au XVIIIe siècle (1988).[1]

The study of traditional Chinese logic

The interest in Chinese grammar is old in Europe. Chinese logic, on the other hand, was very late to attract attention. China was regarded as a predominantly Confucian country, and logic was not seen to play a part in Confucianism. The binary 'logic' of the *I Ching* (*The Book of Changes*), did arouse the early curiosity of logicians like Leibniz, but neither Leibniz nor his successors in the field of formal and philosophical logic over the next 250 years knew anything of the indigenous Chinese logical tradition. In 1962, William and Martha Kneale could still write an outstanding book entitled *The Development of Logic* which manages completely to disregard not only Chinese but also Indian indigenous traditions. I. M. Bochenski in his *Formale Logik* (1956) also gave short shrift to Chinese logic although there was at least an extremely brief section on the Indian traditions. Even in the 1980s the history of logic – like the philosophy of language – is dealt with largely without reference to China.

China has most emphatically been put on the map of the history of technology, and to some extent of science. But when it comes to the history of logical science this has yet to be done, if indeed it can be done.

The pioneering first forays into this difficult terrain were by Alfred Forke (1902) and (1922). P. Masson-Oursel (1906) and (1918) made a valiant attempt at introducing forms of Chinese reasoning to a Western public. But still, from the standpoint of a logician this did not add up to much. Henri Maspero (1928) made a first detailed philological attempt at a coherent interpretation of Mohist logic.

None of these scholars of the history of Chinese logic had any grounding in logic, and not surprisingly their reports aroused little attention outside the narrow circle of sinological specialists.

Hu Shih, on the other hand, by calling his 1922 account of Chinese philosophy *The Development of the Logical Method in Ancient China* adopted so broad a concept of 'logic' that the term seemed to become synonymous with philosophy. Ignace Kou Pao-Koh (1953) made perhaps the most successful and thorough attempt to apply traditional philological methods to Chinese logical texts. Ralf Moritz (1974) takes special account of the social context of logical thought in ancient China but still takes no notice of formal logic or analytical philosophy.

It was during the 1960s, when some scholars with a solid logical background began to write about logic in China, that Chinese logical tradition began to be interpreted in a way that could make sense to historians of logic in general. It is the

---

[1] Cf. also Cheng Yat-Shing (1976), Miller (1952), Crump (1950) and Dew (1965). Since we are mainly concerned with Classical Chinese, we have, in this survey, largely ignored the study of colloquial Chinese. For a useful survey of linguistic studies on Chinese see Serruys (1943).

considerable merit of Janusz Chmielewski to have acted on the insight that in order to study Chinese logic it is useful to know about Western formal logic. Janusz Chmielewski's *Notes on Early Chinese Logic* I to VIII (1962 to 1969) are the first sustained attempt to apply formal logic to Classical Chinese texts. Donald Leslie has pursued similar ambitions with less theoretical rigour and discipline in a number of papers.[1] Chung-ying Cheng, a student of the very distinguished analytical philosopher W. V. O. Quine, has written on Chinese logic from the point of view of modern analytical philosophy.[2]

The problem with all this early analytical work on Chinese logic was that the Chinese texts were in too bad a shape to be properly interpreted, too unreliably attributed and too poorly dated. This problem had to be solved before any serious advance could be made. It is the merit of A. C. Graham that he took up this philological challenge. He demonstrated that only two dialogues attributed to Kungsun Lung can be called genuine.[3] Next, he established and interpreted the text of the main source for Chinese logic, the Dialectical Chapters of the *Mo Tzu* 墨子 in his monumental work *Later Mohist Logic, Ethics and Science* (1978). Without Graham's many pioneering contributions to the study of logic in China, the present survey of language and logic could never have been attempted.

The obvious next step was to apply logical reflection to the corpus thus established. Also in this area A. C. Graham's conceptual and philosophical clarifications published during the last thirty years are essential contributions.

Chad Hansen's recent book *Language and Logic in Ancient China* (1983) is a philosophically very ambitious attempt to describe the special character of language and logic in pre-Han China. Hansen aims to draw philosophically and logically important conclusions from his Chinese logical material.

Hansen claims that 'Chinese philosophers have no concept of truth at all',[4] that they had no concept of a sentence,[5] that they had no concept of propositional knowledge ('knowing that a sentence's true') or of propositional belief ('believing that a sentence is true').[6] Taken together, this would mean that the Chinese could not possibly have had a science which they understood as a body of *sentences or theorems* which they claimed they *knew* were *true*, and which in any case they *believed* to be *true*.[7]

The question of the suitability of the Chinese language for science

There have been widely differing views as to the suitability of Chinese for science. Marcel Granet's spirited and seminal article 'Quelques particularités de la langue et de la pensée chinoises' (1920), written very much in the tradition of Wilhelm von Humboldt, is wonderfully clear in his conclusion that Chinese, lacking grammatical

---

[1] Leslie (1964a) and (1964b).   [2] Cheng Chung-ying (1965), (1970), (1971), (1972) and (1975).
[3] See Graham's recent *Studies in Chinese Philosophy and Philosophical Literature* (1986a).
[4] Hansen (1985), p. 491.   [5] *Ibid.*, p. 517 *et passim.*   [6] *Ibid.*, pp. 500f.
[7] In (*d*,3) and (*d*,9) we shall consider the concepts of 'truth', 'belief' and 'knowledge' in ancient China.

forms, is a picturesque and an essentially inarticulate language. According to Granet, the Chinese language is inherently inappropriate for scientific analysis and precise scientific discourse:

There is no doubt that the progress and the diffusion of the scientific spirit are linked to the existence, in the West, of languages all of which – to different extents – are instruments of analysis which allow one to define classes, which teach one to think logically, and which also make it easy to transmit in a clear and distinct fashion a very elaborate way of thinking. Now I do not think that Chinese as it is written and spoken, in the slightest degree has any of these qualities of the great languages of Europe.[1]

Can a language which suggests rather than defines be suitable for the expression of scientific thought, for the diffusion of science, for the teaching of science? A language made for poetry and composed of images rather than concepts is not only not an instrument of analysis. It also fails to constitute a rich heritage of the work of abstraction which each generation has been able to achieve.[2]

In our section on definition we shall investigate the Chinese tradition for non-suggestive precision and analysis.

Marcel Granet blames the ideographic Chinese writing system for the articulatory poverty of Chinese:

If, in the course of a long historical development, the Chinese language has been able to remain essentially a simple means of picturesque expression it owes this, in my opinion, to the figurative writing which, linking every word to an ideogram, from the start was opposed to the employment of all manner of grammatical forms and derivations in such a way that syntax is almost reduced to the development of rhythm only. Above all this figurative writing has stifled all the inner life of words.[3]

In our section on the semiotics of Chinese characters we shall investigate the question of the influence of the writing system on the conduct of Chinese science. In our section on lexical and functional categories we shall look into the question of grammatical form in Chinese.

Marcel Granet observes that the Chinese language works through musical and picturesque symbolisation, and he concludes:

In this case one must admit that when language cannot translate the operations of thought, these operations must proceed beyond language.[4]

Here we have the core of Granet's thought on the Chinese language. In his view the Chinese language does not properly speaking articulate thought, does not 'translate' it. It only poetically and picturesquely suggests precise thought. Precise thought has to proceed without the support of linguistic articulation by speakers of Chinese. In our section on logical and grammatical explicitness we shall enquire into related aspects of the Chinese language.

---

[1] Granet (1920), p. 150. [2] *Ibid.*, p. 154. [3] *Ibid.*, p. 151. [4] *Ibid.*, p. 153.

Marcel Granet asks rhetorically:

This (Chinese) thought which seems in essence picturesque and musical and which expresses itself in any case through rhythm and concrete symbols, what can it achieve when applied to a domain where precise and distinct formulations as well as explicit judgments are required? What kind of sincerity can there be in a kind of thought which takes not lived experience but tradition as its point of departure? . . . What power would the principles of contradiction and of causality have – without which scientific thought can hardly proceed or be expressed?[1]

In our sections on negation and contradiction in Chinese we shall investigate to what extent the Chinese were concerned with contradiction.

Marcel Granet, clearly influenced by Wilhelm von Humboldt's *Lettre à Monsieur Abel-Rémusat sur la nature des formes grammaticales en général, et sur le génie de la langue chinoise en particulier* of 1827, brings out with admirable verve and eloquence some profound doubts concerning the adequacy of the Chinese language as a medium of science. As a highly intelligent and thoughtful observer of Chinese thought he deserves to be taken seriously. His challenge needs to be answered in philological detail. It will not do summarily to dismiss Granet's intuitions. These intuitions have, in any case, been profoundly influential ever since they were published.

The negative perception of the Chinese language is graphically brought out by J. E. Renan (1823 to 1892):

Is not the Chinese language, with its inorganic and imperfect structure, the reflection of the aridity of genius and heart which characterises the Chinese race? Sufficing for the wants of life, for the technicalities of the manual arts, for a light literature of low standard, for a philosophy which is only the expression, often fine but never elevated, of common sense, the Chinese language excluded all philosophy, all science, all religion, in the sense in which we understand these words. God has no name in it, and metaphysical matters are expressed in it only by round-about forms of speech.[2]

However, one point must be added at this stage in order to avoid all misunderstandings. To the extent that the preceding volumes of *Science and Civilisation in China* have shown that the Chinese were rather good at some parts of science, they have also shown that one can use Literary Chinese to do science. If Marcel Granet had known more about the Chinese scientific tradition he would, I like to think, have expressed himself in a different, less abrasive way. I also believe that if he had known more about the precise syntactic structure of Classical Chinese and the very subtle semantic and syntactic rules governing the use of Classical Chinese grammatical particles, he might have shown a little more respect for the articulatory power of that language.[3]

---

[1] *Ibid.*, pp. 153ff.   [2] Renan (1889), p. 195, as translated in Watters (1889), p. 18.
[3] There is no similar excuse for sinologists writing in the 1980s.

Chinese as a medium for the philosophy of science

There still remains a crucial philosophical point which is not answered so simply. In this case, the issue has been raised most forcefully by Georg Friedrich Wilhelm Hegel (+1770 to 1831):

> When we speak of the Chinese sciences . . . we see that they enjoy very great public admiration and support from the government. . . . Thus on the one hand the sciences are highly honoured and cultivated, but on the other hand these sciences lack the free space of inner reflection and the properly scientific interest that would make it into a scientific endeavour. A free and ideal realm of the spirit has no place here, and what is called scientific here is of an empirical nature and is essentially in the service of what is useful for the state and for the needs of the state and the individuals. The nature of the written language in itself is a great hindrance for the development of the sciences; or rather *vice versa* since the true scientific interest is lacking, the Chinese have no better instrument for the articulation and communication of thoughts.[1]

As usual, Hegel is thinking and writing in a highly philosophical vein. He feels he has an inner realm of (intellectual) freedom (*der freie Boden der Innerlichkeit*) which he suspects is lacking in China. He feels the Chinese did not have that autonomous realm of 'scientific spirituality', that properly scientific intellectual interest which in its concern with truth is indifferent to practical application and which makes an enquiry into a theoretical rather than practical activity. Hegel sees Chinese science as basically utilitarian and pragmatic in purpose, and above all he sees Chinese science as subservient to non-scientific purposes. Hegel sees the written Chinese language as inferior. And he sees this inferiority of the Chinese language as essentially linked with the untheoretical nature of Chinese thinking.

Hegel may have been wrong. His doubts may not have been very well-informed. But I think Hegel's doubts were profound and significant. The Chinese achievements in astronomy, medicine, biology, and even in mathematics, which are laid out in the preceding volumes of *Science and Civilisation in China*, I fear, would not necessarily have removed Hegel's doubts. For none of these achievements affect the crucial observation that the 'Chinese had sciences but no science, no single conception or word for the overarching sum of all of them', as Nathan Sivin (1982) put it in his admirably clear paper 'Why the scientific revolution did not take place in China – or didn't it?' Neither would the history of (basically moral and political) philosophy in China put to rest Hegel's suspicions. These things only go to reinforce the positive judgement with which our quotation begins: the conviction that the admirable intellectual activity in China was strongly utilitarian rather than purely theoretical in purpose.

Hegel's question is of a different order. Do the Chinese have, and is the Classical Chinese language equipped to express, a non-empirical, abstract and theoretical Science with a purely philosophical and cognitive perspective?[2] Here we must

---

[1] Hsia (1985), p. 184.
[2] One may, of course, also ask whether the West ever developed an ultimately non-utilitarian kind of science for its own sake. But the ideal of a science for science's sake was clear enough.

carefully distinguish two questions: 1. did the Chinese have what we might call second-order science, i.e., the enquiry into the nature, status and methods of science and 2. did the Chinese ever develop science for its own sake?

From the early 20th century onwards there have been a number of studies that have a bearing on this subject. Jên Hung-Chün's 'The reason for China's lack of science' (1915) was the first important paper. (It was published in Chinese.) In 1922 there followed Fêng Yu-lan *Why China has no Science – An Interpretation of the History and Consequences of Chinese Philosophy*. Then there was Homer H. Dubs' 'The Failure of the Chinese to Produce Philosophical Systems' (1929) and Derk Bodde's 'The Attitude toward Science and Scientific Method in Ancient China' (1936).

The publication of the first volumes of *Science and Civilisation in China* and associated publications have provoked considerable discussion, notably A. C. Graham's 'China, Europe and the Origins of Modern Science' and Nakayama's 'Joseph Needham, Organic Philosopher' (1973). A very stimulating critical survey of the current state of the art is provided in N. Sivin (1982), which we have quoted above.

The latter half of this Section will be devoted to the very subtle and difficult question not of whether China had sciences, or of whether it had an all-embracing notion of 'Science', but whether China had anything we can interpret and recognise as purely theoretical *logical concerns*[1] during the formative period of Chinese civilisation from Confucius's times (−551 to −479) to Wang Chhung 王充 (+27 to +100) and beyond. Did they discuss rules of argumentation and the structure of scientific propositions? Did they have what we can recognise as *logic*? I shall outline the history of logic in China down to the +7th century when Buddhist logic was practised with great success in China, and to the +17th century when Aristotelian logic was introduced into China – with notably lesser success.

(*b*) THE LANGUAGE

(1) THE TYPOLOGY OF EAST ASIAN LANGUAGES

It is useful to distinguish four major, overlapping, historical varieties of Chinese:

1. Proto-Chinese *c.* −1300 to *c.* −500
2. Classical Chinese *c.* −500 to *c.* −100
3. Literary Chinese *c.* −100 to the early 20th century, overlapping with
4. *Paihua* 白話 or colloquial Chinese

Proto-Chinese and Classical Chinese reflect more or less standardised spoken languages from the periods in question. Literary Chinese largely remained fossilised throughout its history and never followed closely any spoken norm although it was consistently influenced by spoken practice.

---

[1] I shall not be concerned with Chinese philosophy and methodology of science. This is a matter to be discussed by scientists who are familiar with detailed Chinese scientific practice, and with the technical scientific literature.

Our treatment of the Chinese language will concentrate on the Classical Chinese language of the formative period of Chinese culture from about −500 to about −100, and on what we call Literary Chinese, which, basing itself on Classical Chinese as a model, was the slowly evolving more or less standardised medium of written communication throughout China from −100 onwards.

Before we turn to a detailed consideration of Classical and Literary Chinese, it will be both healthy and methodologically essential to put these varieties of Chinese in a proper historical and comparative perspective.[1]

The typology of East Asian languages

From a typological point of view East Asian languages may be divided into isolating ones which have very few traces of inflections, and agglutinative languages in which whole clusters of morphemes are 'glued-onto' the words they modify.[2] This obvious division might lead one to suppose a genetic relation between the isolating languages on the one hand and the agglutinative languages on the other. However, matters are not as simple as that. There has been complex interaction between these languages and we need sophisticated criteria to reconstruct the genetic relationships, if indeed the genetic paradigm is appropriate to explain the complex interrelationship of the languages concerned.

How, then, are we to decide whether two languages are genetically related or not? The classical comparative method worked out for Indo-European languages establishes sound-laws for correspondences of shared vocabulary and determines the nature and degree of shared features.

Using such sound-laws many scholars have tried to establish among the East Asian languages three genetic groups: 1. the Altaic (including Uighur, Mongol, Tungusic, Manchu, and possibly Korean and Japanese); 2. the Sino-Tibetan (including the Sinitic languages often referred to as Chinese dialects, as well as the very large number of Tibeto-Burman and Karen languages); 3. the Austronesian-Thai (including such languages as Vietnamese, Thai, Malay, and Tagalog).

The Sinitic languages

The Sinitic languages (or Chinese dialects) and some adjacent languages have one prominent feature in common: their non-composite words or morphemes all have only one syllable. We therefore call them monosyllabic languages. Both north and south of the Sinitic languages there are many languages with large numbers of morphemes containing more than one syllable. We call these languages polysyllabic. Historically, there has been profound influence from the northern and southern polysyllabic East Asian languages upon the Sinitic languages.[3] The variety within

---

[1] Cf. Vol. 1, pp. 27ff. For a survey of more recent research see vol. 2 of Sebeok (1967). I owe a special debt of gratitude to Søren Egerod for advice on this Sub-section.

[2] Yü Min 俞敏 has pointed out to me that this typological difference is very much one of degree.

[3] See Hashimoto (1976a), Hashimoto (1976b), Hashimoto (1986) and Egerod (1983), pp. 37–52.

what are known as 'Chinese dialects' is no smaller than that of languages in Europe.[1] Moreover, the diversity of languages spoken within the Chinese territory is vastly greater than the variety of languages spoken in Europe.[2]

Word order in the Sinitic languages tends to be fixed rather than free. Thus Propertius expresses a profoundly Chinese sentiment using profoundly un-Chinese word order when he says: *Fortunae miseras auximus arte vias*: FORTUNE'S MISERABLE WE-ADD ART WAYS. Chinese word order would have to be more like the English translation: 'We artificially add to fortune's miserable ways.'[3] Predominantly, but by no means completely, Chinese word order obeys what one might call 'a principle of adjacency', which says that no constituent of a sentence tends to modify any other constituent that is not immediately adjacent to it.[4]

All – or almost all – of the East Asian monosyllabic languages turn out to have evolved tones which serve to distinguish otherwise like-sounding or homophonous syllables from each other.[5] These tones or syllabic intonation patterns increase the number of distinct morphemes which can be expressed through monosyllables. Tones are particularly useful in monosyllabic languages, but it turns out that tonality is not a basic unchangeable feature of natural languages.[6] The history of East Asian languages shows that it is easily transferred from one genetic language group to another.

Northern versus Southern Sinitic

Among the monosyllabic tonal languages, the Sinitic languages (or Chinese dialects) are the most important and best known. These may broadly be divided into Northern Sinitic and Southern Sinitic languages. This division certainly antedates the phonological system reconstructed by Karlgren as Ancient Chinese (also known as the Chhieh Yün language named after the *Chhieh Yün* 切韻 Dictionary of +601). E. G. Pulleyblank, in his monumental recent work on the history of Chinese phonology, prefers to interpret this basic division as a contrast between the dialects as spoken in different capitals or cultural centres of the Chinese empire.[7]

The contemporary Chinese languages (or dialects) represent the result of migration – especially from north to south – which has led Northern Sinitic languages far

---

[1] See S. Egerod (1983), pp. 37ff.   [2] See Chang Kun (1967).   [3] Propertius II. vii. 32.

[4] An exception to this law is the final *erh i* 而己 in Classical Chinese which often refers to elements inside the sentence at the end of which it stands. See Harbsmeier (1986). Yü Min has kindly pointed out to me a very nice modern Chinese exception: *tsou-chhu ta mên chhi* 走出大門起 'went out of the door', where *chhi* 起 which is an unstressed form of *chhü* 去 'go' and modifies *chhu* 出 'to go out, leave'.

[5] The four tones of Modern Standard Chinese are historically developed from other phonetic features, quite possibly the following: the first high tone developed in words with an originally unvoiced initial consonant; the second rising tone developed in words with an originally voiced initial consonant; the third tone developed in words with a final laryngal stop (like the Danish (stød)) and the fourth tone developed in words which ended in an h-like sound. See Kennedy (1952b), Haudricourt (1954), and E. G. Pulleyblank (1964).

[6] We note that ancient Greek was a tonal language: what we tend to read as stress in Classical Greek is known to have been not stress but a tonal pattern somewhat like the Modern Standard Chinese rising second tone, while the circumflex accent is thought to have been pronounced somewhat like a Modern Standard Chinese short second rising tone immediately followed by a short fourth falling tone.

[7] Pulleyblank (1984).

south. The great Han dynasty migrations followed rivers and valleys through Hunan and Kuangsi to Indo-China and Kuangtung, and these same routes have been followed throughout history. Fukien was colonised from this direction as well as from the Lower Yangtze River and from the coast.

South Sinitic, on the other hand, has spread northward as a standard way of reading the Classical Chinese language.[1] Henri Maspero called his 1920 work on ancient Chinese *Le Dialecte de Tch'ang-ngan sous les T'ang*, but Chou Tsu-Mo has shown it to belong rather to the Lower Yangtse region.[2] Literary Chinese, the medium of most writing in traditional China, functioned as a bearer of the Southern Chinese tradition especially from Thang times onwards. That is why traditional Chinese poetry is more appropriately read in Cantonese than in the Peking dialect. Cantonese pronunciations are much closer to Thang dynasty Literary Chinese pronunciation.[3]

Modern Standard Chinese turns out, in this scheme of things, as a Northern Sinitic language with considerable accretions of loan words from Literary Chinese.[4] More recently a wealth of loan words have been incorporated from European languages – especially from English – mostly via Japanese.[5]

Structurally, Modern Standard Chinese has a fair amount in common with its Northern and Western neighbours,[6] but like all other Sinitic languages it has also received considerable influence from the Literary Chinese which was introduced to all parts of China together with literacy. In modern times there has been a strong tendency towards 'Westernising' grammatical patterns in Chinese.[7] For example, I count 130 'isms' like *Ma-Lieh-chu-i* 馬列主義 'Marxism–Leninism' in the recently published *A Reverse Chinese-English Dictionary*, (Peking, 1986), whereas traditionally China was mercifully free of all manner of 'isms'. Similarly, I count over 160 'isations' like *La ting hua* 拉丁化 'Latinisation' which differ markedly from indigenous Chinese formations in *-hua* 化 'ise' like *tsao hua* 造化 CREATE TRANSFORM 'good luck' and Buddhist loan translations like *tso hua* 坐化 SIT TRANSFORM 'die while sitting'. Cases of Westernised syntax in the modern writer Pa Chin's work are conveniently presented for the Western reader in Kubler (1985).

A typical Southern Sinitic language, modern Cantonese (with its many (sub)-dialects), turns out to be structurally and lexically closely related to Thai. In fact

[1] The Literary Chinese as described in the classic *Chhieh Yün* 切韻 (+601) rhyming dictionary is phonologically speaking clearly a variety of what we call Southern Sinitic more closely related to Japanese *gōon* 吳音 than to the Japanese *kanon* 漢音, as Pulleyblank has successfully shown. For a brief description of the correlation between change of linguistic standard and change of location of the capital city see Pulleyblank (1984), pp. 1–4. Pulleyblank minimises the effect of other languages, whether Altaic in the North or Austric in the South on the various Sinitic langauges. For the opposite view see Ballard (1979) and Hashimoto (1976).

[2] See Malmqvist (1968).

[3] Not surprisingly, the Northern Chinese have traditionally called themselves Han people (*Han jen* 漢人) after the Han dynasty, whereas the Southern Chinese prefer to call themselves Thang people (*Thang jen* 唐人) after the Thang dynasty.

[4] P. Demiéville (1950) and Egerod (1972).   [5] See Kao Ming-Khai *et al.* (1984).

[6] Particularly interesting in this connection is the Dungan language spoken in Far Eastern parts of the Soviet Union. See Dragunov (1936), Kalimov (1968), and more recently the very interesting study Imazov (1982). A fascinating sampling of Dungan sentences will be found in Salmi (1984).

[7] The most sensitive account of this remains Wang Li (1959), vol. 2, pp. 299–383.

there are certain varieties of Cantonese which, from a linguistic point of view, could perhaps be described as dialects of Thai which happen to have come to employ Chinese characters as a writing system.[1]

Archaic Chinese of the −1st millennium (including what we call Classical Chinese) turns out to be based on an ancient variety of Northern Chinese.[2] The most palpable evidence for this is the paucity in Archaic Chinese of sentence-final particles which is paralleled in Modern Standard Chinese, and which is in absolute contrast to the profusion of sentence-final particles in Southern Sinitic languages such as Cantonese (and Thai).

One might expect that at least the ancient languages of the −1st millennium were 'pure' Chinese in the sense that the basic vocabulary consisted of only Sinitic words. Such, however, is not the case. There certainly are quite a number of basic words for which we cannot establish regular cognates in other Sinitic languages. These are most plausibly explained as loan words. In some cases the source of these loan words may even seem to be Indo-European: *mi* 蜜 'honey', which in the oldest reconstructed pronunciation had a final t-sound, is generally taken to be a loan word from Indo-European, and related to Sanskrit *madhu*, Greek *methy*, μέϑυ, 'mead', and is also related to Russian *mjed*, English *mead*, Danish *mjød*. The horse is not indigenous to China, and neither is the word *ma* 馬 'horse', which in fact is generally assumed to be an Indo-European loan word related to the English *mare* and the German *Mähre*. The wild goose *yen* 雁 is often taken to be related to Greek *chēn* χῆυ 'goose' and German *Gans*.[3]

We mention these details to illustrate the fact that from the very beginnings of the written Chinese languages these have included non-Sinitic words. Sinitic languages in historical times never evolved in splendid isolation from surrounding languages. The indications are that at all known historical stages, the Chinese languages were structurally and lexically influenced by very complex linguistic surroundings. Chinese languages may be isolating languages, but they were not isolated languages.

Having said that, the fact remains that Literary Chinese has remained surprisingly reluctant to absorb large numbers of foreign morphemes or words. One only has to think of the vast number of Latin and French borrowings in English to appreciate the significance of the contrast with Chinese.[4]

---

[1] In varying degrees modern Sinitic languages show signs of substrata, i.e., that they represent varieties of Sinitic spoken by people who originally spoke another type of language which has left its traces in the language they speak. It has been suggested that even Archaic Chinese of the −1st millennium may not be an exception. See e.g., Egerod (1967) pp. 91–129; Hashimoto (1976).

[2] Even in its earliest known forms Chinese seems to have involved loan words not generally attested in Sino-Tibetan languages. For Austro-Thai loan words see Benedict (1975).

[3] See Forrest (1948), p. 136. For a more recent treatment of early Chinese loan words see Ulving (1969). Chang (1988) takes up the old theme of Joseph Edkins' classical study (1871) and surveys recent studies in the field. By far the most sophisticated work on the problem of the relations between Indo-European and Chinese has been done by E. G. Pulleyblank in a series of hitherto unpublished lectures delivered over the past 20 years, the latest being the John Fletcher memorial lecture at Harvard in April 1987 in which Pulleyblank argued that Chinese and Indo-European may have had a common linguistic ancestor located in Central Asia.

[4] See e.g., Serjeantson (1935).

When a vast number of basic French words were introduced into English there were complaints and real fears that English was losing its identity. In our times there are similar fears in France. There never were such fears and there never was reason for such fears in traditional China.[1] Even at the height of Sanskrit influence during the Thang dynasty, when poets like Han Yü 韓愈 (+768 to +824) could be extremely worried that Buddhism was infesting and replacing Chinese tradition, there was no tendency to replace or extensively supplement the basic vocabulary of Chinese with Sanskrit morphemes. Pai Chü-I 白居易 (+772 to +846) may have had reason for his vision that 'people's homes all turn into Buddhist temples',[2] but people's daily speech was never anything like as pervasively Sanskritised as English speech was Frenchified in medieval times.

The predominant practice in China was to produce loan words by paraphrasing meaning rather than by imitating sound. Thus the Chinese say *tien-shih* 電視 'lightning/electric-look' for 'television', unlike the Japanese who say *terebi*, and like the Germans who say *Fernsehen* ('distant-look'). Phonetic loans remain comparatively few in Chinese. *Chhan* for Sanskrit *dhyāna*, (Japanese *zen* 禪) 'meditation' is a familiar example of this. What was phonetically borrowed from Sanskrit were mostly specific technical Buddhist terms. And when Buddhism ceased to be an all-powerful foreign influence, the majority of the borrowed Sanskrit loan words disappeared from the language. Of the thousands of loan words from Sanskrit that have put in an appearance in Literary Chinese texts only a few hundred, such as *tha* 塔 (Sanskrit: *stūpa*) 'pagoda' survived as part of the general vocabulary into modern times.[3]

Insofar as Classical Chinese defined a very large part of the basic linguistic and logical conceptual framework for later Chinese thought, it deserves our special attention. This is not to deny the fact or to underestimate the importance of the fact that important modifications and innovations occurred in later times. None the less the present section on language and logic will pay special attention to the formative phase of the Classical Chinese language. The subsequent section on language and science will deal in much greater detail with later Literary Chinese sources for Chinese history of science.

(2) Spoken Chinese and the Semiotics of Chinese Characters

Characters and words

Aristotle maintained: 'What is spoken is a symbol (*symbolon*) of states (*pathēmata*) in the soul; what is written is a symbol of what is spoken.'[4]

---

[1] The modern invasion of English loan words is an entirely different matter. It is indeed a cause for profound concern among many Chinese in our time.

[2] *Liang Chu Ko, Pai Chü-I chi* 兩朱閣, 白居易集, vol. 1, p. 74.

[3] For an interesting but incomplete survey of loan words in Modern Chinese see Kao Ming-Khai *et al.* (1984). For rich bibliographic information on loan words in Chinese see Yang (1985), pp. 111–22.

[4] *De Interpretatione* 16A3.

The ancient Chinese said comparable things. The realist philosopher Han Fei 韓非 (died −233) has an anecdote about two unknown individuals, Wang Shou 43and Hsü Fêng, that vividly illustrates the primacy, from a certain point of view, of the spoken over the written medium:

Wang Shou was travelling along with written materials (*shu* 書) on his back. At a big cross-roads he caught sight of Hsü Fêng. Said Hsü Fêng: 'Conduct consists of actions. Actions arise from circumstance. The person who knows (*chih chê* 知者) has no constant pattern of conduct. Books consist of sayings. Sayings arise from knowing. Therefore the knowing person does not keep written materials. Why are you travelling along with these things on your back?' At this point Wang Shou burnt his written materials and danced round the bonfire.[1]

Yang Hsiung 揚雄 (−53 to +18) asks:

When our words are unable to express our thought (*yen pu nêng ta chhi hsin* 言不能達其心), and when our writing is unable to express our words (*shu pu nêng ta chhi yen* 書不能達其言), is this not a great difficulty?[2]

What writing is supposed to do here is to represent spoken words. But in so doing writing does represent what is in one's mind:

Therefore words are sounds of the heart/mind (*ku yen hsin shêng yeh* 故言心聲也). Writing is a picture of (what is in one's) heart/mind (*shu hsin hua yeh* 書心畫也).[3]

Wang Chhung 王充 (+23 to +100) writes in his autobiographical chapter:

Through words one makes clear one's intention (*chih* 志). One is afraid that words may disappear, and therefore one writes them down with characters.[4]

Chêng Hsüan 鄭玄 (+127 to +200) makes it clear in his explanation of the origin of phonetic loan words that to him what was written down (*shu* 書) were words of the spoken language. He looked upon Chinese as a logographic script in the sense that characters were designed to represent spoken words.[5] Ko Hung 葛洪 writes in *c.* +320:

What emerges from the mouth are words. What is written down on paper is writing (*shu*). Writing stands for words, and words are notations for things.[6]

Chhên Khuei 陳騤 concurs in his *Wên Tsê* (+1170):

Through words we get across (*tsai* 載) facts; through writing (*wên* 文) we note down words.[7]

---

[1] *Han Fei Tzu* 21.15.1, in Liao (1939), vol. 1, p. 219. Cf. also Chuang Tzu's story about Duke Huan of Chhi reading a book in his hall, who is lectured by wheelwright Pien on the primacy of living speech, the spoken word. (*Chuang Tzu* 13.6)

[2] *Fa Yen* 法言 8 ed. Wang Jung-Pao, p. 246, cf. Von Zach (1939), p. 23.   [3] *Ibid.*

[4] Wang Chhung, *Lun Hêng* 王充, 論衡 ed. *Chung-hua-shu-chü*, p. 1688, cf. Forke (1911), Part 1, p. 72. Cf. Chiang Tsu-1 (*1965*), p. 46.

[5] See Hung Chheng (*1982*), p. 132.   [6] *Pao Phu Tzu*, ch. 43, ed. *WYWK*, p. 748.

[7] Ed. Liu Yen-Chheng, p. 12.

The most telling kind of evidence is found in passages not at all explicitly concerned with the matters at hand, but with the way of reading texts. Chu Hsi 朱熹 (+1130 to +1200) is reported to have said:

Generally, when one looks at books, one must first read them until one is familiar with them, one must bring it about that it is as if the words all come from one's own mouth. Then one (must) continue to reflect subtly and bring it about that it is as if the meanings all come from one's own heart. Only then can one get the point.[1]

Chu Hsi reads the Classical Chinese texts as 'sayable Chinese'. The reader imagines himself saying the same words, not writing the same strokes of the brush. This is significant.

At the opening of an imperial edict of +1269 I came across this statement:

I believe that one writes characters in order to note down words, and one uses words in order to report things.[2]

From a certain point of view the primacy of the spoken word versus the written character is obvious in exclamations like *wu hu* 惡呼 'alas' which is already common in the Book of History. For this phrase there is a wide variety of graphs as there is indeed in all sorts of other onomatopoeic expressions, i.e., words that plainly imitate the sounds of this world.

One might say these are exceptional cases of no general typological significance, but take a common expression *wer i* 逶移, 逶迤, 逶迆, 'to writhe (gracefully), be graceful', which is attested already in the Book of Songs[3] for which Morohashi's great Chinese-Japanese dictionary (vol. 11, p. 638) provides eight alternative written forms, and for which I have found some more variant forms. It is almost as if it didn't much matter to the Chinese what graph the scribe employed as long as it was clear which word of the spoken language was intended. Compare the case of the English word 'through', for which we have no less than 350 orthographic variants recorded from the +13th to the early +15th century in England alone, with another 100 or so to be added from various dialect sources.[4]

But whereas in the case of English various ways of spelling 'through' reflect dialectal variation as well as scribal uncertainty or indifference, the situation is different in Chinese. In Chinese there can be an important suggestive nuance when the scribe uses the radical for 'path, to walk' (suggesting a winding road), or the insect radical and the word for 'snake' (suggesting a writhing snake), or the mountain radical on top (visualizing a difficult winding mountain path). The graphs offer glimpses of the range of imagery conjured up by the word in the writer's mind. The writing offers more than the word. It does have an autonomous graphic effect. Chinese characters often involve a kind of graphic poetry. And for certain Chinese

---

[1] *Chu Tzu Yü Lei*, ed. Chung-hua-shu-chü, 1986, vol. 1, p. 168.
[2] *Yüan Shih*, ed. *SPPY*, ch. 202, p. 1381, quoted in I. T. Zograf (1984), p. 11.　　[3] *Shih Ching* 18.1.
[4] See MacIntosh, Samuels, and Benskin (1986), vol. 4, pp. 96–101. These ways of spelling the word 'through' represent not only different scribal conventions but also dialectal differences on a scale unrecorded in China with the Chinese writing system.

writers this poetry of the graphs was important. In any case it has always fascinated Westerners.

Western views of Chinese characters

In the West, as opposed to China itself, an 'exotic' view of Chinese characters became current since the +15th century to the effect that these characters, quite unlike other writing systems, and rather like the mathematical symbols '1', '2', '3', etc., stand directly for concepts and only incidentally have pronunciations (pronunciations like English 'one', in Russian 'odin' or in Greek 'heis'). Following H. G. Creel we can call this the ideographic conception of the Chinese characters.[1] According to this conception, Chinese characters would turn out to constitute a kind of natural *Begriffsschrift*, a natural and culture-specific version of Leibniz's *characteristica universalis*, or an 'alphabet of thought', as Leibniz liked to say.

The opposite view, namely that Chinese characters stand for pronunciations of morphemes, or words of the spoken language, and mean what they mean because the morphemes have the meanings that they have in the spoken language, has been argued in the work of the Swedish linguist Bernhard Karlgren and the American scholar Peter A. Boodberg. Following an article by Y. R. Chao (1940), we may call theirs a logographic conception of Chinese characters. This view is by now quite generally accepted by specialists in Chinese linguistics.[2] This logographic view does not, of course, in any way deny that Chinese characters have important graphic etymologies and that in addition these characters have autonomous æsthetic dimensions of their own that may serve important literary purposes.

Some comparisons between Greek and Chinese may be useful at this point. We can trace the history of the pronunciations of Greek words in the spelling of the *Iliad* and the *Odyssey*, and we can trace variant pronunciations of the same words in written dialect material that has come down to us from ancient Greece. In China the evolution of the orthography of Chinese characters tells us precious little[3] about the history of the pronunciation of a given word. When it comes to dialects, it is in the nature of the Chinese script that a given word, though pronounced differently in different dialects, will normally be written with exactly the same character in all dialects.

---

[1] Creel (1936) and (1938). The term 'ideographic' seems to go back to the very scholar (Champollion) to whom we owe the deciphering of the ancient Egyptian script and the demonstration that the Egyptian script is *not* ideographic in the sense just outlined. Champollion's crucial discovery was that the Egyptian hieroglyphs can only be properly understood when one attributes to them phonetic values. (For the story of the discovery of Egyptian in the West see Erik Iversen (1963).) The idea of ideography is, of course, much older than this. Cf. Plotinus, *Enneads* V.8.6, ed. Paul Henry and Hans-Rudolf Schwyzer (*Plotini Opera*, vol. 2, 1959), p. 391.

[2] See e.g., Chao (1940), Boodberg (1957), Serruys (1943), p. 172ff., and Boltz (1986). Sampson (1985) gives the general linguistic context. DeFrancis, *The Chinese Language*, Honolulu 1984, ch. 8: 'The Ideographic Myth', gives some historical background.

[3] We do learn a certain amount about the pronunciations of Chinese characters in the course of Chinese history by observing which characters were used as phonetic elements in other more complex characters, and which were used as simple phonetic loans to represent other characters: such phonetic loans must indicate more or less close similarity of pronunciation of the characters involved.

Whereas the internal structure of a word in Greek will tend to be evident in the phonetic representation of that word, in ancient Chinese the internal morphemic structure of words written with one character, which in some cases we can try to reconstruct, is hidden under the blanket of the character.[1] This has meant that the Chinese morphemes written with characters had to be treated as unanalysable even when they did contain some internal structure. Grammatically significant slight distinctions in pronunciation as between *advise* and *advice* may have been present, in ancient Chinese pronunciation, but they were unfortunately not visible in the writing system. Again, such pairs as *practise* versus *practice*, where an original difference in pronunciation has been neutralised but is preserved in the more conservative writing system, would be indistinguishable in Chinese writing. All such cases would look in the Chinese writing system more like the case of *(to) promise* versus *(the) promise*, except that there was no inflexion in Chinese to help disambiguate between 'nominal' and 'verbal' uses.

Homophonous but etymologically entirely distinct words, rather like the English *sun* and *son*, or *right*, *write*, and *rite* may be written by the same character, be thus graphically indistinguishable, and create the quite spurious impression that Chinese words are grotesquely ambiguous. Examples of this sort are common. But there are also notorious cases where a single Chinese character is used to write two entirely different words with quite distinct meanings, as in the case of *shih* 石 'stone', which is also read as *tan* 'hectolitre of dry grain'.

What is ambiguous in all such cases is the writing system, not the words of the language. Axel Schuessler's new *Dictionary of Early Zhou Chinese* (1987) takes account of this situation and tries to be a dictionary of words rather than of characters.

It was the reconstruction of the ancient pronunciations of Chinese characters by B. Karlgren and others that has enabled us to carry the analysis to the inside of the unit written by a Chinese character. And the process of this systematic reconstruction started by Karlgren is still continuing. The most recent contribution being the important book by E. G. Pulleyblank, *Middle Chinese* (1984).

We can safely conclude that Chinese characters did hide a range of important features which phonetic spelling tends to reveal, and that they discourage the analytical perception of the internal structure of Chinese morphemes. Phonetic reconstruction of ancient pronunciations can certainly not make up for this disadvantage: for one thing these competing reconstructions are no more than working hypotheses,[2] and for another the fact that two characters are reconstructed as having the same pronunciation does not under any circumstances signify more than the inability of researchers to identify a difference by the means of reconstruction so far available.

Thus the fact that *jen* 人 'man' and *jen* 仁 'humaneness, goodness' are reconstructed as having identical pronunciations does not at all, in my view, allow us to

---

[1] Cf. Axel Schuessler (1976) and S. Egerod (1971).

[2] Reconstructions have been proposed in Baxter III (1986), in Light ed. (1968), pp. 1–34, Bodman (1968), in Light ed. (1968), pp. 34–199, Karlgren (1952b), Li Fang-Kuei (*1971*), Pulleyblank (1962), Schuessler (1987), and Wang Li (*1958*).

conclude that these two crucially distinct words really were pronounced exactly the same way. On the contrary, it would seem more likely that our reconstructions so far leave out certain features of ancient pronunciations, than that Classical Chinese was as full of such homophones as the reconstructions so far available would make it appear.

The typology and evolution of the Chinese writing system

Like other known writing symbols, many Chinese characters began as pictures of things.

Confucius said: 'When we look at the character *chhüan* 犬 (for a dog) it looks as if one had drawn a dog.'[1]

But it is profoundly significant that the name of the scribe in ancient times was not a designation of a painter or artist but that of a diviner: *shih* 史. The pictures and designs of the scribes/diviners of antiquity turned into writing to the extent that they were associated with fixed words of the spoken language.

From around −4800 onwards we have pottery inscribed with marks, and it must be emphasised that these marks were not, so far as we know, connected with divination. These marks have been considered by a number of distinguished scholars including Yü Hsing-Wu 于省吾 as an early writing system,[2] but there is no evidence that the marks represented meanings, words, sentences or phonetic values. We simply know too little about their function.[3] There are, however, most puzzling exceptions, like the ancient character for *tan* 旦 'dawn' depicting a mountain, and above it a horizon with a sun on top.[4]

The earliest evidence we have of writing in China dates from not much earlier than −1200. We have a large number of oracle bone inscriptions from this period.[5] The evidence we have suggests that Chinese writing was first used in the service of the ancient practice of divination. The inscriptions we have are records made by diviners for their successors, not ritual communications with some spirits. One can say that the Chinese script evolved along much the same lines as Mesopotamian and Egyptian writing did:

1. By the pictographic strategy pictures or pictographs were used to denote the words for what was depicted. This strategy can never have been very successful, since there are many words the meaning of which is not readily drawn. That problem was tackled by the phono-pictographic strategy.
2. By the phono-pictographic strategy the pictographs were used both for the words the meanings of which they depicted and also for other similar-sounding

---

[1] *Shuo Wen Chieh Tzu*, ed. Tuan Yü-Tashai, p. 473.
[2] Yü Hsing-Wu (*1972*). For a detailed survey of the early evidence see Cheung Kwong-yue (1983).
[3] Cf. Boltz (1986), pp. 430ff. Chang (1983), pp. 84ff., plausibly maintains that the most ancient inscriptions were potter's marks.
[4] Yü Hsing-Wu (*1972*). Cf. also Aylmer (1981), p. 6.
[5] A convenient, thematically organised anthology of these inscriptions is in Chang Tsung-tung (1970).

words. This raised increasing problems of ambiguity, since one given pictograph could stand for a series of more or less homophonous words. Such problems were reduced by the logographic strategy.

3. By the logographic strategy semantic elements were added to the pictographs in order to specify which of the homophones of a given pictograph was intended. These additional elements are generally known today as 'radicals', and their use to disambiguate graphs has been increasing until it became fairly, but never completely, standardised in Han times.

Historically, all three strategies coexisted: a given word could be represented either by a pictograph, a phono-pictograph or by a logographic character with a disambiguating element.

The next and fourth strategy would have been to select for each syllable of the Chinese language of the time one standard pictograph or phono-pictograph to represent that syllable. That would have led to a syllabic 'alphabet', but for interesting reasons on which it is tempting to speculate the Chinese never did quite take this step. Chinese writing thus always retained its logographic character: in Chinese one writes not syllables but words, and – as the language developed many poly-syllabic words – morphemes.

The standardisation of this logographic script remained incomplete. We count no less than nine variants for 'spring' in the early inscriptions: the picture of a bud seems to have been the only obligatory element. Added to the bud were such things as the sun, various numbers of either grass-pictures or tree-pictures.[1]

Chinese characters as indicators of pronunciation

Of the 9,353 characters purportedly defined in the early dictionary *Shuo Wên Chieh Tzu* 說文解字 (postface +100) at least 7,697 are claimed by Hsü Shen 許慎 (died *c.* +149) to contain phonetic elements indicating – more or less precisely – the pronunciation of the characters in question. It is therefore important to realise that Chinese writing, though it is not simply phonetic, none the less contains many phonetic elements.[2]

Let us look more closely at the modern *Hsin-hua Tzu-tien* 新華字典 of 1971 which contains 8,077 common characters and gives their pronunciations in Modern Standard Chinese.[3] 1,348 of these constitute elements recurring in other characters

---

[1] Chang Tsung-tung (1972), p. 22.

[2] On the other hand the choice of one phonetic element rather than another has often been determined by mnemonic 'semantic' considerations. As far as possible, one chooses a suggestive pictogram which is easiest to remember because it not only represents the right kind of sound but because it originally has a suggestive meaning. Thus even when an element is clearly phonetic, we are still entitled to look for a semantic reason for this phonetic element rather than another. For example, in *mên* 悶 'depressed, tightly closed' the element *mên* 門 'gate' is phonetic, but it is at the same time highly suggestive. Cases of this sort are common, and many of them are already noted in *Shuo Wên Chieh Tzu* 說文解字.

[3] We disregard the fact that different dialect pronunciations would give a somewhat different picture from that presented here.

as phonetic elements. They are, in a sense, indicators of pronunciation. No less than a further 6,542 characters contain such phonetic elements. Only 187 characters contain no hint whatever of pronunciation in the graph.

However, there are a number of important factors limiting the value of characters as indicators of pronunciation. One point is that the phonetic recurrent elements generally do not provide any hint of the tone in which a syllable is to be pronounced. Even leaving aside the question of tones, 68 (16%) of the 415 syllables which occur in Modern Chinese cannot be indicated by existing phonetically recurring elements and will thus always have to be learned separately.

Another serious problem is that only 52% of the characters with a recurring phonetic element have their pronunciations (regardless of tone) indicated exactly when we consider modern standard Peking pronunciation. Another 27% have at least the correct rhyme, but a changed initial consonant. The remaining 21% fail to correspond exactly even in rhyme and thus offer little help for remembering the pronunciation today, even when their is a certain phonetic similarity.[1]

Finally, of the recurring phonetic elements *c.* 10% have more than one phonetic value, which reduces their value as constant and reliable indicators of pronunciation.

A look through Bernhard Karlgren's *Analytic Dictionary of Chinese and Sino-Japanese* or the more up-to-date but less attractively arranged *Grammata Serica Recensa* will confirm that among Chinese characters we find a large number of phonetic series containing recurring phonetic elements and their reconstructed ancient pronunciations are conveniently lined up for us to compare. We also find a considerable number of phonetic loans where a character with one meaning is used for a homophonous or near-homophonous word with an entirely different meaning. Moreover, when reading ancient Chinese texts we frequently find that Chinese writers make use of *ad hoc* phonetic loans.[2]

Chinese characters were constructed *and used* extensively with reference to their pronunciations. Note the frequency of onomatopoeic words from most ancient times onwards. Chinese characters are and always were a device to note down spoken words or morphemes. They are not devices to depict meaning independently from the spoken language. A character, irrespective of its graphic composition, has the range of meanings of the word/morpheme it represents.

Because characters do not directly stand for meanings as such but for words of a living spoken language, a dictionary could be written of +1st century 'dialects' that purported to describe the varieties of language(s) spoken in Chinese territory, and this 'dialect dictionary' could be written *with Chinese characters*. If characters stood directly for meanings, then etymologically unrelated but synonymous morphemes in the dialects should regularly have been written with the same standard characters. In general, this turns out not to be the case. Hence the need for and the possibility of a 'dialect' dictionary written entirely in Chinese characters.

---

[1] For a detailed analysis of these data see Chou Yu-Kuang (*1980*), pp. 1–13. Chou treats the modern simplified characters without reference to their history.

[2] See Karlgren (1963) and (1964).

When the Japanese used Chinese characters to write their language, they did indeed write their own etymologically quite unrelated morphemes with semantically appropriate Chinese characters, the *kanji* 漢字. The *kanji* were conventional marks used to write down words of the Japanese language. To a very limited extent the same thing has also happened in the transcription of Chinese dialect words. Thus for the Taiwanese word *bêq* 'want to, will' we sometimes find the near-synonymous standard Chinese character pronounced *yao* 要. But in general it is clear that Chinese characters stand for etymologically and not just semantically related morphemes of the spoken language.

The advantages of Chinese characters

There is one significant advantage of Chinese characters for a language like Chinese, and this has to do with homophones, i.e., words of the same pronunciation. In English we have pairs like *morning/mourning* or *sun/son* which are unambiguous in writing while they are ambiguous in speech. On the other hand we have pairs like the verb *permit* versus the noun *permit* which are unambiguous in speech but ambiguous in writing. In English, phenomena of the first kind are not common while in Chinese they are endemic. Homophones pronounced *i* in the fourth tone can be counted by the score, and dozens of them are actually quite common. Under these conditions it is psychologically natural that the scores of neatly distinct characters with which these morphemes are written will be intimately linked with the identity of the morpheme itself. The Chinese character is drastically closer than the Greek spelling to the essence of a morpheme represented by it.[1]

All the ancient Chinese linguists have perceived a profound link between the hexagrams of the *Book of Changes* and the Chinese characters. Many characters show an ambition to provide emblems which make the true nature of the things designated as transparent as possible so that ideally one can approach the mysteries of the universe through the study of characters. Here lies a profound difference between the Chinese and the Near Eastern invention of writing which is spelt out in L. Vandermeersch (1988).

This explains the extraordinary Chinese fascination with calligraphy,[2] a phenomenon which is hard to understand for the outsider, except on a purely æshetic level. But Chinese calligraphy has a deeper metaphysical dimension. It touches the nerve of morphemes and aspires to reach the profound cosmic essence of things. At the same time it expresses a person's cultural personality and identity. The Chinese calligraphic tradition has a cultural depth quite unparalleled in the West.[3]

---

[1] The Chinese customarily refer to their words as *tzu* 字 'character', although they have a more westernised term *tzhu* 詞 for 'word'. Thus the Chinese today can be heard to say things like 'I didn't say a character' instead of 'I didn't say a word'.

[2] During the +4th century already, we are told, 58 styles of calligraphy were known. By the +6th century the number had risen to 120. See von Rosthorn (1941a), p. 142.

[3] The literature on Chinese calligraphy is vast. See Fu (1977), which contains a useful bibliography on pages 308 to 310.

And this is where the pictographic and thought-provoking internal structure of Chinese characters does come into our account. For the literate person such structures are associated psychologically with the morpheme as a kind of graphic etymology of the morpheme as well as to the essence of things. Of course, like all kinds of etymology and metaphysical resonance, this may be present to a greater or lesser extent in different individuals or at different times. But the result is that it becomes not entirely unnatural to say that the characters stand in a more or less direct relation to their meaning, to what the culture perceives as the essence of things. That is why the ideographic conception of Chinese characters has its poetic force and attractiveness.

E. F. Fenollosa and Ezra Pound have done more than anyone else in the West to explore this poetic potential of the Chinese characters.[1] A very fine presentation of the connection between the graphic semantics of Chinese characters and archeological finds in China is presented in Cecilia Lindqvist (1991), which is currently being translated into many European languages.

From a historical rather than literary point of view the advantage of Chinese characters lay in the fact that every literate dialect speaker would be able to read most of them in his own dialect and would thus to some extent regard them as his own.

The Chinese characters became the focus of cultural self-identification among the Chinese when any unified phonetic writing system might have been divisive and difficult to learn for non-speakers of the standard language. As has often been noted, the Chinese characters were eminently suitable as a *scriptura franca* of the empire and were thus to some extent adapted to their historical environment. It is for this reason − and not for reasons of traditionalist inertia among the Chinese − that the Chinese characters still seem to have a stable future in that country.[2]

Chinese dialects are not much less diverse than European languages of the Indo-European family of languages. What made the speakers of these diverse languages feel that they spoke 'the same language' was their effective (and beautiful) common writing system which constituted a link not only with other parts of China, but also to the common literary heritage.

Written and spoken Chinese

Until now we have concentrated on characters and we have identified these as certainly representing morphemes or words of the ancient spoken language. But there is a quite separate and more controversial further question: to what extent did the the Classical Chinese sentences written with these characters represent the current

---

[1] Cf. Fenollosa (1936) and Pound (1952a, 1952b and 1954). A. Fang (1953), and Kennedy (1964, in 'Fenollosa, Pound and the Chinese character') have commented on Fenollosa and Pound from a sinological point of view.

[2] On the inconvenience of characters see e.g., Lü Shu-Hsiang (*1980*), p. 37 and Lü Shu-Hsiang (*1984*), p. 127. It is profoundly misleading when Charles Aylmer closes his eminently useful booklet *Origins of the Chinese Script* (Aylmer (1981), p. 22) with the remark that 'It (the Chinese script) is in fact ideally suited to the language it represents.'

speech of the time? Essentially it is the question whether Classical Chinese is a natural language or just a language-related notational system. M. Granet thinks it is the latter:

The great difficulty for a systematic study of Chinese stems from the disparity between the spoken and the written language. This disparity is so strong today that one can ask oneself whether the written language is a true language at all, whether it is an exact representation of the thought (the only difference compared with our languages being that this representation is achieved by a graphic representation and not a phonetic one), or whether it is just a mnemotechnic notation which permits the reconstruction of a verbal expression for the thought.[1]

First of all we must be careful not to distort the issue by asking simply whether Classical Chinese written sentences completely represent the spoken Chinese language or not. No written language completely represents the corresponding spoken language. There are plenty of German words I can think of which one would never write and for which there is no recognised spelling.[2] Within modern China, this phenomenon is particularly common in the Min and Yüeh 'dialect' areas. There are colloquial constructions which we would hardly ever use in writing. We must assume that such constraints were even more real in ancient China: written Chinese never was a completely faithful record of the speech at any time. This much should, I think, be uncontroversial.

Moreover, the relation between the spoken and the written language has evidently not been the same throughout the ages. The *Iliad* was still recited, and for the first time written down in 'Homeric Greek', long after anything like 'Homeric Greek', if indeed it ever was a current spoken language, had become a matter of the past.[3] Similarly one must assume that the *Book of Songs* of the Chinese was written down at a time when its language already sounded archaic or was at least obsolescent. The crucial point that the Homeric poems and the *Book of Songs* have in common is that both, though written in a sometimes formulaic, somewhat artificial language, were evidently based on oral poetry which was only incidentally – almost literally *post festum* – written down and almost certainly first performed by illiterate people. Bards could be blind even after the invention of writing because they did not need to read.

Indeed, at least as late as the −3rd century it appears that the texts of the *Book of Songs* were known and understood by less learned scribes by their sounds, not their characters, as the phonetic way of writing quotations in the famous *Lao Tzu* 老子

---

[1] 'La grande difficulté d'une étude systématique du chinois provient de la disparité de la langue parlée et de la langue écrite. Cette disparité, de nos jours, est assez forte pour qu'on puisse se demander si la langue écrite est bien une langue véritable, c'est à dire une figuration exacte de la pensée (avec cette seule différence avec nos langues que cette figuration est obtenue par une représentation graphique et non phonétique), ou si elle n'est qu'un système de notations mnémotéchniques permettant de reconstituer l'expression verbale de l'idée.' (M. Granet (1920), p. 99. Henry Rosemont (1974), argues that written Classical Chinese was a notational system only very indirectly related to a spoken language.

[2] Cf. also Harper (1881) for an early account of the similar situation in Chinese.

[3] It is worth reminding ourselves that to some extent Homeric Greek is an artificial artistic language which is not thought to have been identical with any particular dialect of ancient Greek. However, it was surely close enough to what was being spoken to be easily comprehensible when heard.

manuscripts recently discovered would suggest.[1] In general, the profusion of phonetic loan characters throughout the epigraphic evidence accumulated through archeological discoveries must indicate that texts were remembered primarily not as graphic form but as spoken sound. The proliferation (and irregularity) of phonetic loans throughout even our printed texts is the strongest proof we have of the primacy of the spoken over the written forms of texts. In a predominantly illiterate society this is not in the slightest surprising: indeed anything else would be anthropologically and historically extraordinary.

From the −12th to the −11th century, the time from which we have our first extensive written documents in the Chinese language, the purpose and style of writing was sacral and formal: records were kept of communications with the gods. This naturally restricted the kinds of things that would be written down, but it also restricted the kinds of linguistic forms that would be included in the texts. From the −10th century onwards there followed a process of increasing secularisation of writing, and by the −5th century the written language was used to note down a wide variety of historical, philosophical and anecdotal material, although even at that stage writing something down was still a solemn thing to do.

Especially from the −5th to the −2nd century there is evidence that the written literature to various extents, though presumably never completely, represented spoken standards of the time.[2] The dialogues in texts like *Lun Yü* or the *Mêng Tzu* probably come as close to the spoken language of the time as Plato's almost contemporary dialogues.

It ought to be clear to any unbiased reader that the dialogues of *Lun yü*, *Mêng-tsi* and *Chuang-tsi*, the dramatically narrated episodes of the *Tso chuan,* etc., are the purest possible reproduction of a spoken language. We can positively hear the speakers, with all their little curious turnings, anacoluthic sentences, exclamations, etc.[3]

We do in any case have some direct evidence that what people said could be written down:

The emperor says nothing in jest. When he speaks the clerk will write it down.[4]

One might, I suppose, insist that what the scribe writes down is a translation into 'notationese' of what the emperor intended to convey rather than what he literally said. But we note that Confucius, for one, is *recorded* as first saying something and then claiming that he was only joking a moment ago.[5]

Even substandard colloquial Chinese seems to make its occasional appearance in our early sources. For example in the story about Chi Mao-Pien who was noted

---

[1] Cf. *Lao Tzu China Pên Chüan Hou Ku I Shu Shih Wên* 老子甲本卷後古佚書釋文, p. 22a, bamboo strip no. 339f.
[2] Chmielewski (1957), p. 74, says of pre-Chhin Chinese: 'The former is represented in the literary monuments of the Chou era and reflects to a considerable extent the living linguistic usage of the time.' I agree with this judgement.
[3] Karlgren (1929), pp. 177–8.
[4] *Lü Shih Chhun Chhiu* 18.2, ed. Chhen Chhi-Yu, p. 1156, and *Shuo Yüan* 1.11, ed. Chao Shan-I, p. 8.
[5] *Lun Yü* 17.4, in Lau (1983a), p. 171.

for his fondness for dirty language about others (*to tzu* 多訾). His language is so vulgar that one of the characters used in *Lü Shih Chhun Chhiu* 呂氏春秋 to represent his speech has rarely entered any traditional Chinese dictionary.[1]

But let us leave jokes and possible vulgarisms aside and consider the grammatical particles of Classical Chinese instead. These will, as I argue, and as the Chinese through the ages have seen it, originally represent words from the spoken language. Chêng Hsüan 鄭玄 (+127 to +200) describes the particle *chü* 居 as a dialect word: 'It is an auxiliary particle used in the area between Chhi and Lu.'[2] Essentially the uses and functions of grammatical particles must have been taken (though possibly adapted) from the contemporary colloquial language. In the literature of the −5th to −2nd century the usage of particles strikes one as strictly idiomatic and we find few grammatical anachronisms. When we get to the −1st century, the situation changes.

A. Conrady (1926) has identified a colloquial particle *yen* in the Book of Songs which clearly went out of use around the −6th century. We know little of the colloquial language of the time, but Conrady's arguments suggest that *yen* 言 was a colloquial particle at the time. Again, the sentence-final modal particle *i* 已 has been shown to have been well established as a colloquialism around the time of Confucius.[3]

Bernhard Karlgren has attempted to identify dialects in Classical Chinese. Certainly, the language of the *Tso Chuan* 左傳 is distinct from that of the *Lun Yü* and the *Mêng Tzu*. Such differences are perhaps most naturally explained as reflecting different dialectal varieties of spoken Chinese or dialects, especially since − as we have seen − such differences in dialect are mentioned in our ancient texts.[4] Some eminent scholars like H. Maspero and A. Forke have seen the differences as a matter not of dialect but only of written style,[5] but this notion is less than plausible.

Finally, there is an interesting piece of evidence first brought up by Bernhard Karlgren in 1929, the case of stuttering as recorded in *Shih Chi* 史記 (−1st century):

'I cannot get the words out my mouth.' he replied. 'But I know it will n-n-n-ever do (*chhên chhi chhi chih chhi pu kho* 臣期期知其不可)! Although Your Majesty wishes to remove the heir apparent, I shall n-n-n-ever obey such an order (*chhên chhi chhi pu fêng chao* 臣期期不奉詔).'[6]

---

[1] 'The heir is so wicked that his looks exceed those of NN (*thai tzu chih pu jên kuo NN shih* 太子之不仁過頤涿視)' (*Lü Shih Chhun Chhiu* 9.3, ed. Chhen Chhi-Yu (1984), p. 490. Wilhelm (1928), p. 109 rises to the occasion by translating '*er hat es dick hinter den Ohren*'). The parallel in *Chan Kuo Tshe* no. 139, ed. Chu Tsu-Keng, p. 479, cf. Crump (1970), pp. 170ff., is little help.) NN is clearly described as a name referring to a wicked person by Kao You's commentary. Since the character in question is not in any traditional dictionaries, it is obviously tempting to explain it as a variant of an existing character. But if it is really a variant, why does this miswritten character occur exactly in the direct speech of someone notorious for his dirty speech? Even more important: if the character had been understood as an epigraphic variant of another character, why have none of the lexicographers, who used the *Lü Shih Chhun Chhiu* extensively, added this character to their treasury of variant characters? The most plausible explanation seems to be that the character was in fact felt to represent a vulgarism, and was omitted on the same grounds as a number of other characters representing exceedingly vulgar words.
[2] *Li Chi Chi Chieh* ed. Sun Hsi-Tan, vol. 2, p. 55.
[3] See Harbsmeier (1986).   [4] Bernhard Karlgren (1929) argues this point.
[5] Cf. *Orientalische Literaturzeitung* 1928, p. 514, and *Journal Asiatique* 1928, p. 150.
[6] *Shih Chi* 96, ed. Takigawa, p. 5; cf. Watson (1958) vol. 1, p. 260. The passage was first discussed in B. Karlgren (1929), p. 178.

B. Karlgren (1929), p. 178, comments: 'It ought be evident to anybody that the stutterings *k'i k'i* . . . would not be inserted into a "literary, non-colloquial" sentence.' However, it must be said that it is not inconceivable that the stutter could be introduced into a sentence written in a different language. B. Watson's English translation of our passage quoted above provides instant exemplification. Stuttering could just be represented by Chinese chartacters, but other speech defects such as that of Alcibiades reported by Aristophanes in *holāis Theōlon? tēn kephalēn kolakos echei,* ὁλᾷς θεωλόν; τὴν κεφαλήν κόλακος ἔχει, 'Do you see Theoros? He has the head of a crow (flatterer)',[1] are not so easily rendered by characters.

It is the coherence of the detailed picture we can reconstruct of the historical evolution of colloquial Chinese grammar which makes it so plausible to assume that our Classical Chinese sources are written in what was close to the spoken language of the time.[2] Alain Peyraube (1988) provides a detailed case in point: the evolution of the dative construction in Chinese. Wang Li (1958) and Ōta Tatsuō 太田辰夫 (1958) are standard handbooks on Chinese historical grammar which will illustrate our point in a more general way.

What we call Literary Chinese is a language which is to varying degrees removed from the spoken Chinese of its time. Chinese writers since the historian Ssuma Chhien 司馬遷 (*c.* −145 to −89) have occasionally commented on the gap between the written and the spoken medium. In his time the written Chinese language was beginning to become an autonomous medium only partly determined by the colloquial language. But Bernhard Karlgren still insists: 'I go so far as to say that I believe even in the Han period the written language was not far removed from the colloquial.'[3] Many of the strict idiomatic rules which had governed the use of grammatical particles until then ceased to be obligatory. Literary Chinese became increasingly (but never completely) an autonomous written medium.[4] In a felicitous phrase, Henri Maspero called Han dynasty Chinese *'une sorte de koinē littéraire'*.[5] The concept *koinē* is particularly apt because it indicates that what was written was 'a second language' not identical with the local dialects/languages spoken in China at the time.

For example, a poem like *Khung Chhüeh Tung Nan Fei* 孔雀東南飛 (Southeast Fly the Peacocks), probably of the +5th or +6th century[6], is plain folklore and colloquial

---

[1] Plutarch, *Lives,* ed. Loeb, vol. 4, p. 4. The Greek ought to have been *horais Theōron* ὁρᾷς Θεώρὸν. In Alcibiades's speech *korax* 'raven' becomes *kolax* 'flatterer'.

[2] By studying the history of Chinese literature we can trace the development of the early colloquialisms of the Book of Songs, the colloquial language of Confucius, of Mêng Tzu (−4th century), and then the more standardised common language of the −3rd century as represented in *Han Fei Tzu,* the language of the commentators of the +2nd century, the language of the colloquial Buddhist texts from the +3rd century onwards, of the popular ballads like *Khung Chhüeh Tung Nan Fei* 孔雀東南飛 of the Zen Buddhist colloquial texts and the poetry from the +8th century onwards, the colloquial texts of Chu Hsi's conversations of the +12th century, the colloquial plays of the +13th century, and then the colloquial novel from the +15th century onwards down to the 20th century and to early Modern Standard Chinese.

[3] B. Karlgren (1929), p. 178. The distinguished German scholar Alfred Forke, on the other hand, believed that China never had a written language directly based on the colloquial language.

[4] Cf. Harbsmeier (1981), p. 9. An important change seems to have occurred between the composition of the *Huai Nan Tzu* 准南子 and the *Shi Chi* 史記.

[5] *Journal Asiatique,* vol. 222 (1933) p. 47.

[6] See Mei Tsu-lin (1982) for the dating of this poem, and Fränkel (1969).

in style. One strongly senses in this ballad the rhythms and patterns of oral literature, and one can demonstrate the presence of many universal features of oral poetry all over the world. It is extremely unlikely that a poem like *Khung Chhüeh Tung Nan Fei* should have been deliberately concocted in a language which was not currently spoken, used in oral poetry and naturally understood at the time.

In the Buddhist colloquial Chinese literature of the +3rd to +5th century we also find a wide range of popular usages and particles that were, and long remained, unacceptable in formal Literary Chinese.[1] By that time a fairly clear stylistic gap had opened up between respectable Classical Chinese texts and unrespectable colloquial literature of various kinds. An incidental remark in the autobiography of Emperor Hsiao 蕭繹 (+508 to +555) suggests that whatever the differences between colloquial and written Chinese, he was able to understand books in classical Chinese when they were read out to him:

Since the age of 14 I have suffered a chronic eye disease which has made me increasingly blind, so that after a while I could not read for myself any longer. For the last 36 years I have employed servants to read aloud for me. To quote the words of Tsêng Tzu: 'Reciting poetry, reading prose, I live with the men of old. Reading prose, reciting poetry, I hope together with the men of old.'[2]

One may say that the separation of the written from the spoken language over the ages contributed to the problems of illiteracy in ancient China and in any case made Chinese books less accessible than they might have been, and that it made the reading of books more of an elitist concern than it need have been. To some extent one may also feel that it failed to encourage the sort of spontaneous creativity which was at the heart of the Renaissance in Europe. Colloquial texts (like the *Khung Chhüeh Tung Nan Fei* 孔雀東南飛) largely remained a subcultural phenomenon in China until the 20th century. This was not the case after the Renaissance in Europe.

It turned out that there were also certain advantages attached to the stabilisation of the written language and its relative isolation from the changes of colloquial idiom. For this stabilisation over the centuries meant that the literature of all ages since the −6th century continued to be read by many right down to the 20th century. Through the stabilisation of Literary Chinese, the distant past was made to look more familiar and accessible to the present. After all, the Chinese would be reading sentences in the ancient books which they might literally use in the next letter they would write. In this way the literary heritage could be felt to be a natural living part of the present.

In the context of science this meant that Chinese scholars throughout the ages have been able to draw with natural ease on achievements of the past. The literary language played the crucial role that Greek and (later) Latin used to play in Europe, that Biblical Hebrew plays among orthodox Jews, and that Classical Arabic to a

---

[1] See Gurevich (1974), which is the standard monograph on the subject, and Zürcher (1984), which adds fascinating material but seems unaware of Gurevich's earlier work.
[2] Kuo Têng-Fêng (*1965*), p. 182.

considerable extent continues to play today in the Islamic world. Literary Chinese was the medium of the all-important cultural continuity in traditional China. The effective loss of our Latin and Greek past should remind us vividly of the importance of such a medium of historical continuity.

(3) TRADITIONAL CHINESE COMMENTS ON LANGUAGE

Before we discuss our modern views on the ancient Chinese language, let us listen to what the ancient Chinese themselves had to say on the subject of their own language. It will emerge in our section on lexicography that by Han times, at least, the Chinese were able to analyse and classify their characters in a remarkably systematic way. But did they generally reflect on the nature and purpose of (their) language?

In pre-Han texts we find a few graphological comments on the structure and significance of Chinese characters, as when the *Tso Chuan* 左傳 commentary (+4th century) (Duke Hsüan, 12th year) interprets the character *wu* 武 to be composed of the symbol for 'spear' and for 'to stop'. These attempts at popular graphic etymology are not at all common in early China.

The clearly predominant traditional Chinese interest in language was social, not theoretically linguistic in a narrow sense. There is no Classical Chinese word for linguistics.[1] Language tended to be considered from what we today might be tempted to call a 'socio-linguistic' point of view.

Already Confucius seems to have recognised a cognitive function for language:

If you do not know about words (*yen* 言), you do not have what it takes to know about men (or: others).[2]

The important thing that language allows us to do, according to Hsün Tzu, is setting up distinctions between the different things in the world.

What is it that makes man into man? It is the fact that he makes distinctions (*pien* 辨).[3]

The result is expressed in an ancient proverb:

From words one knows things.[4]

As a philologist one heartily agrees. But what exactly, one wonders, would constitute 'knowing about words' to the ancient Chinese? The question is put in the book *Mêng Tzu* and the answer there makes it clear that Mêng Tzu was in any case not especially referring to what we understand by linguistics in any narrow sense.

'What is "knowing about words"?'
'To know from biased words what someone is blind to, to know from immoderate words what someone is ensnared in, to know from heretical words where someone has strayed from the right path, to know from evasive words where someone is at his wit's end.'[5]

---

[1] Chang Ping-Lin 章炳麟 (1868 to 1936) was the first to coin the phrase *yü yen wên tzu chih hsüeh* 語言文字之學. See Li Jung (*1971*), p. 1.
[2] *Lun Yü* 20.3, see Lau (1983a), p. 205.   [3] *Hsün Tzu* 5.24, see Dubs (1929), p. 71.
[4] *Tso Chuan*, Duke Chao, 1.2 (end); see Legge (1872), p. 576.   [5] *Mêng Tzu* 2A2, see Lau (1983c), p. 59.

The Chinese did not have a distinct abstract notion of language as opposed to speech, talk, words, no division between *dialektos* 'language' versus *logos* 'word, speech'.

The essence of man, that which is peculiar to man, is declared to be language in a passage from one of the earliest textual commentaries in Chinese literature:

That which makes man into man is speech (*yen* 言). A man who cannot speak, how can he count as a man?

That which makes speech (*yen*) into speech is truthfulness (*hsin* 信). Speech that is not truthful, how could that count as speech?

That which makes truthfulness into truthfulness is the Way (*tao* 道). If truthfulness is not in accordance with the Way, how can it be truthfulness?[1]

Man is here conceived and almost defined as a speaking animal. His ability to speak truthfully and in accordance with the Way is what makes him properly human. It is by no means a coincidence that such a view of language is advocated in a systematic commentary on an older text: the *Ku Liang* 穀梁傳 and the *Kung Yang* 公羊傳 commentaries constitute coherent evidence of how seriously the Chinese took the matter of language, of diction, of formulation. The *Kung Yang* commentary relentlessly asks why the older Annals, on which it comments, use one word rather than another, why the text says what it does say, and why it says it in one way rather than another. For the study of early linguistic sensitivities the *Kung Yang* and *Ku Liang* commentaries provide crucial early evidence.[2]

The Book of Rites (*Li Chi* 禮記) takes a slightly different perspective on language:

The parrot can speak, but it is not distinct from (*pu li* 不離) the birds. The *hsing-hsing* 猩猩-ape can speak, but it is not distinct from (*pu li*) the wild animals. Now when a man has no sense of propriety, even if he can speak, does he not have the mentality of a wild animal? It is because the animals have no sense of propriety that father and son copulate with the same female. Therefore the sages instituted the rites to educate men, and they made men understand that because they had a sense of propriety they belonged to a different category (*pieh* 別) from birds and beasts.[3]

Thus the *Li Chi* insists that propriety, not language, is what makes man into man.

Definitions of language

The Chinese have certainly used more of their analytical skills on the analysis of propriety than on that of language. The commonest definitions of language involve communication:

Confucius said: 'An ancient book says that words serve to give adequate expression to the (communicative) will (*yen i tsu chih* 言以足志); written style serves to give adequate expres-

---

[1] *Ku Liang Chuan*, Duke Hsi, 22.4.
[2] For a close linguistic study of this early tradition see Malmqvist (1971).
[3] *Li Chi chi-chieh*, ed. Sun Hsi-Tan, *WYWK*, vol. I, p. 10.

sion to words. If you do not speak, who will understand your (communicative) will (*chih* 志)? If your words have no elegant style, they will not reach far.'[1]

Words (*yen* 言) are the means by which one makes plain (*yü* 喻) one's communicative intention (*i* 意).[2]

Words (*yen*) are the means whereby what is of ourselves is communicated to others. Hearing is the means whereby what is of others is communicated to oneself.[3]

As for words (*yen*), these are the means by which we dredge out our chests/minds and bring out (*fa* 發) our true feelings (*chhing* 情).[4] Sentences (*tzhu* 辭) are the means by which men make themselves understood to others (*jên chih so i tzu thung yêh* 人之所以自通也).[5]

Language is seen as a means of communication also within the individual personality in the following intriguing passage, which is quite unique for the attention it pays to internal psychology and the rôle of language:

> The heart/mind (*hsin* 心) is such that it hides a heart/mind (*hsin*). Within the heart/mind (*hsin*) there is another heart/mind (*hsin*). As for this heart/mind (*hsin*) of the heart/mind (*hsin*), thought/communicative intention (*i*) precedes speech (*gen*). Only when there is articulated sound (*yin* 音), does it (i.e., thought) take shape. Only when it takes shape is there speech. Only when there is speech is there internal control (*shih* 使). Only when there is internal control is there order. When one does not order things, there is bound to be chaos.[6]

The limitations of language

Words are conceived as a means to an end. Ultimately, this end tended to be practical. When the end is reached the means cease to be important. The traditionally dominant conception is clearly stated in the opening phrase of the book *Fa Yen* 法言 by Yang Hsiung 揚雄:

> In study, practising (the thing learnt) is the most important thing. Expressing it in words is secondary.[7]

In any case, we have a whole chapter of an ancient book on the theme:

> When sages make themselves understood to each other, they do not rely on words.[8]

For further discussion on this subject the reader is referred to the Section on logical and grammatical explicitness, (*c*,5).

Words are not in themselves important symbols to the ancient Chinese mind, as they certainly were to the ancient Indian mind. They were not regarded in China as complete representations of what they transmit. They are conventional tools only.[9] This is a point of view shared by Confucians and the Taoists.

---

[1] *Tso Chuan*, Duke Hsiang, 25.7fu2, see Legge (1872), p. 517. For the Chinese ideals of style, see our Section (*b*,7).
[2] *Lü Shih Chhun Chhiu* 18.4, ed. Chhen Chhi-Yu, p. 1177; cf. Wilhelm (1928), p. 300. Cf. also *ibid.* 18.5, ed. Chhen Chhi-Yu, p. 1185; cf. Wilhelm (1928), p. 304.
[3] *Huai Nan Tzu*, ch. 20, ed. Liu Wen-Tien, p. 20b.
[4] *Shuo Yüan*, ch. 8, ed. Chao Shan-I, p. 218.   [5] *Shuo Yüan*, ch. 11, ed. Chao Shan-I, p. 301.
[6] *Kuan Tzu* 49, ed. Tai Wang, vol. 2, p. 101f.; cf. Kuo Mo-Jo *et al.* (1957) (*Kuan Tzu Chi Chiao*), p. 788.
[7] *Fa Yen*, ch. 1, ed. Wang Jung-Pao, p. 3a.
[8] *Lü Shih Chhun Chhiu* 18.3, ed. Chhen Chhi-Yu, p. 1167; cf. R. Wilhelm (1928) p. 297.
[9] Also, Chinese poetry is justly famous for its economy of diction and powers of suggestion rather than its comprehensiveness of articulation or representation.

Therefore when names are sufficient to point out (*chih* 指) realities and when sentences (*tzhu* 辭) are sufficient to bring out what he is aiming at, the gentleman leaves off.[1]

Of course, there is a difference in style and in the extent to which humorous and poetic sensitivity get into the picture in some texts that have come to be regarded as Taoist:

With nets one captures fish. Once one has captured the fish, one forgets the net. With traps one captures hares. Once one has captured the hares, one forgets the trap. With speech one captures communicative intention (*i* 意). Once one has captured the communicative intention (*i*), one forgets the speech. How can I find someone who forgets about speech and speak to him?[2] Consummate words dispense with words. Consummate action dispenses with action.[3]

The *Chuang Tzu* is in profound doubt whether words are an altogether effective means of capturing/expressing meaning:

Saying is not blowing breath. Saying says something. The only trouble is that what it says is never quite fixed. Do we really say something? Or have we never said anything? If you think it different from the twitter of fledgelings, is there proof of the distinction?[4]

We have here a theoretical insistence on the indeterminacy inherent in the Chinese language. Words are not conceived as an eternally adequate representation of reality, as they were by the writers of Sanskrit. For the Taoists words are only pointers to a reality which they do not truly reach.[5] This perception of the nature of language is, I feel, endemic in Chinese literary culture, even when that culture was predominantly Confucian. It comes out in the absolutely crucial, and very popular, Chinese notion of the 'picture beyond the picture' in Chinese art, and of the 'intended meaning beyond the words (*yen wai chih i* 言外之意)' in poetry. If my view of Chinese grammar is correct, this way of thinking may even have affected the character of the Chinese grammatical system.

The *Chuang Tzu* revels in the insufficiencies of the written as against the spoken word, and in the insufficiencies of the spoken word as a vehicle for intended meaning.[6] That famous passage is not unique:

Sentences (*tzhu* 辭) are only the outward manifestation of intended meaning (*i* 意). It would be paradoxical if one were to concentrate on the outward form and neglect the intended meaning (*i*). Therefore, the people of old, when they had got hold of the intended meaning (*i*) discarded the words (*yen* 言). When one listens to words, this is because one glimpses the intended meaning (*i*) by means of the words.[7]

Quite generally, language was regarded as secondary to action, a point of view by no means alien to the Greek tradition. The autonomous theoretical and

---

[1] *Hsün Tzu* 22.49, see Dubs (1928), p. 292. Cf. *Shuo Yüan*, ch. 11, ed. Chao Shan-I, p. 302.
[2] *Chuang Tzu* 26.48, see Watson (1964), p. 302.   [3] *Chuang Tzu* 22.84, see Watson (1964), p. 247.
[4] *Chuang Tzu* 2.23, see Graham (1964), p. 52.
[5] Could this be the import of the sophist Hui Shih's 惠施 enigmatic saying 'pointing does not reach (*chih pu chih* 指不至)', quoted in *Chuang Tzu* 33.76?
[6] Cf. Vol. 2, p. 122, *Chuang Tzu* 2.23, and Graham (1981), p. 52.
[7] *Lü Shih Chhun Chhiu* 18.4, ed. Chhen Chhi-Yu, p. 1179; cf. Wilhelm (1928), p. 303.

non-pragmatic value of words counted only in so far as they were connected to action (*hsing* 行), use (*yung* 用). This is a recurrent theme. The disrespect for discourse as such is graphically expressed in the following saying attributed to Confucius:

The Master said: 'Thus when the Way prevails in the world action flourishes. When the Way does not prevail talking flourishes.'[1]

An alternative view of language

It was a commonly held conviction that language had to conform to ancient norms, and occasionally we find references to semi-divine origins of nomenclature, as of most human inventions:

The Yellow Emperor assigned the correct designations to the various things (*chêng ming pai wu* 正名百物).[2]

The *Chhun Chhiu Fan Lu* 春秋繁露 is widely regarded as an early Han document. However, Göran Malmqvist has accumulated evidence which inclines him to regard the book as a whole as a product of Six Dynasties times around the +6th century, although it does contain earlier material.[3] Whatever its precise date, the *Chhun Chhiu Fan Lu* provides an interesting account of names that deserves our attention[4]. Its sacral view of names is worth studying, if only as providing a useful contrast to what I see as the dominant tradition:

Names represent the intentions of Heaven as expressed by the sages (*ming tsê shêng jên so fa thien i* 名則聖人所發天意).[5]

Even the Taoists could claim things like:

Words that are not in harmony with the former kings cannot be considered as the Way (*tao* 道).[6]

Linguistic conventions and varieties

In spite of occasional references to the Yellow Emperor as the originator of the right use of names, the contrast to biblical ideas of a divinely sanctioned nomenclature of things, or to Indian ideas of an eternally right holy language is striking. Hsün Tzu's pragmatic view of language is so basically sensible that it almost strikes us as banal, because we have got so used to the idea of linguistic conventions:

Names do not have an unchanging proper use. One designates by names because of conventions. Only when the convention is fixed and a usage is established does one speak of the

---

[1] *Li Chi*, ed. S. Couvreur, vol. 2, p. 507.   [2] *Li Chi*, ed. S. Couvreur, vol. 2, p. 269.
[3] G. Malmqvist has unfortunately not yet published his findings.
[4] Hsü Fu-Kuan (*1975*) (Liang Han ssuhsiang shih) vol. 2, p. 240 maintains that the *Chhun Chhiu Fan Lu* takes the names back to the stage where they were magical spells of primitive societies. Cf. also Kuan Han-Heng (*1981*).
[5] *Chhun Chhiu Fan Lu*, ed. *SPTK*, 10.1b3.   [6] *Huai Nan Tzu*, ch. 20, ed. Liu Wen-Tien, p. 16b.

'proper use' of a name . . . There is no fixed correspondence between names and realities. Only when the convention is fixed and a usage is established does one speak of a (correct) designation for a reality.[1]

Compare Aristotle: *to de kata synthēkēn, hoti physei tōn onomatōn ouden estin, τὸ δὲ κατὰ συνθήκην, ὅτι φύσει τῶν ονομάτων οὐδέν ἐστιν.* '(Names have their meaning) by convention, for none of the names are by nature.'[2] Hsi Khang 嵇康 (+223 to +262) echoed Hsün Tzu – and, I suppose Aristotle – when he said in his essay arguing that music is beyond pleasure and grief:

This is because the mind is not tied to what is said. Words are sometimes not sufficient for bringing out (*chêng* 證) what is on one's mind. . . . Words are not things that are naturally fixed once and for all (*fu yen fei tzu jan i ting chih wu* 夫言非自然一定之物). In the Five Regions customs differ, and the same thing has different appellations (*hao* 號).[3]

The conventions were recognised to change with time:

Ancient ordinances (or terms, *ming* 命) are often not comprehensible as modern speech.[4]

Conventions, of course, can also differ in different parts of a country. The variation of the languages/dialects spoken in China were recognised and commented on at an early stage:

The people of the Five Regions[5] speak mutually incomprehensible languages, and their predilections are different. Those who act as translators and who interpret the desires between these different groups are called *chi* 寄 in the east, *hsiang* 象 in the south, *ti shih* 狄氏 in the west, and *i* 譯 in the north.[6]

Mêng Tzu, too, noticed the way in which one's language depends on one's environment:

Mêng Tzu said to Tai Pu-Shêng: 'Do you wish your kind to be good? I shall speak to you plainly. Suppose a counsellor of Chhu wished his son to speak the language of Chhi. Would he have a man from Chhi to tutor his son or would he have a man from Chhu?'

With one man from Chhi tutoring the boy and a host of Chhu men chattering around him, even though you caned the boy every day to make him speak Chhi,[7] you would not succeed. Take him away to some district like Chuang and Yüeh for a few years, then even if you cane him every day to make him speak Chhu, you will not succeed. Now, you have placed Hsüeh Chü-Chou near the king because you think him a good man. If everyone around the king, old or young, high or low, was a Hsüeh Chü-Chou, then who would help the king to do evil? But if no one (else) around the king is a Hsüeh Chü-Chou, then who will help the king to do good? What difference can one Hsüeh Chü-Chou make to the King of Sung?[8]

---

[1] *Hsün Tzu* 22.25, see Watson (1963), p. 144.   [2] *De Interpretatione* 16a28.
[3] Tai Ming-Yang (*1962*), p. 211.
[4] *Lü Shih Chhun Chhiu* 15.8, ed. Chhen Chhi-Yu, p. 935; cf. Wilhelm (1928), p. 230.
[5] North, south, east, west and centre.   [6] *Li Chi*, ed. S. Couvreur, vol. 1, p. 296.
[7] For the verbal use of *Chhi* 'speak Chhi dialect', compare, e.g., the German verb *berlinern*.
[8] *Mêng Tzu* 3B6, see Lau (1983c), p. 125.

Language is described as a matter of convention or habit (*hsi* 習) in a rather interesting piece of early psycholinguistic observation:

Now a madman may gnaw at himself, oblivious that his is not the flesh of an animal raised for food. He may eat dirt, unaware that it is not millet or rice. None the less, a madman of Chhu will speak the Chhu language, while a madman from Chhi will speak the Chhi language. That is a matter of habit. Now the effect of habit on man is such that . . . it is more adhesive than glue or lacquer.[1]

Hsün Tzu summarises with his customary lapidary precision:

Children from Kan, I, Yüeh, or Mo all utter the same sounds at birth, but when they grow older, they develop differing customs. Instruction (*chiao* 教) brings this about.[2]

During the Han dynasty, a whole dictionary of 'dialects' was compiled, the *Fang Yen* 方言 by Yang Hsiung 揚雄 (−53 to +18). In their cultivation of the art of definition and in their lexicography, the Chinese have shown a most extraordinary attention to semantic detail and documentation which is quite unparalleled by Western dictionaries of the same period. Definition and lexicography in China deserve their separate Sub-sections, to which we shall turn below.

Chêng ming 正名 '*the right use of names*'

Confucius recommends discipline in the use of words:

The gentleman, in his relation to words, is in no way arbitrary, that is all.[3]

From an anachronistically scientific point of view one may look upon this as a programme for terminological discipline, and in Confucian historiography such terminological discipline was indeed important. Confucius, like most other philosophers in pre-Chhin China, was interested in language within the context of the political and social context of his time.

Confucius has an unforgettably humorous remark about things not living up to their names:

A horn-gourd that is neither horn nor gourd! A pretty horn-gourd indeed! A pretty horn-gourd indeed![4]

The ancient Chinese generally had a keen perception that categories set up by language (particularly moral and administrative categories) have an important function in organising and regulating public and private human behaviour.[5] The use of this language is supposed to symbolise, and to induce, orthodoxy in thought and in behaviour. The setting up of such a code of speaking was called *chêng ming*

---

[1] *Han Shih Wai Chuan* 4.26, ed. Hsü Wei-Yü, p. 158; cf. Hightower (1951), p. 152.
[2] *Hsün Tzu* 1.3; see Köster (1967), p. 2.   [3] *Lun Yü* 13.3, see Lau (1983a), p. 121.
[4] *Lun Yü* 6.25. I quote the inspired translation in Waley (1938), p. 120.
[5] By the way: the regulated official language of the Communist Party in the German Democratic Republic was popularly known as *Parteichinesisch*.

'setting up of right names, right use of names'. It is under this heading *chêng ming* that Hsün Tzu discusses logic from a Confucian point of view, as we shall see in the Section on Hsün Tzu's logic below.

In ancient Chinese literature the predominant concern was not the relation between names and objects, but between names and behaviour, between what one says and how one performs. The Confucians, with their theory of the right use of names (*chêng ming*) were particularly interested in such concepts of social roles as 'father', 'son', 'elder brother', 'younger brother', 'ruler', 'minister', 'husband', 'wife' etc., which embodied values and moral norms that were obligatory for those who filled these rôles. A father should live up to the title of a father through paternal care, the son should live up to the title of son through filial piety, the ruler should live up to the title of a ruler through efficient government, and so forth. Words for social rôles were interpreted as specifying standards or ideals of social behaviour. Understanding these words, then, constituted the understanding of the principles governing proper behaviour. Hence the Chinese fascination for the definition of moral terms which we shall see in our Section on definition (*b*,4) below.

A rather late anecdote will give an idea of some wider Confucian thoughts on the right use of names:

Confucius was sitting with one of the Chisun family. The Chisun's minister, Thung, said: 'If the prince should send someone to borrow a horse, should it be given him?'

Confucius said: 'I have heard that when a prince takes a thing from his subject, that is termed "taking". One does not speak of "borrowing".'

The Chisun understood and said to the minister Thung: 'From now on when your prince takes a thing, call it taking. Do not speak of borrowing.'

Confucius used the expression 'borrowing a horse' correctly and as a result the proper relation between prince and subject was established. The *Lun Yü* says: 'What is necessary is to use names correctly.'[1]

In its extended form represented in the above story it represents an attempt at a Confucian *Sprachregelung*, a mandatory use of language, which in different transmutations has indeed remained extraordinarily influential in China to this day.

Hsing ming 刑名 *'title and performance'*

Another variety of the social concern with language was the legalist theory of 'performance and title (*hsing ming*)': But whereas the Confucians were, in principle, traditionalists upholding the values inherent in linguistic tradition, the legalists saw in the normative use of language an important lever of political control. Like their Confucian counterparts, they insisted on the binding character of titles (*ming* 名), but the definitions were no more simply traditional. They were to be formulated according to changing political needs. By defining the duties of all members of

[1] *Han Shih Wai Chuan* 5.34, ed. Hsu Wei-Yu, p. 200; cf. Hightower (1951), p. 190. The quotation is from *Lun Yü* 13.3, the *locus classicus* of ancient Confucian logic, which is very probably a fairly late addition to the text. (Cf. Creel (1974)), p. 116ff.

society according to their titles the legalist ruler hoped to define and control social practice. What used to be, for the Confucians, a traditional moral obligation imposed on the people by the moral connotation of key words involving social roles and social behaviour, became, at the hands of the legalists like Shen Pu-Hai 申不害, a political obligation imposed by the ruler's duty-defining decrees according to the changing exigencies of the time.[1]

(4) THE ART OF DEFINITION

There has been much discussion on the alleged vagueness of the ancient Chinese language. Indeed, it does look as if the ancient Chinese did not have a standard abstract word meaning 'definition'.[2] However, Hsün Tzu's term *chih ming* 制名 'management of names' may seem to come close enough to our notion of establishing networks of interrelated clarifications of terms.

Significantly, the ancient Chinese did have a standard way of referring to the making of relevant distinctions: *pieh thung i* 別同異. The Chinese tended to be interested in definitions not in a Socratic way and for their own sake as descriptions of the essence of things, and they were rarely interested in definition as an abstract art in the Aristotelian manner.[3] They were interested in clarifications of the meanings of terms in their literary contexts and as instrumental in controlling the physical and social environment. The Chinese were more concerned to find useful distinctions than to find ingenious definitions.[4] They commonly took a great interest in the clarification of the meanings of terms through examples and illustration. The extent to which such attempted clarifications amount to formal definitions, on the other hand, is often in doubt.

Hsün Tzu takes up the matter of creating a terminology for systematic analysis and asks three questions:

If today there were true rulers, they would in some cases stick to the old names but would in other cases create new ones. In this one must be careful to investigate the following

---

[1] For a detailed discussion of Shen Pu-Hai and *hsing ming* see Creel (1974).

[2] Compare, however, the notoriously ill-understood word *chhi* 氣, for example in the *Hsün Tzu*. The precise force of this remains to be investigated.

[3] Cf. Graham (1986a), p. 378: 'The definitions of Chinese philosophy are therefore conceived as presenting, not what is essential to being X, but what is indispensable to being called "X". In either case, however, there is the same exclusion of the accidental.'

[4] A little episode from the *Mêng Tzu* will illustrate our point:

King Hsüan of Chhi asked about ministers.
  Mêng Tzu asked: 'Which kind of ministers are you asking about?'
  'Are there different kinds of ministers (*chhing pu thung hu* 卿不同乎)?'
  'Yes, there are ministers of royal blood and those of families other than the royal house.'
  'What about ministers of royal blood?'
  'If the ruler made serious mistakes they should remonstrate with him. But if repeated remonstrations fell on deaf ears, the ruler should be dismissed.'
  The King looked appalled.
  'Your Majesty should not be surprised by my answer. Since you asked me, I dared not answer with anything but the truth (*pu kan i pu chêng tui* 不敢以不正對).
  As soon as the King had calmed down did he ask about the ministers from different clans.
  'When the ruler makes mistakes they should remonstrate. If they remonstrate repeatedly without being listened to they should resign.' (*Mêng Tzu* 589; Lau (1983c), p. 219).

questions: Why are names necessary? How do we distinguish 'the same' from 'different'? What is the most important thing in the management of names (*chih ming*)?[1]

The most important thing in the management of names, for Hsün Tzu, is simply that (relevantly) similar things be called by the same name, and that (relevantly) dissimilar/different things be called by different names.[2] As far as Hsün Tzu is concerned, one can manage names X and Y as long as one can distinguish between X and Y.[3]

Our question is whether or not the Chinese were able to ask for definitions and conceptual clarifications when these were needed, and whether they were able to provide such definitions when appropriate.

We note first that the art of definition is by no means limited to the so-called literary texts. The newly discovered legal texts from the −3rd century make sustained and extensive use of legal definitions of terms as well as phrases relevant to their precise legal code. Moreover, their definitions were plainly stipulative in the sense that they did not purport to describe how a term generally tends to be used, but rather how it is henceforth to be interpreted in a well-defined context of legal discourse. They prove beyond doubt that at the lower end of literacy widespread use was made of pithy legal definitions designed to facilitate the application of the law. The art of definition is not a privilege of the literary elite.

Often, such legal definitions explained ordinary expressions through technical legal terms:

What does 'neighbourhood' (*ssu lin* 四鄰) mean?
'Neighbourhood' means the members of one's mutual responsibility group (*wu jên* 伍人).[4]

Or again, they might explain a legal technical term in concrete terms:

When a son accuses his father or mother, or when a male or female slave accuse their master, this is not a matter for the public court. One should not take up the case.

What does 'not a matter for the public court' mean? When the ruling person on his own authority kills, amputates or brands his children, male or female slaves, this is called 'not a matter for the public court' and should not be taken up as a legal case.[5]

The case of the legal definitions is particularly important because these belong to the non-literary world of public administration and thus show that at this basic level of everyday life definition played a crucial and pervasive part in Chinese culture. And even when technical terms were not explicitly defined, they were used, in the legal texts, with impressive consistency. This suggests that the technical precision of Mohist vocabulary, far from being marginal in ancient Chinese life, was part of everyday administrative practice.

---

[1] *Hsün Tzu* 22.12, see Köster (1967), p. 289.   [2] *Hsün Tzu* 22.21ff., see Köster (1967), p. 290.
[3] For a more detailed treatment of Hsün Tzu's logical thought see below Section on 'Explicit logic'.
[4] *Shui Hu Ti Chhin Mu Chu Chien* 睡虎地秦墓竹簡, p. 194; Hulsewé (1985) D82.
[5] *Shui Hu Ti Chhin Mu Chu Chien*, p. 196, Hulsewé (1985) D86f. Cf. also D87–9, and D166–90 for an extended series of legal definitions.

Turning now to more literary evidence, the ancient Chinese philological commentaries from the −3rd century onwards are rich sources of definitions of terms. In many cases whole sets of related terms are defined.[1] The commentators also provide paraphrases of knotty passages and sentences. As far as the art of systematic glossography goes, China must be said to possess one of the richest traditions in the world.

But let us look more closely at the art of definition in the earliest literature. First, a typical conceptual clarification from the military sphere:

I have heard it said that military action within the state is called internal strife (*luan* 亂). Military action from the outside is called an incursion (*khou* 寇).[2]

It turns out that we can trace the art of definition of moral terms back to Confucius himself, but I shall start with an example probably wrongly attributed to him in the book *Chuang Tzu*.

'Allow me to ask what you mean by "benevolence" (*jên* 仁) and "duty" (*i* 義).'
Confucius said: 'To delight from your innermost heart in loving everyone impartially, these are the essentials (*chhing* 情) of benevolence and duty.'[3]

Confucians were naturally interested in the definition of moral terms:

To give of one's wealth to others is called 'generosity' (*hui* 惠). To teach others benevolence is called 'conscientiousness' (*chung* 忠). To find the right man for the empire is called 'benevolence' (*jên*).[4]

There was also a distinct interest in the concept of 'man':

What is it that makes man into man? It is the fact that he makes distinctions (*pien* 辨) . . . That by means of which man is man is not just that he has two feet and no proper fur. It is because he has the capacity to set up distinctions. The *hsing hsing* 猩猩-ape, as far as its physical shape is concerned, also has two legs and no hair (on its face), but the gentleman will drink his soup and eat his meat. Therefore that by means of which man is man is not just that he has two feet and no hair (on his face). It is because he has the capacity to make distinctions.[5]

Note how carefully Hsün Tzu ensures that his definition distinguishes between the most closely similar cases: man and ape. They are physically alike, but intellectually different. Hsün Tzu's attitude to definition at this point strikes us as a very 'scientific' one.[6]

Wang Chhung defines the concept of man in terms that remind us of the traditional Western use of *genus* and *differentia*:

---

[1] Many glosses are also found in the non-commentarial literature. Characteristically, such early glosses were often by words with similar sounds, as Wang Li (*1981*), p. 1, points out.
[2] *Tso Chuan*, Duke Wên 7, ed. Yang Po-Chün, p. 563; cf. S. Couvreur (1951b), vol. 1, p. 486.
[3] *Chuang Tzu* 13.49, see Graham (1981), p. 128.   [4] *Mêng Tzu* 3A4, see Lau (1983c), p. 107.
[5] *Hsün Tzu* 5.24, see Dubs (1928), p. 71. Cf. Read (1976), no. 402, for a description of the *hsing hsing* ape.
[6] For further examples of definitions of man see the Sub-section on ancient Chinese comments on the Chinese language.

Among the 360 naked (unfurry) animals man ranks first. Man is a creature (*wu* 物). Among the ten thousand creatures he is the most intelligent one.[1]

The standard request for a definition of *X* is of the form *ho wei* 何謂 *X* 'what is called *X*?' This formula is the standard in legal texts. Mencius, when challenged on the meaning of benevolence and trustworthiness, comes up with a strikingly elegant and coherent answer:

'What does one call "good", and what does one call "trustworthy"?'
'That which one ought to desire is called "good", and having it within oneself is called being "trustworthy".'[2]

Both definitions *de verbo* and *de re* can coexist in the same passage:

Tzu Chang asked: 'What must a gentleman be like before he can be said to have won through?'
The Master asked: 'What do you mean by winning through?'
Tzu Chang replied: 'What I have in mind is a man who is sure to be known whether he serves in a state or in a noble family.'
The Master said: '. . . The term "winning through" describes a man who is straight by nature and fond of what is right, sensitive to other people's words and observant of the expression on their faces, and always mindful of being modest. Such a man is bound to win through whether he serves in a state or in a noble family. On the other hand, the term "having a reputation" describes a man who has no misgivings about his own claim to benevolence when all he is doing is putting up a façade of benevolence which is belied by his deeds. Such a man is sure to be known, whether he serves in a state or in a noble family.'[3]

Very commonly, in China as elsewhere, definition is by examples:

'What does one call "from Heaven"? What does one call "from man"?'
'That oxen and horses have four feet, this is called "from Heaven". Haltering horses' heads and piercing oxen's noses, this is called "from man".'[4]

The Chinese were more interested in the question how a thing should be described than in the question how a term should be defined.

To consider right as right and wrong as wrong is called 'knowledge'. To consider right as wrong and wrong as right is called 'stupidity'. To call right 'right' and wrong 'wrong' is called 'being straight/correct'.[5]

Clearly, the extent to which various Chinese authors have wished to, or needed to, define their terms has varied. With the flourishing of competing systems of thought during the Warring States period the need for definitions increased. The chapter on Explaining Lao Tzu (*Chieh Lao* 解老) in the book *Han Fei Tzu* 韓非子 provides some of the finest examples of the art of definition in ancient China.

---

[1] *Lun Hêng*, ch. 43, p. 1392, see Forke (1911), vol. 1, p. 528.   [2] *Mêng Tzu* 7B25, tr. Lau (1983c), p. 297.
[3] *Lun Yü* 12.20, see Lau (1983a), p. 115.   [4] *Chuang Tzu* 17.51, see Graham (1981), p. 149.
[5] *Hsün Tzu* 2.12, see Köster (1967), p. 13.

'The Way' is that wherein all things are so (*jan* 然), what all principles of things observe.[1]

'The principle' is the pattern which makes things what they are.[2]

'Humaneness' (*jên* 仁) refers to loving others dearly from the bottom of one's heart, to rejoice in other people's good fortune and to dislike other people's misfortune, and to do all these things undeliberately without being able to stop oneself, without seeking a reward for it.[3]

'Duties' (*i* 義) are the service of minister to ruler, of inferior to superior, the distinction between father and son, noble and mean, the interactions between acquaintances and friends, the differentiation between close and distant relatives, between those inside and outside the clan. That the servant should serve the ruler is fitting. That the inferior should cherish his superior is fitting. That a son should serve his father is fitting. That a lowly person should honour a nobleman is fitting. That acquaintances and friends should help each other is fitting. That relatives are counted as insiders and non-relatives as outsiders is fitting. 'Duty' (*i*) refers to the fittingness (*i* 宜) of these things.[4]

'Ritual' (*li* 禮) is that by which one gives external expression to one's real internal feelings. It is the external embellishment of all the duties . . .[5]

The Confucian theory of the proper use of names was closely associated with definition. Hsün Tzu's chapter on the subject opens with an interesting sequence of definitions, which even pay careful attention to the fact that many terms have more than one usage:

These are the common names that apply to man. That which is as it is from the time of birth is called man's 'nature'. That which is harmonious from birth, which is capable of perceiving through the senses and of responding to stimulus spontaneously and without effort is also called 'nature'. The likes and dislikes, delights and angers, griefs and joys of nature are called 'emotions'. When the emotions are aroused and the mind makes a choice from among them, this is called 'thought'. When the mind conceives something, and thought and the body put it into action, this is called 'conscious activity'. When the thoughts have accumulated sufficiently, the body is well trained and the action is carried to completion, this is also called 'conscious activity'. When one acts from considerations of profit, it is called 'business'.

When one acts from considerations of duty, it is called 'conduct'. The means of knowing within man is called 'intelligence'. His intelligence tallying with something is called 'knowing'. The means of being able within man is called 'ability'. His ability tallying with something is called 'being able'. Injuries to one's nature are called 'sickness'. Unexpected occurrences with which one meets are called 'fate'. These are the common names that apply to man, the names that have been fixed by the kings of later times.[6]

Even the Taoist book *Chuang Tzu* contains a series of terminological clarifications which show a probing concern with the clarification of the meanings of interrelated terms.

The 'Way' is the layout of the Power. 'Life' is the radiance of the Power. The 'nature' of something is its resources for living. What is prompted by its nature is called 'doing'. A doing

---

[1] *Han Fei Tzu* 20.27.1, see Liao (1939), vol. 1, p. 191.   [2] *Ibid.* 20.27.4, see Liao (1939), vol. 1, p. 191.
[3] *Ibid.* 20.3.1, see Liao (1939), vol. 1, p. 171.   [4] *Ibid.* 20.4.1, see Liao (1939), vol. 1, p. 171.
[5] *Ibid.* 20.5.1, see Liao (1939), vol. 1, p. 172.   [6] *Hsün Tzu* 22.4, see Watson (1963), p. 139f.

which is contrived is a 'misdoing'. To 'know' is to be in touch with something. 'Knowledge' is a representation of it. As for what knowledge does not know, it is as if we are peering in one direction.

That which prompts on the course which is inevitable is what is meant by 'Power'. The prompting being from nowhere but oneself is what is meant by 'ordered'.[1]

For the scientifically-minded authors of the Mohist dialectical chapters, the art of definition came to occupy a central place in their methodology.[2] A. C. Graham has argued that the items A1 to A87 from the Mohist canons constitute what aspires to be a *system* of definitions. Even the principles of application of the system are carefully defined in ways that remind one of modern technical definitions in logic:

The 'standard' (*fa* 法) is that in being like which something is so.[3]
The 'criterion' (*yin* 因) is that wherein it is so.[4]

A number of concepts are defined in terms of a few basic ones like those of benefit (*li* 利) and harm (*hai* 害).

'Benefit' is what one is pleased to get.[5]
'Harm' is what one dislikes getting.[6]
'Loyalty' is to be energetic in sustaining responsibility when one deems something beneficial.[7]
'Filial piety' is benefiting one's parents.[8]
'Achievement' is benefiting the people.[9]

'Desire' and 'dislike' are defined in terms of benefit and harm:

'To desire' is directly weighing the benefits. 'To dislike' is directly weighing the harm.[10]

In the context of Chinese science it is particularly interesting to note the way in which the Mohists used interlocking formal definitions for the coherent clarification of concepts like that of a circle.

'The same in length' is exhausting each other when laid straight.[11]
'The centre' is the place from which (straight lines to the circumference) are the same in length.[12]
'Circular' is having the same lengths from a single centre.[13]

Many of the definitions are exasperatingly short, and they do not constitute discursive explications of the terms they are concerned with. They read like cues for memory, but they do have a rigid formalised syntax and are by no means just strings of key words. One of the reasons for their economy of terms is that the definitions were designed to minimise the use of undefined terms. Thus the fewer terms the Mohists used in their definitions the neater these definitions would fit into the pattern of interlocking definitions where all defining terms were in turn defined in the system.

---

[1] *Chuang Tzu* 23.70, see Graham (1981), p. 190.   [2] Graham (1978), pp. 261–336.
[3] *Ibid.*, A70.   [4] *Ibid.*, A71.   [5] *Ibid.*, A26.   [6] *Ibid.*, A27.   [7] *Ibid.*, A12.   [8] *Ibid.*, A13.
[9] *Ibid.*, A35.   [10] *Ibid.*, A84.   [11] *Ibid.*, A53.   [12] *Ibid.*, A54.   [13] *Ibid.*, A58.

The system of Mohist definitions as we have it today does not realise these high ambitions in a satisfactory way. But the Mohist canons are full of conceptual analyses and explications that show a very inquisitive attitude towards the problem of the systematic clarification of meanings of words and expressions in the context of a formal description of reality.

While it is true that the school of Mohist logicians declined, and that with it the art of Mohist definition disappeared, it is only proper to point to the lexicographical tradition as the context in which the arts of definition and clarification of meaning have flourished throughout the centuries. Here is a quotation which – were it not for the Chinese transcriptions – one might think was taken straight out of Aristotle. Indeed, for all we know, this definition may indeed have been produced during Aristotle's life-time, although we have no way of being sure.

> Bipeds with feathers are called *chhin* 禽 'birds'. Four-footed creatures with fur are called *shou* 獸 'beasts'.[1]

The important point here is that a non-furry tetrapod, or a non-four-footed furry animal would not count as *shou*.

The art of definition in Chinese dictionaries will be the subject of our next section. But before we leave the history of definition in ancient China, we must turn to a unique system of Confucian definitions.

A glossary of ethical terms by Chia I 賈誼

Within the human and social sphere, the Confucians developed a distinct interest in definitions, especially definitions of moral terms. By far the most impressive system of such definitions that has come down to us is that by Chia I (−200 to −168). Curiously, it has never been translated into any European language, and it is not even mentioned in the standard histories of Chinese philosophy. Here is a first partial translation of this central document in the history of the art of definition and of ethical theory in ancient China.

> He (King Huai of Liang(?)) said: 'I have often heard the term "the Way" (*tao* 道), but I still do not know the reality (that corresponds to) the Way (*tao*). May I ask what "the Way" (*tao*) refers to?' Chia I[2] replied: 'The Way is that which we use as a guideline to deal with things. Its root is called emptiness (*hsü* 虛), and its branches are called arts (*shu* 術). . . .'
> 'May I ask about the constituents (*thi* 體) of moral appraisal?' Chia I replied:
> '1.[3] The parents loving and benefiting the son is called kind-heartedness (*tzhu* 慈). The opposite of kind-heartedness is heartlessness (*yin* 嚚).
> 2.   The son loving and benefiting his parents is called filial piety (*hsiao* 孝). The opposite of filial piety is impiety (*nieh* 孽).

---

[1] *Erh Ya*, ed. *SPTK*, 2.12b.
[2] The way of calling Chia I 賈誼 'Mr Chia (*Chia chün* 賈君)' is worth noting, since it may be significant for the dating of the chapter. Cf. *Hsin Shu* 7.1, ed. Chhi Yü-Chang, p. 919f., where Chia I seems to use the term to refer to himself.
[3] The numbering of the definitions is added by the translator.

3. Loving and the benefiting coming from within is called loyalty (*chung* 忠). The opposite of loyalty is disobedience (*pei* 倍).

4. The heart being aware and in sympathy with other people is called generosity of heart (*hui* 惠). The opposite of generosity of heart is called hostility (*chhou* 儱).

5. The elder brother respecting and loving the younger brother is called brotherly affection (*yu* 友). The opposite of brotherly affection is cruelty (*nüeh* 虐).

6. The younger brother respecting and loving his elder brother is called brotherly piety (*thi* 悌). The opposite of brotherly love is nonchalance (*ao* 敖).

7. In one's contacts with others to be careful about appearances is called respectfulness (*kung* 恭). The opposite of respectfulness is rudeness (*hsieh* 媟).

8. In one's contacts with others to be serious and correct is called reverence (*ching* 敬). The opposite of reverence is irreverence (*man* 慢).

9. In words and actions to be consistent is called moral firmness (*chên* 貞). The opposite of moral firmness is moral duplicity (*wei* 偽).

10. One's promises being effective and one's words correct, that is called trustworthiness (*hsin* 信). The opposite of trustworthiness is unreliability (*man* 慢).

15. Caring for all and showing no partiality is called impartiality (*kung* 公). The opposite of fairness is partiality (*ssu* 私).

16. To be straightforward and unwarped is called correctness (*chêng* 正). The opposite of correctness is crookedness (*hsieh* 邪).

17. Seeing oneself with other people's eyes is called proper judgement (*tu* 度). The opposite of proper judgement is biased judgement (*wang* 妄).

18. Judging others by comparison with oneself is called fairness of judgement (*shu* 恕). The opposite of fairness of judgement is arbitrariness of judgement (*huang* 荒).

21. In one's actions to be in accordance with principle is called moral charisma (*tê* 德). The opposite of moral charisma is moral vindictiveness (*yüan* 怨).

22. To give rein to principle and be pure and calm is called proper comportment (*hsing* 行). The opposite of proper comportment is improper comportment (*wu* 污).

25. In one's heart to love all men properly is called benevolence (*jên* 仁). The opposite of benevolence is cruelty (*lei* 戾).

26. In one's actions to do the completely fitting thing is dutifulness (*i* 義). The opposite of dutifulness is brutishness (*mêng* 懵).

39. To be deeply aware of disaster and good fortune is called wisdom (*chih* 知). The opposite of wisdom is stupidity (*yü* 愚).

40. To notice things swiftly and to investigate them profoundly is called intelligence (*hui* 慧). The opposite of intelligence is childishness (*thung* 童).

47. In sorting things to be clear and discriminate is called discrimination (*pien* 辯). The opposite of discrimination is indiscriminate talkativeness (*na* 訥).

48. To understand all subtle matters is called sharpness of mind (*chha* 察). The opposite of sharpness of mind is obfuscation of mind (*mao* 旄).

49. To move with conviction and be awe-inspiring is called moral authority (*wei* 威). The opposite of moral authority is moral disorientation (*hun* 圂).

50. To govern and control a people so that they do not go against one's regulations, that is called strictness (*yen* 嚴). The opposite of strictness is laxity (*juan* 軟).[1]

---

[1] In all, there are 56 such pairs of definitions. These definitions invite comparison with those in Aristotle's *Nicomachean Ethics* and *Eudemian Ethics*. Aristotle treats fewer terms but is much more analytical in his approach to those terms he does treat.

All these qualities belong to the range of what is good. They are what is called the Way (*tao* 道). Thus he who keeps to the Way is called a knight (*shih* 士). He who rejoices in the Way is called a gentleman (*chün tzu* 君子). He who knows the Way is called enlightened (*ming* 明). He who practises the Way is called talented (*hsien* 賢). He who is both talented and enlightened is called a sage (*shêng jên* 聖人).'[1]

This did not remain the last long list of philosophical definitions. Chhên Chhun 陳淳 (+1159 to +1223), a disciple of the famous Neo-Confucian philosopher Chu Hsi 朱熹 (+1130 to +1200) even wrote what amounts to a Dictionary of Philosophical Terms known as the *Pei Hsi Tzu I* 北溪字義.[2] Chhên Chhun frequently distinguishes between different meanings of a given philosophical term and goes into considerable detail about each of the terms he discusses. His *dictionnaire philosophique* provides as clear a proof as one could hope for that Neo-Confucians were interested in conceptual questions and definitions. They were certainly not just wallowing in edifying but undefined and ultimately indefinable generalities. There was a genuine concern for the clarification of concepts, although the pursuit of this clarification was not often conducted in the sort of systematic critical spirit cultivated in Greece in the tradition of Socrates.

As Yü Min has reminded me, it is worth reflecting that there was a strong Buddhist influence on many of the founders of Neo-Confucianism, which may have affected the intellectual habits of conceptual clarification.

Often, Chhên Chhun starts out with a traditional gloss, but then goes on to ask penetrating questions:

Nature (*hsing* 性) is nothing other than principle (*li* 理). But why in that case do we call it '*hsing*' and not '*li*'?[3]
Passion (*chhing* 情) and nature (*hsing*) make up a conceptual pair. Passion (*chhing*) is the movement of nature (*hsing*).[4]

In all, 43 terms are given careful contrastive and systematic conceptual attention, of the kind that Plato would have respected, and that Aristotle tried to provide, for corresponding Greek concepts. The existence of this book, and the high respect it enjoyed in the Neo-Confucian tradition refutes once and for all the idea that Confucian thinking was merely 'picturesque' and did not aim for conceptual clarity.

We may find that the conceptual clarity actually achieved was insufficient, but, as observers of Greek philosophy, we will most certainly want to say the same about Aristotle and Plato, as recent philosophical research on these philosophers has shown at many points.[5]

---

[1] Chia I, *Hsin Shu*, ch. 8, ed. Chhi Yü-Chang, p. 919, ed. *SPTK*, 2.31bff.
[2] A useful new edition of this work was published by Chung-hua-shu-chü in 1983. There is also a recent translation of the book by Wing-Tsit Chan (1986).
[3] Chhên Chhun, *Hsi Tzu I*, p. 6.
[4] Chhên Chhun, *Pei Hsi Tzu I*, p. 14. For another instance of a conceptual pair (*tui* 對) see *ibid.*, p. 28.
[5] See the volumes of the Clarendon Aristotle Series edited by J. L. Ackrill at the Clarendon Press, Oxford, for memorable pieces of the sort of philological analysis of ancient texts that have served as a model for my work on Chinese language.

Quotation marks

In many of the above translations we have seen a proliferation of quotation marks. In ancient China there was no typographical or epigraphic device equivalent to the modern Western quotation marks. Moreover, the question marks in the passage just quoted, and in many others, are not made explicit in the original. One might therefore be tempted to think that the ancient Chinese did not have any linguistic devices to indicate the nuances expressed by quotation marks in English.

Consider the following spoken sentences:

A. *Poverty is a common phenomenon.*
B. *Poverty is a common word.*
C. *Poverty is a subtle concept.*
D. *Poverty is really only in the mind.*

In written English we tend to mark off the uses of the word 'poverty' in B, C and D with quotation marks. There are no such devices in ancient Greek or ancient Chinese.

From a logical point of view, the important point is not whether ancient Chinese (or Greek) inscriptions use quotation marks, but whether the ancient Chinese (or the Greeks) did or did not devise a method or convention which enabled them to make explicit the distinctions we indicate by quotation marks (some of which are covered by the Sanskrit *iti*).

The idiom *suo wei X chê* 所謂 . . . 者 'the so-called *X*' gets close to the nuances B and C which we express with quotation marks. Another device the Chinese have used increasingly, from the time of Confucius onwards, is the combination of the particles *yeh chê* 也者 after the word or phrase to be marked out. This usage had the obvious disadvantage that while the end of the quotation marks was very clear indeed, the beginning was by no means physically obvious in all instances. The result was that the Chinese have shown a strong predilection for placing the words that they put into quotation marks at the beginning of their sentences. However, it turns out in practice, that even when a phrase *X yeh chê* occurs in the middle of the sentence, it is usually quite easy to ascertain where the 'quotation marks' are meant to begin.

The case is different with the quotation marks for direct speech, indicated by the word *yüeh* 曰, which is often translated as 'to say'. Actually, the syntactic behaviour of this word *yüeh* is so restricted that it is often better understood as an initial quotation mark or a colon. The word is rarely preceded by a negative or followed by anything other than direct speech or items of a list. In any case, however one decides to interpret *yüeh* it will sometimes remain uncertain where the direct speech that follows *yüeh* ends. Thus while *yeh chê* fails to mark the beginning of the phrase put into quotation marks, *yüeh* fails to mark the end, except in the occasional cases where *yün* 云 'to say' is used to mark the end of a quotation which begins with *yüeh*.

As we have seen above, the practice of marking off a term to be defined by *yeh chê* was already current in the time of Confucius. In the *Kuo Yü* 國語, a historical text, there is only a sprinkling of three examples, where quotations from the *Book of Songs* are marked off with *yeh chê* and two cases where the concept of a ruler is discussed in a way that invites us to use quotation marks even in English, although there is no distinction between use and mention involved. (It is useful to keep in mind that in English, too, the use of quotation marks is not limited to the marking of the mentioning versus the use of a term.)

The irreverent philosopher Chuang Tzu is fond of the use of the Chinese equivalent of the quotation marks for subtle literary purposes (I count over 20 instances). A. C. Graham (1981) consistently renders the formula by quotation marks in English, a new practice which does seem to improve our understanding of the stylistic subtleties in the passages involved.

'Knowledge' (*yeh chê*) is a tool in competition.[1]

The subtle point here is that Chuang Tzu does not really believe that what people call 'knowledge' in fact *is* knowledge. He manages to indicate this by a device which is remarkably like our non-standard use of quotation marks, the particles *yeh chê*.

In the more theoretically orientated *Hsün Tzu*, the combination *yeh chê* becomes quite common (I count over 30 instances). When Hsün Tzu introduces a concept as such, he commonly puts it into 'quotation marks' to draw attention to the fact that he is referring not to a thing which instances the concept but either to the form of words (mention versus use) or to a concept as such.

In the *Kungsun Lung Tzu* 公孫龍子, an example of a book of logic, we find that the particle *chê* 者 by itself is used where we should expect quotation marks distinguishing the mentioning of a term from the use of it:

'Horse' (*ma chê* 馬者) is that by which we name the shape. 'White' (*pai chê* 白者) is that by which we name the colour.[2]

Without the *chê* this logical observation would be much less transparent. After all, horses are not the sorts of things which we use to designate shapes. Kungsun Lung is concerned to make it plain that he is talking about the word *ma* 馬 'horse'. It is crucial to his argument that he is mentioning the word *ma* 'horse' and not using it. The use of *chê* alone where we would use quotation marks remained common in ancient Chinese philosophical literature. The problem is that there are so many other functions of the particle *chê* on its own.

At other times, when making paradoxes, the Chinese sophists cheerfully exploit the non-transparent use of words to designate not their meanings but themselves, as in the famous paradox:

Shan chhu khou 山出口 MOUNTAIN COME-OUT-FROM MOUTH 'Mountain' comes out of the mouth.[3]

---

[1] *Chuang Tzu* 4.6, see Graham (1981), p. 67.
[2] *Kungsun Lung Tzu*, ch. 2, beginning, ed. Luan Hsing, p. 15.
[3] *Chuang Tzu* 33.75, see Graham (1981), p. 284.

Compare Aristotle's comment:

Anthrōpos gar kai leukon kai pragma kai onoma estin. ἄνθρωπος γὰρ καὶ λευκὸν καὶ πρᾶγμα καὶ ὄνομά ἐστιν. 'For *anthrōpos* (man) is something white, a thing, and a name'.[1]

Kungsun Lung and Aristotle seem to have been struck by the same logical feature, which we make explicit through quotation marks, but characteristically Kungsun Lung exploits his observation to make a very short paradox which is easily misunderstood and is indeed deliberately provocative.

At least by Han times, the formula *yeh chê* had apparently entered everyday speech, if the following dialogue is anything to go by:

Prince Chien went out on a tour of Chheng-fu and encountered Duke Kan of Chheng in a (hemp)field.
'What's this?' asked the Prince.
'It's a (hemp)field.'
'What's a "(hemp)field" (*yeh chê*)?'
'What you grow hemp in.'
'What's "hemp" (*yeh chê*)?'
'What you make clothes from.'[2]

The first question is about a thing, it is *de re*: 'What is this?' The second and third questions are not about this thing any more, they are about the Chinese words (*de verbis*) for '(hemp)field' and 'hemp', which the Prince apparently does not understand, and which are then defined for him. The Chinese construction *yeh chê* here can be naturally translated into quotation marks. But the question whether we have 'use' or 'mention' is a debatable one.

(5) DICTIONARIES IN TRADITIONAL CHINA

One way of reacting to and commenting on one's own language is by writing a dictionary of it. A civilisation's awareness of its own language may be conveniently observed in the dictionaries it produces.

Lexicography proper in China is historically and essentially linked to the tradition of writing commentaries or glosses on old texts. Such glosses could be of three kinds: those concerned with meaning, those concerned with the shape of the characters, and those concerned with the pronunciation of the characters. Correspondingly, we find in China three kinds of dictionary. The semantic dictionary, of which the *Erh Ya* 爾雅 (*c.* −2nd century) is the illustrious precursor, the pictographic-cum-semantic dictionary of which the *Shuo Wên Chieh Tzu* 説文解字 (postface +100, presented to the throne in +121) is the great model, and the primarily folk-etymological or phonetic dictionary of which the *Shih Ming* 釋名 (+2nd century) is the distinguished earliest example. There were other types of dictionaries, like the fascinating Dictionary of Local Languages (*Fang Yen* 方言 −1st century) but these fall into a class by themselves.

---

[1] *Sophistici Elenchi* 1.14, 174a8f.　　[2] *Shuo Yüan* 18.32, ed. Chao Shan-I, p. 558.

One kind of dictionary is conspicuous indeed through its rarity: the dictionary of foreign languages. The Chinese had two things in common with the Greeks: 1, they were fascinated by their dialects; 2. they did not seriously respect foreign languages as interesting in their own right.

In this section we shall look at the historical development of Chinese dictionaries from late Chou times down to the period when the Western impact made itself felt in Chinese lexicography. We shall pay special attention to the semantic aspects of lexicography, the way the Chinese solved the complex problems of the description of meaning. Fukuda Jōnosuke (1979) and Liu Yeh-Chhiu (1983), as well as the beautifully detailed Chhien Chien-Fu (*1986*) are the main handbooks I have used in this brief survey on a subject which deserves a monograph in its own right.[1]

Pre-history of lexicography: the character primers

The modern Chinese word *tzu-tien* 字典 is of fairly recent origin, deriving as it does from the authoritative status of the *Khang Hsi Tzu Tien* 康熙字典 (1716). The earliest Chinese bibliography that has come down to us, the bibliographic section of the *Han Shu* 漢書 features a number of word-lists under the heading *hsiao hsüeh* 小學 'little studies'. With the help of this bibliography we can trace back the pre-history of the dictionary almost 2,500 years to the late Chou dynasty. Tradition even ascribes the earliest list of words to the court historian of King Hsüan of the Chou dynasty, who reigned from −827 to −781. It is hard to know what to make of this tradition today, but it would appear that the word-list *Shih Chou* 史籍, ascribed to the period around −800, did take some definite form not later than during early Warring States times, i.e., some time in the −5th century.[2]

The *Shih Chou* was a work by a scribe for the use of those who aspired to become scribes. It was a (rhymed!) primer of writing for the select who aspired to learn that arcane art in those distant times. By the −3rd century writing had become something of a political issue with China's First Emperor insisting on the standardisation and modernisation of the script, the result of which was the small seal style (*hsiao chuan* 小篆). Li Ssu 李斯, the right-hand man of the First Emperor, produced a new primer of standard characters, as did two other courtiers of his time. During the Han dynasty these three works were combined to produce one standard handbook, the *Tshang Chieh Phien* 倉頡篇 comprising apparently no less than 3,300 characters in all.[3]

---

[1] The important tradition of the phonetic and phonological description of Chinese falls outside the scope of the present section.

[2] Cf. Wang Li (*1981*), p. 7. A short Akkadian word list, from central Mesopotamia, is ascribed to the −7th century. It would appear that the development was almost contemporary in both cultures.

[3] It may be useful to compare the vocabulary of some early books: The Spring and Autumn Annals, attributed to Confucius, use about 950 different characters, the three commentaries to this work combined *c.* 4,000, the Book of Songs no less than 3,000, the *Lun Yü* 論語 2,200, the *Mêng Tzu* 孟子 a little over 2,000, the *Lao Tzu* 老子 little more than 800, the *Hsün Tzu* 荀子 about 3,000, the *Mo Tzu* 墨子 about 2,750, the *Kuo Yü* 國語, surprisingly, only about 1,500.

The lexicographer Yang Hsiung 揚雄 (−53 to +18), of whom we will hear more in due course, wrote a commentary on this work, and those who have suffered through the extraordinarily recondite vocabulary of the great poet Ssuma Hsiang-Ju 司馬相如 (−179 to −117) will not be surprised to learn that this poet actually produced his own extended list of characters, the *Fan Chiang Phien* 凡將篇, which unfortunately, like so many of the early practical character books, was lost when it became superseded by later practical character books.

The only one of these word lists to survive in a form that allows us to get some idea of its original shape is the *Chi Chiu Phien* 急就篇 by a certain Shih Yu who flourished around the years −40 to −33.[1] The work consists of 2,168 characters, few of which are repeated in the text.[2] It was thus suitable, and widely used, as a text to copy by way of practising calligraphy. Indeed, its very survival in many versions is due to this fact.

The text begins with a list of 130 family names. Next there is a section on embroidery, food, clothes, social status, instruments, and many other things, and finally on social ranks. Taken together with its learned commentaries by scholars such as Yen Shih-Ku 顏師古 (+581 to +645), this work has retained a certain usefulness until today.

The books we have so far surveyed are more than purely mnemonic word-lists, but they really still belong to the pre-history of Chinese lexicography. We shall now turn to some of the most important works of lexicography proper. And the first work we shall have to consider is essentially a compilation of traditional glosses on various Chinese words.

The 'Erh Ya': thesaurus of early Chinese glosses

The dictionary of glosses *Erh Ya* 爾雅, on which the first commentary was written during early Han times,[3] often looks like little more than a sparsely annotated word-list. Within the Chinese tradition it occupies a place similar to that of Yâska's *Nirukta*[4] (*c.* −5th century) in ancient India, although it is much less analytically ambitious than the latter. It is easily the most respected word-list in Chinese cultural history. The origins of this book are traditionally placed in distant antiquity. Even Confucius himself is said to have had a hand in its compilation. With such (spurious) credentials, the book became one of the Thirteen Classics during the Sung dynasty. This says perhaps less about the intellectual importance of the *Erh Ya* than about the importance attached to the knowledge of characters in Chinese culture.

The first three chapters of the *Erh Ya* are lists of semantically close or related words belonging to common usage. Each of these series of semantically related

---

[1] See Wang Li (*1981*), pp. 8ff., for a general account of the *Chi Chiu Phien* 急就篇. The book owes its survival to the fact that it was used for calligraphy practice.

[2] The authoritative bibliographic handbook, *Ssu Khu Chhüan Shu Tsung Mu Thi Yao* 四庫全書總目提要 (ed. Chung-hua-shu-chü, Peking 1980, vol. 1, pp. 102–3) wrongly claims there are no repetitions at all.

[3] See Wang Li (1981), p. 12.

[4] The *Nirukta* is a treatise dealing with Vedic words and with their etymological explanations. See Cardona (1976), pp. 270–3 for a survey of the literature on this work.

terms is then followed by a 'heading' (or rather: a 'tail') indicating the general category the (unknown) compiler had in mind.[1] Thus the first three books are a kind of mini-thesaurus of the basic vocabulary, a proto-Roget[2] of the ancient Chinese language. The principle of the thesaurus was thus discovered early in China and it continued to be used with great proficiency.

The last sixteen chapters of the *Erh Ya* are more specialised: they constitute an encyclopaedic survey of the vocabulary within the main areas of Chinese civilisation.[3]

We have seen that the first three chapters of the *Erh Ya* provide synonym groups and brief glosses, but already at this stage there is an awareness of the fact that one word may have several meanings, as when the dictionary quite properly includes the word *chhung* 崇 among the words in the semantic field of *kao* 高 'high', but adds that *chhung* can also mean *chhung* 充 'fill up'.[4] There are also instances where the dictionary first introduces a series of characters under one 'heading', but then goes on to elaborate distinctions by further subdividing the members of the series.[5]

Synonymy is expressed in the following way: *X wei chih* 謂之 *Y, Y wei chih X* '*X* is called *Y* and *Y* is called *X*', in the section on architecture which is a treasury of detailed terminological information.[6] Considerable skill is shown in the mutual definitions of kinship terms: the precision of the terminology and the meticulous conciseness of the definitions are quite remarkable, especially when they are relational:

If someone calls me an uncle, I call him a nephew.[7]

One senses that those who wrote this section felt entirely safe on their home ground.

There is no doubt that the *Erh Ya* documents a profound, comprehensive, and analytical interest of the Chinese in their own language during the centuries when it was taking shape. The *Erh Ya* is devoid of all undue emphasis on cultural or moral terms. It is entirely concerned with everyday terminology and with the terminology of the science of the day, the so-called *san ming* 散名 or 'miscellaneous names'. One might have thought that such a concentration on *san ming* of no moral consequence would have left the *Erh Ya* unrespected by Confucian orthodoxy. But this just goes to show that our accustomed ways of thinking of Confucian mentality are perhaps not quite adequate. There definitely *was* a place for the *Erh Ya* in the Confucian scheme of things, just as there was a place for science. After all, did not Confucius himself

---

[1] These categories were not always unambiguous or well-chosen. For example, under the category *yü* 予, (1) verb: give; (2) first person pronoun: I, the *Erh Ya* 爾雅, confusingly, lists words of both meanings. Since most Chinese characters are highly ambiguous, a single character is seldom enough to indicate a meaning, but the *Erh Ya*, like the early glosses on which it is based, refuses to make explicit which meaning is intended. Commentators like Chêng Hsüan 鄭玄 (+127 to +200) disambiguated the glosses by expanding them into binomes. Such disambiguation through the making of binomes became the standard practice in colloquial Chinese as it evolved from Han times onwards.

[2] Cf. Roget (1852), often reprinted and revised since. There are lists modelled on Roget for French, Hungarian, Dutch, Swedish, German, and Modern Greek. Cf. Dornseiff (1970), p. 37.

[3] Cf. Vol. VI (1), pp. 186ff., and Vol. VII (2) pp. [135]ff.

[4] *Erh Ya*, ed. *SPTK*, 1.2b.　　[5] *Erh Ya*, ed. *SPTK*, 1.3a. Cf. Liu Yeh-Chhiu (*1983*), p. 31.

[6] *Erh Ya*, ed. *SPTK*, 2.1a.　　[7] *Erh Ya*, ed. *SPTK*, 1.15b.

recommend the Book of Songs on the grounds that it provided insight into *niao shou tshao mu chih ming* 鳥獸草木之名 'the names of birds and beasts, shrubs and trees'?

Careful readers of *Science and Civilisation in China* will have noticed that many a fascinating story of scientific and technological analysis and achievement takes its beginning in the Confucian classic *Erh Ya*. It is a fact of profound historical significance that this dry book so full of scientific or proto-scientific definitions became a *Confucian* classic.

Already in the bibliographic section of the *Han Shu* 漢書, the first comprehensive bibliography that we possess, the *Erh Ya* is listed in a place of honour, right after the crucially important *Hsiao Ching* 孝經 or Classic of Filial Piety. In the bibliographical chapter of the *Thang Shu* 唐書, our encyclopaedia is even promoted to a place of honour after the *Lun Yü* 論語 of Confucius. The scholar Chhien Ta-Hsin 錢大昕 (+1728–1804) summed up current feelings when he said: 'If you want exhaustively to understand the import of the Six Classics you must take the *Erh Ya* as your starting-point.'[1] Until the twentieth century, the *Erh Ya* remained standard fare in the Chinese literary curriculum.

A later supplement to the *Erh Ya*, the *Hsiao Erh Ya* 小爾雅, survives today as ch. 11 of the apocryphal book *Khung Tshung Tzu* 孔叢子. The *Hsiao Erh Ya* is already mentioned in the bibliographical section of the *Han Shu*. The book has often been considered as unreliable for this reason. However, quotations from it by the leading commentators from Tu Yü (+222 to +284) onwards have tended to be in accordance with the text as we have it in ch. 11 of the *Khung Tshung Tzu*. The chances are that this book originated in late Han times and was later considered to be the original work quoted in the *Han Shu* section on bibliography.[2] By and large, the *Hsiao Erh Ya* does not redefine terms already defined in its famous predecessor. It is small in size and would not, perhaps, be worth mentioning if it did not contain the earliest reference in Chinese linguistics to the phenomenon of the fusion word,[3] one character standing for the fusion of two others. This phenomenon is noticed in the opening definition of ch. 3 of the *Hsiao Erh Ya* 小爾雅.[4]

A much larger elaboration on the *Erh Ya* was the *Kuang Ya* 廣雅 of Chang I 張揖 (*c.* +220 to +265) which was over 18,000 characters long and in the version that has come down to us expounds the meanings of 6,913 characters.[5] This work is

---

[1] Chhien Ta-Hsin 錢大昕 (*1989*), ch. 33, *Yü Hui Chih Lun Erh Ya Shu* 與晦之論爾雅書, in Liu Yeh-chhiu (*1983*), p. 61. Given the Chinese passion for writing commentaries, it will not come as a surprise to anyone that the commentaries on the *Erh Ya* should be many. The earliest commentary of which we know the title dates from the −2nd century, and the great scholar to whom we owe the preservation of so much of ancient Chinese literature, Liu Hsin 劉歆 (died −23) wrote a commentary on the work in three *chüan* 卷. The early commentarial literature of a dozen or so works was summarised and – as it so often happened – superseded by the standard work by the philologist Kuo Phu 郭璞 (+276 to +324) which has come down to us in a somewhat truncated form. There is an extremely convenient Harvard Yenching Index including a critical text of the *Erh Ya*.

[2] Cf. Hu Chhêng-Kung (*1936*) and Liu Yeh-Chhiu (*1983*), p. 40.

[3] My wife points out to me a rather interesting case of a fusion word in English: *bloody* for 'by Our Lady'. Thus here as so often elsewhere, the difference between Chinese and English turns out to be less than absolute. Compare also *zounds* for 'God's wounds'.

[4] *Khung Tshung Tzu* 孔叢子 11, ed. *SPTK*, 3, p. 66a, where *chu* 諸 is defined as *chih* 之 + *hu* 乎.

[5] The *Kuang Ya* is most conveniently available in the *SPPY* edition profusely annotated by Wang Nien-Sun.

important because it preserves a large number of early glosses on characters that have not been preserved elsewhere. However, as the ratio of comment per character shows, the explanations – though valuable – are sparse in the extreme.[1]

'Shuo Wên Chieh Tzu' 説文解字

In spite of its fame, the *Erh Ya* must have remained of limited use as a dictionary. For one thing, it was not easy to find a given character in the book. For another, even after finding it, the definitions found were often far too vague to be useful. The real breakthrough, a practical dictionary of eminent usefulness to anyone who wants to understand ancient Chinese texts, came with the *Shuo Wên Chieh Tzu* by Hsü Shen 許慎 (*c.* +55 to *c.* +149).[2] Hsü Shen (who, incidentally, was born in Ten Thousand Year Village) held various ritual positions under the Han dynasty. He was famous for his commentaries on the classics, which are unfortunately lost. Hsü Shen undoubtedly incorporated many of his commentarial glosses into his dictionary.

The *Erh Ya* had still remained essentially in the tradition of the earlier character-lists like the *Shih Chou* 史籀. The *Shuo Wên Chieh Tzu* was essentially a systematic repository of philological glosses. In the postface, Hsü Shen reports that his dictionary contains no less than 9,353 characters. In addition there are about one thousand graphic variants, so that in Hsü Hsüan's 徐鉉 (+920 to +974) recension the dictionary contained 10,710 characters.[3] The sheer bulk of the dictionary is impressive: there are 133,441 characters of comment, as one of the many loving admirers of the book has calculated.[4]

In an ancient book of such bulk it was crucial that material should be so arranged that the reader could readily find what he was looking for. Hsü Shen attempted to solve this problem by arranging the more than 9,000 characters under 'only' 540 graphic radicals. These radicals were recurring graphic elements in Chinese characters, and in the typical case they would indicate the semantic class to which a word belongs. Thus any designation of a variety of fish is likely to have the character for fish, *yü* 魚, on the left side of the graph. Hsü Shen would ask his readers to look for all characters containing the *yü* under this radical. Of course readers would already have followed something like this practice in *Erh Ya*, but what Hsü Shen did was to generalise the method to include all characters he discussed. They were either declared to be one of the 540 radicals or subsumed under one of these radicals.

The radicals were themselves arranged in semantic groups. Thus in chapter 8 of the dictionary there is a sequence of 36 radicals, all of which are either transformations of the man-radical (*jên* 人) or otherwise related to the human body. Within each of

---

[1] During Sung times two important works in the tradition of *Erh Ya* appeared: the *Phi Ya* 埤雅 by Lu Tien 陸佃 (postface dated +1125) and the *Erh Ya I* 爾雅翼 by Lo Yüan 羅願 finished in +1174, but first printed in +1270. Again, during Ming times there was the *Phien Ya* 駢雅 by Chu Mou-Wei 朱謀瑋 (preface dated +1587), and the very large *Thung Ya* 通雅 in 52 *chüan* by Fang I-Chih 方以智 which was finished some time before +1579. These are just some of the hightlights of the lexicographic *Erh Ya* tradition in China.

[2] The best Western treatment of this most important work remains Roy Andrew Miller (1953).

[3] I owe these figures to Li Jung 李榮 and to Yü Min 俞敏, both from Peking.

[4] Ting Fu-Pao (1930), first series, pp. 245f.

the 540 radical groups Hsü Shen again grouped together what he felt were semantically related characters, starting often from the more general and moving towards the more specific. As Roy Andrew Miller has pointed out, the arrangement of a character under one rather than another radical was often inspired by cosmological speculation rather than linguistic reflection.[1] Under each character, Hsü Shen concentrated on the explanation of two features: the internal graphic structure and the graphically relevant meaning of the character concerned. In both of these areas he made decisive contributions.

Developing earlier ideas of Liu Hsin 劉歆 (died +23), Pan Ku 班固 (+32 to +92) and others, Hsü Shen divided the Chinese characters into six classes (*liu shu* 六書) according to their internal structure and this division has remained canonical until the 20th century.[2] His explanations of the graphs remain, of course, puzzling in many instances, but they still form the indispensable point of departure for discussions by modern scholars. However, by some curious reversal of history, modern archaeology has given us a much more detailed picture of early Chinese epigraphy than was available to Hsü Shen who was so much closer in time to the evidence. On the basis of the oracle bone inscriptions from around the −11th century, available to us but unknown to Hsü Shen, we can in many cases improve on the analyses provided by him.

It is a measure of the tremendous spirit of scientific accuracy that Hsü Shen aimed for, that among his many precise editorial conventions he had a standard device for indicating that he felt he had no good explanation for a given character. He would write the word *chhüeh* 闕 in the place where the explanation should have come. This convention is all the more remarkable in the face of the exasperating tendency among later Chinese commentators to pretend omniscience, to present even the most conjectural guesses with an air of certainty and to make up far-fetched explanations when they did not understand a passage. Hsü Shen's humility greatly increases our confidence in what he does say, although we often may be inclined to disagree with him.

For the study of the small seal characters introduced during the −3rd century and for the study of early Han glosses on Chinese characters, the *Shuo Wên Chieh Tzu* 說文解字 is a quite indispensable tool. However, it had its serious weaknesses. Already the specialist scholar Hsü Hsüan, who in +986 published a new, corrected and improved version of the *Shuo Wên Chieh Tzu*, complains in his preface: 'When one is looking for a character one often has to leaf through a whole volume.'[3]

Hsü Shen had given no phonetic glosses,[4] and as the *Shuo Wên Chieh Tzu* was proving so popular, it was found convenient to add these to his book from other sources

---

[1] Miller (1973), p. 1223.   [2] Cf. Vol. I, p. 32.   [3] *Shuo Wên Chieh Tzu Chuan Yün Phu*, preface.

[4] Phonetic glosses were given at the earliest stage simply by providing a (near-)homophonous other character. This was often difficult (in the case of rare pronunciations), and sometimes impossible (when there were no homophonous known characters). The crucial step was taken in what is known as *fan chhieh* 反切, where the initial consonant of one character was combined with the rhyme of another. The latter system was apparently invented, perfected, and further developed under an increasing influence from Sanskrit. Cf. e.g., Miller (1973), pp. 1232ff. and Wang Li (*1981*), pp. 55ff. for an orientation on the history of phonetics in China.

to be discussed presently. Apart from a fragment of a Thang dynasty manuscript, it is only these supplemented versions of Hsü Shen's dictionary dating from the +10th century which have come down to us.[1]

Let us now turn to the nature of the definitions in *Shuo Wên Chieh Tzu*. Hsü Shen distinguishes general terms, which he calls *tsung ming* 總名, from others. Thus he defines *chhê* 車 as 'a general term for a conveyance with wheels',[2] and *niao* 鳥 'bird' as 'general term for long-tailed pteroid'.

He defines tools by their uses, as when he defines *fu* 斧 'axe' as 'that wherewith one hacks'.

Fascinatingly, he sometimes seems to define what we are inclined to regard as verbs by nouns, as when he defines *ming* 鳴 'to sing, of birds, etc.' as 'the sound of birds (*niao shêng* 鳥聲)'. Or does he really think of *shêng* 聲 as verbal here? Again, Hsü Shen defines *ku* 故 'reason' as 'to cause something/someone to do something (*shih wei chih yeh* 使為之也)'.

He defines the historically (and especially graphologically) primary meaning of a graph, that is to say, he is really writing a historical dictionary. For those looking for definitions of grammatical particles in the *Shuo Wên Chieh Tzu*, this can be most frustrating. As when Hsü Shen defines the object pronoun *so* 所 as follows: '*So* is the sound of felling a tree.' And as if to rub salt into our wounds, he even goes on to quote a passage from the *Book of Songs*, where the word is indeed used in that way.[3]

Hsü Shen's failure to define the current meaning of *so* is symptomatic: his primary interest is in the explanation of the graph and its relation to the original sense of the graph. Hsü Shen sets out to explain the semantics of graphs. Hence the title of his great work, 'Explaining Graphs and Expounding Characters'.

The problem is not restricted to grammatical particles. When he is explaining the word *hsin* 新, meaning 'new', one might expect him to explain this word as 'not old', but he says no such thing. For the graphic structure of the graph suggests another basic meaning which had all but disappeared by Hsü Shen's time, the gathering of firewood, hence 'firewood', a meaning which is expressed with a character written with the grass radical on top, as *hsin*.

Again, when explaining the character *chhing* 輕 'light', Hsü says it means 'a light cart', which may be true enough etymologically, and it certainly would explain the presence of the character for a cart on the left of *chhing*, but it does not help very much as an explanation of current usage. The point is that it does not really help to consider the *Shuo Wên Chieh Tzu* as a dictionary in our terms. We have to read and use the book on its own terms. This means that it would be totally misguided to treat the definitions of the *Shuo Wên Chieh Tzu* as a reflection of Han dynasty usage or even

---

[1] During the Chhing dynasty a number of very important commentaries on the *Shuo Wên* were published which are an essential aid to our understanding of the dictionary. These may all be conveniently consulted in Ting Fu-Pao (*1930*) (Shuo Wên Ku Lin), which also contains a singularly useful index to the characters treated in the *Shuo Wên*.

[2] Cf. Vol. VII (I) KR, p. 47.

[3] Consider the word *yu* 猶 'like'. Hsü Shen defines it as 'a species of large ape (*chüeh* 玃) while being fully aware that this is not the meaning that would get one anywhere reading ancient Chinese texts.

of the usage in the Classical literature Hsü Shen was familiar with. Hsü Shen was only interested in such usages in so far as these serve his purpose, which is that of explaining graphs.

We need therefore not be worried by the fact that he very often defines one character X by another character Y, and on the same page goes on to define the character *Y* by the character *X*, as when he defines *i* 意 'meaning' as *chih* 志 'intention', and on the same half page goes on to define *chih* 'intention' by *i* 'meaning'.[1] His is not a learner's dictionary of Classical Chinese, and as we are all raw learners of the language, this limits the usefulness of his dictionary in practical terms. It is really most useful for those who understand the words but wish to understand the graphs.

In fact, it takes a tremendous ability for abstract thought to be able to disregard, as consistently as Hsü Shen does, current usage and to concentrate so entirely, as he does, on the aspects that are relevant to his theoretical purpose.

There is a clear system in this: Hsü Shen is not explaining words, he is explaining graphs in terms of *the meaning relevant for a satisfactory explanation of the graph*. In other words, Hsü Shen is consciously and consistently doing the *semiotics of the Chinese graphs*. It is an extraordinary thought that so long before semiotics was recognised in the West as a crucially important philosophical and philological discipline, it was coherently practised in China and consistently applied to a large body of Chinese characters. Even the literal translation of the title of Hsü Shen's book makes it quite plain what he sets out to do: 'Explaining Graphs and Expounding Characters'. There are occasional excursions or deviations, as when Hsü Shen explains the final particle *i* 已 quite rightly as *yü i tzhu* 語已辭 'a word indicating that the speech has ended'.[2]

Hsü Shen's graphic tradition was followed by a long line of distinguished graphic dictionaries, for example the *Liu Shu Ku* 六書故 by Tai Thung 戴侗, first published in +1320 but considered to be of Sung provenance. This book is divided into nine sections: mathematics, astronomy, geography, human affairs, animals, plants, artificial objects, miscellaneous, problematic characters. A valiant attempt is made to base the explanations of characters not on the standardised small seal script introduced during the brief Chhin dynasty (−221 to −207), as Hsü Shen had done, but on earlier inscriptions.

The study of the *liu shu* 六書, or six categories of characters, has become a large industry in China.[3]

'Yü Phien' 玉篇 *and 'Lei Phien'* 類篇

In our context, we are more interested in the semantic side of lexicography than in the graphic etymology of characters. In Chin times (+265 to +420), Lü Chhên 呂忱

---

[1] *Shuo Wên Chieh Tzu* in Tuan Yü-Tshai (*1981*), p. 502. [2] *Shuo Wên*, sub verbo *i* 已.
[3] *Shuo Wên Chieh Tzu Ku Lin* ed. Ting Fu-Pao (*1930*), with its supplement published in 1932, reprints 288 books as well as 544 articles on *Shuo Wên*. Since then, a number of important works have appeared, notably Ma Hsü-Lun (*1957*).

produced an expanded edition of the *Shuo Wên Chieh Tzu* explaining 12,824 characters. For many centuries this work maintained its status as the equal of the *Shuo Wên Chieh Tzu*, but by the +12th century the book only survived in part and gradually fell out of use, being eclipsed by the mammoth *Yü Phien* (+548)[1] by the child prodigy Ku Yeh-Wang 顧野王 (+519 to +581). This dictionary is impressive in sheer bulk: there are 290,770 characters of text defining in all 16,917 characters, and in addition there are over 400,000 characters of commentary, if we are to believe the preface of the work.

However, the discovery in Japan early in this century of an earlier version of somewhat less than a fifth of the *Yü Phien* known as the *Yü Phien Ling Chüan* 玉篇零卷,[2] which is considerably more detailed in its semantic descriptions than the current edition we have, suggests that what we have today is a boiled-down, abbreviated and doctored version of Ku Yeh-Wang's original work.

The *Yü Phien* changes the number of radicals from the *Shuo Wên Chieh Tzu* slightly and rearranges the characters with less emphasis on the six categories (*liu shu*) set up by Hsü Shen. In general, the *Yü Phien* concentrates on phonetics and meaning rather than on graphic structure.

An avowed successor to the *Yü Phien*, still very much in the tradition of the *Shuo Wên Chieh Tzu*, but placing more emphasis on pronunciation and meaning rather than the shape of characters, was the *Lei Phien* 類篇, compiled between +1039 and +1066 and finally edited by the great historian Ssu-Ma Kuang (+1019 to +1086). It comprised no fewer than 31,319 characters, but even this number was surpassed by the contemporary more phonetically orientated dictionaries like the *Kuang Yün* 廣韻 and *Chi Yün* 集韻, to be discussed below.

'Shih Ming' 釋名 *and etymological glossography*

Folk etymology is extremely common in early brahmanic Indian literature, and it is well known in ancient Greece. It features prominently, for example, in Plato's dialogue *Cratylus*. In fact, Plato's contemporary Mencius did occasionally use etymologising glosses to explain why and how a word came to mean what it means, but the golden age of etymologising glossography was the Han dynasty. A book like the *Pai Hu Thung* 白虎通 'Comprehensive Discussions in the White Tiger Hall' introduces folk etymologies in almost every chapter and gives a vivid picture of the range of traditional Chinese folk-etymological fantasy and imagination, and of its function in the thought of the time.[3]

The folk-etymolgically orientated dictionary *Shih Ming*, containing etymologising glosses for somewhat over 1,500 characters, was written around +200 by Liu Hsi 劉熙. The material is arranged in a manner strongly reminiscent of Roget's

---

[1] The current editions of the work include additions dating from +674 and from +1013. Our present editions are based on that of +1013 by Chhên Phêng-Nien 陳彭年.

[2] This is reprinted in *Tshung-shu chi-chhêng* 叢書集成, vols. 1054 to 1057.

[3] See Tjan (1952).

Thesaurus of the English Language, and in the tradition of the *Erh Ya* 爾雅, under 27 headings as follows:

1. Heaven, 2. Earth, 3. Mountains, 4. Streams, 5. Hills, 6. Roads, 7. Provinces, states, 8. Body parts, 9. Attitudes, body actions, 10. Age and youth, 11. Kinship, 12. Language, 13. Food and drink, 14. Colours and stuffs, 15. Headdress, 16. Clothing, 17. Dwellings, 18. Curtains, carpets, 19. Writing, 20. Classics, 21. Tools, 22. Musical instruments, 23. Weapons, 24. Wheeled vehicles, 25. Boats, 26. Diseases, 27. Mourning rites.[1]

This compares favourably with one of the most comprehensive synonym-and-homonym dictionaries, the *Vaijayant ī* (+12th century), in which 3,500 stanzas were organised into eight sections: 1. heaven, 2. atmosphere, 3. earth, 4. nether world, 5. general, 6. disyllables, 7. trisyllables and 8. sundry.[2]

The *Shih Ming* 釋名 is famous for its 'paronomastic[3] glosses', i.e., glosses that are similar in pronunciation to the original character and at the same time close in meaning. The word-play of these glosses comes out in examples like gloss 325:

Tsu 足 *(tsiuk/tsiwok)* 'the foot' is *hsü* 續 *(dziuk/ziwok)* 'to continue'. This means it is a continuation of the leg.[4]

It is a reflection of Liu Hsi's preoccupation with phonetic description that many of his glosses are chosen for sound resemblance at the expense of similarity in meaning. Moreover, the explanations appended to the paronomastic glosses are not essentially explanations of the characters but rather explanations of why Liu Hsi chose the gloss he did for the character at hand. This does, of course, make the *Shih Ming* a crucially important document for a historical study of the Chinese sound system, but from our semantic point of view the dictionary is of limited interest.[5]

Chhieh Yün 切韻, *Kuang Yün* 廣韻 and *Chi Yün* 集韻[6]

For our purposes the rhyme dictionaries are of interest not for their information on rhymes but because they often include precious semantic glosses. These dictionaries were, by and large, not intended as works of scientific phonological analysis but rather as practical handbooks for those who needed to know what rhymes and what does not rhyme in order to pass their examinations. The *Chhieh Yün* 切韻 rhyme dictionary of +601, which distinguishes 32 initial consonants and 136 finals, is remarkable especially for its highly sophisticated methodological introduction.[7] This famous rhyme dictionary, which has been crucial for the reconstruction of the

---

[1] Bodman (1954), p. 11.   [2] Vogel (1975), p. 323.

[3] Bodman (1954) consistently misspells this word as 'paranomastic' in his book, and this practice has in fact had considerable influence on later sinological writing.

[4] The bracketed pronunciations are the forms reconstructed for Archaic and Ancient Chinese respectively in Karlgren (1952b).

[5] The most convenient edition to use is that finished by Wang Hsien-Chhien 王先謙 and his collaborators in 1896 and reprinted in the *Wan-yu wên-khu* 萬有文庫 collection no. 177–8.

[6] For the history of the rhyme dictionaries see especially Chao Chheng (*1979*). Liu Qiuye, p. 218.

[7] See Henri Maspero (1920).

pronunciation of ancient Chinese, represents the pronunciations in the present-day Nanking area and not, as previously assumed, the area of Chhang-an, the capital of the Sui dynasty under which it was compiled.

In this section we shall concentrate on those rhyme dictionaries that have important semantic components, the *Kuang Yün* and the *Chi Yün*, both of which may be regarded as semantic dictionaries with a phonetic component. Before we turn to these landmarks of traditional Chinese lexicography, a brief survey of the predecessors of these dictionaries is in order.

The earliest rhyme dictionaries that we know of are the *Sheng Lei* 聲類 (+3rd century) by Li Têng and the slightly later *Yün Chi* 韻集 by Lü Ching 呂靜. Much more famous than these was the extant *Chhieh Yün* dictionary (preface +601, an improved edition dated +677) in which the characters are categorised under tones, and originally under a total of 193 rhyme categories – which we have discussed above. Unfortunately, only fragments of these early rhyme dictionaries survive.

The earliest important rhyme *dictionary* that survives today is Chhên Phêng-Nien's 陳彭年 *Kuang Yün* 廣韻 (first edition +1007), which is intended as an extended version of the *Chhieh Yün*. It is arranged according to tones and rhyme categories. There are semantic and phonetic definitions of 26,194 characters which are divided into 206 rhyming categories. The pronunciations of 3,877 syllables are distinguished. There are a massive 191,000 characters of text. The special strength of this dictionary lies in its quotations. It does not limit itself to the Confucian classics, but actually quotes freely and abundantly from narrative literature and informal literature.[1]

The *Chi Yün* (+1067) was finished by a team under the historian Ssuma Kuang (+1019 to +1086). Ssuma Kuang 司馬光 became famous for his superb and vast comprehensive history of China, the *Tzû Chih Thung Chien* 資治通鑑. Among mankind's dictionaries of its time, the *Chi Yün* was an achievement of similarly gigantic dimensions. The *Chi Yün* defines no fewer than 53,525 characters. There are thus considerably more definitions than in the *Kuang Yün* but the definitions are somewhat shorter. For a long time this remained the most comprehensive reference book on Chinese characters and, quite arguably, it remained for several centuries the biggest dictionary in the world.

'Fang Yen' 方言

The book *Fang Yen* (Local Languages), attributed to Yang Hsiung 揚雄 (−53 to +18),[2] must surely be the earliest dictionary of local languages in any country. It is common to regard the *Fang Yen* as a 'dialect dictionary', but it would appear that many of the linguistic phenomena covered in that work actually belong to different

---

[1] By far the most useful edition of the *Kuang Yün* is Chou Tsu-Mo (*1960*).

[2] See Serruys (1959), which remains the standard work on this dictionary. Yang Hsiung's authorship of the *Fang Yen* has been doubted by some from Sung times onwards. However, the prevailing opinion is that the famous poet and scholar was indeed the author of the dialect dictionary. (Liu Yeh-Chhiu (*1983*), p. 45.)

languages rather than different dialects. Yang Hsiung apparently derived the idea for his dictionary from the long-standing interest in local folklore at Chinese imperial courts. By his own account he spent at least 27 years collecting material for his work from visitors at the capital Chhang-an, and like the author of the *Shuo Wên Chieh Tzu* 説文解字, Hsü Shen (died +149), he seems to have worked on his *magnum opus* until he realised that he did not have much longer to live, i.e. shortly before +18.

The arrangement of the material discussed largely follows that of the *Erh Ya*, but in this case each character is followed by a string of equivalents in various dialects. According to Ying Shao 應劭 (+2nd century), the dictionary contained 9,000 characters, but the current edition has closer to 12,000 characters, a discrepancy which indicates later additions to Yang Hsiung's 揚雄 work as it was current in late Han times.

Among the dialect words, Yang Hsiung distinguished five kinds of linguistic characterisations:

1. of the common language;
2. of a certain dialect area;
3. of a certain group of dialect areas;
4. an obsolete dialect form;
5. a local variation on a common language expression.

Interestingly, Yang Hsiung included a number of expressions from non-Chinese minority languages which he describes as *ti* 狄, *man* 蠻, and *chhiang* 羌 .

For example, Yang Hsiung first enumerates 12 words for *ta* 大 'big', and then continues to explain which of these words are used where. Or under the word *chu* 豬 'pig' he goes on to list nine words for 'pig' and their area of use. In each case, Yang Hsiung is careful to note some semantic nuances relating to certain dialect forms. The great philologist Kuo Phu 郭璞 (+276 to +324) wrote the first detailed commentary on the *Fang Yen* that has come down to us, and Chou Tsu-Mo (*1951*) provides an exemplary index to the book, which contains by far the most convenient critical edition of the *Fang Yen*.

In China, the *Fang Yen* remained unparalleled and unsurpassed until the flowering of Chhing philology from the +18th century onwards. Both in sheer size and in coherent dialectological design the *Fang Yen* also remained unparalleled in the West for a long time. What the *Fang Yen* demonstrates without a shadow of doubt is that Chinese characters stand not for ideas or thoughts but for words of the spoken language.

'Ching Tien Shih Wên' 經典釋文 and 'I-chhieh Ching Yin I' 一切經音義

From the bibliographic sections of the dynastic histories it would appear that the Thang dynasty was an age of lexicographical megalomania: we hear of one dictionary in no less than 360 *chüan* 卷, a bulk that was quite unprecedented in Chinese history.[1] A long book might have thirty *chüan*, but over three hundred *chüan* was beyond the imagination even of the bibliophile. For later bibliophiles like ourselves it is a matter of profound regret that none of these early lexicographical mammoths has come down to us. None the less it is significant that the Chinese were sufficiently interested in lexicography to produce some of their most voluminous early books in that area.

Consider the *Ching Tien Shih Wên* 經典釋文 by Lu Tê-Ming 陸德明 (+550 to +630), which is thirty *chüan* long and concentrates on the phonetic as well as semantic explanation of the characters found in thirteen Chinese classics. Strictly speaking this is not a dictionary but a systematic collection of glosses. In some cases Lu concentrates on the phonetic glosses, in others on the semantic glosses. But whatever he concentrates on has remained invaluable lexicographical evidence for scholars of Chinese literature ever since, not so much because of the originality of his own observations, but because of the care with which he collected earlier glosses and thus saved them from oblivion.

The arrangement of the characters and phrases in the *Ching Tien Shih Wên* is according to their glossed occurrence in the classics. This makes the book extremely inconvenient to consult as a dictionary. However, much of the contents of the *Ching Tien Shih Wên* are accessible in the large modern Chinese-Japanese dictionary by Morohashi.

What Lu Tê-Ming did for the Confucian Classics, the monk Hsüan Ying 玄應 did around the year +645 for the Buddhist works in his *I-chhieh Ching Yin I* 一切經音義 in 25 *chüan*, systematically excerpting glosses from no fewer than 454 Mahayâna and Hinayâna texts. Hsüan Ying glossed both Sanskrit-derived technical terms and ordinary Chinese characters and arranged his glosses, as Lu Tê-Ming had done shortly before, according to the provenance of the glosses he quotes. The glosses on ordinary Chinese characters occupy about half of the book.

The Empress Wu 武則天 of the Thang dynasty was megalomaniac in her architectural taste, and she had similar inclinations when it came to dictionaries. Under her direction the *Tzu Hai* 字海 was compiled in no less than 100 *chüan*, and the bibliography in the *Hsin Thang Shu* 新唐書 'The New Thang History' even mentions a work entitled *Yün Hai Ching Yüan* 韻海鏡源 by a certain Yen Chên 顏真 in 360 *chüan*.[2] This must have been, by any standards, a very large dictionary, and by the standards of the time it must have been simply gigantic. Unfortunately, unlike the large *I-chhieh Ching Yin I*, which is much less than a tenth of its size, it is lost.

Around +807 the monk Hui Lin produced another *I-chhieh Ching Yin I*, this time in 100 *chüan*. Hui Lin excerpted from no fewer than 1,300 Buddhist works, and his glosses are taken from a vast range of lexicographical and commentarial literary works now lost to us. These earlier works have actually later been partly reconstructed on the basis of the ample and carefully attributed quotations in Hui Lin's large collection of glosses.

---

[1] For some details on this dictionary, of which we have a detailed description, see Liu Yeh-Chhiu (*1983*), p. 179.   [2] *Ibid.*, p. 80.

Other efforts at systematic glossography fade in comparison with this, but mention must be made of Juan Yüan's 阮元 *Ching Chi Tsuan Ku* 經籍纂詁 (+1798), which is a systematic collection of pre-Thang semantic glosses on Chinese characters, a work rightly admired by later Chhing dynasty philologists.

'*Tzu Hui*' 字匯 (+1615)

The dictionaries we have discussed so far may look impressive when described, and they are awe-inspiring when one tries to use them, but they all have one major defect: they remained infuriatingly impractical. The breakthrough came during the Ming dynasty, and was due to the scholar Mei Ying-Tso 梅膺祚 (+1570 to +1615). He arranged the 33,179 characters, which he set out to explain, under 214 headings (*pu* 部) which have proved (in essentially unchanged form) to be useful and efficient mnemonic aids for the retrieval of characters down to the present day. Even more important was Mei's recognition that the order of characters under each radical was essential for rapid use of the dictionary. He introduced the natural and efficient rule that characters under each radical are arranged by the number of strokes needed in addition to the radical. In addition to this, Mei included tables of content in each *chüan* with references to page-numbers, thus greatly facilitating ease of access. (One wishes the compilers of large books in China down to the twentieth century had learned that lesson from Mei Ying-Tso!) With these major innovations, there was for the first time a dictionary that could be used rapidly and efficiently. So conscious was Mei of ready accessibility in lexicography that he even included lists of troublesome characters that were not easily categorised under his radicals.

Mei's dictionary remains a mine of information on phonology, and it became the most important source of later standard works such as the *Khang Hsi Tzu Tien* 康熙字典, which we shall presently discuss. In fact, Mei's work was surreptitiously copied even before it was published. Chang Tzu-Lieh 張自烈 (+1564 to +1650) based his *Chêng Tzu Thung* 正字通 (preface +1670) on Mei's dictionary, introducing a number of minor corrections, minor rearrangements and new mistakes.[1] This was only the first of a long line of popularised editions of the *Tzu Hui* 字匯, and the very fact that the *Tzu Hui* was so widely popularised testifies to the public interest in its new practical approach.

The *Tzu Hui* itself was first printed in Japan (+1671), where several more popular editions of it appeared.

'*Khang Hsi Tzu Tien*' 康熙字典 (+1716)

By Western standards, the *Khang Hsi Tzu Tien*, or 'Character Classic of the Khang Hsi Emperor', might still have to count as a plagiarisation of the *Tzu Hui*, but this

---

[1] For an interesting discussion of the fate of this book see Goodrich (1976), pp. 1062f.

plagiarisation was imperially sponsored, and it must be admitted that it involved considerable improvement in reliability and detail, and also in scope: there are 47,035[1] characters defined in the Khang Hsi Dictionary, 10,000 more than in the *Tzu Hui* or its popularisations including the *Chêng Tzu Thung* 正字通. In fact the quality of the Khang Hsi Dictionary is such that most serious scholars of Classical Chinese will at least have it on their shelves – not out of some kind of piety to the Khang Hsi emperor, but because under that man's tutelage a remarkable monument of traditional Chinese lexicography was achieved. The Khang Hsi Dictionary was a collective effort by thirty scholars over the years +1710 to +1716.

The cultural importance of the Khang Hsi Dictionary can be gauged by the fact that the modern Chinese word for 'dictionary' *tzu-tien* 字典 actually derives from the title *Khang Hsi Tzu Tien*. The Khang Hsi Dictionary became synonymous with the general concept of a dictionary, and in fact it has all but replaced its pioneering predecessors (excepting the *Shuo Wên Chieh Tzu* 説文解字) on modern scholars' bookshelves. One consequence of the official character of the Khang Hsi Dictionary was that it concentrated entirely on the educated literary language and omitted non-literary meanings of characters. This was a matter of policy, and it constitutes an unfortunate shortcoming of the book.

On the other hand, the Dictionary aimed at, and boasted of, total reliability when it came to its references to the Classical literature. Owing to the prestige of the Khang Hsi emperor, this image was preserved until Wang Yin-Chih 王引之, the philologist (+1766 to 1834), in the true spirit of Chhing dynasty *khao chêng* 考證 textual scholarship, published a list of 2,588 entries containing mistakes. Since then it has been the Chinese philologists' delight to point out in exquisite detail that even this impressive list of errata is far from complete.[2]

In spite of all its shortcomings, the Khang Hsi Dictionary has enjoyed a quite unique popularity in China. When, in 1882, a popular edition was published in Shanghai, 40,000 copies were sold within a few months, and the second edition of another 60,000 copies again sold out within a few months.[3]

'Phei Wên Yün Fu' 佩文韻府 *(+1726)*

The dictionaries proper we have discussed so far are essentially single character dictionaries. Combinations are covered only incidentally. However, the meaning of a combination of characters in Chinese is by no means simply the sum or product of the components. There are, in short, idioms. If one wishes to look up a combination of two characters, with both of which one may be quite familiar, all the traditional dictionaries we have so far discussed are generally quite inadequate for the purpose. This gap is filled by the *Phei Wên Yün Fu*, compiled between + 1704 and +1711 under the direction of Chang Yü-Shu 張玉書 (+1642 to + 1711) and others. In the 1983 reprint edition by the Shanghai Ku-chi-chhu-pan-shê the index to the work

---

[1] I must record that Li Jung (*1985*), p. 3 counts 47,021 characters.
[2] See Wang Yin-Chih (1831).   [3] Cf. Fang Hou-Shu (*1979*), part II, p. 226.

alone covers *c.* 1,200 large pages in small print and gives one access to 13,000 double pages of detail on Chinese combinations of characters arranged by the rhyme of the last characters in the combinations.

The *Phei Wên Yün Fu* was based on two earlier compilations of the same type, the *Yün Fu Chhün Yü* 韻府群玉 and the *Wu Chhê Yün Jui* 五車韻瑞. It is a measure of the scholarly meticulousness of this work that the material taken over from these sources is clearly marked as such and distinguished explicitly from the original additions by the editors of the *Phei Wên Yün Fu.*

The weak points of the *Phei Wên Yün Fu* are none the less evident enough: essentially this is not a dictionary but a mammoth collection of basic material for any future dictionary that would have to include combinations. By and large, the semantic explanations in *Phei Wên Yün Fu* are all quotations of earlier glosses. When no such glosses were available, there normally are no explanations of the combination in question. In a dictionary of that kind there would have to be glosses on each combination, not just examples of how it is used. The meanings of combinations would have to be made explicit in each case, not assumed as self-evident from the context when no semantic gloss happens to have been found.[1]

Considering the tremendous variety of meanings attributed to Chinese words in traditional Chinese dictionaries, one might be excused for getting the impression that Chinese is a semantically rather diffuse language when compared with, say, Greek. A look through a standard dictionary like Mathews's *Chinese-English Dictionary* (1931) will certainly leave one with this impression when one compares it to Liddell and Scott's *A Greek English Lexicon.* But the comparison is quite inappropriate. Mathews's dictionary covers 3,000 years of linguistic history, whereas Liddell and Scott's dictionary is overwhelmingly concerned with the 305 surviving classical authors, who are divided into nine principal eras in Greek literature, which they list as running from Homer, *c.* −900, to Lucian, *c.* +160. It is true that 165 authors later than Lucian are cited, and some of them are even as late as Theodorus of Byzantium, who flourished in +1430. But for comparison with Chinese literature one must imagine a mammoth Liddell and Scott covering not only Classical Greek, but all later forms of Byzantine Greek, medieval colloquial Greek, as well as semi-Turkic and Americanised modern colloquial Greek. Such a mammoth dictionary, if presented in a historically indiscriminate, summary, way without attribution to sources, would present a most confusing picture of the meanings of Greek words, quite as confusing I should think, as the picture emerging in Mathews's dictionary of the semantics of Chinese words.

The curious fact is that we still do not have a specialised dictionary of pre-Buddhist Classical Chinese, which would be comparable to Liddell and Scott's

---

[1] The final breakthrough in lexicography never came during Classical times in China. It was for the Japanese scholar Morohashi Tetsuji to incorporate the material from the Chinese dictionaries and from their Japanese counterparts into the best dictionary of Classical Chinese until today, the *Dai Kanwa Jiten* 大漢和辭典, which collects all the earlier glosses, and where every combination is explained and interpreted. Needless to say, the *Dai Kanwa Jiten* could not have been compiled without the use of its Chinese predecessors, notably *Phei Wên Yün Fu* 佩文韻府.

dictionary. Couvreur's great *Dictionnaire classique de la langue chinoise* does not limit iself to pre-Buddhist usage. Even the very careful Wang Li (*1979*), covering far too long a time span for our purposes, still presents Chinese words with a range of meanings that is not significantly larger than that which Liddell and Scott attribute to Greek words.

Glossaries of foreign words

Since Chou dynasty times the Chinese had intepreters of various languages attached to their various courts. Surprisingly, they did not produce detailed and reliable dictionaries of foreign languages of anything like the lexicographical standards they applied to their own language until modern times. The contrast with Europe is striking. Leaving aside a number of minor glossaries of foreign languages printed in the late +15th century, we find the gigantic *Calepini Octoglotton* (+1502)[1] by Ambrogio Calepino (+1435 to +1511), which gives the meanings of Latin words first in Latin and then in seven other languages. Calepino's work became so popular that *calepin* became a current word for 'dictionary'. By comparison, the *Pentaglot Lexicon*[2] published under the auspices of the Chhien Lung emperor (+1736 to +1795) strikes one as vastly inferior not only in scope but also in philological quality. The extraordinary achievements of European lexicographers of foreign languages up to 1889 may be found conveniently surveyed in the bibliography assembled in the 11th edition of *Encyclopædia Britannica*, vol. 8, p. 190ff. We find nothing remotely like this fascination for foreign languages in traditional China at that time in spite of the fact that the Chinese-speaking people lived in an exceptionally rich and varied linguistic environment.

In this the Chinese were rather like the Indians who wrote in Sanskrit. There are hardly any significant foreign language dictionaries into Sanskrit, whereas there is a considerable lexicographical tradition of Sanskrit in India. The comparison with Sanskrit lexicography in India is indeed instructive. The major works of Indian lexicography are listed in Vogel (1975). In his general characterisation of Sanskrit dictionaries Vogel notes that 'the classical dictionaries are generally limited to nouns and indeclinables'.[3] Vogel reports that traditional Sanskrit dictionaries are of two types: they list synonyms, or they list homonyms. They are essentially systematic word-lists. There are no sizeable traditional Sanskrit dictionaries that give discursive definitions of the meanings of all current Sanskrit words.[4] The earliest fragment of a glossary that survives is dated to the +6th or +7th centuries. The art of lexicography of Sanskrit flourished particularly during the +11th to +14th centuries, but apparently no major lexicographical effort was made in traditional times towards the writing of dictionaries that explained foreign languages in Sanskrit.

---

[1] The book was reprinted and added to many times. The largest edition (Basle, 1590) treats eleven languages, among them Hungarian. However, the best edition is the seven-language version (Padua, 1772).

[2] This work was photographically republished by Raghu Vira as vol. 19 of the Sata-Pitaka Series, New Delhi 1961, but was first described already by J. P. Abel-Rémusat in his *Mélanges Asiatiques* (1825), vol. 1, p. 153.

[3] Vogel (1975), p. 304.   [4] Vogel (1975), p. 305.

Such Chinese interest in foreign languages as there was, was entirely practical. There is, as far as I know, no deep *intellectual* curiosity with respect to foreign languages. A few people were interested to see that somebody in China *knew* foreign languages, but nobody was particularly concerned to understand them or learn more about them. In this the Chinese were rather like the ancient Greeks.

An early survey of the study of foreign languages is Abel-Rémusat's '*De l'étude des langues étrangères chez les chinois*' of 1826. Paul Pelliot (1948), pp. 207–90 remains the best overall introduction to the main institution responsible for foreign language instruction, the Ming dynasty Four Barbarians' Office (*ssu i kuan* 四夷館) and its successor in the Chhing, the Four Translations Office (*ssu i kuan* 四譯館).[1]

By far the most important source for Chinese glossaries of foreign languages is the *Hua I I Yü* 華夷譯語 (Chinese Barbarian Translations), an early version of which was published in +1389.[2]

The pioneering detailed work on the all-important *Hua I I Yü* was Wilhelm Grube's *Die Sprache und Schrift der Jučen*, published in 1896 and carefully lining up 871 phonetic and semantic glosses on the *Ju-chên* 汝真 language. Lewicki (1949) remains a standard account of the Chinese achievements in Mongolian lexicography.[3] For Uighur we now have Ligeti (1966), for Malay we have Edwards and Blagden (1931), which lists 482 expressions divided into seventeen semantic categories. For Cham we have Edwards and Blagden (1939), with its 601 words and phrases divided into the same categories as in the case of Malaccan Malay. For Vietnamese we have Gaspardonne (1953) and Chhen Ching-Ho (*1953*). For Japanese the best reference work is Ōtomo and Kimura (1968). For what the Chinese called *hui hui* 回回 or Muslim language we have Tazaka ('Kaikaikan yakugoto goshaku' in *Tōyō Gakuhō* 30 (1943), pp. 96–131 and 232–96; 30 (1944), pp. 534–60; and 33 (1950), pp. 132–45. For Tibetan there is Nishida. For Thai and related languages an early European treatment is Müller (1892).[4]

All these vocabularies are of interest to students of historical phonology and comparative East Asian linguistics. From our lexicographical point of view they are disappointingly limited in scope and show little semantic perceptiveness on the words that they do treat. Chinese culture, like Greek culture, did receive outside stimuli, but it remained generally introspective as far as lexicography was concerned.

One might expect that the Chinese might have developed a sophisticated intellectual interest in Sanskrit, since we have such an enormous body of translations from the Sanskrit. The classic work on Sanskrit-Chinese glossaries is Prabodh

---

[1] See *Ssu I Kuan Khao* by Wang Tsung-Tsai, preface 1580. Copies kept in *Tōyō Bunko* and Leningrad Library no. 130). See also Devéria (1896). For a readable account of the regulations and the history of these institutions see also Wild (1946).

[2] The earliest version is that of Huo Yüan-Chieh (*floruit* 1389). *Hua I I Yü*, reprint Kyōto 1937. Cf. Fuchs (1931) and Miller (1954).

[3] See however already Julius Klaproth's *Abhandlung über die Sprache und Schrift der Uiguren. Nebst einem Wörterverzeichnisse und anderen Uigurischen Sprachproben, aus dem kaiserlichen Übersetzungshofe zu Peking*, Berlin 1812.

[4] For bibliographic detail see Yang (1974), pp. 113–21.

Chandra Bagchi's *Deux lexiques Sankrit-Chinois* (Paris, 1929 and 1937), which discusses in detail four glossaries,[1] none of which shows any detailed philological interest in Sanskrit philology or grammar. R. van Gulik summarises his findings in his monograph *Siddham, The Study of Sanskrit in China and Japan* (Delhi, 1956) as follows:

Thus one arrives at the conclusion that the study of Sanskrit as we understand the term, namely as leading to a thorough knowledge of Sanskrit grammar and vocabulary such as will enable one to read and translate independently Sanskrit texts, never flourished in China.

Here the reader will ask how the huge Buddhist Canon in Chinese came into being, as so few Chinese had a sufficient knowledge of Sanskrit to be able to translate Indian sutras.

The answer is that the stupendous task of translating into Chinese the hundreds of sutras that constitute the bulk of the Canon was performed mainly by Indian and Central Asiatic monks.

We shall see in our Section on Buddhist logic that there were notable exceptions to this rule, like Hsüan-Tsang 玄奘,[2] but the general disinterest of the Chinese in Sanskrit lexicography as well as grammar is plain enough. However, the ideal of enabling the Chinese to read Sanskrit was not absent in the Chinese tradition. I-Ching 義淨 writes in the preface to his One Thousand Words in Sanskrit:

If one reads this together with the Siddham Parivarta, one will be able to translate in one or two years.[3]

Whatever Chinese interest there was in Sanskrit practically, it disappeared from the +15th to the early 20th century. There was no competitor in China to Japan's Ji-un (+1718 to 1804), who showed a sustained interest in Sanskrit philology and achieved a fair understanding of Sanskrit grammar.[4]

Another important language with which the Chinese were in intense and sustained contact was Tibetan. We have an early glossary described in Thomas and Giles (1948), which presents a Tibeto-Chinese word- and phrase-book written on the back of a +9th-century Buddhist text found in Tun Huang. There are in fact two lists containing 154 and 60 items respectively and supplying Chinese glosses in each case.[5] We also have a number of sizeable Buddhist translators' glossaries of technical terms from Tibetan to Chinese. But none of these amounts to a comprehensive dictionary of the Tibetan language. None of them shows an intellectual interest in the vocabulary and structure of the Tibetan language as such.

During the period of Manchu domination in China there was, understandably, a certain practical interest in Manchu. Chinese dictionaries and studies of the Manchu language are surveyed in the Russian monograph Grebenshchikov (1913).

---

[1] *Taishō Tripitaka* nos. 2133, 2134, 2135, and 2136. Van Gulik (1956), p. 35, comments: 'We must be grateful to those Japanese monks (who preserved these glossaries C.H.), for if they had not carefully preserved these books, much important material regarding the ancient pronunciation of the Chinese language would have been lost. But this aspect of the vocabularies falls outside the scope of the present essay. Here we are concerned only with their value as aids in learning Sanskrit for the Chinese of former times. As such their value is practically nil.'

[2] See our Section (*f,*7). Van Gulik was, of course, familiar with these exceptions.

[3] *Taishō Tripitaka* no. 2133.   [4] In van Gulik (1956), pp. 13f.

[5] For more recent comments see Ligeti (1968).

(6) THE ART OF GRAMMAR IN TRADITIONAL CHINA

Background

We have noted that the ancient Chinese paid considerable philosophical attention to the nature of their own language, and that they developed a commentarial lexicographical tradition that shows their profound philological interests.

However, dictionaries as well as commentaries are essentially cumulative. They are typically concerned with individual questions of interpretation within a given passage. They are not concerned with the grammatical system as such. When it came to the study of the pronunciation of Chinese characters, the Chinese linguists showed a considerable systematising interest and ability. They were interested not only in pronunciation of individual characters but in the sound system of their language as a whole.[1] But what about the grammatical system of their language? What was the evolution of Chinese views on the grammatical system of their own language?[2] By and large, one will find that the grammatical system as a whole has been of only marginal interest when compared to lexicography and phonology.

Grammar in the West

In Greece, the study of grammar proper started mainly with the Stoics in the 3rd century.[3] The first complete grammar, the *Technē grammatikē* in 25 sections, was traditionally attributed to Dionysios Thrax (*c.* −100) but is probably a somewhat later compilation. The *Technē grammatikē* defines, for example, the concept of a sentence ('expression of a complete thought'), and of a word ('smallest unit within a sentence'), and sets up eight word-classes (noun, verb, participle, article, pronoun, preposition, adverb, conjunction). It deals in detail with inflection but does not pay the same attention to syntax.

In Rome, Varro (−116 to −27) wrote his famous *De lingua Latina*, of which books 5 to 10, and some further fragments, survive today. From *c.* +500 we have the celebrated Latin grammar by Priscian. Grammar became a part of the *trivium*, the educational curriculum consisting of grammar, logic and rhetoric, which, together with music, arithmetic, geometry and astronomy, was the mainstay of secular education throughout the Middle Ages and beyond. H. Keil (1961), *Grammatici Latini*, published between the years 1855 and 1880, is an impressive monument to the ingenuity and fertility of the Latin grammatical tradition. Nothing remotely similar to this collection could be compiled from traditional Chinese sources.

---

[1] The preface to the phonetic dictionary *Chhieh Yün* 切韻 (+601) shows a quite theoretical interest in what we today would call the phonological system of Chinese.

[2] See Miller (1973).

[3] See Pohlenz (1939). The general story of the early history of grammar is well told in Steinthal (1890) Arens (1955), pp. 1–28, Robins (1967), pp. 1–65, and Hovdhaugen (1982). Borst (1963) provides a detailed comparative perspective on the history of men's ideas on language.

Even more striking is the contrast when we turn to the philosophy of grammar. Anselm of Canterbury's fascinating dialogue *De Grammatico*,[1] or Boethius, *De modis significandi*[2] are only two examples of a rich Western tradition which has no parallel in China.

In India, Pâṇini evolved what long remained the most advanced system of grammatical description in the world as early as *c.* −500. The Sanskrit grammatical tradition is among the most impressive scientific achievements in pre-modern India.[3]

Grammar in traditional China

What corresponds to European grammar in China is a proper part of the science of characters, which we have discussed in detail in our section on lexicography. In so far as we have grammar in traditional China, it is in the form of explanatory glosses on grammatical particles and – much later – dictionaries of grammatical particles. Grammar as a specialised and systematic non-lexicographical exercise was late to develop.[4]

The earliest book we can plausibly describe as a grammar of the Chinese language, the *Ma Shih Wên Thung* 馬氏文通 was published in 1904[5] and was written (under clear and acknowledged Western influence) by Ma Chien-Chung 馬建忠 (1844 to 1900) and his brother Ma Hsiang-Po 馬相伯 (1840 to 1939),[6] both of whom were well versed in Latin and were reputed to have taught Latin to Liang Chhi-Chhao 梁啟超 (1873 to 1929) and Tshai Yüan-Phei 蔡元培 (1868 to 1940). We shall now investigate the evolution of grammatical reflection and analysis in China before the impact of Western linguistics in that part of the world.[7]

How could the ancient Chinese be brilliant philologists and fine linguists[8] without sorting out grammar? Why did they fail to produce their own grammatical system to account for their own language? Hu Shih 胡適 (1891 to 1962) comments:

How is it that Chinese grammar study developed so late? I think there are three reasons. First, Chinese grammar is basically very easy, so people felt no need for grammar study. Intelligent people could 'grasp it (directly) with their minds', while dullards, having laboriously to use the dullards' method of 'A book if read a thousand-fold, its meaning will itself

---

[1] See Henry (1964) and (1967).   [2] See Pinborg and Roos (1969).   [3] See Staal (1972).

[4] I must report as a curiosity, though, that I have stumbled across the combination *chü fa* 句法 which is the Modern Chinese word for syntax, already in the sayings attributed to Chu Hsi 朱熹 (+1130 to +1200) in *Chu Tzu Yü Lei* 朱子語類, ed. Chung-hua-shu-chü, Peking 1986, p. 3330. The meaning of *chü fa* here seems to be something like 'the art of making sentences in poetry'.

[5] The preface of this work is dated 1898 which is usually given as the date of publication. However, the first volume was published only in 1900, and the complete work appeared in 1904.

[6] Since the work was published soon after Ma Chien-Chung's death, his brother Ma Hsiang-Po saw it through the press. There has been considerable controversy on the authorship of the *Ma Shih Wên Thung* 馬氏文通, notably Chu Hsing (1980). Cf. Peverelli (1986), pp. 59f. The most recent studies (not mentioned by Peverelli) are Wu Kuo-1 (1982) and Lü Shu-Hsiang and Wang Hai-Fên (1986). See also Lin Yü-Shan (1983), pp. 41f.

[7] For the history of Chinese grammar from *Ma Shih Wên Thung* onwards we now have Peverelli (1986), which provides a useful survey, as well as Kung Chhien-Yen (1987).

[8] For a convenient biographic and bibliographic dictionary of the practitioners of these arts in traditional China see Chhen Kao-Chhun (1986).

unfold', did not conceive of the short-cut of grammar study. Second, education in China was for the most part limited to a very small number of people, so no one considered the inconvenience of the greatest number, so there was no necessity to study grammar. Third, the Chinese language and orthography have stood alone for thousands of years with no chance for comparison with other types of highly-developed language and orthography. Only Sanskrit came into contact with Chinese fairly early; but the grammar of Sanskrit is too difficult and too far removed from Chinese grammar to serve as a basis for comparison. Other languages that came into contact with Chinese were each looked down upon by the Chinese and so were not able to induce comparison. Having such comparison, the Chinese did not develop the concept of grammar.[1]

A few hundred years earlier Andreas Müller wrote:

Lingua Sinica nulla indiget GRAMMATICA. *Sufficit lexicon.*[2]

This is evidently an exaggeration, but it hints at an important factor. One reason why the Chinese did not (need to) develop systematic grammar is that in an analytic language like Chinese the lexicon of words plus the lexicon of grammatical particles taken together go a long way towards accounting for what it takes to understand the texts − a much longer way than would be possible in more synthetic languages like Greek or Sanskrit. In Greek, there is an obvious and pervasive need for an analysis of the cases (nominative, genitive, dative, accusative), for the endings of tenses and their aspects (present, imperfect, aorist, perfect, pluperfect, future), for agreement (as in the English 'I go', 'he goes'), and many other things. All these can only be accounted for in a grammar. They cannot very well be treated in a dictionary. In Chinese, what roughly corresponds to the genitive in Western languages, must in principle be discussed in the dictionary under *chih* 之. In general a great deal of the grammar of Classical Chinese can be formulated as an extended dictionary entry under the various grammatical particles of that language. Grammars could therefore take the form of dictionaries of grammatical particles, which is exactly what happened.[3]

The grammar of English is not as different from that of Chinese as one might think. Speakers of English used to be teased for speaking a language without a grammar. Sir Philip Sidney (+1554 to +1586), in his *The Defence of Poesie*, put up a splendid defence of English. It reminds me of what was said about Chinese by seventeenth-century European observers. Sir Philip answered the charge that English had no grammar as follows:

Nay, truly, it hath that prayse, that it wanteth not Grammar: for Grammar it might have, but it needes it not: beeing so easie of itself, and so voyd of those cumbersome differences of Cases, Genders, Moodes, and Tenses which I think was a peece of the Tower of *Babilon's* curse, that a man should be put to schoole to learne his mother-tongue.[4]

---

[1] *Hu Shih Wên Tshun* 胡適文存 1.3, p. 67, cf. Cikoski (1970), p. 7.

[2] 'The Chinese language does not need grammar. The dictionary suffices.' Müller (1672), p. 13.

[3] By comparison, specialised dictionaries of Greek grammatical particles were relatively late to develop. Cf. Denniston (1934).

[4] Sidney (1868), p. 70.

'Full words' and 'empty words'

The most fundamental and important grammatical distinction in Classical Chinese is that between 'full words' and 'empty words'. We shall first discuss the early history of the division and then turn to the problems of subdivisions within the two main categories.

The distinction between categorematic and syncategorematic in medieval grammatical and logical theory[1] is closely related to that between empty and full words in Chinese.[2] A comparison shows that the Western distinction has attracted much more sustained systematic philosophical attention than its Chinese counterpart.[3]

Already Hsü Shen 許慎 (died ca. +149) in his *Shuo Wên Chieh Tzu* 説文解字 distinguished between what he called particles (*tzhu* 辭 or *yü* 語) on the one hand, and full lexical words *tzu* 字 on the other. He defines *ning* 寧 'would that, rather' as 'particle of intention', *chieh* 皆 'all' as 'particle of universality', *ko* 各 'each' as 'particle of difference', the final particle *i* 巳 as 'sentence final particle'. By the time Chêng Hsüan 鄭玄 (+127 to +200) wrote his commentary on Hsü Shen's great dictionary he draws a consistent line between *ming* 名 'names', and *tzhu* 詞 'particles'. But as

---

[1] Cf. especially William of Sherwood's *Syncategoremata*, edited by J. R. O'Donell (1941).

[2] It will be useful to compare the distinction between categorematic and syncategorematic terms as made by William of Ockham:

> There is still another distinction holding both between vocal, and between mental terms. Some are categorematic, others are syncategorematic terms. Categorematic terms have a definite and fixed signification, as for instance the word 'man' signifying all men, and the word 'animal' signifying all animals, and the noun 'whiteness' signifying all whitenesses.
>
> The syncategorematic terms, on the others hand, such as 'all', 'no', 'some', 'the whole', 'besides', 'only', 'in so far as', do not have a fixed and definite meaning, nor do they signify things distinct from the things signified by categorematic terms. Rather, just as in the system of numbers zero standing alone does not signify anything, but when added to another number gives it a new signification, so a syncategorematic term does not properly speaking signify anything, but when added to another number gives it a new signification. In the same way the categorematic term does not properly speaking signify anything, but rather when added to another term it makes it signify something or makes it stand for some thing or things in a definite manner, or has some other function with regard to a categorematic term. William of Ockham (1957), p. 51.

[3] William of Sherwood (+1210 to *c.* +1270) wrote even before his namesake of Ockham:

> In order to understand anything, one must understand its parts, and so in order that the statement may be fully understood, one must understand the parts of it. Now its parts are of two kinds: principal and secondary. The principal parts are the substantival name and the verb, for they are necessary for an understanding of the statement. The secondary parts are the adjectival name, the adverb, and conjunctions and prepositions, for a statement can exist without them.
>
> Some secondary parts are determinations of principal parts with respect to the things belonging to them; these are not syncategoremata. For example, when I say 'white man', the 'white' signifies that some thing belonging to 'man' is white. Other secondary parts are determinations of principal parts in so far as they are subjects or predicates. For example, when I say 'every man runs' the 'every', which is a universal sign, does not signify that some thing belonging to 'man' is universal, but that 'man' is a universal subject. (Secondary parts) of this sort are called syncategoremata. They cause a great deal of difficulty in discourse and for that reason they are to be investigated. Cf. Kretzmann (1968), p. 13 and Boehner (1966), pp. 19ff.

I quote these medieval definitions so that Chinese explanations may be seen in their proper comparative perspective.

the Sung dynasty scholar Liu Shu-Tshai 劉叔才 points out: 'Without pronouncing these (grammatical particles) one cannot exhaust the meaning.'[1]

The standard terms *shih tzu* 實字 'full characters' and *hsü tzu* 'empty characters' first become widely current during Sung dynasty times.[2]

The relationship between empty and full words

In sentences there are auxiliary words (*chu tzhu* 助字), just as in ritual and music there are masters of ceremonies. If in ritual or music there are no masters of ceremonies, then the ritual will not proceed properly and the music will not be pleasant. If in sentences there are no auxiliaries, then they will not flow smoothly (*pu shun* 不順).[3]

Lu Chiu-Yüan 陸九淵 (+1139 to +1192), the famous Confucian philosopher, advises: 'When it comes to what meanings tend towards, there is again the distinction between empty (*hsü* 虛) and full (*shih* 實). In the case of empty characters (*hsü tzu* 虛字), one will only discuss the meaning of the characters (*tzu i* 字義). In the case of full characters (*shih tzu*), one will discuss the objects referred to.'[4] And Chou Po-Chhi, a Yüan dynasty scholar, suggests an important historical perspective 'The empty characters of the present time were all full characters of ancient times.'[5]

One aspect of this is that the Chinese use characters with full lexical meanings as loan characters for grammatical particles, as Chou Po-Chhi 周伯琦 himself points out in a different place.[6] Another aspect is raised in Yüan Jên-Lin's 袁仁林 fascinating book *Hsü Tzu Shuo* 虛字説 (+1747). 'On the whole the ancients created the characters taking concrete things as their point of departure. Afterwards those who made the (written) language in each case loaned the full characters to write empty words in order to convey their meaning.'[7]

Yüan Jên-Lin 袁仁林 even provides explanatory accounts of some particles. Thus he writes: 'Every pot has a bottom and a lid. When the lid is put on top to cover things, its position is such that it embraces the rest. Now, when we use the full word *kai* 蓋 "lid" as an empty word: "on the whole, probably", it still has something of its full meaning.'[8]

The anonymous *Hsü Tzu Chu Shih* 虛字助釋 (1820) lists up the lexical meanings of 39 grammatical particles, but without focusing on the link between the lexical and grammatical functions discussed.[9]

[1] Cf. Lin Yü-Shan (1983), p. 28.

[2] Chêng Tien and Mai Mei-Chhiao (1965), p. 91, gives a rich selection of early quotations. Chhên Huan (+1786 to +1863) speaks of the 'empty *meaning* (*hsü i* 虛義)' of a character in his commentary on the *Shih Ching* 136.1.

[3] Chhên Khuei, *Wên Tsê*, ed. Liu Yen-Chhêng, p. 27. Cf. also Liu Chih-Chi 劉知幾 (+661 to +721), who distinguishes between sentence-initial and sentence-final particles and then continues: 'If you leave these out, the language is insufficient. When you add them, then the paragraphs and the sentences (*chang chü* 章句) achieve completeness.' (Chêng Tien and Than Chhüan-Chi (1988), p. 97).

[4] Translated from Chêng Tien and Mai Mei-Chhiao (1965), p. 95.

[5] *Ibid.* [6] *Ibid.*, p. 100. [7] *Ibid.*, p. 100. [8] *Hsü Tzu Shuo*, p. 2b.

[9] Chou Yin-Thung (1983) systematically documents the non-grammaticalised usages of almost 100 common grammatical particles in Classical Chinese.

The subcategorisation of empty words

Hsü Shen 許慎 (died +149) defines ten words as distinct kinds of particles (*tzhu* 詞).[1] In *Wên Hsin Tiao Lung* 文心雕龍 (*The Literary Mind and the Carving of Dragons*), Liu Hsieh 劉勰 (+465 to +522) subdivides particles into four classes: those which do not add to the meaning of a sentence (*wu i wên i* 無益文義) like *hsi* 兮, sentence initial particles, intra-sentential particles, and sentence-final particles.[2] Liu Tsung-Yüan 柳宗元 (+773 to +819) distinguished further between *i tzhu* 疑詞 'doubtful (final) particles' like *hu* 乎 and *chüeh tzhu* 決詞 'definite (final) particles' like *yeh* 也.[3] These, however, were isolated remarks rather than systematic attempts at subcategorising Chinese grammatical particles.

An alternative distinction between empty and full words

At least from Ming times onwards there emerges a new distinction *within* the lexical items between 'full (*shih* 實)' nouns on the one hand, and 'empty (*hsü* 虛)' verbs and other non-nominal words on the other. It is this latter distinction between nominal and non-nominal words, which came to play an important part in discussions of parallelism in poetry, and it is this latter distinction which actually entered classical dictionaries like the *Tzhu Hai* 辭海 (1948), p. 426, where we read *sub verbo shih tzu* 實字 'full character':

In ancient times the nouns were considered as full words and all other characters as empty words. Chu Hsi 朱熹 (+1130 to +1200) finds this distinction less than water-tight: Vital energy (*chhi* 氣) is a solid/real/full (*shih* 實) thing. 'Contract (*yüeh* 約)' is a half-empty half-full character. These two cannot function as corresponding elements in parallelism.[4]

The rationale behind this distinction between the nominal and the non-nominal is in the philosophical distinction between substance (*thi* 體) and function (*yung* 用). Thus Yen Jo-Chü 閻若璩 says:

Every character can be used for a substance (*thi*) or for a function (*yung*). For example, *chên* 枕 'pillow' in the upper tone is used for the substance and is a full (*shih*) word (meaning 'pillow').[5] Read in the falling tone, *chên* refers to the function (*yung*) and is an empty (*hsü* 虛) word (meaning 'use as a pillow').[6]

A Ming dynasty compilation on parallelism in poetry explains: characters which correspond to a form and substance count as full. Characters which do not correspond to a form and substance count as empty. Those which appear to correspond

---

[1] Ancient dictionary entries on grammatical particles are conveniently assembled in Chêng Tien and Mai Mei-Chhiao (1965), pp. 281–91.

[2] *Wên Hsin Tiao Lung*, cf. 34, ed. Lu Khan-Ju (1981), vol. 2, p. 185; tr. Shih (1983), p. 189.

[3] *Ta Tu Wên-Fu Shu*, quoted after Lin Yü-Shan (1983), p. 28.

[4] *Chu Tzu Yü Lei*, translated from Chêng Tien and Mai Mei-Chhiao (1965), p. 103.

[5] These different readings for characters in different functions go back to at least the Han dynasty.

[6] Chêng Tien and Mai Mei-Chhiao (1965), p. 95.

to a form and substance but do not so correspond, count as half-empty. Those which appear not to correspond to a form and substance, count as half-full. [1]

'Dead words' and 'living words'

The handbook on parallelism continues:

The full characters are all dead characters. Only among the empty characters are there both dead and living characters. The dead characters refer to things that are naturally so, like high and low, large and fine. Living characters refer to a bringing about, like 'fly', 'dive', 'be transformed'. Empty characters must correspond in parallelism to empty characters, full characters to full characters. Similar remarks apply to half empty and half full characters. It is particularly important that dead characters must not correspond in parallelism to living characters, and living characters must not correspond to dead characters. If one does not pay attention to this, then the style goes wrong. [2]

This passage shows clearly how the system of rules for parallelism could lead to classifications of characters that can be interpreted as grammatical in the light of later developments.

Ancient comments on word order

On several occasions the *Kung Yang* 公羊 commentary brings up questions of word order and explains these in terms of the psychology of perception. Thus, when a sentence reads in Chinese SIX FISH-HAWKS RETURN FLY, the commentary asks:

Why does SIX come first and FISH-HAWKS afterwards? SIX FISH-HAWKS RETURN FLY reports something that is seen. One looks and sees there are six things. One looks more closely and finds there are six fish-hawks. One investigates more slowly and finds they are flying back. [3]

In another, perhaps even earlier, commentary we find a discussion of the sentence WILD-GOOSE NORTH HEAD-FOR 'the wild goose head northwards' versus NINTH MONTH DISAPPEAR WILD-SWAN WILD-GOOSE:

Why is WILD-GOOSE named first and HEAD-FOR afterwards? That is because you see the wild goose first and only after that you ascertain the direction it is taking. . . . Why is DISAPPEAR named first and WILD-SWAN WILD-GOOSE afterwards? This is because you see that they are gone and only afterwards you ascertain that what is gone are the wild swans and the wild geese. [4]

For the sentence BROWNISH THE *SO-SUI*-PLANT 'the *so-sui* 莎隨 plants have turned brownish' we have from the same early source the following disquisition:

Why does it first say BROWNISH and then mention the plant? It is because the brownish colour is what you see first. [5]

---

[1] *Tui Lei* 對類, cf. Chêng Tien and Mai Mei-Chhiao (1965), p. 110. [2] *Ibid.*
[3] *Kung Yang Chuan*, Duke Hsi, year 16. Cf. also Duke Yin, year 1; and Duke Hsi, year 1.
[4] *Ta Tai Li Chi*, ch. 47, ed. Kao Ming (1975), p. 59. [5] *Ibid.*, p. 61.

The philosopher and commentator Chu Hsi 朱熹 (+1130 to +1200) notes an important semantic distinction which he thinks is marked by word order.

Wei chi 謂之 CALL IT means to name it. *Chih wei* 之謂 IT CALL means simply that it is it.[1]

What Chu Hsi is saying seems to be that when in Classical Chinese we have *wei chih X* this should be read as 'we call it *X*', whereas when we see *X chih wei* it just means 'is (an) *X*'. The point is debatable, but clearly we have here caught Chu Hsi out as he is trying to do Classical Chinese grammar.

Grammatical comments in the traditional Chinese commentaries

From earliest times Chinese philological commentaries are rich in concrete observations on grammatical usages of given grammatical particles. These observations tend to be descriptive rather than explanatory, and the grammatical descriptions are never justified or argued for. One hardly finds commentators considering alternative grammatical glosses and deciding between them. Moreover, the grammatical particles were never seen as part of a grammatical *system*. Without conscious and reasoned choice between alternative grammatical descriptions and without the establishment of a grammatical system, Chinese grammatical observations in traditional times remained – from a modern point of view – *ad hoc* and disconnected comments. But precisely because they were so untheoretical, these comments give us precious clues on traditional Chinese natural intuitions concerning their grammatical particles. Disconnected and *ad hoc* as they may be, traditional grammatical glosses deserve our close attention. And it does stand to reason that grammar, after all, is more *ad hoc* and disconnected in nature than some theoretical linguists (like Noam Chomsky) would have us believe.

In some instances Chinese grammatical glosses are contrastive, as when the *Kung Yang Chuan* 公羊傳 notes that in two entirely parallel phrases one uses *êrh* 而 'and then' and the other uses *nai* 乃 'and then':

At midday (then) one could bury the princess (*Jih chung êrh kho tsang* 日中而可葬).[2]
When the sun went down (then) one could bury the princess *Jih hsia chih nai kho tsang* 日下之乃可葬).[3]

The *Kung Yang Chuan* goes on to ask the perfectly pertinent question: why is *êrh* used in one case and *nai* in the other. This question is answered perhaps most clearly in the *Ku Liang Chuan* 穀梁傳 commenting on the same passages:

Erh is a slow particle. . . . *Nai* is a (more) urgent particle.

Thus *nai* is taken to mean 'then immediately', whereas *êrh* is just 'then'. The negative *fu* 弗 is explained as a stronger variant of *pu* 不.[4] The poet and philosopher Liu

---

[1] *Chu Tzu Yü Lei* 38, ed. Wang Hsing-Hsien, p. 3280.
[2] *Chhun Chhiu*, Duke Hsüan 8, ed. Yang Po-Chün (1984), p. 695; cf. Couvreur (1951 b), vol. 1, p. 595.
[3] *Chhun Chhiu*, Duke Ting 15, ed. Yang Po-Chün (1984), p. 1600; cf. Couvreur (1951 b), vol. 3, pp. 592f.
[4] Ho Hsiu 何休, *Kung Yang Chuan Chieh Ku* 公羊傳解詁, Duke Huan, year 10. For further evidence on the comparison between *pu* 不 and *fu* 弗 in traditional Chinese grammar see Chêng Tien and Mai Mei-Chhiao (1965), pp. 165ff.

Tsung-Yüan 柳宗元 (+773 to +819), even employs a notion that comes close to that of a 'grammatical rule' when he complains that one of his correspondents 'in his use of empty characters did not conform to the rules and orders (*lü ling* 律令)'.[1]

Dictionaries of grammatical particles in traditional China[2]

The first specialised dictionary of grammatical particles was the *Yü Chu* 語助 by Lu I-Wei 盧以緯 (preface +1324),[3] which treats over one hundred grammatical particles and provides often contrastive glosses on their usage in an almost colloquial style for beginners. Classical particles are often glossed by colloquial equivalents. One special feature of this work is the fact that it also defines a number of combinations of particles.

In +1711, the scholar Liu Chhi 劉淇 produced the first coherent and comprehensive dictionary explaining 476 grammatical particles, the *Chu Tzu Pien Lüeh* 助字辨略, which contains abundant illustrative examples from Classical literature from pre-Chhin times down to the Yüan dynasty. The annotated editions of this book remain singularly useful today.

Yüan Jên-Lin's 袁仁林 *Hsü Tzu Shuo* 虛字説 (Explanation of Grammatical Particles, preface dated +1747) is only 68 pages long, but full of careful and original observations on the grammatical functions of both individual grammatical particles and combinations of these. The attention Yüan Jên-Lin paid to combinations of grammatical particles is remarkable for its time. Yüan Jên-Lin had a profound influence on the authors of the first proper grammar of Literary Chinese, Ma Chien-Chung 馬建忠 and Ma Hsiang-Po 馬相伯.

Wang Yin-Chih's 王引之 much larger and less erratic *Ching Chuan Shih Tzhu* 經傳釋詞 (Explanations of the Grammatical Particles in the Classics and Commentaries), preface +1798, remains a popular grammatical handbook to this day. Wang sets up no less than 52 categories of grammatical particles. In his introduction he complains that since Han times disproportionate attention has been given to 'full' lexical items, and hardly to grammatical particles. He proposes to make up for this imbalance in a comprehensive way.[4]

All the above dictionaries of grammatical particles are essentially collections of glosses and make no attempt at a systematic analysis of the characters they discuss or of the grammatical system as a whole. The first grammar of Chinese was published in 1904. However, there are a number of works which include interesting remarks on grammar and which deserve special study. Notable among these is

---

[1] *Fu Tu Wên-Fu Shu* 復杜溫夫書 ed. *WYWK*, vol. 5, p. 13.

[2] An outstandingly useful compilation on traditional Chinese grammatical theories is Chêng Tien and Mai Mei-Chhiao (1965). On pp. 320ff. of this work one will find a bibliographic survey of traditional Chinese books related to grammar. Hu Chhi-Kuang (1987), too, supplies a well-documented survey.

[3] See Chhen Wang-Tao (1980).

[4] Yü Min (1987) provides detailed criticism of Wang Yin-Chih's classic. Yü Min's book is a major contribution to methodology in Chinese syntax and provides a wealth of original analyses based on a comparative approach to the study of ancient Chinese grammar: he takes into account not only evidence from Chinese dialects (which provide crucial hints), from Tibetan and from Japanese, but he also provides stimulating comparisons with Sanskrit, German, English, Latin and Greek grammar.

Chhên Khuei's 陳騤 *Wên Tsê* 文則 (preface dated +1170, new edition by Liu Yen-Chhêng 劉彥成, Peking 1988), which contains a section on the use of empty words.[1]

Moreover, Watters (1889), p. 97, has drawn attention to one early 19th-century work that does contain a systematic and coherent account of Chinese grammatical categories. This work is said to have already been freely used by Joseph Edkins in his grammar of the colloquial Chinese, and it deserves our attention.

Pi Hua-Chên's 畢華真 *'Lun Wên Chhien Shuo'* 論文淺説 *and* *'Lun Wên Hsü Shuo'* 論文續説[2]

Pi Hua-Chên (writing in the early 19th century) begins by dividing words (*tzû* 字) into full (*shih* 實) and empty (*hsü* 虛) ones. The characteristic feature of the full words is that by themselves they describe something which may be thought of and spoken of. The concepts they designate must have features which may be designated (*yu tuan kho chih chê* 有端可指者). Moreover, they must be designed to combine into sentences (*phai tzhu pien chü* 排詞編句).

Among the full words, Pi distinguishes between primary words (*mu tzu* 母字, lit. 'mother words') which directly refer to things, and secondary words (*tzu tzu* 子字, lit. 'son words') which refer to features of things.

Two full words, we are told, can be either co-ordinate (*phing lieh* 平列) or they are subordinate (*tshê lieh* 側列) with a head (*chu* 主) and a modifier (*pin* 賓). The head is said to come last.

We now turn to the empty words in Pi's system. Among these he describes the rigid empty words (*tai hsü tzu* 呆虛字), intransitive or stative verbs, which are designed to describe circumstances (*i hsieh chhu wu chih chhing chuang* 以寫出物之情狀) and which do not require a complement. A sentence with such a predicate is called a simple sentence of the first order (*tan tshêng tai chü* 單層呆句).

Next Pi describes the living empty words (*huo hsü tzu* 活虛字), which express a human action (*i hsieh chhu jên shih* 以寫出人事) or which combine what precedes with what comes afterwards (*i lien chui shang hsia* 以聯綴上下), and sentences involving these are called second-order rigid sentences (*shuang tshêng tai chü* 雙層呆句). Finally, there are two categories of empty words which come close to grammatical particles (*khou chhi yü chu hsü tzu* 口氣語助虛字 and the *khung huo hsü tzu* 空活虛字). Rosthorn (1898) does not manage to give a coherent picture of the distinction.

From Arthur von Rosthorn's description this does indeed appear to be a fascinating book, quite different from the grammatical literature we have discussed so far. One is at times tempted to assume some Western influence, but on the other hand one would expect such influence to have manifested itself more obviously and to

---

[1] *Wên Tsê Chu-i* 文則注譯, ed. Liu Yen-Chhheng, p. 27.
[2] Cf. *Yen Hsü Tshao Thang Pi Chi* 衍緒草堂筆記. The present account of this little book has to rely entirely on Rosthorn (1898), since I have not been able to gain access to the original Chinese work. Pi Hua-Chên 畢華真 is mentioned in Hummel (1975), p. 641, as active in 1809 and interested in music, but without further detail. It would be interesting to know whether Pi Hua-Chên had any connections with Westerners.

have aroused the suspicion not only of Rosthorn, but even more of Watters and Edkins.

Phonology

We have seen that the treatment of grammar was generally atomistic and *ad hoc* in nature: what we have are essentially glosses and collections of glosses on grammatical particles in particular contexts.

One might be tempted to conclude that the Chinese were quite generally incapable of systematically organised theoretical work within the field of linguistics. But the case of Chinese phonology shows abundantly that this is not true.

There is no need here to give a detailed account of phonological sciences in China. The literature on this subject is vast.[1] However, from the point of view of the history of science, we must emphasise that in the field of phonology the Chinese did approach a systematic theoretical method, which we have found lacking in the study of grammar. This important achievement deserves a separate monograph in its own right, and I shall not attempt to summarise it here.

(7) THE ART OF LITERACY IN TRADITIONAL CHINA

The ancient Greeks as well as the Romans cultivated and celebrated the art of rhetoric and particularly of public speech. This was an integral part of the social fabric of public life in these societies. Admittedly, the ability to speak persuasively to rulers is commended in such works as *Chan Kuo Tshê* 戰國策 (−3rd century), and the wit of the quick riposte was widely admired in ancient China and celebrated in such works as the *Shih Shuo Hsin Yü* 世説新語 (+5th century). But the overwhelming emphasis in traditional China was not so much on public delivery of speech as on the written memorial. Inspired calligraphy counted for more than spirited oral performance. The cultural emphasis is scriptural.

Here we have a most profound contrast between ancient China and ancient Greece. There is no proper parallel in China to the ritualised sporting contest between arguments that was so important in Greek intellectual history. What mattered in Chinese rhetoric was generally to give the right advice to a ruler. Figures like Kungsun Lung, who were rhetorical sophists, remained marginal in Chinese intellectual history, whereas the conversationalist Socrates became an all-important philosophical figure.

There are a few stylistic features of written Literary Chinese that deserve our brief attention because they were intimately connected with Chinese ways of communicating and of thinking. First, there is the ideal of the brief, concise and pregnant style. Then there is the habit of allusion to earlier literature. Finally, there is the

---

[1] See Pulleyblank's monumental study *Middle Chinese* (Toronto 1984) and T. A. Sebeok ed., *Current Trends in Linguistics* vol. 2 (The Hague 1967) for a competent recent survey of the field.

pervasive habit of stylistic parallelism. We shall deal with these stylistic features in that order.

Brevity

The dominant ideal in Chinese style and in Chinese philosophy was economy of language.[1] Meng Tzu has given perfect expression to the ideals of good style both in prose and in poetry throughout the ages:

When one talks about what is close at hand but one's words have a far-reaching import, one's words are good.[2]

A sparing use of language is commended by Confucius:

The people of Lu were rebuilding the treasury. Min Tzu-Chien said: 'Why not simply restore it? Why must it be totally rebuilt?'
The Master said: 'This man does not speak. And when he does, he hits the mark'.[3]

It is this hitting the mark (*chung* 中) that is Confucius's main concern in language and that became proverbial in later Confucian literature. Intellectual garrulousness (*ning* 佞) was one of his important targets for attack, and one might well suspect that what he was attacking there was what in the West evolved into discursive theoretical philosophy.

The Taoists carried the ideal of brevity to its logical conclusion, and they commanded wide sympathy outside their own Taoist circles.

Consummate words do away with words.[4]

Sages were said to understand each other without recourse to words.

When the sages make things clear (*yü* 喻) to each other, they do not rely on words.[5]

And this is necessary because, as Chuang Tzu points out:

What can be said and discussed are crude things. What can be reached by thought are subtle things.[6]

Speaking of the beginnings of time, a Taoist text of the −2nd century states:

Words were brief (*lüeh* 略), but they were in accordance with the inherent principles of things.[7]

The pragmatically inclined book *Mo Tzu* 墨子 recommends:

When it comes to words (*yen* 言), do not aim for quantity but for wisdom; do not aim for stylistic polish (*wên* 文) but for perspicacity.[8]

---

[1] The philosopher Immanuel Kant was among those who were impressed by the conciseness of Chinese. He comments: '"Z. E. Guten Morgen, mein Herr!" wird durch *ein* Zeichen ausgedrückt.' E.g., '"good morning, sir" is expressed by one character.' (A. Hsia (1985), p. 99).
[2] *Mêng Tzu* 7B32, see Lau (1983c), p. 299.   [3] *Lun Yü* 11.14, see Lau (1983a), p. 101.
[4] *Chuang Tzu* 22.84, see Watson (1964), p. 247.   [5] *Lü Shih Chhun Chhiu* 18.3, ed. Chhen Chhi-Yu, p. 1167.
[6] *Chuang Tzu* 17.23, see Graham (1981), p. 146.   [7] *Huai Nan Tzu* ch. 8, ed. Liu Wen-Tien, p. 1b.
[8] *Mo Tzu* 2.13, see Mei (1929), p. 8.

It remained the high ambition of Chinese stylists to capture what was important and essential by rigorously omitting what was not. Confucius said:

When sentences (*tzhu* 辭) convey (what they are intended to convey) one stops.[1]

Unfortunately, this stylistic ideal was cultivated to an extent that makes many texts incomprehensible to outsiders and even more to outsiders who are also latecomers like ourselves. There was a deliberate – and often politically prudent – practice of expressing things in such an indirect way that one could not be held responsible for what one had said.

Therefore the gentleman is subtle (*wei* 微) in what he tells others.[2]

The Sung philosopher Chu Hsi 朱熹 (+1130 to +1200) summed up the traditionally dominant perception:

One should not set store by wordiness.[3]

Chhên Khuei 陳騤, writing about +1170, coined the phrase:

In writing simplicity is of the essence (*Wên kuei chhi chien* 文貴其簡).[4]

He provides a useful illustration of the virtue of brevity (*chien* 簡). He quotes the *Book of History*: 'You are wind. The people below are grass.' He compares a longer and more explicit version in *Lun Yü* 12.20: 'The virtue of the gentleman is like wind; the virtue of the small man is like grass. Let the wind blow over the grass and it is sure to bend.' To hammer home his point he adds a third version quoted by Liu Hsiang 劉向 (−79 to −8) which uses thirty-two characters to make the same point.[5]

Chhên Khuei is preoccupied with the pregnancy (*hsü . . . i* 蓄 . . . 意) of good style. There should be more to words than what is directly said in them. Explicitness in a text strikes him as vulgar.[6] Unfortunately, such explicitness can be of great educational and scientific value. We shall discuss the problems of semantic and grammatical explicitness in detail below.

Allusion

In any language, words and phrases may evoke literary antecedents. Sermons, for example, not only contain quotations from the Bible, they are also full of allusions to biblical texts. Few poets in world literature can have been more allusive than T. S. Eliot or Ezra Pound. Few prose writers anywhere will have been more allusive than James Joyce.

The first thing one has to understand about Literary Chinese style is that by force of a dominant convention it tends to be saturated with allusions to a large corpus of literature with which the reader is assumed to be thoroughly conversant. In this, Chinese literature resembles traditional Christian religious literature.

[1] *Lun Yü* 15.41, see Lau (1983a), p. 159.
[2] *Han Shih Wai Chuan* 4.33, ed. Chhen Chhi-Yu, p. 163; cf. Hightower (1951), p. 158.
[3] Quoted in *Li Chi Chi-chieh*, ed. Sun Hsi-Tan, *WYWK*, vol. 1, p. 6.
[4] Chhen Khuei, *Wen Tsê*, ed. Liu Yen-Chhêng, p. 12.   [5] *Ibid.*, p. 12.   [6] *Ibid.*, p. 15.

Consider, for a moment, the style of Theophilus Sigfridus Bayer (+1694 to +1738), one of the founding fathers of sinology. Bayer was in the habit of alluding to Greek and Latin authors without indicating his sources. This was difficult enough. But in addition many of his printed pages include passages in Arabic, Persian and Syriac.[1] It is worth noting that the widespread display of foreign language skills in traditional Western style has no parallel at all in China. This difference is culturally significant.

Memorisation as part of literary culture

In traditional China, the Chinese classics were memorised in vast quantities by any aspiring scholar/reader. This detailed familiarity with the classics among writers and readers alike created an esoteric literary high culture which is of difficult access to those who do not share the required memorised background. When the writer of Literary Chinese uses a phrase or even a relatively rare character which recurs in one of the classics, he will not only effortlessly associate the *locus classicus*, but can quite properly assume that his readers will naturally perceive the resonance.

Western observers have often complained about the inordinate burden placed upon the Chinese by obligatory memorisation of a wide range of texts. The central place of memorisation in Literary Chinese education is seen as a symptom of the supposed lack of creativity displayed in that culture.

This view is mistaken. When, on August 7, +1582, Matteo Ricci arrived in Macao, he was determined to teach the Chinese not only Christianity but also the art of memorisation, which he continued to believe was much more satisfactorily developed in Europe than in China. In +1596 he drafted his Chinese Treatise of Mnemonic Arts (*Chi Fa* 記法) to help the aboriginal Chinese on the way towards the sort of mnemonic skills he thought he possessed and which he knew were commonly cultivated in the Europe of his time.[2]

Frances A. Yates's classic work *The Art of Memory* describes in detail how the skill of memorisation from Greek and Roman antiquity onwards was regarded as a central discipline within rhetoric as taught in the educational institutions of the time. Memory is of paramount importance in all societies where there is no easy access to books and written information. China is in no way an exception. In Ricci's time a detailed memory of things was known as a *thesaurus eloquentiae*, and by all accounts Ricci had as colossal a memory as any Chinese of his time.[3] Moreover, Ricci was not at all unique in a European context. If an enormous memory is indeed a burden on the creative mind, Erasmus of Rotterdam, Montaigne, or François Rabelais certainly carried more than their fair share of that burden.

---

[1] See Lundbæk (1986), p. 4.
[2] Spence (1986), pp. 1–23. For memorisation techniques in China see Vol. 4, pt. 2, pp. 48 and 528, Vol. 4, pt. 3, p. 583, Vol. 5, pt. 4, p. 261 and Spence (1986), p. 156ff.
[3] Spence (1986), p. 5.

Erasmus finishes his *Encomium morias* as follows:

> Vetus illud, MISŌ MNAMONA SYMPOTAN
> novum hoc: MISŌ MNAMONA AKROATĒN
> Quare valete, plaudite, vivite, bibite, Moriae celeberrimi Mystæ.[1]

The published work of Erasmus shows that he had a truly astonishing memory. And, as his *Colloquia familiaria* show, he was very far from being a dry scholastic sort of man.

Even men like Montaigne and Rabelais, who despised and ridiculed memorisation, had, by our standards, remarkably good memories and were full of literary traditionalisms. It so happens that these were among the most unconventional writers in European intellectual and literary history. It is perhaps superfluous to stress this, but it needs repeating in our society with its cult of the new: traditionalism of allusion and comprehensive memorisation of traditional heritage are not an impediment to creativeness, innovation, anti-traditionalism, or even flippancy. The work of Rabelais should be sufficient to prove this point.

The argument that the Chinese habit of memorisation, the allusive traditionalism of Literary Chinese style constituted an impediment to creativity, innovativeness and scientific progress loses all its force as soon as one looks into the crucial era when the breakthrough came in the West, the Renaissance. If anything impeded the Chinese, it was not their traditionalism as such but the style in which they cultivated it.

The Renaissance was far from untraditional. It was steeped in scholastic learning and scholastic allusions. The point is that the writers of the European Renaissance used this tradition in new creative ways. And this creative way of using the tradition became the dominant literary trend.

The special character of allusions in Literary Chinese

In Literary Chinese texts the literary allusions often not only add to one's aesthetic appreciation of a passage, they are essential even for the grammatically correct parsing and the elementary understanding of passages. In English one will grammatically understand a sentence even when one fails to recognise a biblical or other allusion that occurs in it and gives the sentence a special force or a specific reference. In Literary Chinese, the texts as they stand often do not begin to make any sense until you get the allusions.

If, for example, you read that someone's 'mind was on study', it is important to realise that this most likely should be translated as 'was fifteen years old' and is a reference to *Lun Yü* 2.4 'At fifteen I had my mind on study'. Conventional allusions of this type abound in many – but not all – types of Literary Chinese.

---

[1] 'There is an old saying: I hate the fellow-drinker with a good memory. Here is a new one: I hate a listener with a good memory. Therefore farewell, clap your hands, live and drink, you distinguished initiates of Folly.'

Such allusions create, among Literary Chinese texts, a hierarchy of abstruseness, rather like the degree of Latinity used to create a hierarchy among English texts. There seems to be no general reason why the use of allusions should be more obfuscating in the traditional Chinese context than the use of Latin in traditional European societies.

However, since the gradual modernisation of the Chinese educational system from the 1920s onwards, familiarity with the classics has markedly declined in China, just as it has at the same time in the West. The recognition of allusions used to be a natural and effortless thing. Today, it is turning into more of a speciality of the few who still possess something of the traditional command of the texts.[1]

What used to be trivially obvious allusions understood as effortlessly as our *veni, vidi, vici* by the leisured literati are slowly becoming learned *apperçus* even to Chinese scholars.

Allusions of all types play a greater part in the recommended Literary Chinese style than generally in, say, contemporary English or in ancient Greek, or indeed in Classical Chinese.[2] This is a measure of the extent to which Chinese literary culture perceived itself as embedded in a tradition. Writing a sentence in Literary Chinese tended to be a dialogue with the past. One can only understand this dialogue if one is familiar with both sides: the Literary Chinese text one is reading and the Classical and Literary Chinese heritage that text relates to. The writer in Literary Chinese expressed himself through the medium of the tradition of which he felt he was a part. He spoke to those who were also part of this tradition. His discourse was not directed towards those who lacked what he considered the essential educational background.

The comparison with the Christians' allusions to the biblical tradition is instructive. Such references often express a commitment to and reaffirmation of the basic values of that tradition. Nothing would be further from the truth than a suggestion that the allusive conventions of theological and religious discourse have prevented theology from being inventive or creative. Thomas Aquinas remained entirely a medieval traditionalist, and yet I doubt that European intellectual history has seen any more creative and original systematiser since Aristotle than Thomas Aquinas with his *Summa theologiae* and his *Summa contra gentiles*. Having to use traditional, scholastic and allusive Latin did not prevent Thomas Aquinas from applying what I am tempted to see as a scientific, critical, method across a vast range of human knowledge in often original ways.

Moreover, the overwhelming dominance of Christianity did not prevent the growth, within its monastic centres, of a most admirable tradition of natural science largely unencumbered, or at least unimpeded, by the allusive style of religious discourse. We used to look upon the Middle Ages as a period of narrow religious preoccupation, but we have long since had to recognise that a uniquely rich culture of

---

[1] See Fang (1953) for a vivid picture of the pitfalls and the burdens this presents for the modern scholar.

[2] They also play a much larger part in modern Chinese than they do in, say, English, German, or Russian. Yang (1985), pp. 135–92 has a comprehensive bibliography on Chinese proverbs and traditional sayings.

varied intellectual endeavour had developed under this superficial appearance. The logicians of the 'dark Middle Ages' have turned out to have been in many ways vastly more enlightened and more advanced theoreticians than their 19th-century detractors. In earlier volumes of this series we have made a similar discovery about seemingly Confucian-dominated, unscientific China.

Lofty moralism and precious literary allusiveness may have preoccupied missionaries and the Sinologues in their wake. But from the −4th century onwards we find an almost totally unornate, unallusive hard-headed scientific style in China. The book *Kuan Tzu* 管子 contains a great deal of unornate straightforward expository prose. The book *Mo Tzu* 墨子 is stylistically most pedestrian, even pedantic much of the time. It subordinates style to requirements of clarity of exposition and argument. The Dialectical Chapters of the book *Mo Tzu* are algebraic almost to the point of obscurity in their technical use of language. The logician Kungsun Lung 公孫龍 (*c.* −320 to −250) writes with the same sort of technical rigour that we find in Anselm of Canterbury's (+1033 to +1109) *De grammatico*. Wang Chhung 王充 (+27 to +100) explicitly declares himself, in the preface to his *Lun Hêng* 論衡, to be an adherent of plain, unallusive style, although we should emphasise that in point of fact his style has considerable literary qualities of its own.

Thus we have, on the one hand, a 'protoscientific' stylistic tradition which appears indifferent to – perhaps even defiant of – the aesthetic demands of Literary Chinese. On the other hand we also have writers who use exquisite Literary Chinese to expose and oppose the traditionalism Literary Chinese may seem to stand for.[1]

Even within the literary and non-scientific tradition not everyone indulged in allusions all the time or to the same extent. Yü Hsin 庾信 (+513 to +581) used so many of them that one suspects that he defeated even the seasoned pre-modern scholar on many points.[2] Thao Yüan-Ming 陶淵明 (+365 to +427) and Pai Chü-I 白居易 (+772 to +846), on the other hand, deliberately tried to avoid allusion in an effort to create a fresh and natural style. Many other scholars like Han Yü 韓愈 (+768 to +824) repudiated and attacked the excessive use of allusion in Classical Chinese. Such common traditional Chinese reactions only go to show that as modern observers we are not the only ones to find much Literary Chinese writing ridden with ossified allusions and clichés.

The Chinese have had their fair share of literary hacks, and easy is the task of finding allusions ineptly or inaptly used in the vast treasury of Classical Chinese literature. But more fruitful and much more exciting is the challenge of fathoming the

---

[1] The traditions of the Literary Chinese jest book and much of Chinese informal literature (*pi chi* 筆記) use well-turned Classical Chinese phrases full of allusions to make outrageously rude and thoroughly sarcastic comments about almost everything that is holy or taboo in Chinese culture. The *Liao Chai Chih I* 聊齋志異 (Stories From a Chinese Studio) by Phu Sung-Ling 蒲松齡 (+ 1640 to + 1714) is justly famous for its superb terse Classical style, which is rich in allusions. At the same time it is also notoriously rich in improprieties and crude expressions, so much so that even the best of modern Chinese translations admit that they do not dare to follow the original in this supposedly more liberal new age! There is nothing inherently conformist about using Literary Chinese.

[2] See Graham (1980) for a thorough analysis of allusion in one of Yü Hsin's works.

depth of semantic subtlety that is often encapsulated in the apt and agile deployment of what at first sight might look like a stale cliché. The reader of Classical Chinese prose and poetry often finds that what at first sight was a hackneyed cliché turns out upon closer inspection to be a rather subtle turn.

Let us now, finally, turn to the use of allusion as a riddle for the reader or as a display of superior learning by the writer, a phenomenon which corresponds to the use of Latin or Greek words in modern European literature. Traditional Chinese letters and prefaces (usually added as postfaces at the end of traditional books) illustrate the cases we are interested in. The phenomenon is beautifully summarised in a lively passage from a book published in 1895:

One day in China, when I had replied in Chinese to a Chinese letter, my Chinese professor told me: 'This will not do. This is not profound enough. It is too clear. I shall write a letter for you of which he will understand nothing.' When I objected that I was precisely hoping that my correspondent should understand me, my Chinese professor replied: 'That does not matter. He should much rather understand nothing than understand everything.' There you are! It is the same with the prefaces to Chinese books. The author does not want them to be understood. The more obscure he is, the less intelligible he is, the more he will be admired by the ordinary reader. All the more so, if one in a thousand is able to grasp all the allusions with which the author has studded his piece. And the reader does not want to understand. No more than our peasants consider that a preacher preaches well if he preaches intelligibly.[1]

This is an account of a culture clash. Chinese prefaces are indeed often by far the most difficult part of a book. When it comes to Pre-Han Chinese literature we certainly do well to heed the observation by Ko Hung 葛洪 (+283 to +343):

Moreover, the many obscurities in ancient books are not necessarily because men of the past wanted to be difficult to understand. In some cases, language has changed over periods of time. Sometimes the dialect was not the same. Books have been all disarranged and put in incorrect order. Buried and hidden over a long period of time, the slips and bindings have rotted and broken apart and much has been lost. Some sections have been added at the end of the wrong book, and some books exist only in remnants and with lacunae. Some have missing chapters or sentences. In this case it is difficult to figure them out. It may appear as if they were obscure.[2]

Ko Hung 葛洪 concludes:

When it comes to speech, we call it rhetorically successful if it is easy to understand; then why should we call a book good if it is hard to understand?[3]

---

[1] Un jour, en Chine, ayant répondu en Chinois à une lettre chinoise, mon professeur chinois me dit: 'Cela ne va point. Ce n'est pas assez profond (*shen*); c'est trop clair. Je vais vous écrire pour lui une lettre dont il ne comprendra rien.' Sur ma remarque que je voulais justement que mon correspondant me comprisse il répliqua: 'Cela ne fait rien; vaut mieux qu'il n'y comprenne rien, que qu'il y comprenne tout.'

Eh bien! c'est la même chose avec les Préfaces des livres chinois. L'auteur ne veut pas qu'on comprenne. Plus il est obscur, moins il est intelligible, d'autant plus il sera admiré par le lecteur ordinaire. Tout au plus si un entre mille peut saisir toutes les allusions dont l'auteur a farci sa pièce. Et le lecteur chinois ne veut pas le comprendre, pas plus que nos paysans n'estiment un prêtre qui prêche intelligiblement. Schlegel (1895), p. 3.

[2] *Pao Phu Tzu* 30, ed. *WYWK*, p. 629; cf. Sailey (1978), p. 163.   [3] *Ibid.*, p. 632; cf. Sailey (1978), p. 166.

On the other hand the great scholar Chang Ping-Lin 章炳麟 (1868 to 1936)
wrote obscure prose, which was so studded with allusions that Lu Hsün 魯迅 (1881
to 1936), a man of quite considerable learning, liked proudly to declare that he
understood little of it.[1] But then: who would claim he understood *Finnegans Wake*?
Here, as so often, the difference between China and the West is a matter of degree.

Rhythmic euphony and parallelism

Rhythmic regularity and the recurrence of rhythmic patterns are a well-known
stylistic feature in Latin rhetoric as they are in many other languages. What distinguishes the Literary Chinese style is a strong and pervasive tendency towards pleasing rhythmic patterning throughout their sentences. We can call this a principle of
rhythmic euphony. The rhythm must be regular enough to aid with the proper
scanning of the text, but it must avoid being mechanically monotonous.[2] Above all,
the rhythm must be predictable enough to assist in the correct parsing of a text.[3]

Parallelism, which is common in Classical Chinese, is endemic in Literary
Chinese. No one has explained this more eloquently and more graphically than Liu
Hsieh 劉勰 (+465 to +522):

When living creatures are created, their limbs are created in pairs. In the mysterious workings of Nature nothing stands alone. The mind creates literary language, and in doing this it
organises and shapes all sorts of thoughts, making what is high supplement what is low, and
spontaneously producing parallelism (*tzu jan chhêng tui* 自然成對).[4]

Note the way Liu Hsieh stresses the spontaneous, natural way in which parallelism enters Chinese diction. He returns to this subject throughout his detailed and
theoretically pioneering chapter entitled *Li Yü* 儷語 (Matched Words), of which we
have just quoted the beginning. Liu Hsieh discusses first the historical development
in Chinese literature, then the typology, and finally the stylistic dangers involved in
the use of parallelism.

In many other languages sentence patterns will not be repeated in successive
sentences. If one sentence has a certain structure there is no reason to assume that
the next will follow suit. Each sentence comes with its own fresh and new syntactic
pattern. In Classical and Literary Chinese the situation is different. Here parallelism is a significant aid to comprehension. If you see that a passage begins with
a six-character phrase, you expect that it might well continue with either one or
three more similar phrases. A sequence of sentences in Literary Chinese is often a
series of repeated rhythmic and syntactic patterns. To some extent this syntactic

---

[1] This is unfortunate for our particular mission, because Chang Ping-Lin took a certain interest in logic.

[2] The preferred rhythm is the four-beat rhythm: for example in Hsün Yüeh's 荀悦 book *Shen Chien* 申鑑 (*c.*
+200) we find that no fewer than 52% of all prosodic groups are made up of four-syllable clauses, 17.3% are three-
syllable groups, and 15.5% are five-syllable groups. This leaves no more than about 15% of the prosodic groups
for all numbers of syllables below three and above five syllables.

[3] One has to remember that Literary Chinese texts were nearly always intended as unpunctuated texts. See
our Section on the sentence.

[4] *Wen Hsin Tiao Lung*, ch. 35, ed. Lu Khan-Ju (1981) p. 189; tr. Shih (1983), p. 190.

repetitiveness compensates for the absence of morphology to make syntactic relations palpable and explicit in Literary Chinese.

But there is a far more profound aspect to parallelism in Classical and Literary
Chinese: this parallelism represents the characteristically Chinese tendency to see
the world as a harmoniously patterned whole in which certain patterns of reality
repeat themselves in all spheres. The patterns of physical reality are found again in
social reality. The patterns of social reality again repeat themselves in the micro-
cosm of the self. The patterns of the past repeat themselves in the present. The patterns of the human sphere repeat themselves in the patterns of the sphere of the
gods, ghosts and spirits. The universe is seen as an ordered pattern of patterns, and
this vision of the universe is represented in a language dominated by parallelism, an
ordered pattern of linguistic patterns.

During the Han dynasty, when the systematicity of the Chinese view of the
universe and of social reality was first made fully explicit, the use of parallelism
dramatically increased and then remained a central part of the Chinese stylistic
repertory. Linguistic parallelism is a linguistic reflection of the intellectual habit of
parallel patterning, 'correlative thinking'.[1]

The definition of parallelism

How then do we define parallelism in Chinese? Let us begin by quoting the
European classic on the subject of parallelism, *La loi du parallélisme en style chinois,
Demonstrée par la préface du Si-Yü Ki* by Gustave Schlegel:

In two parallel or juxtaposed phrases the rules of Chinese style demand that all the parts
correspond to each other: the subject to the subject, the verb to the verb, the noun to the
noun, the adjective to the adjective, the adverb to the adverb, the place name to the place
name, the sign of the genitive to the sign of the genitive, the object to the object, etc., etc.[2]

The subtitle of Schlegel's book is instructive: *La traduction de cette préface par Feu
Stanislas Julien défendue contre la nouvelle traduction du Père A. Gueluy.* Schlegel analyses a
text only 836 characters long on no less than 203 pages, and shows through this *tour
de force* how the 'law' we have just quoted is necessary for an adequate understanding of it. Again and again Schlegel shows how disregard for the law of parallelism
prevents the unfortunate Father Gueluy from getting the grammar and meaning of
sentences right. Convincingly, Schlegel shows how parallelism is not only pleasing
to our sensibilities, but literally essential to our understanding of texts.

Tchang Tcheng-Ming, in his thorough study *Le Parallélisme dans les Vers du Cheu
King* (Paris and Shanghai, 1937), has shown the ancient and popular origins of parallelism in the ancient Chinese *Book of Songs.* Thus parallelism is in no way a late
pedantic literary invention in China. It has its deep popular roots.

The conditions under which words may count as parallel in two sentences have
exercised lovers of Chinese poetry a great deal, because parallelism not only is a

---

[1] Cf. Graham (1986b).   [2] Schlegel (1895), p. 1.

pervasive feature of Literary Chinese prose but also forms an integral obligatory part of the so-called *lü shih* 律詩 or 'Regulated Verse'.[1] Roughly, two sentences count as parallel if they contain the same number of words in the same pattern. Words or phrases *X* and *Y* count as parallel in such sentences when in some sense they correspond to each other. This correspondence will normally involve that *X* and *Y* play grammatically similar rôles, but more specifically it may take the following forms:

1. *X* and *Y* are identical and have the same meaning
2. *X* and *Y* are synonymous
3. *X* and *Y* are the opposite of each other
4. *X* and *Y* are similar in a relevant respect
5. *X* and *Y* are dissimilar in a precise relevant respect
6. *X* and *Y* look as if they correspond to each other in one of the above five ways but do not in fact correspond (in that case we have only superficial formal parallelism).[2]

Parallel prose phien wên 駢文

In general it must be said that the proverbial *loi du parallélisme* is not a law but a strong stylistic habit. There is, however, one style of Literary Chinese in which parallelism is almost literally obligatory. This is the *phien wên thi* 駢文體 'Parallel Prose Style', which is said to have had its origins in the +2nd century and flourished well into Thang times.[3] It is remarkable, though, to see how much of such texts as the *Huai Nan Tzu* 淮南子 (−2nd century) is dominated by parallelism.

Let us take some examples of the Parallel Prose Style. The *Pei Shan I Wên* 北山移文 'Proclamation on the North Mountain' by Khung Chih-Khuei (+447 to +501) is a stylistic *tour de force* which consists of 124 lines and exactly 62 couplets. There are quite a few quadruplets (parallel pairs of parallel couplets). There is also a sprinkling of introductory phrases such as 'therefore', 'when it comes to' and the like. But there is not a single triplet or quintuplet of parallel clauses.[4]

The *Yü Thai Hsin Yung Hsü* 玉台新詠序 Preface to 'New Songs from the Tower of Jade' has been laid out to consist of 105 lines. With some few exceptions they all make structurally parallel couplets.[5]

Yü Hsin 庾信 (+513 to +581) probably went further than anyone else in his use of allusion and parallelism. His famous *fu* 賦-poem *Ai Chiang-nan Fu* 哀江南賦 'The Lament for the South' consists of no fewer than 520 carefully paired lines. The text of this complex poem is beautifully laid out with a translation in Graham (1980), pp. 50–103.

---

[1] For an interesting recent book-length treatment of the subject see Fu Phei-Han (1986).
[2] Our account is based on Hightower (1965).
[3] See Hightower (1965) for an introductory survey of some prominent features of this style of literature in two famous examples of the genre.
[4] See *ibid.*, pp. 75–6, which conveniently displays the text to bring out its structure.
[5] *Ibid.*, pp. 77–91.

We are not saying that Literary Chinese was generally like the Proclamation on the North Mountain or the Lament for the South. What we are saying is that parallelism was natural and endemic to Literary Chinese style in a way that is alien to all Indo-European language I know. Parallelism is almost part of the grammar of Literary Chinese.[1]

Parallelism in Literary Chinese is typically between pairs of sentences. Quite regularly one pattern of parallelism may be maintained over four sentences or two couplets. Occasionally, one gets three couplets maintaining the same structure. Triplets or quintuplets of parallel sentences, on the other hand, are not part of the convention, although there is no strict rule against them.

Interlocking parallel prose

Apart from the strict parallelism we have so far discussed, there is a more complex form of 'interlocking prose style' discussed in some detail in Wagner (1980). A typical example of this runs as follows:

> Whoever does anything to things will ruin them;
> whoever lays hold of things will lose them.
> Therefore the sage
> because he does nothing to things, does not ruin them;
> because he does not lay hold of things, does not lose them.[2]

Patterns of interlocking resumptive parallelism such as these are common in early and later Chinese literature.

Rhythmic requirements and parallelism sometimes interfere with semantic requirements. One occasionally suspects that the presence of a character in a sentence is due to purely rhythmic considerations and adds nothing to what is said. In scientific texts, for example, we often find a strong tendency towards rhythmic regularity, which leads to uncertainty whether certain characters are heavily significant or purely euphonic and present only for the sake of rhythm. This is true, for example, for the texts of Buddhist logic from the +7th century, which we shall discuss below.[3] The logically disconcerting effect is that the technical vocabulary is occasionally varied for the sake of rhythm, at the expense of terminological consistency.

With rhymes, the phenomenon is well known in Western literature, and immortalised in the German poem by Christian Morgenstern:

> Ein Wiesel
> saß auf einem Kiesel
> inmitten Bachgeriesel.
> Du fragst, weshalb:
> Das Mondkalb
> verriet es mir im stillen:

---

[1] Almost, but not quite. Because, as we have seen, two lines may count as parallel on only superficially formal grounds.   [2] *Lao Tzŭ* 64, Lau (1984), p. 95.   [3] See Sub-section (*f*,7).

das raffinier-
te Tier
tat's um des Reimes willen.

In scientific texts, on the other hand, we should like to be sure that whatever is said is not just said for reasons of euphony.

Rhyme

The ancient *Shih Ching* 詩經 'Book of Songs' (−9th to −6th century) and the *Chhu Tzhu* 楚辭 'Songs of the South' (*c.* −3rd to +2nd century) have regular rhyme patterns. These are not evident in the changed modern pronunciations of the words involved, but they have proved useful in the reconstruction of ancient Chinese pronunciation. A book like the *Lao Tzu* 老子 employs a looser system of semi-rhymes and contains a fair amount of unrhymed prose.[1] This mixture of poetry and prose is characteristic of much early Chinese writing. The classic of Taoist prose *Chuang Tzu* 莊子 (−4th to −3rd century) has recently been shown to contain a great many more semi-rhymed passages than were generally recognized.[2] Significant parts of the *Huai Nan Tzu* 淮南子 (−2nd century) are rhymed.[3] The *Mo Tzu* 墨子 (−5th to −3rd century) and the *Mêng Tzu* 孟子 (−4th century), by comparison, are texts with few rhymes. Occasionally, one has to assume that the choice of a word in rhymed texts is due to the rhyme rather than the semantics. But generally this does not in my experience pose insurmountable problems in the interpretation of Literary Chinese scientific texts.

(*c*) LOGICAL FEATURES OF THE CLASSICAL CHINESE LANGUAGE

(1) NEGATION AND THE LAW OF DOUBLE NEGATION IN CLASSICAL CHINESE

In standard logic we have sentential negation, which applies to whole sentences, i.e., has sentences as its scope.

Not–p

is understood to mean that the proposition p is not true. The scope of the sentential negation is a whole proposition. Thus the sentential negation corresponds roughly to a phrase like 'it is not the case that . . .'.

Aristotle developed an abstract notion of negation, *apophasis*, which he defines:

An affirmation is a statement affirming something of something. A negation (*apophasis*, ἀπόφασις) is a statement denying something of something.[4]

---

[1] See Karlgren (1932). [2] Chang Tsung-Tung (1982).
[3] Cf. Chou Tsu-Mo (*1958*). Ames (1983) finds rhymes in almost half of chapter 9 of that book.
[4] *De Interpretatione* 6, 17a25f.

All natural languages I know of have devices to deny or negate sentences, and Classical Chinese is certainly no exception. On the other hand, few cultures have had occasion to formulate an abstract concept of negation, and in this respect traditional China does not form an exception either.

In Classical Chinese literature we find no abstract notion of negation as such, and even when one looks for a word for 'to deny, to negate' the verb *fei* 非 to disapprove, to criticise', which comes closest to this meaning, is not often used in the technical sense. However, we do have comments like the following where *fei* is used to mean 'to deny, reject' as opposed to *li* 立 establish, assert, put forward':

The explanations put forward by Confucius were rejected by Mo Tzu. The explanations put forward by Mo Tzu were rejected by Yang Chu. The explanations put forward by Yang Chu were rejected by Mêng Tzu.[1]

And as we shall see in detail below, the Later Mohists argued along the following lines:

Rejecting denial is self-contradictory. Explained by 'he does not reject it (i.e., the denial itself)'.[2]
Whether denials are admissible or not does not depend on how many or few there are. Explained by 'deserving rejection'.[3]

Abstract disquisition of this sort is rare in China, and when it occurs, it is important from a logical point of view to realise that the notion of negation discussed is the psychologising one of a negative act of judgement, not the purely logical one.[4] On the other hand there is a surprisingly rich flora of negatives in the Classical Chinese language.

The main negative particles in Classical Chinese

1. *Pu* 不 negates a verb or a verbal predicate in a narrative sentence and always precedes the verb phrase it modifies in the pattern (SUBJECT) *pu* PREDICATE[5] which we tend to understand along the lines of '(the SUBJECT) does not PREDICATE'.

Chün tzu pu chhi 君子不器 GENTLEMAN NOT VESSEL; 'The gentleman does not act as a tool/ vessel.'[6]

Pu is a verbal negation, and even if what follows is lexically more likely to function as a noun, it must be taken verbally. However, the verbal predication can be judgemental rather than narrative:

Fu tzu pu wei yeh 夫子不為也 MASTER NOT BE-ON-SIDE *yeh*; 'The Master is not on his side.'[7]

---

[1] *Huai Nan Tzu*, ed. Liu Wen-Tien, ch. 13, 10b.   [2] Graham (1978), B79.
[3] *Ibid.*, B78.   [4] For this distinction see, e.g., Ritter and Gründer (1984), vol. 6, p. 666.
[5] Elements in brackets are optional and may not be present.
[6] *Lun Yü* 2.12; cf. Lau (1983a), p. 13, who misconstrues the essential grammatical point: 'The gentleman is no vessel.' This would have to be *chün tzu fei chhi yeh* 君子非器也 according to the grammatical rules of Classical Chinese.
[7] *Lun Yü* 7.15; cf. Lau (1983a), p. 59.

The logical scope of *pu* can be more than one verb phrase:

If your lordship wishes to see it he must for half a year not (*pu*) enter the harem and he must not (*pu*) drink wine or eat meat.[1]

2. *Fei* etymologically possibly a contraction of *pu wei* 不為 NOT BE, negates the whole predicate in the pattern: (SUBJECT) *fei* PREDICATE *yeh* 也, which we tend to understand along the lines of '(the SUBJECT) is by no means correctly classified by the PREDICATE', where the final particle *yeh* marks the judgemental mode of the statement.

> *Fei wu thu yeh* 非吾徒也 NOT MY DISCIPLE *yeh*; 'He is not my disciple.'[2]
> *Tzu fei wo yeh* 子非我也 YOU NOT ME *yeh* 'You are not (identical with) me.'[3]
> *Fei wo yeh* 非我也 NOT I *yeh*; 'It wasn't me (i.e., it was not my doing).' (Compare *wu wo* 毋我 below.)[4]

Since *fei* is very common before what we translate as nominal predicates, it is often called a nominal negation.

When *fei* comes before a predicate that we ordinarily might take as narrative or descriptive, we are forced by the rules of Classical Chinese grammar to read that predicate as non-narrative and classificatory.

> *Kuan Chung fei than* 管仲非貪 KUAN CHUNG NOT BE-AVARICIOUS; 'Kuan Chung was by no means (correctly classifiable as) avaricious.'[5]

We are not grammatically free to take the verb after *fei* as straightforwardly narrative 'he went on to act avariciously (on that occasion)'. On the other hand, we are not forced to take *fei than* 非貪 NOT BE-AVARICIOUS as a nominal predicate 'be not (identical with) avarice', although we are obviously free to take it that way when the context demands it.[6]

3. *Wei* 未 '(temporally:) not yet, (logically:) not quite' refers to the whole predicate.[7] (Compare the sentence-final particle *i* 已 '(temporally:) by then, (logically:) under these conditions'.)

> (SUBJECT) *wei* PREDICATE TRANSLATES INTO
> (The SUBJECT) does not yet/quite PREDICATE.
> *Wei to yü ti* 未墮於地 NOT-YET FALL TO GROUND; 'It has not yet fallen to the ground.'[8]
> *Wei shan yeh* 未善也 NOT-QUITE GOOD *yeh*; 'He is not quite good.'[9]

4. *Fu* 弗 '(often contrastively[10]) 'refuse/fail to . . . (the object)' always precedes verb phrases, and very often transitive verbs with their object understood.[11] *Fu ju yeh*

---

[1] *Han Fei Tzu* 32.12.32; cf. Liao (1939), vol. 2, p. 36.   [2] *Lun Yü* 11.17; cf. Lau (1983a), p. 101.
[3] *Chuang Tzu* 17.89; cf. Graham (1981), p. 123.   [4] *Lun Yü* 11.11; cf. Lau (1983a), p. 99.
[5] Cf. *Han Fei Tzu* 36.15.23; cf. Liao (1939), vol. 2, p. 155.
[6] Harbsmeier (1981) In, pp. 17ff., I argue that pre-verbal *fei* 非 is nicely conveyed by the German *nicht etwa* ('it is not as if'). A less cumbersome English gloss would be 'by no means'. For an excellent treatment of other usages of *fei* see Pulleyblank (1959) and also Yen (1971).
[7] Harbsmeier (1981), pp. 42ff.   [8] *Lun Yü* 19.22; cf. Lau (1983a), p. 196.
[9] *Lun Yü* 15.33; cf. Lau (1983a), p. 157.
[10] Compare the originally emphatic *ne pas* in French, which has lost its emphatic force in contemporary speech.
[11] See Harbsmeier (1993). For an interesting alternative treatment of this particle, see Ting Sheng-Shu (*1935*).

弗如也, NOT BE-UP-TO *yeh*; 'You fail to come up to him, i.e., you are not as good as he.'[1]

5. *Mo* 莫 none (of the subjects)' has the subject as its scope.

Min mo kan pu fu 民莫敢不服 PEOPLE NONE DARE NOT SUBMIT; 'No people dare to fail to submit.'[2]

6. *Wu so* 無所, LACK THAT-WHICH 'none (of the objects)' has as its scope the object of the verb which it precedes:

> (SUBJECT) *wu so* VERB
> (The SUBJECT) VERBS nothing.
> *Chün tzu wu so chêng* 君子無所爭 GENTLEMAN LACK THAT-WHICH FIGHT; 'The gentle-man does not compete over anything.'[3]

7. *Wu* 無 is a negative verb 'to lack, there are none'.

8. *Wu* 毋 'make sure that not, don't!'.[4] *Wu* is preverbal. When it precedes a pronoun, it will force this pronoun into a verbal rôle:

Wu wo 毋我 MAKE-SURE-THAT-NOT I; 'He avoided being self-centred.'[5]

9. *Wu* 勿 'avoid to, don't' is a prohibitory negation which includes an object:

Chün wu thing 君勿聽 RULER SHOULD-NOT/IT LISTEN-TO; 'You should not listen to this.'[6]

10. *Wei* 微 'if it were not for, but for' is restricted to pre-nominal position and to sub-ordinate position.[7]

Wei Kuan Chung wu chhi pei fa tso jen i 微管仲吾其被髮左衽矣 BUT FOR (*wei*) KUAN CHUNG WE PRESUMABLY LET-DOWN HAIR LEFT FOLD-CLOTHES; 'But for Kuan Chung we should presumably be wearing our hair down and folding our robes to the left.'[8]

11. *Fou* 否 'such is not the case, no' is a sentential negation which functions as a whole sentence by itself.

Kennedy (1952a), p. 2, lists altogether 18 negatives and speculates on their pho-netic relations. Mulder (1959) continues this line of research.[9] In view of such an abundance of subtly distinguished negatives, it is curious that Liu Kia-hway could maintain the idea that in Chinese '*l'affirmation et la négation se distinguent à peine*'.[10] If this were so, the Chinese would indeed be a singularly illogical people. The mind boggles at the thought of what legal system the Chinese could have developed if they really hardly distinguished between the affirmation and the negation of a statement.

But Liu Kia-hway (1961) is desperately and demonstrably wrong. A quotation attributed to Confucius himself will conveniently illustrate the point:

---

[1] *Lun Yü* 3.6; cf. Lau (1983a), p. 39.   [2] *Lun Yü* 13.4; cf. Lau (1983a), p. 123.
[3] *Lun Yü* 3.7; cf. Lau (1983a), p. 21.   [4] Cf. Lü Shu-Hsiang (*1958*), pp. 12–35, and Graham (1952).
[5] *Lun Yü* 9.4; cf. Lau (1983a), p. 77. Note the verbal use of the pronoun 'I'.
[6] *Han Fei Tzu* 22.4.4; cf. Liao (1939), vol. 1, p. 229.
[7] For an admirable treatment of this particle see Yen (1978).
[8] *Lun Yü* 14.14; cf. Lau (1983a), p. 137.   [9] Cf. also Unger (1957).   [10] Cf. Liu Kia-hway (1961), p. 9.

As for the Way, there are only two alternatives: to be humane or to be inhumane (*jên yü pu jên* 仁與不仁), that is all.[1]

How can one possibly say about a moral culture, at the heart of which lies this division between *jên* 仁 'be humane' and *pu jên* 不仁 'be inhumane', that affirmation and negation are hardly distinguished? How could the Master have committed himself more emphatically to the absolute distinction between a moral term and its opposite? I suggest that it is only the remoteness of Chinese culture from our own that makes it at all possible publicly to attribute such exotic attitudes to the Chinese as Liou Kia-hway does.[2]

The scope of negatives

The problems of the scope of Classical Chinese negatives cannot here be treated in detail. But let us consider a sentence like

Ho pu chhu thu 河不出圖 RIVER NOT BRING-FORTH CHART; 'The river did not bring forth its chart.'[3]

The English translation may be understood in many ways according to intonation. For example:

1. It was not the river that brought forth its chart.
2. It was not the chart that the river brought forth.
3. It was not bringing forth that the river did to the chart.

In Classical Chinese, readings one, two and three would be expressed in quite different sentences like the following:

1'. *Chhu thu chê fei ho yeh* 出圖者非河也 BRING-FORTH CHART THAT-WHICH NOT-BE RIVER *yeh*.
2'. *Ho so chhu chê fei thu yeh* 河所出者非圖也 RIVER THE-OBJECT-WHICH BRING FORTH NOT-BE CHART *yeh*.
3'. *Ho chih yü thu chê fei chhu yeh* 河之於圖者非出也 RIVER'S BE-RELATED-TO CHART THAT-WHICH NOT BRING-FORTH *yeh*.

When reading Classical Chinese texts, one gets the impression that some of the burden that is carried by intonation and that goes unmarked in written English is generally carried by explicit syntax in Classical Chinese. It is not at all easy to find any clear cases where a sentence like *ho pu chhu thu* 河不出圖 obviously invites a reading along the lines of readings one to three above.

The system of Chinese negation shows considerable subtlety and precision. Moreover, when it comes to the problems of so-called 'cumulative negation', Classical

---

[1] *Mêng Tzu* 4A2(3); cf. Lau (1983c), p. 139.
[2] For a sensible discussion of other adherents of Liu's view, see Leslie (1964), p. 2ff.
[3] *Lun Yü* 9.9; cf. Lau (1983a), p. 79.

Chinese shows distinctly greater logical discipline than, say, Classical Greek or Modern English.

Double negation and cumulative negation

Battling Billson[1] expresses the proposition 'I am not going to fight tonight in Oddfellows' Hall' as follows:

'I ai*n't* fighting at *no* Oddfellows' Hall', he replied. '*Not* at *no* Oddfellows' Hall, *nor no*where else I'm *not* fighting, *not* tonight, *nor no* night.' He pondered stolidly, and then, as if coming to the conclusion that his last sentence could be improved by the addition of a negative, added '*No*.'

Plato does only slightly worse (or should we say better?) in the following:

Aneu toutou **oudeis** *eis* **ouden oudenos** *an hymōn* **oudepote** *genoito axios.*

Literally: 'Without this no one for nothing to one of you will never become worthy', i.e., 'no one will ever become worthy of any of you for any purpose without this.'[2]

In Middle High German illogical cumulative negation is obligatory in sentences like *nu* **en***kan ich es* **niemanne** *gesagen*, literally 'now not can I it no one say', i.e., 'now I cannot tell anyone'; the indomitable translator of Rabelais, Johann Fischart, asks: *Kan* **keiner kein** *Liedlin?*[3] is there no one who knows no song'. In modern Russian we have to say *on* **nichto ne** *skazal*, literally: 'he nothing not said', i.e., 'he said nothing'. In modern French we say *on* **ne** *le voit* **nulle** *part*, and this evidently does not make the French illogical people. On the contrary, I remember that I have heard it said that French is a very logical language. If that is so, Classical Chinese seems to be more logical still when it comes to negation.

There certainly are no sayings in Chinese that on the face of it commit such inexactitudes of logical articulation as we find in the saying 'all that glitters is not gold', which does not improve when one translates it into French: *tout ce qui brille n'est pas d'or*. As if gold did not glitter! One can easily imagine a Chinese scholar writing a treatise wondering about what it does to the mind when one is condemned to think in an exotic Far-Western language, like English or French, that involves such a muddle surrounding negation. How, the Chinese may wonder, do they keep their negated thoughts straight?

Consider the French, who are so justly proud of their language. We have *se déranger de rien* 'to make a fuss about nothing', *cela se réduit à rien, beaucoup de bruit pour*

---

[1] Wodehouse (1986), p. 184. Shakespeare commits logical atrocities like 'You may deny that you were not the cause.' (*Richard III* i.3.90) and 'First he denied you had in him no right.' (*Comedy of Errors* iv.2.7).

[2] See Jespersen (1929), p. 311ff., for a survey of the phenomena of redundant double negation in Modern English and other European languages. Kühner and Gerth (1892), vol. 1, pp. 203–23, provides a wonderfully well-documented study of the phenomenon in Classical Greek. There are in fact some rules for when negatives cancel one another, and when not in Greek, but unfortunately they are by no means consistently observed.

[3] F. Von der Leyen (1942), vol. 3, p. 85.

rien, ce n'est pas pour rien, where *rien* means 'nothing'. But we say *il n'a rien vu, il est incapable de rien dire*. Molière said *diable emporte, si j'entends rien en médecine*, and one used to be able to ask *a-t'on jamais rien vu de pareil?*, where *rien* means 'something' or 'anything'.[1]

Multiple negation in Chinese

Many Indo-European languages are infested with illogical cumulative negation.[2] As Janusz Chmielewski pointed out a long time ago, Classical Chinese is entirely free of it. I have yet to find a single instance in which negation is not construed strictly according to logic. Anyone who, in Classical Chinese, were to use a form of words literally corresponding to 'I ain't got no money' would certainly be taken to mean that he has money, as logic indeed would encourage one to think. In this respect, then, the ancient Chinese were far more logical than their Greek counterparts. Consider the following Chinese sentence involving three negations in close proximity:

Fei fu neng wu chhu yeh NOT (*fei*) NOT (*fu*) CAN NOT (*wu*) DISCARD *yeh*. Séraphim Couvreur's Latin paraphrase runs like this: *non quod non possit non dimittere* 'Not as if he could not avoid to discard (the ritual clothing).'[3]

There certainly is no illogical cumulative negation here, but the important point is that there *could* never be any such thing according to the grammatical rules of Classical Chinese. And the structural organisation of the negations is grammatically as well as logically crystal-clear.[4]

Even three negations in a row do not work cumulatively. They work strictly according to logical syntax:

A ruler may be shrewd and wise, but NOT-QUITE (*wei* 未) LACK (*Wu* 無) NOT (*pu* 不) KNOW (i.e.: it is not quite that there is nothing he does not know).[5]

We shall see later on that there were, in ancient China as in ancient Greece, those who maintained a relativistic and deliberately paradoxical stance according to which contradictories can be concurrently true. In China, this view was known as the *liang kho chih shuo* 兩可之説 maintained by some sophists.[6] But the very fact that,

---

[1] *Le petit Robert*, new edition 1981, p. 1717.

[2] The language of my own country, Norwegian, forms an exception. Negation is as strictly logical in our language as it is in Classical Chinese. Modern Chinese, on the other hand, has an unsettling idiom like *hao pu* GOOD NOT, where *hao pu jê-nao* GOOD NOT BUSY is synonymous with *hao jê-nao* GOOD BUSY and means 'very busy'. On the other hand we have – astonishingly – the pair *hao jung-i* GOOD EASY and *hao pu jung-i* GOOD NOT EASY, which both mean 'not at all easy'. I have not the slightest idea how to explain these fascinating 'illogical' facts about Modern Chinese. And I do not wish to imply that Classical Chinese was in any sense a perfectly logical language either.

[3] *Li Chi, Tseng Tzu Wen*, ed. Couvreur (1951), vol. 1, p. 446.

[4] One exception is the phenomenon known to linguists as negation-raising in sentences like 'I don't think she will come', which means 'I think she won't come'. In Harbsmeier (1981), p. 31ff., I have shown that similar phenomena do exist in Classical Chinese.

[5] *Lü Shih Chhun Chhiu* 17.5, ed. Chhen Chhi-Yu, p. 1092; cf. Wilhelm (1928), p. 277.

[6] See our Sub-section on Têng Hsi 鄧析 below.

in China as well as in Greece, such views were regarded as outrageous and flippant, shows that such relativistic attitudes went against what was common sense at the time.

Indeed, Janusz Chmielewski has suggested that the very feeling of logical outrage towards the *liang kho chih shuo* sparked off, as a response, Later Mohist Logic,[1] and one might add that similar feelings of outrage at the claims made by early Greek sophists were an important factor in the development of logic in Greece.

For more theoretical discussion of the closely connected notion of a contradiction, see below.

(2) LOGICAL SENTENCE CONNECTIVES

Apart from negation, some of the basic concepts of propositional logic are those of implication (roughly rendered by the English 'if'), disjunction (roughly rendered by 'or'),[2] equivalence (roughly: 'if and only if'), conjunction (roughly: 'and'). Let us look briefly at the representation of each of these logical concepts in Classical Chinese. We shall pay particular attention to the question whether the ancient Chinese were able to make explicit counterfactual conditionals such as 'if the world were different from the way it is, people would also be different.' The possibility of making such hypotheses that run counter to known fact is crucial for the development of critical and systematic general scientific explanations.

Implication

In standard propositional logic an implication is a complex sentence of the form:

(A) *p* materially implies *q*.

We call *p* the antecedent and *q* the consequent in this formula. Now in standard propositional logic a sentence like that in (A) is taken to mean something like 'it is not the case that *p* is true and that *q* is false.' According to the concept of material implication there need be no semantic connection whatever between the antecedent and the consequent of a true implication.

Strict implication, on the other hand, does demand a strict semantic relation between antecedent and consequent:

(B) *p* strictly implies *q*.

is taken to mean something like 'it is impossible in any conceivable world for *p* to be true and for *q* at the same time to be false, as long as *p* and *q* mean what they do mean'.

---

[1] See Chmielewski (1969), Part VI, pp. 35ff. Chan Chien-Feng (1979), pp. 30ff., was the first to argue this point in detail.

[2] 'Or' turns out to be a logically ambiguous word. We shall refer to parts of this ambiguity below.

With this distinction in mind, consider now a typical Classical Chinese clause with *tsê* 則 'rule, principle; on that pattern, then as a rule, then':

If you study without thinking, then (*tsê*) you are lost. If you think without studying, then (*tsê*) you are in danger.[1]

Here we have two generalisations stated in two conditionals. Confucius argues that study and reflection must go together, given that the world is as it is. He maintains that there is a real link between the antecedents and the conclusions in the world. This world being as it is, you will get lost if you study without reflection, and you will get into trouble if you just indulge in reflection without study.

By writing sentences like '*p tsê q*', the ancient Chinese established a relation between different statements *p* and *q* about the world. The general force of *tsê* is that given *p* one would expect that *q*. The nature and the intensity of this expectation can vary according to context. But by using the particle *tsê* the ancient Chinese identified what they saw as (more or less) regular patterns of co-occurrence or concomitance in this world. It was by establishing such patterns of concomitance that they attempted to orientate themselves in the welter of appearances and things and to articulate regularities in their world.

When used by a logician, the meaning of *tsê* may harden into the rigidity of strict implication.

If you hear that something you do not know is like something you do not know, then (*tsê*) you know them both.[2]

When used in a narrative context, the meaning of *tsê* may very occasionally soften to become a purely descriptive indicator of contemporaneity.[3]

Classical Chinese distinguishes between a general implication of the sort 'whenever *p* then *q*', which tends to be expressed in sentences like '*p tsê q*', and particular implications of the sort 'if (on a given occasion) *p* then *q*', which tends to be expressed in sentence patterns like '*ju* 如/*jo* 若 *p, q*.'[4]

The concept of a necessary condition 'unless p, not-q' is expressed as follows:

If you do not know language, you do not have the means whereby to know men.[5]

Here knowledge of language is said to be a necessary condition for knowledge of men.

Early writers were aware of the transitivity of the relation expressed by this formula:

If names are not correct, then (*tsê*) speech will not be in accordance with things. When speech is not in accordance with things, then (*tsê*) tasks are not fulfilled. When tasks are not fulfilled, then (*tsê*) ritual and music will not flourish. When ritual and music do not flourish,

---

[1] *Lun Yü* 2.15; cf. Waley (1938), p. 91. [2] *Mo Tzu*, ed. Graham (1978), p. 443.
[3] E.g., *Han Fei Tzu* 36. 13.6.
[4] Harbsmeier (1981), pp. 245ff., details the contrast but overstates the regularity with which it is maintained.
[5] *Lun Yü* 20.3; cf. Lau (1983a), p. 305.

then (*tsê*) punishments and fines will not be adequate. When punishments and fines are not adequate, then (*tsê*) the people have nowhere to seek refuge.[1]

The force of this argument is that the correct use of names is a necessary condition for a tolerable life of the people. We shall look more closely into this sort of argument, consisting of a chain of implications, in our Sub-section on the sorites.

Counterfactual conditional sentences

If one wishes to explain why the world is the way it is, one needs to be able to imagine the world to be different from the way it happens to be. Counterfactuals can either be implicit ('If this is not a counterfactual, I am a Dutchman') or they can be explicit ('If he had been born in Holland, he would be a Dutchman.') Implicit counterfactuals are no less counterfactual than explicit ones.

Alfred Bloom has argued concerning modern Chinese:

Yet, despite Chinese grammatical precision in expressing both the degree of likelihood of the premise of implicational statements and the distinction between if-then and if-and-only-then interpretations of the relationship of the premise to its consequence, the Chinese language has no distinct lexical, grammatical, or intonational device to signal entry into the counterfactual realm, to indicate explicitly that the events referred to have definitely not occurred and are being discussed for the purpose only of exploring the might-have-been or the might-be.[2]

Bloom maintains:

Historically-speaking, the fact that Chinese has not offered its speakers incentives for thinking about the world in counterfactual and entificational ways is likely to have contributed substantially to sustaining an intellectual climate in which these modes of thinking were less likely to arise.[3]

Bloom's account raises very interesting questions:[4] is it true that there was no way of making explicit counterfactuals in Classical Chinese? And is it true that the ancient Chinese were notably sparse in their use of counterfactual reasoning? Such counterfactual reasoning is crucial for the intellectual development of a people. Imagining the world as being different from the way it actually is, is an important strategy of scientific explanation.

Wang Chhung 王充 uses counterfactual reasoning to demonstrate the unacceptability of traditional superstitions:

If Yao and Kao Tsu had really been the sons of dragons, then, since it is in the nature of offspring to be like their parents, and since a dragon can fly on the backs of clouds, then it follows that Yao and Kao Tsu should have been able to mount clouds and fly.[5]

---

[1] *Lun Yü* 13.3; cf. Lau (1983a), p. 121.

[2] Bloom (1981), p. 16. Curiously, Bloom does not mention the crucial *yao pu shih* 要不是 'if it had not been that' which is frequently used in modern Chinese to make counterfactual sentences with a negative antecedent. For a lucid account of *yao pu shih* and other related matter see Eifring (1988).

[3] *Ibid.*, p. 59.   [4] Cf. the review by Garrett (1985) and also Wu (1987).

[5] *Lun Heng*, ch. 15; cf. Forke (1911), vol. 1, p. 320.

It turns out that for counterfactual conditionals of the sort:

(C) if (contrary to fact) it were the case that p, then q

Classical Chinese writers regularly use patterns like '*shih* 使 *p*, (*tsê* 則) *q*' or, much more rarely, '*chi* 藉 *p*, (*tsê*) *q*'. Chinese counterfactuals about a contingent event in the future cannot *per definitionem* be strictly counterfactual: 'If he came, I would simply send him home again' speaks about a remote hypothesis, or about an inconceivable but strictly speaking possible future event. In just the same way counterfactual *shih* 使, when followed by a clause describing a contingent future event, will introduce a remote or 'unrealistic' possibility.

If (*shih*) the Way could be handed up (to a ruler), everyone would hand it up to his ruler. If (*shih*) the Way could be presented, everyone would present it to his parents. If (*shih*) the Way could be reported, everyone would report it to this brother. If (*shih*) the Way could be bequeathed, everyone would bequeathe it to his descendants. But it is impossible to do any of these things, and the reason is this. . . .[1]

The combinations *chia shih* 假使 and *jo shih* 若使 always mark counterfactual clauses.[2] Hsün Tzu's combination *chia chih* 假之 'let us assume this as an arbitrary hypothesis'[3] is not so much counterfactual as theoretically hypothetical.

The counterfactual particle *wei* 微 'if it had not been for, but for', on the other hand, like its English paraphrases, is entirely limited to counterfactual usage: the noun mentioned after *wei* must refer to something that is presupposed to have been non-existent.[4]

If it had not been for your assistance (*wei fu jên chih li* 微夫人之力), I would not have got to this state.[5]

Sometimes the noun phrase after *wei* is a nominalised sentence:

If the Master had not removed my lid for me (*wei fu tzu chih fa wu fu yeh* 微夫子之發吾覆也), I would never have understood the Great Integrity of Heaven and Earth.[6]

At other times the nominalisation, if present, is in any case totally unmarked:

But for the fact that your Royal Highness mentioned it (*wei thai tzu yen* 微太子言), I would have requested permission to go along.[7]

Counterfactual *wei* 微 can even come after the nominal subject: 'If your Highness had not left (*chün wei chhu* 君微出), made clear your resentment and rejected me, I would (still) certainly be guilty.'[8]

---

[1] *Chuang Tzu* 14.47; cf. Watson (1964) p. 161. Harbsmeier (1981), pp. 272ff., analyses more than fifty counterfactuals in *shih* 使 like this one.

[2] See Harbsmeier (1981), p. 277.　　[3] *Hsün Tzu* 22.68 and 23.30.

[4] For a sound treatment of *wei* 微 see S. L. Yen (1978).

[5] *Tso Chuan*, Duke Hsi 30, ed. Yang Po-Chün, p. 482; cf. Legge (1872), p. 217. Yang Po-Chün's note on this passage contains further examples of our pattern.

[6] *Chuang Tzu* 21.38; cf. Watson (1964), p. 227.

[7] *Chan Kuo Tshê*, no. 473, ed. Chu Tsu-Keng, p. 1651; cf. Crump (1970), p. 556. Yen (1978), p. 475, misunderstands this sentence.

[8] *Chan Kuo Tshê*, no. 471, ed. Chu Tsu-Keng, p. 1639; cf. Crump (1970), p. 549. Our interpretation follows Fêng Tso-Min (*1983*), p. 980, which is also in accordance with the oldest commentary.

In collocation with *sui* 雖 'even if', *wei* 'if it had not been for' makes unambiguously counterfactual concessive clauses 'even if it had not been the case that':

Even if it were not so (*sui wei* 雖微) that the former great official had done it, dare I do anything but approve of your command?[1]

It turns out that explicit counterfactual reasoning was frequent in ancient China. The lack of verb forms like the subjunctive in Chinese did *not* lead to a lack of explicit counterfactuals.

Inference

The conditional particle *tsê* 則 'then', which can mark the relation between an antecedent and a consequent in conditionals, is among the most frequent words of the Classical Chinese language. Curiously enough, the word does not occur even once in the works which we can confidently attribute to the logician Kungsun Lung 公孫龍 (*c.* −320 to −250). By contrast the word *ku* 故 'therefore', which marks inferences from premises to a conclusion, occurs frequently in Kungsun Lung's work.[2]

What, then, is the difference between conditionals like '*p tsê* 則 *q*' on the one hand and inferences like '*p ku* 故 *q*', '*p ku yüeh* 故曰 *q*', '*p shih i* 是以 *q*', or '*p shih ku* 是故 *q*' on the other?[3] The consequent in conditionals is not asserted to be true, while the conclusion of the inference is indeed maintained to be true (against the background indicated in the premises). A chapter from the *Lao Tzu* will illustrate our point: Lao Tzu begins with a series of conditionals where the truth of the second clause is predicated only conditionally upon the truth of the antecedent. He then continues with a series of inferences where the truth of the conclusions is asserted, and what precedes the *ku* are grounds adduced to justify the conclusion.

> (If you) bow down, then (*tsê*) you are preserved;
> (If you) bend, then (*tsê*) you will be straight;
> (If you) are hollow, then (*tsê*) you will be full;
> (If you) are worn out, then (*tsê*) you will be renewed;
> (If you) have little, then (*tsê*) you will be enriched;
> (If you) have a lot, then (*tsê*) you will get confused.
> Therefore (*ku*) the sage embraces the One and is a model for the empire.
> He does not show himself and therefore (*ku*) he is conspicuous.
> He does not consider himself right and therefore (*ku*) is illustrious.
> He does not brag and therefore (*ku*) has merit.
> He does not boast and therefore (*ku*) endures.
> Precisely because he does not contend, therefore (*ku*) no one in the empire is in a position to contend with him.[4]

---

[1] *Tso Chuan*, Duke Chheng 16.7; cf. Legge (1872), p. 398. Cf. also Pulleyblank (1959), p. 187.
[2] Cf. Cheng (1975).   [3] For useful discussion on this point see Cheng (1975).
[4] *Lao Tzu* 22; cf. Lau (1984), p. 33.

Ku yüeh 故曰 can be used in strictly logical contexts, as when we read in the *Kungsun Lung Tzu* 公孫龍子:

'Horse' is that by which we name the shape. 'White' is that by which we name the colour. Naming the colour is not (the same as) naming the shape. Therefore (*ku*) I say 'white horse' is not (the same as) 'horse.'[1]

In other contexts the meaning of *ku* is a vague 'thus', as in the summarising formula *ku yüeh* 故曰 'thus it is said . . .', which often introduces apt traditional quotations that are in no way logically implied but much more loosely illustrated or suggested as true by what precedes.

In spite of the loose usages of both *ku* and *tsê* (which we would expect in a natural language) the distinction between the non-assertive, hypothetical force of *tsê* and the assertive (inferential) force of *ku* is fairly clearly maintained throughout all of early Chinese literature.

Disjunction

The standard way to ask whether something is fish or fowl in Classical Chinese is to ask two questions: 'Is it fish? Is it fowl?' The juxtaposition is quite enough to make the disjunctive point, but quite often the particle *chhi* 其 'his, its' is used in this sort of alternative question regarding the same subject. The disjunctive or alternative nature of the question may be brought out explicitly by the word *i* 抑 'or rather?', Latin: *an?* 'or?'. The interesting point is that Classical Chinese has no current equivalent for the declarative *vel* 'or', and the exclusive *aut* '(either) or'. *I* (as well as the 'alternative' anaphoric *chhi*) is entirely limited to alternative questions of this sort. *I* 'or' never occurs in declarative sentences.

In pre-nominal position, we very occasionally find the particle *jo* 若 used to mean 'or':

Please let the ruler or the heir apparent come (*chhing chün jo ta tzu lai* 請君若大子來).[2]

During Han times, moreover, we begin to find the particle *huo* 或 'some' used to mean 'or'. The earliest example that I have found, from the −2nd century, comes from the *Huai Nan Tzu* 淮南子:

Generally, when people think or plan, they all first consider something as right and only then proceed to doing it. That they are right or wrong (*chhi shih huo fei* 其是或非) is what distinguishes the stupid from the clever.[3]

---

[1] *Kungsun Lung Tzu*, ch. 2, ed. Luan Hsing, p. 15.

[2] *Tso Chuan*, Duke Ai 17, fu 2; cf. Couvreur (1951b), vol. 3, p. 731. In *Tso Chuan*, Duke Hsiang 13.3, we have a similar case of *jo* 若 within the scope of *chhing* 請. There is a puzzling isolated case of *ju* 如 being used as a disjunctive particle:

These can either be understood or they can be not understood. They can either be seen or they can be not seen (*ju kho chih ju kho pu chih. ju kho chien ju kho pu chien* 如可知如可不知。如可見如可不見). (*Lü Shih Chhun Chhiu* 16.6, ed. Chhen Chhi-Yu, p. 1002; cf. Wilhelm (1928), p. 252.)

[3] *Huai Nan Tzu*, ch. 9, ed. Liu Wen-Tien, p. 33a; tr. Ames (1983), p. 207.

How, then, *does* one ordinarily say 'This is fish or fowl' in pre-Han times (and after)? Formal logic teaches us that this proposition is adequately rendered without any notion of disjunction. In fact, disjunction may be systematically defined in terms of implication and negation, both of which we have already discussed with respect to Classical Chinese. In formal logic '*p* or *q*' is understood to be exactly synonymous with 'if not-*p* then *q*'. The ancient Chinese availed themselves of just this logical equivalence. Instead of 'This is fish or fowl' they said something like 'If this is not fish then it is fowl.' The pattern for a declarative disjunction in Classical Chinese is '*fei* 非 *p tsê* 則 *q*' ' if not–*p* then *q*'. This is indeed a current pattern in ancient Chinese literature. It is current exactly where we would expect a disjunction.[1]

Conjunction

The English 'and' is ambiguous. It can stand between nouns ('John and Bill are Americans'), between adjectives ('a nice and attractive girl'), between verbs ('he loved and adored her') or between sentences ('John is an American and Bill is British'). In Classical Chinese nominal conjunction has to be expressed by particles like *yü* 與 'and, with' and *chi* 及 'and, with', whereas sentential conjunction is expressed either (as in English) by simple collocation of the co-ordinate sentences, or by particles like *êrh* 而 'but, and then, and thus'.

In English we say things like 'You do that again and I shall report you to the police', where 'and' is logically speaking a marker of implication. Such logically confusing, purely conditional, usage for *êrh* is rare but not unattested in the early Chinese literature.

(3) LOGICAL QUANTIFIERS

Next to negation and sentence connectives, the only remaining crucial ingredient in modern standard logic are the quantifiers, words like 'all', 'some', 'no(ne)'. One may ask, for example, whether all the quantified categorial propositions in Aristotelian logic can be naturally and painlessly expressed in Classical Chinese or not. Conversely, we must ask whether all the subtle quantificational patterns in Classical Chinese can be naturally and painlessly expressed in Classical Greek. Or more generally: are quantificational strategies like those we find in modern logic parochial to some languages like Greek or are they universal and equally applicable to very different languages like Chinese?

Aristotle's logical *Organon* remained for a very long time the finest achievement in the field of logic, although in more recent times the contribution of the Stoics has been increasingly appreciated for its considerably wider range of philosophical

---

[1] Lahu as well as a wide variety of other East Asian languages use 'if' and 'not' to express 'or'. It would be interesting to find out whether any languages express 'if *p* then *q*' as 'not-*p* or *q*'. I know of no such language. For some structural limitations created by the absence of one single word for 'or', see the Section on logical and syntactic complexity.

analysis. Let us begin with the Aristotelian classification of the relations between the subject term and the predicate term and see whether these logical patterns, which involve quantification, can be naturally rendered into Classical Chinese.

1. *SaP* 'all *S* are *P*' becomes '*S chieh* 皆 *P yeh* 也'.
2. *SeP* 'no *S* are *P*' becomes '*S mo* 莫 *P yeh*'.
3. *SiP* 'some *S* are *P*' becomes '*S huo* 或 *P yeh*'.
4. *SoP* 'some *S* are not *P*' becomes '*S huo fei* 或非 *P yeh*'.

Given that we can translate each of these four basic relations between *S* and *P* that Aristotle envisages, and that the Chinese had ways of expressing conjunctions and implications, we can translate all the 256 forms of the categorial syllogism, from 'Barbara' to 'Celarent', into Classical Chinese. Aristotle might just as well have discussed not Greek but Chinese sentences. He could have expounded exactly the same theory. Syllogistics is not in principle and for linguistic reasons parochial to Greek and other Indo-European languages like Greek. And that, of course, is *one of* the reasons why this form of logic could spread so easily among speakers of very different languages.

Universal quantifiers

Universal quantifiers are – roughly speaking – words or constructions which do the job that 'all' typically does in English. Just as we find many different and non-synonymous negations in Classical Chinese, so we find many different and non-synonymous words for 'all'. At this point we shall summarise some basic distinctions among universal quantifiers.[1]

The most important general observation is that while quantification in Greek is through adjective-like words (like *pantes* 'all', *pas* 'every'), in Chinese the predominant strategy is to use adverb-like words (like *chieh, chien, chin, chou, fan, fan, hsi, ko, pien* 皆, 兼, 盡, 周, 汎, 氾, 悉, 各, 遍) in phrases like *ku chih jên chieh jan* 古之人皆然 ANTIQUITY'S MAN ALL BE-LIKE-THIS 'the men of old were all like this.'[2]

When there is an object as well as a subject in a phrase, then the question arises which noun is quantified by an adverbial quantifier. In a sentence like *bai hsing chieh ai chhi shang* 百姓皆愛其上 HUNDRED FAMILIES ALL LOVE THEIR SUPERIOR, the *chieh* 皆 will refer to the subject, so that we have to translate 'the hundred families (i.e., the citizens) all loved their superiors'[3] and we cannot translate 'the hundred families loved all their superiors'.

If, on the other hand, we had a sentence like HUNDRED FAMILY ALL (*KO* 各) LOVE THEIR SUPERIOR, we must translate 'the hundred families *each* loved their superiors (not other people's superiors)',[4] and we cannot simply translate 'they all loved their superiors'.

---

[1] A detailed account of quantification in Classical Chinese will be found in Harbsmeier (1981), pp. 49–176.
[2] *Lun Yü* 14.40; cf. Lau (1983a), p. 145. [3] *Hsün Tzu* 10.76; cf. Köster (1967), p. 122.
[4] Cf., e.g., *Mo Tzu* 14.11.

If we read HUNDRED FAMILY ALL (*CHIEN* 兼) LOVE THEIR SUPERIOR, we have to translate 'the hundred families loved all their superiors (each in their own way)'[1] and we cannot treat *chien* as synonymous with *chieh* 皆 or *ko* 各.

Fan 凡 'all' refers indiscriminately to all objects, never to the subjects.

Pien 遍 'all' refers indiscriminately to all objects wherever they may be in relation to the speaker, and never to the subjects.

Chou 周 'all objects universally' comes in logical contexts like:

Someone has to universally (*chou*) love (all) men, only then does that count as loving men. But one does not have to universally (*chou*) not love (all) men to count as not loving men. Someone does not have to universally (*chou*) ride (all) horses in order to count as riding horses, but only if he universally (*chou*) not-rides (all) horses can one say that he does count as not riding horses.[2]

In passages like these a subject quantifier like *chieh* 'all' would obviously be impossible, but it also seems semantically significant that the Mohists, who were extremely familiar with object quantifiers such as *chien* 'all objects, each in their own way' chose not to use that quantifier in this technical context: they wanted a logical object quantifier that was free from all connotations, and they used *chou* as such a technical term.

They might have used *chin* 盡 'the whole lot'. Indeed, *chin* is the only universal quantifier which the Later Mohist logicians defined:

Chin 'exhaustively' means 'none is not so (*mo pu jan* 莫不然)'.[3]

Chin 'exhaustively' refers to the object, unless that object is unquantifiable. A sentence like HUNDRED FAMILIES EXHAUSTIVELY (CHIN) LOVE THEM would come to mean 'the hundred families loved the whole lot of their superiors'. Sometimes the object is not quantifiable, for example because it is a place word like 'here':

Yüeh's treasures are exhaustively (i.e., all) (*chin*) here.[4]

But we do have sentences where *chin* 'exhaustively' refers to the subject simply because there is no object to which it could refer:

The ten thousand things are exhaustively/all (*chin*) the way the are.[5]

Universal quantification can be expressed periphrastically. We find the combinations 'none not (*mo pu* 莫不)', 'none is not (*mo fei* 莫非)', 'there is none who not (*wu pu* 無不)', 'there is none who is not (*wu fei* 無非)'. These again are syntactically distinct. *Mo fei*, *mo pu* and *wu fei* invariably refer to the subject, while *wu pu* may refer to the object.

In the pattern SUBJECT NONE (*MO* 莫) VERB OBJECT the negative universal quantifier 'none (*mo*)' will always refer to the subject so that we have to translate the pattern as 'none of the subjects verb the objects'.

---

[1] ALL (CHIEN) LOVE WORLD'S MAN 'he loves all the people of the world (equally).' (*Mo Tzu* 28.18; cf. Mei (1953), p. 153.)

[2] *Mo Tzu* 45.24; Graham (1978), NO 17.   [3] *Mo Tzu*, ed. Graham (1978), A47.

[4] *Mo Tzu* 15.24; cf. Mei (1953), p. 84.   [5] *Chuang Tzu* 2.78; cf. Graham (1981), p. 59.

Note that *mo* unlike the English 'no' cannot precede the noun it quantifies and is thus not an 'adjectival' modifier in any sense at all.

By contrast, *chu* 諸 'all the (members of a definite set)', *chung* 眾 'the whole crowd of' and *chhün* 群 'the whole flock of' do immediately precede the nouns they modify.

The various younger brothers (*chu ti* 諸弟) of the King of Chhu made representations that he should pardon the man.[1]

These three words are not synonymous, but they are very close in meaning and in their crucial structural property, which is that they must immediately precede the noun which is their scope.

Fan 凡 'speaking generally of' is limited to whole subjects or topics of sentences and it cannot be used to quantify an embedded noun as in the example we have just noted about the younger brothers of the King of Chhu.

Existential quantification

The existential quantifiers in Classical Chinese are 'adverbial' like *huo* 或 'some'. The pattern

SUBJECT HUO VERB OBJECT

means 'some subjects verb the object'. If one wants to speak of 'some objects' one needs to use a periphrastic pattern:

SUBJECT *YU SO* 有所 VERB

which means 'the subject verbs some objects'.

(4) LEXICAL AND GRAMMATICAL CATEGORIES

Since the sixteenth century Europeans have often expressed their profound amazement at the fact that the Chinese language seemed to have no fixed word classes. They reported with consternation that the Chinese freely used nouns as verbs and verbs as nouns. Chinese was found to be a most extraordinary language in this respect and quite unlike any of the Indo-European languages we are used to.[2]

Ma Chien-Chung 馬建忠 wrote in 1898:

Characters have no fixed meaning. Therefore they have no fixed word classes. If you want to know their word classes, you must first see what the meaning of the context is like.[3]

Ma Chien-Chung supplies an excellent example of what he means:

MEN NONE MIRROR-THEMSELVES IN FLOW WATER BUT MIRROR-THEMSELVES IN *STILL* (1) WATER. ONLY *STILL* (2) CAN *STILL* (3) THE-VARIOUS *STILL* (4).
Nobody will mirror himself in flowing water: he will mirror himself in still water. Only what is still can make still all things that are still.[4]

---

[1] *Shuo Yüan* 12.3, ed. Chao Shan-I, p. 325.   [2] Cf., e.g., E. Clark and H. Clark (1979).
[3] Ma Chien-Chung (*1904*), p. 9.   [4] The quotation is from *Chuang Tzu* 5.10; cf. Graham (1981), p. 77.

We might be tempted to say that the word *chih* 止, which we have literally glossed as *STILL*, is used here (1) as an adjective 'still', (2) as an intransitive verb 'be still', then (3) as a verb 'make still', and finally (4) as a noun 'still things'.

Chhên Chhêng-Tsê 陳承澤 objects in 1922:

> From my point of view, three of the classes Mr Ma brings up represent derived usages (*huo yung* 活用) of *chih*. The word class of *chih* should be 'intransitive verb'.[1]

According to Ma Chien-Chung, lexical items in Classical Chinese do not belong to word classes, they only have functions characteristic of word classes in interpreted sentences. According to Chhên Chhêng-Tsê, lexical items in Classical Chinese do basically belong to word classes but they can occasionally be used in special functions characteristic of other word classes.

The Chinese case is not as strange as might appear at first thought. We shall begin our enquiry by looking at word classes in English. To begin with, it is healthy to consider such sentences as 'But me no buts!' which show a greater flexibility in the use of grammatical particles than would be acceptable in Classical Chinese.[2] J. Marshman, in his grammar of 1814 already mentioned the English word 'sound', which he claimed can function as an adjective, a transitive verb and a noun.[3]

Marshman and others thought that what defines the grammatical character of a word in a sentence is the syntactic context in which it appears, primarily word order.[4] But how does that context itself acquire its firm structure in the first place? Questions of this order have rarely been raised, but they are decisively important. The answer must be two-fold: firstly, the grammatical particles of the Chinese language impose more definite grammatical structure on their surroundings; secondly, not all Classical Chinese words show the same degree of grammatical flexibility so that the grammatically/functionally more stable words impose a grammatical frame on the less stable ones. In addition, the idiomatic structure of the Classical Chinese language helps to remove the ambiguity of potentially ambiguous phrases.[5]

Words like the English 'sound' and 'round' are fairly special precisely because they are grammatically so multivalent. But Elizabethan English as used by poets like Shakespeare provides an excellent starting-point for a discussion of word classes in Classical Chinese. For Elizabethan English shows a grammatical flexibility in the use of words which is profoundly different from modern English, and which is strongly reminiscent of ancient Chinese practice. By looking at Elizabethan

---

[1] Chhên Chhêng-Tsê (*1957*), p. 20. Cf. Yang Shu-Ta's footnote 3 in Ma Chien-Chung (*1904*), p. 8.

[2] See Simon (1937), p. 116.

[3] Kenneth Robinson has pointed out that Marshman's example is unfortunate: the adjective *sound* is related to German *gesund* and the noun derives from the quite unrelated Latin *sonus*. Thus we just have two homophonous words *sound*. The word 'round' would have been a better example. We happily use this word as a preposition, as a verb, as an adjective, and as a noun. Often only the syntactic context tells us what kind of 'round' we are dealing with.

[4] It is significant that Julien (1870) called his grammar *Syntaxe nouvelle de la langue chinoise fondée sur la position des mots*. Note, however, that word order alone does not distinguish a transitive verb from an adjective: both the adjective and the transitive verb directly precede the noun.

[5] Cf. sentences like 'Oxford City Polytechnic academic staff dislike London University library catalogue card photocopying service charges', which pose no problem in English.

examples we can gain a pretty good first sense of what is going on in Chinese. Thus we may sensibly start our enquiry into Chinese word classes with Shakespeare.

Elizabethan English versus Classical Chinese

Shakespeare asks in the opening lines of Sonnet 108:

> What's in the brain that ink may *character*
> Which hath not figured to thee my true spirit?

Shakespeare feels free to use adjectives as transitive verbs, as in Sonnet 6 'Which *happies* (=makes happy) those that pay the willing loan', or *Macbeth* ii.4.4: 'Threescore and ten I can remember well;/ Within the volume of which time I have seen/ Hours dreadful and things strange; but this sore night/ Hath *trifled* (=made trifling) former knowings.' He can also use adjectives as intransitive verbs, as in *Coriolanus* v.i.6: 'Nay, if he *coy'd* (=be coy) to hear Comenius speak, I'll keep at home'.[1]

A nominalised expression can be used verbally, as in the memorable phrase from the *Lover's Complaint*, 'Who, young and simple, would not be so *lovered*?'. Ben Jonson has such 'Chinese' usages as '*Year'd* but to thirty' (*Sejanus* i.1).

Shakespeare himself even obliges us by using 'king' verbally in *King John* ii.1.371: '*Kinged* of our fears, until our fears, resolved,/ Be by some certain king purged and deposed'. In Classical Chinese this sort of usage of the word for 'king', *wang* would also have been morphologically marked.[2] Shakespeare writes in *King Lear* iii.6.117: 'How light and portable my pain seems now,/ When that which makes me bend makes the King bow -/ He *childed* as I *fathered*'. We may compare *Lun Yü* 12.11: 'Let the father father and the son son (*fu fu tzu tzu*). . . . When the father does not (behave like a) father and the son does not (behave like a) son (*tzu pu tzu fu pu fu*), then even if there be grain, would I get to eat it?' Even 'god' is not immune from such grammatical licence, as in *Coriolanus* v.3.11: 'This last old man . . . Lov'd me above the measure of a father;/ Nay *godded* me indeed.'

The point we need to stress is the *liberty* Elizabethan writers had – or took – when it came to using words in unusual grammatical functions. The verb 'to king' did not at that time need to be lexically sanctioned to be perfectly usable. Moreover, the word did not enter the Elizabethan dictionary as a verb just because Shakespeare used it so. 'Servant', even if it never had been used as a verb, still *could have been* used as a verb. And this possibility, this freedom, is the crucial point about the grammar of Elizabethan and particularly Shakespearean English that concerns us here.[3]

Any reader of Elizabethan English is impressed with the considerable amount of what one might call 'categorial grammatical anarchy' in that language. The English language thrived on such categorial anarchy. Indeed, the period of this

---

[1] Cf. also 'violenteth' meaning 'act violently' in *Troilus and Cressida* iv.4.4: 'The grief is fine, full, perfect, that I taste,/ And *violenteth* in a sense as strong/ As that which causeth it. How can I moderate it?'

[2] See our discussion of derivation by tone change below.

[3] Cf. incidentally: 'My wife, mother, child, I know not. My affairs/ are *servanted* to others.' (*Coriolanus* v.2.78).

anarchy shows up the English language at its best. The grammatical and categorial rigidity introduced into the English language at later stages did not mark grammatical progress but quite arguably a decline in the expressive power and vivacity of the English language.

In Literary Chinese, for its part, words never lost their functional suppleness. In Classical and Literary Chinese the considerable categorial licence the Elizabethans allowed themselves was, so to speak, institutionalised and integrated into the very core of the standard grammatical system itself. A certain amount of categorial anarchy became a stable part of Literary Chinese grammar down to the 20th century.

Chinese writers, like Elizabethan writers, would, of course, differ considerably in the extent to which they made use of the freedom they had. But in principle a writer of Literary Chinese feels free to use any word we would normally translate as a noun N as a verb meaning 'cause to be an N', 'regard as an N' or even, when appropriate, 'act as an N'. Similarly, a writer of Literary Chinese would quite generally feel free to use a verb V as a noun 'V-ness', or as an adjective 'V-ing', or as a transitive verb 'cause to V', 'consider as V-ing', etc..[1]

The Chinese have allowed themselves the luxury of using words in a variety of syntactic functions in which these words can be understood to make good sense. The fundamental contrast may appear to be that the Elizabethans marked such *ad hoc* changes of grammatical or syntactic function not only by word order but also by morphological means such as endings (as in 'king*ed*', 'lover*ed*') whereas the Chinese characterised the unusual functions their words could have by syntactic means: by word order (like the Elizabethans) and through grammatical particles. This is a gross simplification. Just as in English we have standard pairs such as the noun *re*cord versus the verb re*cord*, and the verb *per*mit versus the noun *per*mit (contrasts which are not recognisable in writing but quite distinct in English pronunciation) so the Chinese – as we have seen[2] – often pronounced one and the same character differently in different syntactic functions. Or, to put the matter more accurately: the Chinese developed different derived and grammatically distinct words which their writing system often failed to distinguish.

Unfortunately, we do not know the precise extent to which this differentiation and word derivation was practised in early forms of the Chinese language, and there is considerable controversy regarding the precise nature of the morphological characterisation of the derived words, but we do know that such functional differentiation existed. In 1946 Bernhard Karlgren wrote:

In other words, to the Chinese scribe around −800 it was self-evident that his language had a fully living system of word-formation (*ordbildningslära*) in which different grammatical functions were expressed through sound variations in the word stem.[3]

---

[1] Traditionally, the ancient Chinese called the use of a word in a grammatical or syntactic function which it does not standardly have the 'living use (*huo yung* 活用)' of a character, and this Chinese expression covers very beautifully also what the Elizabethans were doing when they used sentences like *Othello* iii.4.195: 'I do attend here on the General;/ and think it no addition, nor my wish,/ To have seen me *womaned*.'

[2] Cf. our Sub-section on the typology of the Chinese language.

[3] Karlgren (Från Kinas språkvärld. Norstedt, Stockholm, 1929), p. 77; cf. also Chmielewski (1949), p. 397.

We hasten to add that *not all* differences in grammatical function were expressed through variations in the word stem in Classical Chinese. In fact, *very few* such differences were made explicit in this way in Classical Chinese.

Returning for a moment to Elizabethan English, we find that even in that language not all cases of unorthodox grammatical function are made palpable by the addition of morphemes. Shakespeare often *uses* prepositions as verbs, as in *Romeo and Juliet* iii.2.141: 'I'll *to* him', though this is to be derived historically from ellipsis of the verb, as in: '[Go] to your tents, O Israel.' This is reminiscent of Classical Chinese practice where 'prepositions' function as verbs. *King Lear* i.1.264 goes even further than the ancient Chinese ever would: 'Thou losest *here* a better *where* to find'.[1] Nominalisation of interrogative pronouns in this way is more than even Chinese grammatical flexibility permits. And in English we know that 'here' and 'where' are nouns only from the syntactic position in the sentence.[2]

We shall now turn to a more detailed account of logical and grammatical categories in Classical Chinese. We begin with a brief survey of the substantial literature on the subject.

Bibliographic orientation

The question whether there are word classes in Chinese has been discussed by a number of distinguished sinologists.

Wilhelm von Humboldt wrote to Abel-Rémusat in 1827:

All Chinese words, even if they are linked together in a sentence, are in an absolute state, and they resemble in this the root-words in the Sanskrit language.[3]

Humboldt writes:

None the less I maintain that the Chinese language seems to me not so much to neglect as to dismiss the marking of grammatical categories.[4]

Georg von der Gabelentz, in his grammar of 1881, carefully distinguishes word categories (which he designates by German words like *Hauptwort*, *Eigenschaftswort*, *Zeitwort*) and functional categories (which he designates by Latin words like *Substantivum*, *Adjectivum*, *Verbum*). Gabelentz explains:

---

[1] Compare *so* 所 which means 'place' as well as 'where, the object which'.

[2] Again, only syntactic considerations indicate that 'eye' must mean something like 'appear to the eye' in *Anthony and Cleopatra* i.3.97: 'But, sir, forgive me/ Since my becomings kill me, when they do not/ Eye (=appear) well to you.' In *Henry V* iv.3.63, 'This day shall *gentle* his condition', only the fact that 'gentle' has an object tells us that it is a transitive verb. Shakespeare feels no compunction about using nouns as splendid verbs without further morphological notice in *Cymbeline* v.4.145: 'Such stuff as madmen *tongue* (=will use their tongues), and *brain* not (=will not use their brains)', or in *Measure for Measure* i.1.69: 'How might she *tongue* (=scold) me?' Shakespeare also uses 'path' to mean 'walk'. (*Julius Caesar* ii.1.83)

[3] 'Tous les mots chinois, quoique enchaînés dans une phrase, sont in statu absoluto, et ressemblent par-là aux radicaux de la langue sanskrite.' (Humboldt (1827), p. 16)

[4] 'J'avoue cependant que la langue chinoise me semble moins négliger que dédaigner de marquer les catégories grammaticales.' (*Ibid.*, p. 3)

The category is immediately inherent in the word, whereas the syntactic function will be changing in many words.[1]

Henri Maspero flatly denied that there were functional or lexical word classes in Chinese.[2] Maspero was followed in this – with certain qualifications – by a small minority of Chinese scholars such as Kao Ming-Khai.

Paul Demiéville (1948), p. 148, followed Ma Chien-Chung in maintaining that word classes only exist from the functional, not from the lexical point of view.[3] This view was current in Chinese under the slogan *li chü wu phin* 離句無品 'separate from sentences there are no classes'.

Dobson (1966), p. 29, concludes that essentially the question whether Chinese has lexical word classes is inconsequential.[4]

W. Simon (1937)[5] and Bernhard Karlgren (1961, pp. 73–8) maintained that Chinese does have parts of speech.[6] In this they were followed by Chou Fa-Kao (1964) and most Chinese linguists.

G. Kennedy's 'Word Classes in Classical Chinese' (1956) is a detailed study along Bloomfieldian lines and based on the *Mêng Tzu*. Leonard Bloomfield (1887–1949) scoffed at any 'mentalistic' description of language and insisted on discussing only the observable distribution of linguistic items in sentences. Following this rigid methodology, Kennedy shows in great detail how each grammatical particle constitutes a word class of its own and how lexical items of apparently the same 'word class'

---

[1] 'Die Kategorie ist also dem Worte unmittelbar anhaftend, die Function bei vielen Wörtern wechselnd.' (V.d. Gabelentz (1881), p. 113). Cf. Jakhontov (1968), p. 79, who draws attention to Gabelentz's neat distinction.

[2] 'Rien n'y sépare un nom d'un verbe. Evidemment, quand nous traduisons ou quand nous expliquons du chinois, nous disons de certains mots qu'ils sont verbes, et d'autres qu'ils sont noms, d'autres encore qu'ils sont adjectifs, etc. Mais c'est parce qu'il nous est impossible de penser nos mots français sans leur donner immédiatement des valeurs de noms et de verbes, que nous attribuons aux mots chinois des fonctions distinctes qu'ils n'ont pas. En réalité les mots chinois ne sont ni noms ni verbes, ils sont quelque chose d'indifférencié qui, sans être proprement ni l'un ni l'autre, peut établir dans la phrase, suivant les cas, des relations diverses, si bien que notre langue nous oblige à les répartir entre les noms et les verbes, alors qu'en chinois, ils restent indistincts. Cela ne veut pas dire que ces mots n'ont pas un sens précis, mais seulement que ce sens ne se laisse pas enfermer dans nos cadres grammaticaux.' (Maspero (1934), p. 36)

[3] 'Les parties du discours . . . n'existent pas en chinois que du point de vue fonctionnel. Si l'on peut dire que, dans tel ou tel contexte syntaxique, tel ou tel mot chinois est employé ici comme substantif, là comme verbe ou comme adjectif, c'est exclusivement en ce sens qu'il y fonctionne comme sujet, attribut ou régime, comme prédicat ou comme déterminant.'

[4] 'Il importe peu qu'un mot isolé soit considéré comme un 'nom', un 'verbe', un 'adjectif', ou encore les trois à la fois. Il assume une valeur nominale, verbale ou adjective, en vertu de sa présence dans une matrice dont la valeur est nominale, verbale ou adjective. Nous pourrions dire qu'un mot isolé (c'est-à-dire sous la forme d'une citation) est inclassable, mais qu'en 'distribution', c'est-à-dire dans une matrice, il participe à la valeur grammaticale de celle-ci. Dès lors, 'verbal', 'substantif' et autres termes deviennent les qualités d'un cadre distributionel et non les qualités qu'un mot posséderait intrinsèquement.'

[5] Cf. Maspero's review of this in the *Bulletin de la Société Linguistique de Paris* 3, 1938, pp. 200–13.

[6] 'If the Chinese script writes etymologically related but nevertheless different words with one and the same character, a person who starts from the script will ask whether it is the verbal or nominal meaning of the character in his text, and will see transitions of word-categories where there are in reality different words differentiated by tone.' (Simon (1937), p. 115)

'Summarizing, I should like to say that the theory denying the existence of parts of speech in Chinese was built on the following erroneous suppositions: (i) that parts of speech must necessarily be differentiated in sound; (ii) that the Chinese words pass at random from one category to the other.' (*Op.cit.*, p. 117)

combine differently with the grammatical particles. Kennedy opens his article with the following summary and conclusion:

The present material was originally written to serve as Introduction to A Grammar of Mencius, a project that must now be retitled An Abortive Grammar of Mencius. The project had proceeded on the assumption that word-classes can and must be defined before the relation between words can be grammatically treated. It has now reached the conclusion that in the final analysis word-classes cannot be defined, hence that Chinese grammar must start from different premises.[1]

John Cikoski's thesis *Classical Chinese Word Classes* (Yale, 1970) provides the most stringent attempt to date to define word classes for Classical Chinese. Cikoski limits himself to the *Tso Chuan* (−4th or −3rd century) and tries to set up formal criteria for the distinction between nouns and verbs. Cikoski provides a glossary of about 1,500 words (pp. 151–82). All words are assigned to word classes, and the evidence from the *Tso Chuan* for this assignment is listed in this singularly useful glossary. Cikoski's *Three Essays in Chinese Grammar* (1978) are an elaboration of themes in his thesis.

See Harbsmeier (1979), pp. 155–217, for the argument that from the lexical point of view Classical Chinese words do differ grammatically, but that lexical word classes in Chinese took the form of flexible *functional preference* rather than more rigid grammatical specialisation as in Russian or Greek.[2]

One may summarise some of the lessons to be drawn from all this discussion[3] as follows:

1. Not all Classical Chinese words have the same functional capabilities. For example, there is a marked division between grammatical particles and lexical items in Chinese. Consequently we may safely conclude that there are word classes in Chinese.

2. Chinese words are not grammatically characterised in the same way as, e.g., Latin words. Thus, even if we apply our Latin concepts *verbum* and *nomen* to Classical Chinese, verbality in Chinese is not the same as verbality in English. Strictly speaking we equivocate by applying the same concepts *verbum* and *nomen* to entirely different languages. We may safely conclude that Classical Chinese word classes are not the same as Latin word classes: Classical Chinese does *not* have the same word classes as Latin.

3. There is sufficient similarity between Latin and Classical Chinese for it to be possible to apply a wide range[4] of concepts from Latin grammar intelligently to Chinese. 19th-century Chinese grammars, particularly v.d. Gabelentz's *Chinesische Grammatik*, demonstrate this point beyond doubt.

---

[1] Kennedy (1964), p. 323.

[2] The history of the controversy is treated in great detail in Y. V. Rozhdestvenskij's book *Ponyatie formy slova v istorii grammatiki kitajskogo jazyka. Ocherki po istorii kitajevedenija* (The concept of word form in the history of Chinese grammar. Studies on the history of Chinese grammar), Moskow, 1958.

[3] We cannot consider here the vast literature on word classes in Modern Standard Chinese. For an excellent critical survey of Chinese discussions in this area see Wang Sung-Mao (1983), pp. 1–225. Western surveys include Korotkov (1968).

[4] Ma was not indiscriminate: he certainly did not drag such concepts as that of the *supine* into the description of Chinese.

4. A wide range of grammatical categories of Latin can be lucidly expounded in the medium of Classical Chinese. Ma Chien-Chung's 馬建忠 method in *Ma Shih Wên Thung* 馬氏文通 (1898) followed 19th-century European practice in applying selected Latin grammatical categories to Classical Chinese. It is worth pondering how well Ma managed to articulate his Latin-inspired analyses in the medium of Classical Chinese. We may safely conclude that the principles of Latin grammar are not so alien to Literary Chinese that they cannot be expressed in that language, and that the principles of Latin grammar have not been felt to be so strange by Chinese intellectuals that they could not be naturally understood and creatively applied to the Chinese language.

5. While there may be doubt concerning the classification of Classical Chinese lexical items as such, we may safely conclude that 'nominal', 'verbal' and 'adjectival' grammatical functions within a sentence can be syntactically defined by the use of word order and the possible occurrence of grammatical particles. The question whether these functions are quite 'the same' across different language types does not concern us here.

6. We can set up lexical distinctions between Classical Chinese words on the basis of their syntactic and semantic interaction with grammatical particles, number phrases, and with each other. Thus Classical Chinese lexical items as such are by no means grammatically featureless. The precise nature of these lexicalised grammatical distinctive features is the subject of empirical grammatical research, but their existence is beyond doubt.[1]

Let us, then, look at some of this crucial empirical evidence.

The class of lexical items and the class of grammatical particles

Ma Chien-Chung (*1904*) was surely right when he followed Chinese tradition in dividing characters above all into grammatical particles (empty words (*hsü tzu* 虛字)) and lexical items (full words (*shih tzu* 實字)):

When one composes literature, one uses only empty words and full words, and nothing else. . . . And in the classics and commentaries the full words are easy to gloss, the empty words are difficult to explain.[2]

Sentence-final particles like *yeh* 也 (marking declarative sentences), *i* 矣 (marking narrative and declarative sentences), *hu* 乎 (marking questions or emphatic sentences) are clear examples of grammatical particles.

Some particles have distinct lexical and grammaticalised functions:[3]

Ku 故 'reason' is grammaticalised to 'for that reason, therefore'; *tsê* 則 'pattern, rule' is grammaticalised to 'along that pattern, then as a rule, then';
tang 當 'to face, to fit' is grammaticalised to 'when';

---

[1] Kennedy (1964) and Cikoski (1978) are important efforts in this direction. Our own Section (*f*,2.1) details the lexical subclassification of Classical Chinese nouns.
[2] Ma Chien-Chung (*1904*), p. 11.   [3] Cf. Chou Yin-Thung (*1983*).

chhêng 誠 'sincere, genuine, real' is grammaticalised to 'if really'.

Fan 凡 'vulgar, general' is grammaticalised to 'in general, generally, speaking generally of';

kai 蓋 'to cover' is grammaticalised to 'probably';

tai 殆 'danger' is grammaticalised to 'probably'.

Chin 盡 'exhaust' is grammaticalised to 'exhaustively, the whole lot, all';

pi 畢 'finish' is grammaticalised to 'completely, the whole lot, all';

ping 並 'combine' is grammaticalised to 'combining everything, all'.

Chiang 將 'lead' and *yü* 欲 'wish' are grammaticalised to markers of the future, just like the English 'will'.

The so-called prepositions like *yü* 於 are grammaticalised verbs, or deverbalised particles in the process of deverbalisation and grammaticalisation.[1]

The so-called passive particles are transitive verbs in the process of grammaticalisation. *Chien* 見 'see, face'; *pei* 被 'suffer'; *shou* 受 'receive' are grammaticalised to Chinese passive particles.

We conclude that there is an important distinction between grammatical particles and lexical words in Chinese, but that the borderline between these two word classes is systematically blurred in certain places due to a persistent process of grammaticalisation, which is well known also from other languages.

The nominal, verbal and adjectival grammatical functions

Let us attempt to define nominal, verbal and adjectival grammatical functions in terms not of translation into other languages but in terms of syntactic features of the Classical Chinese language itself.

Nouns in Latin have gender, number and case endings. Classical Chinese words have none of these things. Gender does not come into the picture, number is indicated only when this is communicatively relevant,[2] and the function of case endings is 'taken over' – as in English – by various prepositions and word order. None the less, we may define a word or phrase as nominal in a sentence under a certain interpretation if it can only be negated by *fei* 非, not by *pu* 不, and if it can only be conjoined with another word or phrase by *yü* 與, not by *êrh* 而.

We may define a word or phrase as verbal in a sentence under a certain interpretation if – under that interpretation – it can only be conjoined with another word or phrase of the same category by *êrh* 'and', not by *yü* 'with'.

We may define a word or phrase as adjectival in a sentence under a certain interpretation if – under that interpretation – it occurs in a nominal phrase, precedes a noun or noun phrase, and is not itself nominal.

We may define a word or phrase as adverbial in a sentence under a certain interpretation if – under that interpretation – it precedes a verbal phrase and is neither nominal nor verbal nor a grammatical particle.

---

[1] Cf. Harbsmeier (1979), pp. 165–76.
[2] For a fine study of the category of the plural in Chinese see Kaden (1964).

We conclude that we can define functional notions such as 'nominal', 'verbal', 'adjectival' and 'adverbial' for Classical Chinese words and phrases in terms of word order and co-occurence with certain grammatical particles. The question to what extent what we thus formally define as 'nominal', 'verbal', 'adjectival' and 'adverbial' is felt to coincide with European current conceptions of these terms is irrelevant at this point. Our definitions are language-immanent.

We have ancient evidence that the Chinese recognised grammatical or functional distinctions in the use of their words. We know this because occasionally the Chinese have explicitly marked such functional distinctions through affixes. We shall now turn to the way in which grammatical function may be morphologically and intra-syllabically marked in Literary (and perhaps in Classical) Chinese. We turn to the phenomenon of 'derivation by tone change' which was first properly introduced in Chou Tsu-Mo's 周祖謨 article *Ssu sheng pieh i shih li* of 1945, and which is conveniently presented for us by Gordon Downer.[1]

Explicit grammatical derivation in Literary Chinese

Functional contrasts can be marked by an affix which in many cases came to be realised in Modern Standard Chinese pronunciation as the falling fourth tone. Let us designate this affix by X. We then have pairs of words, many of which are indistinguishable in their written form:

a.   *verb versus marked adverb*
   Fu 復   'repeat' versus *fuX* 'again';
   Ping 并   'stand together' versus *pingX* 'together';

b.   *noun versus marked transitive verb*
   I 衣   'dress' versus *iX* 'wear as dress';
   Wang 王   'king' versus *wangX* 'be king over';

c.   *verb versus marked noun*
   Chhi 騎   'ride' versus *chhiX* 'rider';
   Nan 難   'difficult' versus *nanX* 'difficulty';
   Shih 使   'send, despatch' versus *shihX* 'ambassador, messenger';

d.   *transitive verb versus marked intransitive verb*
   Chih 治   'govern something' versus *chihX* 'be well governed';

e.   *intransitive verb versus marked transitive verb*
   Hao 好   'be pretty, good', versus *haoX* 'consider as good, like';

f.   *verb versus marked grammatical particle*
   Wei 為   'work for' versus *weiX* 'for';

g.   *grammatical particle versus marked verb*
   Yü 與   'associate with > with' versus *yüX* 'take part in'.

[1] Downer (1959).

A number of crucial points need to be made in connection with 'derivation by tone change' as outlined above:

1. All the functional changes which are in this way sometimes marked by our affix *X* or the falling tone go unmarked in a vast number of other cases. The falling tone, then, occasionally makes explicit what pervasively tends to be implicit in Literary Chinese. In this way it provides direct evidence that the functional grammatical contrasts so marked are indeed an inherent part of the Chinese grammatical system and not 'read into' Chinese sentences by outsiders.

2. In the vast majority of the derivations of the type we are concerned with the derived word is written with the same character as the word from which it is derived. This is in significant contrast with other derivations through affixes where the derived word is normally written with a new character. As a result of this curious graphic homonymy such distinguished philologists as Ku Yen-Wu 顧炎武 (+1613 to +1682) and Tuan Yü-Tshai 段玉裁 (+1735 to +1815) could seriously maintain that the derivation by tone change was a late scholastic invention with no basis in early Literary Chinese or Classical Chinese.

3. Most derived tones are lost in modern Standard Chinese pronunciations, whereas other words derived by affixes survive in large numbers. For the case of modern Cantonese, however, compare Kam Tak Him (1977), who lists 48 of Downer's examples of derivation by tone change which survive in contemporary Cantonese.[1]

4. We can assign no specific meaning whatever to the affix *X* other than the excessively vague 'derived meaning'. (See Bodman (1967)). Compare the contrastive function of first syllable stress in English pairs like *permit/permit*, which regularly creates nouns as in the cases of words like *conduct, conflict, conscript, consort, contract, present, progress, record, refuse, transfer*, etc., where stress (not tone), though invisible in the written form of the words, is a reliable indicator of word class in spoken English. Word class can remain unmarked in English, as in the case of *regard* or *return*, but when stress on the first syllable is used as a distinctive feature in this way, it always marks the derived noun, never the derived verb. Denominal verbs like *to canoe, to cement, to harpoon, to referee* never are marked by a shift of stress to the first syllable. The Chinese morpheme *X* is not even semantically regular in this limited sense.

5. The functional versatility of the derived words is more limited when compared with that of other words which are not thus explicitly derived. But if traditional modern notions of when a character is to be read in the derived tone is anything to go by, then we should have to conclude that the derived word *wangX* 王 'be king over' can be re-nominalised to mean 'being-king':

When these three qualifications are provided, the being king follows suit (*san tzu chê pei êrh wangX sui chih i* 三資者備而王隨之矣).[2]

---

[1] Kam Tak Him (1977), p. 199.

[2] *Chan Kuo Tshe* no.57, ed. Shanghai Ku-chi-chhu-pan-she, p. 117; cf. Crump (1970), p. 67. Chu Tsu-Keng's edition of the *Chan Kuo Tshe* (p. 188) emends this passage to *wang tao hsingi* 王道興矣 on the basis of *Thai Phing Yü Lan* 太平御覽 in spite of the fact that *Hsin Hsü* 9.8 concurs with our text and that Kao Yu 高誘 (flourished +212) actually glosses the *sui*, which is in the *textus receptus* and not in the emended version.

Niu Hung-En *et al.* (*1984*), p. 34, footnote 16, maintains that we must read *wangX* in the falling tone. If read in the second tone, the passage would have to be translated along the lines 'the king followed . . .', which does not make sense in the context. In any case, our example is far from unique:

Why should this heart (of mine) accord with being-a-king (*tzhu hsin chih so i ho yü wangX chê ho yeh* 此心之所以合於王者何也).[1]

Here the 'preposition' *yü* 於 makes it clear that the *wangX* which follows it must be taken to be re-nominalised. Conversely, *shih* 使 'to despatch' becomes *shihX* 'ambassador' through affixation, but *shihX* in turn may be re-verbalised into 'act as ambassador'. Thus 'derivation by tone change' does *not* in general create words without functional flexibility.

However, we need to qualify this discussion of derivation by tone change with a strong *caveat*: the direct clear evidence we have that a certain character in a certain Classical text was read in one way and not in another way is far too sparse and is of far too late a date for us to discuss the grammar of the distribution of the diacritic morpheme *X* with real confidence. On the other hand, the extreme scepticism of Ku Yen-Wu 顧炎武, who considers *X* as a late philological invention is squarely refuted by Chou Tsu-Mo 周祖謨 (*1945*), pp. 52ff., who shows that derivation by *X* was recognised in the earliest direct sources on the pronunciation of characters we have, i.e., from Han times.[2]

Lexical categories

Having decided that Classical Chinese has functional categories like 'nominal', 'verbal', etc., and that these were sometimes grammatically marked, we are now in a position to ask a more controversial question: do we have lexical word classes in Chinese?

Our first inclination is to state categorically with Wilhelm von Humboldt:

Chin 金 (metal), *mu* 木 (tree), *shui* 水 (water), *shan* 山 (mountain) and *lin* 林 (forest) are substantives in Chinese in the same way as *homme, arbre, eau, montagne, forêt* are in French.[3]

As early as 1814 Marshman (p. viii) recognised that some words show greater grammatical flexibility than others; that, for example, *jên* 人 'man' is less flexible than *jên* 仁 'humane'. There certainly are a number of Chinese characters which we strongly expect to occur in certain grammatical functions and not in others. The verb *yüeh* 曰, for example, is entirely specialised, introducing quotations or lists and the like. *Yüeh* could never, I dare predict, be found in a nominal or any other non-verbal function.[4]

---

[1] *Mêng Tzu* 1A7.9; cf. Legge (1872), p. 141, Couvreur (1951a), p. 315 and Liang Chêng-Thing (*1973*), p. 17, who all agree that we are dealing with *wangX* 'be-king' rather than *wang* 王 'king'.

[2] Indeed, Downer (1959), pp. 264–7, argues that the derivation by *X* is very likely of pre-Han date.

[3] W.v. Humboldt (1827), p. 110.

[4] We disregard the trivial case of the quotation of the word *yüeh* 曰, which, of course, would be nominal.

In spite of the existence of words like *yüeh* 曰 I think most scholars would agree that for Literary Chinese a thesaurus divided into nouns, verbs and adjectives would be systematically, and to a large extent arbitrarily, repetitive.[1] It would be systematically repetitive because a vast number of words can be seen to have been used in these different functions. It would be arbitrarily repetitive because it is unclear how one is to decide whether a word *could have been* used in one function or another in Classical Chinese.[2] Even those words which we have not yet found used in more than one grammatical function quite plausibly *could have been* used in other grammatical functions, and might indeed tomorrow be found so used in the texts that are being unearthed by archaeologists. We must assume that the ancient Chinese – like the Romans and the Greeks – often *chose not to* use words in grammatical ways that would have been acceptable. But this is where agreement ends. We are in no position to judge what the ancient Chinese *could have* written, whether they *could have* used all their verbs as nouns.

Since we cannot ask them what they *could have* written, we are limited to what they *did* write, and this unfortunately leaves a number of important grammatical questions unanswerable. On the other hand, it so happens that the Chinese *did write* about the flexibility in the use of their words, and their own old testimony, it seems to us, is of special interest. It has not received the attention it deserves in the literature we have surveyed at the outset. In our survey of traditional Chinese grammatical thinking we have found that the current opinion was that characters had what was called *huo yung* 活用 'living uses, or *ad hoc* uses' which deviated from the *pên yung* 本用 'basic uses'.

One description of this situation that has been applied with some success to Modern Standard Chinese is to say that what look like *ad hoc* uses of words in unusual grammatical functions are in fact standard uses of homonymous derived words or grammatical homonyms, as in the case of 'sleep' in 'I can sleep' versus 'I like sleep'. This is acceptable for modern English, because the process of unmarked derivation of grammatical homonyms is not freely productive. It is unacceptable for Elizabethan English because in that language the process of unmarked derivation of grammatical homonyms is fairly freely productive. In Classical and Literary Chinese it is even more freely productive than in Elizabethan English.

---

[1] When Peter Roget constructed his *Thesaurus of English Words and Phrases* (London, 1852), he found it natural to divide his material according to word classes in such a way that under most semantic headings he listed separately verbs, nouns and adjectives belonging to a single semantic field. For modern Chinese the *Thung i tzhu tzhu lin* 同義詞詞林 (Thesaurus of Synonyms, Shanghai, 1983) attempts no such division, because it would lead to an unacceptable proliferation of repetitions in its word list. Whatever our theoretical conclusions on word classes in Modern Standard Chinese, this contrast between the English and Chinese treatments of their lexical inventory remains profoundly significant.

[2] If we happen not to find a supine form of a certain verb in Latin, are we simply to assume that this verb did not *have* a supine form? Such a principle would mean that unreasonably many Latin verbs would turn out 'defective' in all sorts of ways. (We recognise that some verbs lack certain grammatical forms which, from a semantic point of view, they could very well have.) Moreover, such a principle of deciding on defective verbs would fail to distinguish between those important cases where there is semantic reason why a verb does not have a supine form, and the others where we just happen not to have come across an example of a supine form in what we have read of Latin literature. We surely do not wait until we find a rare word in the genitive plural before we declare that it *has* a genitive plural.

One possible criterion for deciding whether a Classical or Literary Chinese word is 'basically' a noun or a verb is to observe its interaction with numbers. Suppose we are in doubt whether a character X 'basically' represents a noun or a verb. If the constructions like *san* 三 X THREE X come to meand 'three Xs' (Modern Chinese: *san-ko* 三個 X) or 'three kinds of Xs' (*san chung* 三種 X), we are dealing with a noun. If the constructions come to mean 'Xed thrice' (*X-lê san-tzhu* 了三次), we have a verb. If the construction comes to mean 'the three X ones' we have what is basically a stative verb. If both the verbal and the nominal interpretations of constructions like THREE X are attested, we can then say we have true homonyms. Again if *shu* 數 X SEVERAL X turns out to mean always 'several Xs', we are dealing with a noun, if it means 'to X several times', X has turned out to be a verb. If no such diacritic constructions are attested at all, our criteria do not bite. Moreover, to the extent that such criteria depend on how we translate into a Western language, they remain suspect.

It is the detailed investigation of such diacritic constructions which promises to deepen our understanding of the inherent system of Classical Chinese lexical categories.

Distinct lexical subcategories

J. S. Cikoski sets up a lexical subclassification of verbs on the basis of their syntactic and semantic potential. His subclassification of Chinese verbs into ergative verbs like *mieh* 滅 'destroy' and direct verbs like *hui* 毀 'destroy' is entirely original and grammatically important: ergative verbs become inherently intransitive (translatable as passive) when there is no object, whereas direct verbs do not. Thus *pu mieh* 不滅 without an object comes to mean 'not be destroyed' whereas *pu hui* 不毀 without an object comes to mean 'does not destroy'.[1] These two words are almost synonymous, but Cikoski has shown them to have clearly and neatly distinct lexicalised grammatical properties which ensure that when we see *pu hui*, we translate 'does not destroy', whereas when we see *pu mieh* we *must* translate along the lines of 'is not destroyed'. This grammatical contrast between *hui* and *mieh* is not affected by the question whether these words may also come to function nominally and come to mean 'destruction'. Being an ergative verb is as rigid and stable a lexical property of *mieh* as any lexical property of a word in Latin or English.

Thus we can define rigid lexical categories like that of a transitive verb (a word which when functioning as a verb always has an object), an intransitive verb (a word which when functioning as a verb never has an object), and a direct verb (a word which when functioning as a verb can be either transitive or intransitive, but which retains its transitive meaning even without an object), and ergative verbs (a word which when functioning as a verb may also be either transitive or intransitive, but which take on a passive meaning when there is no object, as we have seen in the case of *mieh* 滅 above).

---

[1] *Tso Chuan*, Duke Chao 12.2 and Duke Hsiang 20.6.

In a similar spirit, one can define rigid lexical categories for nouns: we can define mass nouns like *jou* 肉 'meat' (words which, when functioning as a noun, may be preceded by a quantificative measure like 'one plate of meat'), count nouns like *ma* 馬 'horse' (words which, when functioning nominally, may be followed by an itemising number phrase as in *ma san phi* 馬三匹 'three horses', or which may be immediately preceded by a number as in *san ma* 三馬 'three horses'). We can also set up a class of generic nouns like *min* 民 'people', which like count nouns can be preceded by itemising particles such as *wan* 萬 'the ten thousand, all' and *chu* 諸 'all the individual', but which cannot be counted individually, so that *ssu min* 四民 comes to mean 'four kinds of people'. Again, this lexical division is not affected by the question whether *min* can or cannot be used verbally in Classical Chinese. For example, *shih* 食 'food' is a mass noun, and the fact that this word is also used verbally for 'to eat' does not affect our lexical classification, which essentially says that if and when *shih* functions as a noun it functions as a mass noun.

The categorial continuum

Consider the word *ta* 大 which basically means 'be big, be great, be large'. If we look at the concordanced literature, we find this word used in the following ways:

1. verbally: 'be great'
2. adjectivally: 'great'
3. adverbially: 'greatly'
4. nominally: 'great size'
5. transitively: 'consider great'[1]

If all words in Classical Chinese were lexically quite indifferent as to the grammatical function in which they occur, then a construction consisting of characters such as *ta jên* 大人 'great man' should be indifferently ambiguous between the following translations:

1. great man
2. consider man as great
3. be greatly/very human
4. greatly human
5. the humanness of size
6. be great and be human

This, however, is not the case. As the historian of Chinese literature V. M. Alekseev remarked, 'No language, including Chinese, is mathematics, and therefore there are certain constraints on such elasticity.'[2] In general, Classical Chinese words may be flexible with regard to grammatical function, but they are *not*

---

[1] Cf. C. Harbsmeier (1979), pp. 188ff.
[2] Alekseev (1978), p. 542. The translation from Alekseev's Russian is my own.

indifferent to the way they grammatically function. Classical Chinese lexical items not only have rigid grammatical properties, as we have just seen, they also show clear 'preferences' for certain grammatical functions rather than other functions that in principle they could also perform. The notion of functional preference as applied to words is not easy. There is no suggestion that words have desires and preferences, rather that speakers who use words have structured expectations concerning the most likely functions a word will perform. We say a word in a language shows a preference for a grammatical function to the extent we feel we can attribute to the native speakers of that language an expectation that the word, *ceteris paribus*, will have that function. *Ta* 大 is 'primarily' or 'basically' an intransitive verb or an adjective, and only under special circumstances does it function as an adverb, a noun, or a transitive verb. *Jên* 人 'man' shows a marked preference for nominal syntactic rôles. Consequently, a reader of Classical Chinese will be in little doubt that *ta jên* 大人 probably will mean 'great man', or, at a double take, 'consider man as great/important'.

The case is different if we replace *jên* 'man' with *jên* 仁 'humane': *jên* 'humane' is almost as flexible as *ta* 'great', and the potential syntactic ambiguity of the hypothetical *ta jên* GREAT HUMANE out of context would, perhaps, be greater. But then, happily, such expressions and constructions never naturally occur out of context or in the abstract. They naturally come, if at all, with a context. Normally, in English as in Chinese, this context painlessly removes the ambiguity of constructions which, taken in isolation, would have been ambiguous.

Moreover, not all syntactically possible readings are in themselves equally current or natural, so that in disambiguating sentences one proceeds from what, in view of the context, is the more natural and current to what is less current.[1] We have to develop the grammatical expectations of native speakers of Classical Chinese. The sense of what is natural and current is acquired only by extensive experience.

We conclude that Classical Chinese lexical items or words *do* have inherent grammatical properties. But these properties do not take the form of rigid and total functional specialisation. Rather, they take the form of conditional regularities and patterns of functional preferences, or of *prima facie* tendencies to play certain grammatical rôles. When one sees a Classical Chinese word, this creates a spectrum or field of syntactic expectations, and these expectations can be stronger or weaker as the case may be. And these fields of syntactic expectation may vary individually and subtly for each word in the lexicon, so that one may end up with a tailor-made special word class for each lexical item. The words will then be distributed in a categorial continuum of syntactic tendencies.

---

[1] It is interesting to compare the comprehension of Modern Standard Chinese homonyms in speech. The context of speech creates certain expectations and when one hears a sound sequence which is ambiguous, one begins by assuming that the most likely word is intended and then proceeds down the hierarchy of probabilities. In just the same way one assumes a Classical Chinese word to have its most likely function unless the syntactic environment creates different syntactic expectations.

Chinese words are not predominantly governed by strict functional laws and subcategorised into discrete word clases in quite the same way as Latin. They are importantly governed by flexible grammatical tendencies and subcategorised into a categorial continuum. Chinese words can have a stronger or a weaker tendency to work like nouns, verbs, adjectives. We cannot simply say that they *are* nouns, verbs or adjectives.

Not only do Classical and Literary Chinese have different word classes from Latin. The very notion of belonging to a word class is different in these two languages. Breaking the rules of scholarly discourse one might almost say: 'It's a different ball-game'.

In a language like Greek, a finite verb will look morphologically like a finite verb and function syntactically as a finite verb, just as according to the rules of chess a bishop will both look like a bishop and move as a bishop.

In Classical Chinese verbs do *not* look like verbs. All we can say about them is that, all other things being equal, we naturally *expect* them to work as Chinese verbs do. But all other things never being quite equal, we also know that Classical Chinese verbs can play the part, for example, of nouns or of adjectives. Classical Chinese lexical items, then, are not very much like chess-men with their well-defined shapes and rôles. They are more like football players: we do expect a defender to be at the back and the goal-keeper to guard the goal, but we are also aware that in the heat of the battle a defender may attack and even the goal-keeper has in fact been known to leave his goal for another player to keep.

The scope of functional flexibility

Under the heading 'Distinct lexical categories' we have explored the rigid sub-classification of Chinese words. Now we shall explore the range of the flexibility of Chinese words. I begin with the standard noun *li* 禮 'ritual', which, when used with an object, becomes a transitive verb meaning 'to treat with proper ritual respect'. The generalisation is that, when used as a single predicate, *li* is classificatory and means either 'be identical with ritual' or 'be in accordance with ritual'. The trouble is that we have phrases like *chün tzu li i shih chhing* 君子禮以飾情 'the gentleman cultivates ritual in order to embellish his real nature (or: emotions)'.[1]

Personal pronouns are specialised (rather like goal-keepers) in most languages: we strongly expect them to fulfill their pronominal/nominal duties wherever they occur. But in Classical Chinese we are told about Confucius himself:

The master refused to do four things: . . . and he refused to be self-centred (*wu wo* 毋我).[2]

Here the pronoun *wo* 我, in defiance of all grammatical expectations is used as a verb presumably meaning something like 'to be self-centred'. The existence of this

---

[1] *Li Chi*, ed. Couvreur, vol. 1, p. 445.

[2] *Lun Yü* 9.4; cf. Lau (1983a), p. 77. Note the non-imperative use of *wu* 毋 to mean 'would not'. The old commentary explains that Confucius 'transmitted what was old and did not himself produce things'. This was the thought expressed by the pronoun *wo* 我 as understood in the earliest commentary we have.

parlance does *not* cause us to enter a new verb *wo* 'be self-centered' into the dictionary of Classical Chinese. We simply note that, in the heat of the communicative battle, even a pronoun can come to partake in the grammatical flexibility which is characteristic of the rest of Classical Chinese lexical items. We only understand *wo* in such a way by some sort of grammatical double-take. It is a part of our understanding of *wo* 我 that we do not ordinarily expect the word to function as a verb. Similarly for the pronouns *shih* 是 'this' and *pi* 彼 'that' which are used verbally in the *Chuang Tzu*.[1]

Compound noun phrases, one would have thought, should be stable in their nominal function. They should certainly not be able to function as verbs. But then, in the heat of the syntactic battle, we find Confucius himself applying the verbal negation *pu* 不 to the compound word for 'gentleman':

Is it not gentlemanly (*pu i chün-tzu hu* 不亦君子乎) not to take offence when others fail to appreciate your abilities?[2]

We take *chün tzu* 君子 as verbal along the same lines as the parallel *yüeh* 悦 'pleasant' and *lê* 樂 'joyful'. Confucius is not the only one to take such grammatical liberties with nominal compounds:

Mu Tzu said: 'The Prince of Chhu is very handsome, but he does not behave like a high official (*pu ta-fu* 不大夫).'[3]

Surely proper names should be immune from such functional aberrations, but one could say in Classical Chinese:

Kung Jo said: 'Are you trying to treat me as they treated King Wu (*êrh yü Wu Wang wo hu* 爾欲吳王我乎)?'
Then (Hou Fan) killed Kung Jo.[4]

The proper name *Wu Wang* 吳王 is used verbally to mean 'treat as they treated King Wu'.

Again one may point out that even the name of Confucius himself is not immune to such grammatical 'impropriety':

As for Yen Hui, if he was not like Confucius (*pu Khung* 不孔), then even controlling the whole world would not be enough to give him pleasure.[5]

Pu Khung 不孔 could be paraphrased as 'not live up to the ideals of Confucius'. The verbal negation *pu* 不 makes it clear that *Khung* 孔 must be taken verbally as 'be like Confucius'. Lao Tzu comes in for similar treatment elsewhere in the same text,

[1] Cf. *Chuang Tzu* 26.37 for *pu pi* 不彼, and *Chuang Tzu* 2.90 for *pu shih* 不是, where in each case the pronoun is preceded by the verbal negation *pu* 不 'not'.
[2] *Lun Yü* 1.1. We follow the translation in Lau (1983a), p. 3.
[3] *Kuo Yü* ch. 5, no.6, ed. Shanghai Ku-chi-chhu-pan-she, p. 195.
[4] *Tso Chuan*, Duke Ting 10; cf. Couvreur (1951b), p. 563: '*Est-ce que tu voudrais me traiter comme on a traité le prince de Ou?*'
[5] Yang Hsiung, *Fa Yen* 2, ed. Wang Jung-pao, p. 74.

where we find *wu tsai Lao pu Lao geh* 惡在老不老也 'why should this be a matter of being in Lao Tzu's style or not being in Lao Tzu's style?'[1] Here the explicitly verbal *Lao pu Lao* 老不老 LAO TZU NOT LAO TZU as a whole is nominalised as the object of the transitive verb *tsai* 在 'to be in, to consist in, to be a matter of'!

The verb *kho* 可 *X* 'may be *X*ed' creates another grammatical frame that must be filled by a verb. In one case this 'verb' turns out to be a sequence of two nouns: *fêng* 鳳 'phoenix' and *lin* 麟 'female unicorn', which are not only verbalised by the context but also passivised after having been verbalised, so that *fêng* comes to mean 'be made into a phoenix' and *lin* 'be made into a unicorn':

Can all birds and beasts be phoenixed and unicorned (i.e., made into phoenixes and unicorns) (*niao shou chieh kho fêng lin hu* 鳥獸皆可鳳麟乎)?[2]

The music of the state of Chêng 鄭 being known as notoriously lewd, we find the place word *Chêng* used for 'be Chêng-ishly lewd' construed along with *ya* 雅 'be elegant and dignified'.[3]

At least place names should be stable nouns. But the facts of Classical Chinese syntax are otherwise:

When you live in Chhu, you act-the-Chhu-way (DWELL CHHU AND CHHU *chü Chhu êrh Chhu* 居楚而楚); when you live in Yüeh, you act-the-Yüeh-way (DWELL YÜEH AND YÜEH *chü Yüeh êrh Yüeh* 居越而越); when you live in China, you act the Chinese way (DWELL HSIA AND HSIA *chü Hsia êrh Hsia* 居夏而夏).[4]

A place name *X* can even come to mean more specifically 'to affect *X* manners':

Phien Chhüeh (the famous physician) was from Lu, and many doctors affect Lu manners (i.e., pretend to be from Lu 扁鵲盧人也而醫多盧).[5]

Place names naturally come to function adverbially, too:

If you make a man from Chhu grow up among Jung barbarians, or if you make a Jung barbarian grow up in Chhu, then the man from Chhu will speak the Jung-way (*Chhu jên Jung yen* 楚人戎言) and the man from Jung will speak the Chhu-way (*Jung jên Chhu yen* 戎人楚言).[6]

Verbal or adverbial uses of proper nouns like the ones we have just surveyed may sound strange, but they are not strange at all once one reflects that in German we have *Berlinern* 'to speak with a Berlin accent', and as I am writing these lines I come across the verbs *philippizō Φιλιππίζω,* 'sympathise with Philip' as well as *makedonizō Μακεδωνίζω,* 'sympathise with Macedonia' in Plutarch's biography of Demosthenes.[7] It seems only proper to say that when later Greek writers felt impelled to use the word *aristotelizō Ἀριστοτελίζω,* for 'to imitate Aristotle', and when Lucian

---

[1] Yang Hsiung, *Fa Yen* 10, ed. Wang Jung-Pao, p. 336. Cf. also *pu I pu Hui* 不夷不惠 NOT PO I NOT LIUHSIA HUI 'he behaves neither like Po I nor like Liuhsia Hui', *ibid.* 17, ed. Wang Jung-Pao, p. 722.

[2] Yang Hsiung, *Fa Yen* 9, ed. Wang Yung-Pao, p. 281.　　[3] *Fa Yen* 3, ed. Wang Jung-Pao, p. 93.

[4] *Hsün Tzu* 8.115; cf. Köster (1967), pp. 86f.　　[5] *Fa Yen* 13, ed. Wang Jung-Pao, p. 473.

[6] *Lü Shih Chhun Chhiu* 4.5, ed. Chhen Chhi-Yu, p. 232; cf. Wilhelm (1928), p. 53.

[7] Plutarch, *Parallel Lives*, ed. Loeb, vol. 7, pp. 34 and 46.

permits himself the use of the word *Platōnikōtatos Πλατωνικώτατος*, for 'most Platonic', the Greeks were following the same grammatical urges as the Chinese were used to, the only difference being that the Greeks (like the Germans) were constrained by the grammatical structure of their language to make their grammatical liberties morphologically explicit, except in the adverbial use of nouns as in the frequent Latin *Caesar imperator iussit* 'Caesar *as the imperator* issued an order', which is closely similar in structure to *shih jên li erh thi* 豕人立而啼 'the boar stood up like a man and cried'[1].

Note that in the construction *Chhu jên* 楚人 'Chhu man' one might ask whether *Chhu* 楚 is an adjective or a noun. This is a grammatically moot point. The question may perhaps make sense to the Westerner, but for the Chinese interpreter there is no structural alternative here. We shall turn to this problem under the heading of structural indeterminacy.

Speculations on grammatical indeterminacy in Classical Chinese

We have seen that word classes are less rigid in Chinese than they are in Latin or Greek. Perhaps something similar is true of grammatical structure. Grammatical structures seem suppler, less rigid things in Classical Chinese than they are in Latin.[2] Perhaps the ascription of grammatical structure to Classical Chinese sentences must even more often be taken *cum grano salis*, with a pinch of salt (note the use of 'pinch' versus '*grano*'!) than is the case in ancient Greek. One does feel that the Romans used a more grammatically regimented language than the Chinese.

Perhaps Classical Chinese, as opposed to Latin, is lexically as well as grammatically a somewhat less rigidly defined, a more openly organic structure in which no morphological straight jacket enforces spuriously clear-cut obligatory divisions where none are required for the efficiency of communication. Chinese grammatical structures are only sufficiently well-defined to articulate meaning, but at the same time flexible enough not to impose rigid structure where such structure does not serve an articulatory or communicative purpose.

Perhaps even the functional distinction between noun and verb is *not always* as clear-cut as our familiarity with Indo-European languages leads us to expect, so that in Chinese we must live with the existence of hermaphroditic verbo-nominal hybrid functions.[3]

The Chinese would, then, not only have held an organicist world view. They would also have expressed this world view in a medium of communication which

---

[1] *Tso Chuan*, Duke Chuang 8.5; cf. Couvreur (1951b), vol. I, p. 143.

[2] Problems of functional indeterminacy arise in many, if not all, languages. In Sanskrit, for example, and particularly in the Vedic language, the distinction between noun and adjective is systematically fuzzy, so that Renou (1965), p. 231, noted: 'On discute sur la question de savoir si tel mot est adjectif ou substantif, nom d'agent ou nom d'action: la décision est souvent arbitraire, parce que la catégorisation n'est pas sentie comme telle.'

[3] Harbsmeier (1985) explores this possibility. I am corrected on a number of points in replies by S. Egerod, Zhu Dexi and particularly by E. G. Pulleyblank, all of which are printed in the same issue of *Early China* (pp. 127–145).

was much more (organically?) supple and less rigid in its articulatory strategies than languages like Latin or Greek.

Grammatical suppleness and flexibility should not be confused with (mechanical) grammatical looseness or lack of precision. Classical Chinese grammar is certainly much more precise than would appear from current descriptions. Organisms are precisely structured subtle things. They are *more* highly structured than a very advanced rigid device. What I am suggesting is that Chinese is highly structured in supple and subtle ways analogous to those of organisms.

The suppleness and flexibility of Chinese words is conveniently illustrated by the phenomenon of the *hui wên* 回文 or Chinese palindrome, a poem which makes good sense when read forwards and when read backwards.[1] The *hui wên* has quite a long tradition in China. It is mentioned by Liu Hsieh 劉勰 (+465 to +522), and the earliest illustrious example of a *hui wên* is that attributed to Su Hui 蘇惠 (+4th century).[2]

From Sung times we have a poem by a certain Chhien Wei-Chih 錢惟治 which must count as a consummate *hui wên*. It consists of 20 characters and it makes sense if you start with the first character, or the second or the third, etc., thus giving us effectively 20 poems. But it can also be read backwards, again starting either from the last, or the last but one character, etc., thus giving us another 20 poems.[3]

The extraordinary grammatical suppleness and flexibility of Chinese words is nowhere more palpably demonstrated than in the palindrome, the *hui wên*. From this point of view the playful *hui wên* deserve our serious linguistic attention.

(5) Logical and Grammatical Explicitness

Already the authors of the *Corpus Hermeticum* (+2nd to +4th century) noted:

Language does not reach the truth. But the mind (*nous*) is powerful. Being guided by language up to a certain point, it is capable of reaching the truth.[4]

European observers have for centuries complained of the especially elliptic and insufficiently explicit nature of Classical Chinese. Wilhelm von Humboldt wrote, in his *Lettre à M. Abel Rémusat sur la nature de formes grammaticales en général, et sur le génie de la langue chinoise en particulier* of 1827:

Dans toutes les langues, le sens du contexte doit plus ou moins venir à l'appui de la grammaire.[5]

---

[1] Cf. Latin *Roma tibi subito motibus ibit amor*. The palindrome has a long history in the West.

[2] Cf. also *Wen Hsin Tiao Lung*, ed. Chou Chen-Fu, p. 50, and p. 60, footnote 36.

[3] Chhien's poem will be found in Hou Pao-Lin *et al.* (1985), p. 146. For a detailed Western treatment of three palindromes see Alekseev (1978), pp. 532–44. The best bibliographic orientation on the Chinese palindrome will be found in Franke (1987) and in Ho Wên-Hui (*1985*).

[4] *Corpus Hermeticum* ed. A. N. Nock and A.-F. Festugière (Paris 1960), vol. 1, p. 100.

[5] 'In all languages the sense of the context has to support grammar.' Humboldt (1969), p. 102. The context-dependence of sentence meanings and the problems of vagueness have been in the focus of both linguistic and philosophical interest in recent times. A fine philosophical treatment of the subject will be found in Scheffler (1979). Much more technical is Ballmer and Pinkal (1983). Recent literature is surveyed in the *Journal of Pragmatics* published in Amsterdam.

Humboldt notes degrees of context-dependence in natural languages:

Dans la langue chinoise, la grammaire explicite est dans un rapport infiniment petit, comparativement à la grammaire sous-entendue.[1]

What are the psycho-linguistic effects of a language in which so little is explicit, and so much understood? Consider the following current Literary Chinese saying:

chih jên chih mien pu chih hsin 知人知面不知心
KNOW MAN, KNOW FACE, NOT KNOW HEART
'When you know a person, you know his face but you do not know his heart.'[2]

The crucial sentence connectives as well as the subjects of these three clauses are not explicit in the Chinese. Tense ('goes' versus 'went') and articles ('a person' versus 'the person'), too, are sometimes indicated by syntactic means, but mostly only implicit in the Chinese language. Here is another proverbial sequence of three verb phrases:

chien i pu wei wu yung yeh 見義不為無勇也
SEE RIGHTEOUSNESS, NOT DO, LACK COURAGE PARTICLE
'If, when you see duty you do not act according to it, that is lack of courage.'

But are all these things that the translation makes explicit really properly understood by the Chinese? It has been suggested, especially by the linguist Franz Misteli (1887), that the indeterminacy of language in cases like the sayings I have just quoted represents an indeterminacy of thought, and that translation falsely attributes articulateness to a way of thinking which was essentially vague and indeterminate, or to use Marcel Granet's term summing up the phenomenon: the language is *pittoresque* in nature. According to Marcel Granet, Chinese writing is so elliptic that it only allows one to reconstruct thoughts expressed without actually translating them into linguistic form.[3]

The guiding principle underlying ellipsis in Classical and Literary Chinese is expounded in the following description of the old Spring and Autumn Annals traditionally ascribed to Confucius himself:

The Spring and Autumn Annals in their use of formulations omitted what was already clear and put down what was not yet clear.[4]

This applies not only to the Spring and Autumn Annals, but to traditional Chinese writing in general. Whatever the audience can understand from the context is preferably omitted in literary style. Explicitness is felt to be vulgar. It is not by chance that there is no word for *scilicet* 'i.e., you should remember, that is to say' in Classical Chinese. What you should know is omitted for the very reason that you should know it.

---

[1] 'In Chinese, explicit grammar plays an infinitely small part compared with "understood" grammar.' *Ibid.*
[2] Brace (1925), p. 12.   [3] Granet (1920), p. 99.   [4] *Chhun Chhiu Fan Lu*, ch. 1, ed. Ling Shu, p. 3.

As we shall describe in more detail below, even arithmetic statements are generally elliptic in this way:

San chhi êrh shih i 三七二十一
THREE SEVEN TWO TEN ONE
'Three times seven is twenty-one.'[1]

The suggestion that a Chinese reader or writer just picturesquely conjoins these numbers in his mind without structuring their relation precisely makes no historical or philological sense. The context makes it sufficiently clear even to the modern reader 2,500 years *post festum* that multiplication is intended. Commentators show that multiplication was understood. The Chinese were *more* numerate than their European contemporaries for a long time. They were certainly advanced practitioners of the art of arithmetic. It just so happens that the Chinese here as so often elsewhere – for us disconcertingly often – communicated their insights in remarkably implicit context-dependent ways.

The ambivalent collocation of numbers became the basis for jokes:

'Thirty of Confucius' disciples wore caps (as signs of adulthood) and forty-two did not!'
'What is your textual basis for this claim?'
'Well, it says in the *Lun Yü* that "cap-wearing youths FIVE SIX"[2] and five times six is thirty. Again it says "of boys SIX SEVEN",[3] and six times seven makes forty-two. All this adds up to seventy-two disciples.'[4]

What seems to me to constitute the particular *génie de la langue chinoise* is that it does not grammatically enforce or institutionalise explicitness or semantic completeness where it is deemed communicatively non-functional given the knowledge of the intended audience. On the other hand, the Chinese have often tended to deem explicitness unnecessary where we would feel it was intellectually important or even crucial, especially in philosophical contexts. Whereas Greek philosophers like (Plato's) Socrates or Aristotle have at least aspired to (though not always achieved) an explicit discourse designed to avoid ambiguity, Chinese philosophers show markedly fewer of such ambitions.

Chinese sentences tend to be quite obviously designed not to *make explicit* but only to *communicate* whatever they are used to convey. This communication is often achieved with an admirable economy of articulatory effort which is hard to reproduce in modern English. The stylistic ideal of Classical Chinese encourages the exploitation of the grammatical possibilities for ellipsis: the ideal is to leave out

---

[1] *Lü Shih Chhun Chhiu* 6.4; cf. Wilhelm (1928), p. 76. In *Ta Tai Li Ji*, ch. 67, ed. Kao Ming, p. 291, I find the extraordinary sentence: *êrh chiu ssu chhi wu san liu i pa* 二九四七五三六一八 TWO NINE FOUR SEVEN FIVE THREE SIX ONE EIGHT which is to be interpreted in terms of the *wu hsing* 五行 'five phases', I do not propose to interpret it in detail at this point.

[2] In *Lun Yü* 11.24, *wu liu* 五六 FIVE SIX means 'five or six'.

[3] In *Lun Yü* 11.24, *liu chhi* 六七 SIX SEVEN means 'six or seven'. Both the quotations are found in the *Lun Yü* as we have it today.

[4] Hou Po, *Chhi Yen Lu* 啟顏錄, ed. Wang Li-Chhi, p. 11. The manuscript containing this joke (Stein no. 610) is dated +723. The traditional number of disciples is indeed 72.

whatever can be understood from the context. Confucius both expresses and exemplifies this ideal at the same time in the following often-quoted saying.

Words (should) get (the point) across and that is all (*tzhu ta êrh i* 辭達而已).[1]

The trouble is that what was sufficient to put across a point to Confucius' audience of disciples and therefore lived up to his ideal may not always be sufficient to put across that point to later interpreters like ourselves.

Plato's Socrates was constantly on his guard against being misunderstood (even by himself!) and compensated for this danger by painstaking explicitness in the statement of problems and solutions. To Montaigne, this made him a tedious person to read about. Confucius, on the other hand, seems satisfied that he knows what he is talking about and sees no reason for elaborate explicitness in his sayings as they have come down to us. The point is that Confucius *could* have been more explicit, but he chose not to be, and his choice was typical of the culture to which he belonged.

The question arises what significance we attach to this implicit form of communication typified by Confucius and encouraged by the system of Chinese grammar.

Ellipsis in Shakespeare

The case of Chinese is probably closer to Elizabethan than to modern English. Consider Shakespeare *The Tempest* i.2.219: 'On their sustaining garments (*scil.* there is) not a blemish,/ But (*scil.* the garments are) fresher than before.' In *Anthony and Cleopatra* iv.2.27 we read 'Haply you shall not see me more, or if (*scil.* you see me, then you will see me)/ A mangled shadow.' Or take *Hamlet* i.1.108: 'This must be known; which being kept close might move/ More grief to hide than hate to utter love.' We may apparently paraphrase this as follows: 'This ought to be revealed for it, by being suppressed, might excite more grief in the king and queen by the hiding of the news, than our unwillingness to tell bad news would excite love.'[2] E. A. Abbott (1913), p. 279, introduces his section on ellipsis in Shakespeare as follows: 'The Elizabethan authors objected to scarcely any ellipsis, provided the deficiency could be easily supplied from the context.' This statement highlights a significant feature of Elizabethan English, and it applies without further qualifications to Classical Chinese.

We can go even further: Whatever is unnecessary for the interpretation of a Classical Chinese sentence in its context can be left out *salva grammaticalitate*, or, more technically expressed: Whatever is pragmatically redundant in Classical Chinese is grammatically optional. We may call this a principle of grammatical economy.

The varieties of Pidgin English show that a great many morphemes in both modern and Elizabethan English are 'pragmatically redundant' or unnecessary in the

---

[1] *Lun Yü* 15.41, cf. Lau (1983a), p. 159.   [2] Abbott (1913), p. 284.

communicative context. The third-person verbal ending in 'He reads', for example, is in this way redundant, but the 's' in 'reads' is clearly grammatically obligatory in standard English and cannot be left out. What the speaker of Pidgin does is
to apply something like the principle of economy to modern English. The speaker
of Pidgin will supply indications of tense, definiteness of nouns, etc., only as and
when they are necessary to make his point. That is just what happens in Literary
Chinese.[1]

It turns out that the Chinese were able to write fairly explicitly when they felt the
need or the urge to do so. Wang Chhung 王充 (+27 to +100) was proud of indulging
in this stylistic 'vice' of over-explicitness and describes it in enthusiastic detail in the
postface to his great *Lun Hêng* 論衡. Later scientific writers were able to make their
texts as explicit as they needed to be. The *I Ching* 易經, at one extreme, leaves room
– and power – to the expert interpreter, and to the imagination of the reader. The
Lun Hêng, at the other extreme, was intended to solve doubtful questions of
scientific, historical and religious fact, and I think the patient reader of Literary
Chinese will agree that the *Lun Hêng* may be wrong on many points, but it is not in
general particularly inexplicit.

Neither does a text have to be at all scientifically minded or non-Confucian to use
an 'elaborated code' in this way. The commentator Chao Chhi 趙岐 (died +201),
explaining Mêng Tzu's moral philosophy, wrote, if anything, *more* explicitly even
than Wang Chhung. His sentences are quite deliberately replete with redundant
characters designed to put the intended meaning of Mêng Tzu beyond any doubt
even for the rawest outsider: even after the passage of 1,700 years a modern scholar
can fairly effortlessly follow most of Chao Chhi's interpretations.[2] And in case one
has overlooked the overall message of a section, Chao Chhi adds a very explicit
summary of the significance of every section in the *Mêng Tzu*. Let us look over Chao
Chhi's shoulder as he is expounding Mêng Tzu's terse prose:

BENEVOLENT *yeh chê* 也者 MAN *yeh* 也. COMBINE AND WORD IT, WAY *yeh*. Lau (1983c),
p. 293, translates: ' "Benevolence means "man". When these two are conjoined, the result is
"the Way".'

Chao Chhi glosses:

nêng hsing jên ên chê jên yeh. Jên yü jên ho êrh yen chih kho i wei chih yu tao yeh 能行人恩者人也。人與
仁合而言之可以謂之有道也. CAN PRACTISE BENEVOLENT GENEROSITY *chê* MAN *yeh*.
MAN WITH BENEVOLENT COMBINE AND SAY IT CAN CALL IT HAVE WAY.

'Someone who can practise benevolent generosity is a man. When you combine "be
benevolent" with "be a man" one may call this "having attained the Way".'

---

[1] The fact that he will also add certain elements which he carries over from his own language need not interest us here.

[2] Whether he can *accept* these interpretations is, of course, a very different question, which we shall not discuss
at this point.

Since the advent of the Jesuits in China in the late +16th century, Western observers have been most interested in rarified matters such as philosophy. In reading Chinese philosophical texts, they have often felt exasperated by the elliptic nature of the Chinese language used in these texts.

Part of this apparent vagueness and diffuseness is surely due to their insufficient understanding of the vocabulary and grammar of Classical and Literary Chinese. Classical and Literary Chinese is bound to look to us hazy and indeterminate in so far as we have only a hazy and indistinct grasp of the grammar of that language. I believe that as we understand more of the intricacies of Classical Chinese grammar and lexicography[1], we shall appreciate better the semantic precision of Chinese sentences. To some extent, vagueness, like beauty, is in the eye of the beholder.

Types of discourse

Besides, one has to take into account the type of Chinese discourse in question. Confucius' remarks in *Lun Yü* belong to an intimate sphere of communication which is as naturally and as properly implicit and heavily context-dependent as a modern private conversation. In the type of discourse used among friends it is natural that many things are understood. Explicit discourse would not be intimate. Explicit discourse is essentially public. The form of discourse we have in the *Lun Yü* tends to be not public but private and intimate. If we read it as if it were public discourse, we would profoundly misunderstand it. It is pointless to complain that private speech is not as explicit as public speech should have been.

Private discourse must not be confused with esoteric discourse. Confucius did not speak in a deliberately obscurantist way. There is no time-honoured Chinese conspiracy to write obscurely. Confucius spoke in a familiar, in a private way. We must try to understand him not as we would try to understand a published or public debate, or an Aristotelian disquisition, but rather as if we were eaves-dropping on a stranger's conversations. We must expect to miss out on many things, and we must strive to compensate for this inevitable loss by reconstructing the context and background as best we can.

Socrates conducted pedagogical and didactic conversations. Aristotle wrote didactical lectures. These Greek philosophers tried to convince those who refused to be convinced by anything other than argumentation. Arguments *ex autoritate* were unacceptable. Plato and Aristotle therefore had carefully to spell out and analyse what they had to say.[2] Wang Chhung did not generally accept arguments *ex auctoritate*, and neither did Chuang Tzu.

---

[1] It is interesting to note that we still do not have a standard and generally acceptable grammar of Classical Chinese. Moreover, we still do not have a standard and generally acceptable Classical Chinese – English dictionary.

[2] Note, however, that arguments *ex auctoritate* were routinely accepted in Europe for example from the +2nd century onwards. Thus Plato and Aristotle only represent one tradition, and a tradition which has not always been the dominant one in Europe.

By contrast, Confucius, Mêng Tzu and even Hsün Tzu are predominantly portrayed as preaching to the converted or presenting their case to men in authority.[1] Hsün Tzu is perhaps the most argumentative and systematic of these. But even he, when he bothers to refute thinkers who deviate from the path of Confucian truth, characterises his opponents' views in ways that can by no stretch of the imagination be called fair, and his counter-arguments amount mostly to no more than intellectual denigration. While we can reconstruct a fair amount of pre-Aristotelian thinking in considerable detail from Aristotle's accounts, Hsün Tzu tells us disappointingly little about those with whom he differed. He was expounding Confucianism to an audience of converted Confucians. He felt no need to do justice to the common opponent. A comparison of Hsün Tzu's *Refutation of the Twelve Philosophers* with Aristotle's opening chapter of the *Metaphysics* immediately bears out our point. Plutarch's vitriolic attacks on the Stoics may often seem to us unfair, but they do enable us to reconstruct a fairly detailed picture of Stoic philosophical views.

The central and ultimate purpose of Confucian philosophising was moral edification. Argumentation entered into the picture only in so far as it was morally edifying. Unedifying 'pure' reasoning was rejected as 'petty sophistry (*hsiao pien* 小辯)' and as 'wanton words (*yin tzhu* 淫辭)'.

Indeterminacy and vagueness

One might complain that *tsui jên* 罪人 GUILT MAN is vague or grammatically underdetermined because it can either be taken to mean 'a man characterised by guilt' (with *tsui* 罪 as a nominal modifier) or it may be taken to mean 'guilty man' (with *tsui* as an adjectival modifier). Here it becomes important to distinguish vagueness from structural indeterminacy.

The logician Charles Peirce has supplied a useful definition of vagueness:

A proposition is vague when there are possible states of things concerning which it is intrinsically uncertain whether, had they been contemplated by the speaker, he would have regarded them as excluded or allowed by the proposition. By intrinsically uncertain we mean not uncertain in consequence of any ignorance of the interpreter, but because the speaker's habits of language were indeterminate.[2]

In fact, *tsui jên* 罪 is not a case of semantic vagueness. The vagueness is not on the part of the Chinese expression. The problem arises only when we feel obliged structurally to match the expression with an English expression. Cases of this sort are extremely common in Classical Chinese.

The fact that we have two alternative ways of translating a Chinese construction is not always good evidence that this construction is vague between those two

---

[1] Notable exceptions are the discussions between Kao Tzu and Mêng Tzu, and between Hsün Tzu and Mêng Tzu on the goodness or badness of human nature. Cf. *Mêng Tzu*, book 6 and *Hsü Tzu*, ch. 23. The book *Mo Tzu* contains a great deal of argumentation, but there again its attack on Confucianism (*Mo Tzu*, ch. 29) is clearly directed at a Mohist audience, thus it does not constitute a proper intellectual dialogue with Confucianism.

[2] Baldwin (1902), p. 248.

readings. In principle, every Chinese concept or construction is something *sui generis* anyway which is not normally semantically or structurally isomorphic with anything in another language. The appearance of vagueness in a different language is often simply a symptom of the ethnocentricity of the person who complains about the vagueness. It is the result of insufficient empathy with the strange language, not a feature of that language itself.

The interesting question is whether Classical Chinese is *in Charles Peirce's semantic sense* a more vague language than Classical Greek. My conclusion is that Classical Chinese leaves greater scope for vagueness of meaning than Classical Greek, but that with certain significant limitations Classical Chinese sentences could be made as specific and determinate as each of the ancient Chinese writers wished to make them.[1] Some constraints on this will be discussed in the section on logical and grammatical complexity below.

(6) LOGICAL AND GRAMMATICAL COMPLEXITY IN CLASSICAL CHINESE

In the present Sub-section we shall look at what may turn out to be some restrictions on what can be lucidly expressed in Classical Chinese. Consider Shakespeare's Sonnet 15:

> When I consider everything that grows
> Holds in perfection but a little moment,
> That this huge stage presenteth naught but shows
> whereon the stars in secret influence comment;
> When I perceive that men as plants increase,
> Cheered and checked even by the selfsame sky,
> Vaunt in their youthful sap, at height decrease,
> And wear their brave state out of memory;
> Then the conceit of this inconstant stay
> Sets you most rich in youth before my sight,
> Where wasteful Time debateth with Decay,
> To change your day of youth to sullied night;
> And all in war with Time for love of you,
> As he takes from you I engraft you new.

For stylistic reasons one would not expect a poem of this kind in Classical Chinese: sentences tend to be much shorter than this in Chinese poetry. No Chinese poet would ever coin any word like Christian Morgenstern's *Kurhauskonzertbierterrassenereignis*, which is the subtitle of his poem *Die Fledermaus*.

---

[1] This is not to deny that there were writers in China who were deliberately obscure in order to ensure an exclusive readership. But this is not a peculiarly Chinese phenomenon. Heraclitus was suspected by Diogenes Laertius to have made sure that he wrote in an inexplicit way 'in order that only the able would approach his work and that he should not be open to disrespect because he was popular/vulgar'. (Diogenes Laertius, *Vitae Philosophorum* 9.6) One certainly has similar suspicions with respect to many writers of Literary Chinese, especially the writers of traditional prefaces. These often do strike one as deliberately esoteric and exclusive.

The question that I am concerned with in this Section is not, however, related to matters of style, but with the more fundamental issue of what is and is not lucidly expressible in Chinese.

Preliminary remarks

Let us begin with a historical perspective. In 1827 Wilhelm von Humboldt emphasised the striking way in which Classical and Literary Chinese style strongly favours much simpler sentences than Classical Latin or Classical Greek tend to do. Classical Greek sentences can run over several pages of printed text. Classical and Literary Chinese writers mostly use short sentences. When their sentences are long, they are made structurally simple and transparent by the use of parallelism.

Wilhelm von Humboldt tried to relate the brevity of sentences to the absence of a morphologically explicit system of word classes or 'grammatical forms':

The grammatical relations have an intimate relation to the unity of the proposition. For these grammatical relations are what expresses the relations of the words to the unity of the proposition. If these relations are conceived with precision and clarity, then they will mark off this unity better. They will make it more palpable. The relations between the words must become more in number and more varied as the sentences become longer and more complicated. It emerges from this that the need to try to mark off the distinction between the categories or grammatical forms in all detail derives from the tendency to form long and complicated sentences. When the phrases that are marked off rarely exceed the limits of the simple proposition, then the mind does not require that one represent the grammatical forms exactly. Neither does it demand that one carry the distinction so far that each of the grammatical categories will appear in all its individuality.[1]

In 1934 Liang Shih-Chhiu wrote in the introduction to his very carefully annotated translation from the Latin of Cicero's *De Senectute*:

The sentences in Latin are too long, and their grammatical organisation is too complex, so that there is simply no way of translating them successfully into Chinese without breaking them up into separate parts.[2]

In 1945 the great linguist Wang Li observed with respect to Modern Chinese:

When we translate from Western languages into Chinese, the sentential form of secondary clauses faces us with very great difficulties. . . . When we have a two-layered secondary

---

[1] Les catégories grammaticales se trouvent en relation intime avec l'unité de la proposition; car elles sont les exposans des rapports des mots à cette unité, et si elles sont conçues avec précision et clarté, elles en marquent mieux cette unité et la rendent plus sensible. Les rapports des mots doivent se multiplier, et varier à proportion de la longueur et de la complication des phrases, et il en résulte naturellement que le besoin de poursuivre la distinction des catégories ou formes grammaticales, jusque dans leur dernières ramifications, naît surtout de la tendance à former des périodes longues et compliquées. Là où des phrases entrecoupées dépassent rarement les limites de la proposition simple, l'intelligence n'exige pas qu'on se représente exactement les formes grammaticales des mots, ou qu'on en porte la distinction jusqu'au point où chacune de ces formes paraît dans toute son individualité. W. v. Humboldt (1964), pp. 64f. (Note the difference between sentence length in Humboldt's French and the English translation.) Cf. Menge (1960), p. 541.

[2] Liang Shih-Chhiu (*1934*), p. 3.

sentence form, then unless we break it up into a kind of parataxis, translation is practically impossible. For example, 'They murdered all they met whom they supposed to be gentlemen', if we translate it literally into *tha-mên sha-hai-lê tha-mên so yü-chien-ti tha-mên i-wei shih shang-liu jên ti i-chhieh* 他們殺害了他們所遇見的他們以為是上流人的一切, then the result is simply an uninterpretable Chinese sentence. Again, when we come across a nonrestrictive clause, our literal translation cannot distinguish it from the restrictive clause. Try to compare 'He had four sons, who became lawyers' with 'He had four sons that became lawyers'. If we use a literal translation, then we have no way of translating them into two different forms.[1]

In 1987 Tso Ching-Chhüan 左景權, discussing ancient Chinese versus Classical Greek from a Chinese point of view, observes with characteristic conciseness:

From ancient times long sentences have been a rare sight (*tzu ku i lai, han chien chhang chü* 自古以來罕見長句).[2]

Tso Ching-Chhüan elaborates:

But Greece shows a preference for a long-winded extended sentence. . . . When the reader meets these kinds of complex sentences, the interpretation is actually not difficult. The secret (of their success in interpretation) is their constant practice of 'logical analysis'. . . . They pick out complete constituent clauses and study separately their relations to other constituent clauses. After doing this for a long while, they develop this habit (of analysis), and even if they wanted to stop practising such logical analysis, they still cannot stop themselves.[3]

I shall not trouble the reader with any of the notorious syntactically complex passages in Demosthenes' speeches, or with any of the Latin writers noted for their elaborately balanced complex sentences. But even Seneca, who is generally comparatively brief and concise in his style, could start a letter (to an unhappy lady Marcia) as follows:

Nisi te, Marcia, scirem tam longe ab infirmitate muliebris animi quam a ceteris vitiis recessisse et mores tuos velut aliquod antiquum exemplar aspici, non auderem obviam ire dolori tuo, cui viri quoque libenter haerent et incubant, nec spem concepissem tam iniquo tempore, tam inimico iudice, tam invidioso crimine posse me efficere, ut fortunam tuam absolveres. . . .
'If I did not know, Marcia, that you were as far removed from womanish weakness of mind as from all other vices, and that your character was looked upon as a model of ancient virtue, I should not dare to assail your grief (the grief that even men are prone to nurse and brood upon), nor should I have conceived the hope of being able to induce you to acquit Fortune of your complaint, at a time so unfavourable, with her judge so hostile, after a charge so hateful.[4]

Note the slightly complex scope of *nisi scirem* 'if I did not know'. This complex scope would be hard to make transparent in Classical or Literary Chinese. In the

---

[1] Wang Li (*1984*), vol. 1, p. 57.   [2] Tso Ching-Chhüan (*1987*), p. 230.
[3] *Ibid.*, p. 230.   [4] Seneca (1958), p. 2.

translations we shall survey below the difficulty in representing the complex scope of verbs of intellectual attitude is a recurrent problem.

One may wonder how important the possibility of a complex scope for such verbs of intellectual attitudes is in intellectual culture.

We have a very considerable number of letters from traditional China. I very much doubt that any one of them would start with a sentence remotely as complex and as full of multiple subordination as Seneca's stylistically innocent opening remarks. Moreover, I wonder whether grammatically they *could have* started with a sentence of this order of non-parallelistic embedded complexity.

In the conduct of legal proceedings, syntactic complexity became ritualised in Greece, as a survey of Demosthenes' famous speech *De corona* will show.[1] Rhetoric of this sort, written or spoken, was certainly not encouraged by ancient Chinese ideals of style, as we have seen in my section on the art of literacy. But my point is that such rhetoric was also made difficult by the currently available grammatical devices as used in Classical and Literary Chinese.

Grammatical and intellectual complexity

One hastens to emphasise that, whatever Wilhelm von Humboldt thought, there is no clear evidence of a correlation between the morphological complexity of a language and the syntactic complexities it may be equipped to articulate efficiently. Moreover, there most certainly is nothing culturally 'advanced' about having a complex morphology.

Neither is there any connection whatever between the intellectually advanced nature of a thought and the length of sentences used to express that thought. An elementary textbook on logic, for example, will not normally contain syntactically simpler sentences than a more advanced textbook on the same subject.

Indeed, I was brought up with the idea that a very complex sentence, if true, probably deserves to be restated more briefly and without recourse to the sort of elaborate syntactic flourishes that were the hallmark of Greek and Latin rhetoric, and which are under consideration in the present Section.

And yet there remains the possibility that the ability of a thinker to pass through a phase of clearly articulated complex sentences on his way towards a perhaps briefer and plainer lapidary final statement of his views may be useful. Complex sentences, though perhaps not necessary for the final formulation of one's views, may be useful in the preliminary process of groping towards such views. It is in this groping and probing posture that we tend to find Socrates as described by Plato in the early dialogues. Is such syntactically complex philosophical groping feasible in Classical Chinese?

One may ask whether the ancient Chinese writers were constrained by the structure of their language to write relatively simply or whether the simplicity of

---

[1] E.g., Demosthenes, *De corona* 181ff. ed. Loeb, vol. 2, pp. 141ff.

Classical and Literary Chinese style is a matter of rhetorical choice. One may ask whether Classical and Literary Chinese is a linguistic tool powerful enough to express, for example, Greek logical and philosophical theories.

The rôle of natural language in science

The spread of modern science all over the world and the translation of scientific insights into vastly different languages all over the world would seem to bear clear and abundant witness that all natural languages can be adapted and supplemented in such a way as to serve as media for the communication of complex scientific insight.

Terminology is a case in point. If a language does not have a given concept, one can always introduce it by stipulative definition. After all, this is how the latinised concepts of English were introduced into English in the first place.[1] In principle, such problems of terminology are unproblematic. It is a question of time, attention, education and effort.

The matter of syntactic and logical complexity is only slightly more subtle. If a language does not have a means to represent a complex embedded logical or syntactic structure, in principle one can introduce such a means. For example by introducing a loan word or a loan construction from a language that does possess the means to present the structures involved.

Such terminological and grammatical borrowing has been happening to a quite remarkable extent in many languages. It sometimes takes the form of terminological as well as syntactic anglicisation. In English it often took the form of latinisation. In the case of Classical Tibetan, it took the shape of extensive – almost pervasive – sanskritisation.

Dimensions of logical and grammatical complexity

Complexity can be viewed along many parameters. For example, there is morphological complexity, which we disregard at this point, and there is modal complexity (try to say *utinam venisset* 'I wish he had come' in Classical Chinese!), which we also leave aside, although it certainly invites philosophical attention. Moreover, there is the philosophically highly interesting distinction between direct and indirect speech as in 'Galileo said the Earth was round' and 'Galileo said: "The Earth is round"', which one cannot reproduce in Classical Chinese. But this is another point that we shall reluctantly leave aside. At this point we shall concentrate on quantification, on sentence connectives and on syntactic complexity in that order.

Quantificational complexity

Quantifiers (words like 'all', 'no', 'some', 'only') in Classical and Literary Chinese are not easily compounded within a single proposition. Classical Chinese quantifiers

---

[1] Cf. Vol. VII, pt. 3.

tend to work adverbially as in 'Emperors are *some* wise' rather than adjectival as in '*Some* emperors are wise'.[1]

When the scope of a quantifier is syntactically deeply embedded, the adverbial nature of Classical Chinese quantifiers can cause systematic problems. For example, one might want to say that 'Confucius taught the Way to the ministers in the states of only some kings'. For a Classical Chinese translator, such a sentence would be something of a logical teaser. It is not easy to see how one can make plain that the scope of the quantifier is just the modifying noun 'kings' embedded within the object.

One might also want to say that

I used to believe that only some, but quite frankly not as few rulers of large and wealthy states as most of you might think, understand the Way.

An ancient Chinese who might want to say this, would be unlucky. He would be unable to express clearly and simply what he might have in mind. On the other hand, if many ancient Chinese had had things of this degree of quantificational complexity on their minds, I have little doubt that they would have developed the means lucidly to express such propositions.

Consider another short example, inspired by the philosopher and logician A. N. Prior.

Scholars only occasionally consult a few of the books which every philosopher would consider any true intellectual writing about any subject of interest to all serious students of human culture should be familiar with and make all his or her students read.

There is no reason why the ancient Chinese should have wanted to say this sort of thing. But it is not clear how they should have gone about saying it if they had wanted to. Speakers of Latin may not have wanted to say any such thing either, but our point is that they could easily have said it, had they wanted to.

Again, take the case of the quantifier 'no'. Classical Chinese had no flexible adjectival quantifier 'no' as it recurs in 'No one imagines that nobody fails to understand that no sensible man reads no books that no sage would read and that no sage would write.' The result is that such sentences, which are difficult in English, seem almost impossible to render into Classical Chinese.

What limits comprehension of such English examples is the degree of logical complexity which the reader has learnt and has got used to grasping without losing track of the overall picture. Informally, but adequately, one might say that comprehension of such complex structures depends on the reader's 'logical boggling point'. The difficulty is mostly not one of grammatical structure but of psychological span of attention and concentration. (To use Noam Chomsky's fashionable terminology: it is a matter of performance rather than competence.)

---

[1] In Modern Standard Chinese we do have adjectival forms like *yu-ti* 有的 'some', but it is amusing that *yu-ti* still is rarely used in object position. A sentence like *yu-ti shu hao kan* 有的書好看 SOME BOOKS GOOD TO READ 'Some books are good to read' is clear and easy, while a sentence like I LIKE SOME PEOPLE would be ungrammatical. One has to say SOME PEOPLE I LIKE.

A perfectly perspicuous and transparent logical formula can come to be felt to be opaque when it exceeds a certain degree of complexity. The reason why we do not understand it has nothing to do with the syntax of the artificial language of logic. It is connected with our psychological ability to grasp the clearly articulated structure which is there. This ability can easily be trained and increased manyfold through exposure to logically complex structures. In other words: the logical boggling point is mobile, not genetically or culturally fixed.

However, some people reach their boggling point sooner than others. And we may ask whether some *peoples*, too, on account of the inherent structure of their languages, tend to reach their logical 'boggling points' significantly earlier than other peoples.

In particular, we may wonder from a psychological point of view about the grammatical and logical 'boggling point' of the ancient Chinese. How much quantificational complexity firstly *did* and secondly *could* the ancient Chinese pile into a Classical or Literary Chinese sentence without affecting comprehensibility?

Note that the quantificational articulatory power of English is not due to the efforts of logicians building it up. Elizabethan English would have served quite as well as Modern English. Bertrand Russell and his logician colleagues did not make a difference. The quantificational articulatory power of a language is not a result of a deliberate building-up of this articulatory power.

The Greeks were presumably linguistically equipped to make very complex sentences long before they could be bothered to construct them.

The peasants in ancient Italy spoke a language (Latin) that was excellently suited to express the logical complexities their Greek neighbours liked to indulge in, although the contact with Greek literacy led to a marked increase in the grammatical precision with which speakers of Latin used their language. Latin writers seem never to have had any difficulty with literally reproducing the syntactic complexities they found in Greek rhetoric. The very literal Latin Vulgate generally makes considerably easier and more transparent reading than the Greek original. (But then, of course, Greek and Latin are historically and typologically closely related languages.) One is tempted to add that in religious and perhaps even in metaphysical contexts transparency is not always an unmixed blessing.

The logically felicitous features of Classical Greek and Latin are not properly viewed as logical achievements of these peoples. We should rather think of them as strokes of linguistic good luck.

Complexity through logical connectives

It has often been noticed that Classical Chinese does not have one word for 'or' that was used in declarative sentences like 'Sages are either arrogant or stupid'.[1] They

---

[1] Classical Chinese does have a word for 'or' in questions like 'Are you German or are you Dutch?'. That word (*i* 抑) is not, however, found in declarative sentences with the meaning 'or'. This is very interesting evidence on the greater need for disjunctive structures in questions than in declarative sentences.

tend to express the same idea by saying something like 'As for sages, if they are not arrogant, they are stupid'. It would be biased to maintain that Classical Chinese did not have a concept of disjunction.[1] They expressed the notion of a disjunction very lucidly, and logically quite satisfactorily, through the notions of negation and implication.

But let us now look at the way this enables the Chinese to handle complex sentences involving multiple disjunction. Consider:

Confucius hated or at least disliked arrogant or stupid ministers who, either publicly or privately, criticised or mocked their kings or rulers.

It is a truth of propositional logic that in principle one could express these disjunctions using only negation and implication. But the psychological fact is that a sentence of this order of complexity in the logical connectives would become most cumbersome in Classical Chinese, if indeed it turns out to be translatable at all. And it is not an insignificant fact that the English sentence is very far from difficult or opaque.

Neither is there anything complicated about a sentence like this: 'I never met a more pleasant, more beautiful, more intelligent, more cultured, or more profoundly attractive woman in my life'. Such disjunctive modifiers are impossible to bring out in pre-Han Classical Chinese, and they remain thoroughly uncongenial in Literary Chinese.[2] I leave it to the patient reader to construct other plausible cases of 'self-embedded' disjunctives like

He must either, publicly or privately, have criticised, or, in letters or publications, attacked his leader.

Here again, in principle there is nothing to prevent the ancient Chinese from trying to build up this constellation of disjunctives using only negation and implication. Our observation is that this would result in a less than transparent Classical Chinese construction. (Our English is bad enough!)

One might object that these concocted examples do not represent what, for example, an ancient Chinese or an ancient Roman might want to say. But now I happen to come across this entirely relevant passage in Cicero's rhetorical exploration of the ultimate good:

Aut enim statuet nihil esse bonum nisi honestum, nihil malum nisi turpe, cetera *aut* omnino nihil habere momenti *aut* tantum, ut nec expetenda nec fugienda, sed eligenda modo *aut* reicienda sint, *aut* anteponet eam, quam cum honestate ornatissimam, tum etiam ipsis initiis naturae et totius perfectione vitae locupletam videbit.

---

[1] In any case, the ancient Chinese did have a current word for 'or' (*i*) in questions like 'Is this a man or a woman?' The curious point is that the Chinese never seem to have thought of simply transferring this word into plain declarative sentences.

[2] Cumulative 'conjunctive' modification is also extremely rare in Classical Chinese: 'I met a pleasant, beautiful, intelligent, cultured and in every way very attractive woman' would have to be paraphrased along the lines of 'I met a woman. She was pleasant, etc., etc.'. When this sort of complex noun phrase is syntactically deeply embedded, serious problems of articulation are often inevitable.

Either he will postulate that nothing is good unless it is honourable, and that nothing is evil unless it is dishonourable, and that the other things are *either* of no (moral) significance whatsoever *or* only in such a way that they are neither to be sought nor to be avoided, but only to be chosen *or* to be rejected, *or else* he will prefer the doctrine which he sees is most adorned with honourableness and also rich in exactly the tendencies of nature and in the overall perfection of life.[1]

The logical structure here is plain in the Latin, but logically complex:

Either . . . (and (. . . or . . . (. . . or . . .))) or else . . .

It is hard to render this sentential form into Classical Chinese. The paraphrase using the 'if not . . . then' pattern, though theoretically feasible, would involve extremely heavy bracketing of a sort that only becomes possible in symbolic notation and is hard to represent in any natural language I know of, and which certainly is unheard of in Latin.

Consider now the complexities that are easily and perspicuously created in Latin or English when disjunctions and implications are combined in a single sentence like:

If, either deliberately or by chance, when you meet a sage or a philosopher, or when you run into a distinguished politician, you slight such people, then you should apologise or at least try to make up for your offence.

Such a mixture of conditional and disjunctive constructions in a single sentence or even clause makes things still more difficult for the translator into Classical Chinese.

By adding other logical connectives we could further compound the problems for Classical Chinese without affecting comprehensibility in Latin or English.

It is a curiously easy task to construct logical/syntactic constellations that are easy to articulate in a language like Classical Latin but apparently impossible to articulate in Classical Chinese.

The converse test is most interesting: can we find complex constellations that are easy to express in Classical Chinese but the logical complexity of which is apparently difficult to render in Classical Latin? I find it impossible to build up such examples. It would be interesting to find one.

Are we to conclude, then, that ancient Chinese thought is non-disjunctive, or that a traditional Chinese would never find he was lacking a single word or expression for 'or'? For once, we can give a definite historical answer: No. For it turns out that from Han times onwards the Chinese did tend to develop a word for 'or', the very same word to which we referred above as the quantifier *huo* 或 'some, probably', which sometimes came to be used from Han times onwards in a meaning that comes close to our 'possibly, or possibly, or'. Moreover, we note in passing that the

---

[1] Cicero, *De finibus bonorum et malorum*, Book 2.38. The translation is my own.

Modern Chinese word *huo-chê* 或者, which is defined in dictionaries as 'or', has its very deep pre-modern roots. *Huo-chê* is not an anglicising modernism.[1]

The Chinese did have use for a disjunctive particle, and since they felt they had use for one, they made one. But they felt this need relatively late, and even in Literary Chinese never quite got into the habit of thinking in disjunctions as the ancient Greeks, particularly Socrates, demonstratively groping for the right word, liked to pretend he did.

Logical equivalence (or the bi-conditional) in Modern English provides an interesting case of comparison. Speakers of English, like the speakers of many other languages, have long cultivated thought without feeling the need for a connective which in recent times analytical philosophers have introduced into Modern English: 'if and only if'.

Similarly, English has lived with the treacherous vagueness in the connective 'or' meaning either 'either p or q, but not both' or 'p or q or both', until we artificially introduced a new word 'and/or' to refer unambiguously to the inclusive meaning of 'or'.

Languages like English and Chinese can and do make such additions to their logical tools as and when they feel the need.

In some cases ancient Chinese is far more logically elegant than English. Thus from at least the +3rd century onward, the Chinese frequently used the logically important identifying 'copula' *chi* 即 which we have to render by clumsy glosses such as 'be nothing other than, be identical with'.

Syntactic complexity

Noam Chomsky (1965), p. 12, has a convenient preliminary classification of syntactic constructions:

A. Nested constructions For example, the phrase 'the man who wrote the book that you told me about' is nested in the phrase 'called (the man who wrote the book that you told me about) up'. Such constructions are generally hard to reproduce in Classical or Literary Chinese.

B. Self-embedded constructions For example, 'who committed suicide' is part of a self-embedded construction in 'who taught the student (who committed suicide) Latin'. The (infelicitously) so-called self-embedding constructions are simply nested constructions in which the superordinate and the subordinate construction are of the same type. It would be an interesting task to find self-embedding constructions in Classical or Literary Chinese.

C. Multiple branching constructions To use Chomsky's example, in 'John, Bill, Tom, and several of their friends visited us last night', 'John', 'Bill', 'Tom' and 'several of their friends' are part of a branching construction. Such constructions pose no problems in Classical or Literary Chinese.

---

[1] Cf. Chao (1976), p. 257, who notes that even in Modern Chinese the preferred way of stating an alternative is through negation and conditional.

D. Left-branching constructions For example, '(((((John's) father)'s brother)'s uncle)'. Such constructions are used in Classical or Literary Chinese. In the *Than Kung* 檀弓 section of *Li Chi* 禮記 we read:

Nan Kung Thao's wife's mother's funeral (*Nan Kung Thao chih chhi chih ku chih sang* 南宮紹之妻之姑之桑).[1]

E. Right-branching constructions For example '(this is (the cat that caught (the rat that ate of (the cheese that ruined the meal))))'. Right-branching constructions of this sort are unknown in Classical or Literary Chinese.

I suppose that English allows us to construct combinations of several of these construction types so that in principle we could get a right-branching construction embedded in a left-branching construction, which is part of a multiple branching construction, which is part of a self-embedding construction, which in turn is part of a nested construction. One of the reasons why we do not get sentences of such degree of complexity – not even in Gibbon's *Decline and Fall of the Roman Empire* – is that we have no practical use for them. Another is that even if we had, our grammatical boggling point would too soon be reached: we would not be able to decode the sentences which in principle can be decoded.

On the other hand we do decode things like the so-called 'zero knowledge proof' in computing science, of which Silvio Micali says: 'It allows me to ascertain whether the other person knows what I want to know, without him having to divulge to me what he knows.'[2] One wonders, whether a speaker of Classical Chinese or a writer of Literary Chinese would have been able to think up or articulate such a sentence.

In abstract terms our general observation about Classical or Literary Chinese is that nesting constructions are exceedingly difficult to make in that language, and that branching operations have to be multiple (i.e., co-ordinate) or left-branching. This brief and dry observation has profound consequences for our account of grammatical complexity in Classical or Literary Chinese.

Statements of conditional belief

One kind of syntactic complexity involves embedded complex sentences, for example conditional sentential objects. One may believe that if Deng Xiaoping had continued to hold power for some more years, the reform policies would have been irreversible, or that if the First Emperor of China had lived longer, the Chhin dynasty would have been firmly established.

The standard pre-Han idiom for 'believe' is *i X wei Y* 以...為...'to take *X* to be *Y*, to believe that *X* is *Y* (ing)'. It is not very clear how a conditional (or a counterfactual) could be fitted smoothly into this particular common pattern. Indeed, I believe that the increasing use of the pattern *i wei* 以為 *S* 'to believe that *S*' particularly from

---

[1] *Li Chi*, ch. 2, ed. S. Couvreur, vol. 1, p. 128. Cf. also Chhen Kuei, *Wên Tsê* ed. Liu Yen-Chheng, p. 27.
[2] *Die Zeit* no. 13, 20 March 1987.

Han times onwards is partly motivated by the fact that in this pattern there is much greater freedom for syntactic complexity in the embedded sentence *S*.

However, even in pre-Han times we do occasionally find *hsin chih* 信之 'he holds it to be reliably true'. I quote what might look like a relevant instance where we have a conditional statement which apparently is the grammatical object of a putative verb at least in D. C. Lau's translation:

Everyone believes that Ch'en Chung would refuse the state of Ch'i were it offered to him against the principles of rightness.[1]

But the case is only apparent. We should translate:

Chhen Chung would refuse the state of Chhi were it offered to him against the principles of rightness, and people regard him as trustworthy.[2]

When there is no anaphoric *chih* 之, the sentential object may contain co-ordinate clauses but apparently not subordinate ones:

Now suppose that all the world held it to be reliably true that the ghosts and spirits can reward talent and punish cruelty, then how should there be disorder in the world?[3]

It would be interesting to see if there were individual beliefs of conditional statements expressed with *hsin* 信. We only have a few cases like this one:

Then his friends believed (*hsin*) that he was ashamed to become Official Recorder.[4]

Let us summarise: The ancient Chinese can and will often express their belief in the truth of conditional statements by making conditional statements. What apparently they tend not to do is explicitly to attribute to each other or to themselves such beliefs that conditional statements are true. Thus they will constantly say things like 'if *P* then *Q*', but in pre-Han they are rarely found to embed this structure as an object in sentences like 'I believe that if *P* then *Q*'. Of course, quotations like 'He said: "If *P* then *Q*"' pose no problem, but then they are different.

This is not a superficial point of idiom connected only with the current pre-Han idiom *i X wei Y* 以 X 為 Y and the less current verb *hsin* 信 'hold to be reliably true'. The verb *chih* 知 'know' is extremely common, but I do not find sentences of the form 'He knew that if *P* then *Q*' in the indexed literature. What we do find are

---

[1] *Mêng Tzu* 7A34; cf. Lau (1983c), p. 279.

[2] Yang Po-Chün 楊伯峻 (*1961*), p. 316, translates the *hsin chih* 信之 as: *pieh jên tou hsiang-hsin tha* 別人都相信他 'the others all believed in him'. There is no reason to take *chih* 之 refer to the sentence rather than to the man: *hsin* 信 'to consider as trustworthy' very commonly applies to persons, and more rarely to sentences. Yang Po-Chün's reading is not only possible, it is also by far the most plausible.

Chao Chhi 趙岐 (ed. Shên Wên-Chuo, p. 927) does not paraphrase our *jên chieh hsin chih* 人皆信之. Legge (1872), p. 963, translates: '. . . and all people believe in him, (as a man of the highest worth)'. Couvreur, *Les quatre livres*, p. 623, gives a literal paraphrase of the traditional Neo-Confucian interpretation: *Tchoung tseu, si non justa via dedisses ei Ts'i regnum tunc non accepisset. (Ideo) homines omnes credunt eum (fuisse sapientem).* 'Chung Tzu, if one had presented him by unjust means with Chhi, would not have accepted the reign. (Therefore) all men believed in him (having been) wise.' Dobson (1963), p. 121, writes: '. . . and so everyone trusts him'.

[3] *Mo Tzu* 31.5; cf. Mei (1929), p. 160.

[4] *Lü Shih Chhun Chhiu* 26.1, ed. Chhen Chhi-Yu, p. 1690; cf. Wilhelm (1928), p. 449.

conditional statements followed by questions like *ho i chih chhi jan* 何以知其然 'how do we know that this is so?'[1]

The case for Modern Standard Chinese is different and much more like that of English. One would like to know exactly when and under what conditions sentences of the sort 'I believe that if *P* then *Q*' became current in China. Note the English: 'I have heard that they believe that if, as you get up and stretch your arms, everything turns black, then you ought to go and see a doctor.' In principle, we can have considerable additional complexity within the sentential object of a verb of belief. And we certainly could make this object much more complex than in the above sentence if we tried hard enough. The evolution of the phrase *i X wei Y* into *i wei X shih Y* 以為 X 是 Y in Chinese and the emergence of a single verb *i-wei* 以為 'believe that' and *jên-wei* 認為 'consider that' significantly increased the range of potential syntactic complexity in the sentential objects of belief. Here as often the Chinese evolved syntactic means to express complexity when they needed it.

Grammatical potential versus grammatical performance

It will be wise (though unfashionable or *unzeitgemäß*) *not* to take a modern language like English as our point of comparison, and not to treat our problem on the basis of hypothetical, artificially constructed sentences. We are interested in the actual use of linguistic and logical complexity in various cultures at various historical times. Adapting Chomsky's term we can say we are interested in historical performance.

Our historical question concerning Classical and Literary Chinese is not whether it could have been made to develop such-and-such means of representing logical complexity. The answer to that question is simple and almost axiomatic: yes, it could have been made to develop any means of representing logical complexity that we can think of, the grammatical potential was there. But our question is of an historical kind: how powerful a tool for the written articulation of logical and syntactic complexity did the ancient Chinese actually develop according to our historical records? We need to know:

1. what in point of fact did the ancient Chinese get into the linguistic habit of doing to express semantic complexity, and
2. what did the explicit grammatical strategies they developed equip them to do within that framework?

It is one question to ask what buildings some people might in principle have built. (In principle they might surely all have been able to build the Empire State Building.) It is another question what building techniques and building styles a people actually did develop and cultivate. (It is historically significant that the peoples of ancient times, for example, never built anything like the Empire State Building.)

---

[1] Other verbs expressing propositional attitudes deserve attention: for example the word *khung* 恐 'to fear' and its near synonyms. I do not remember seeing these with complex sentential objects, but the matter needs to be investigated in detail.

The current view in general linguistics seems to be that all languages in principle provide equally powerful means of communication. This may be correct. But historically it is about as interesting as the (correct) observation that the ancient Chinese might have built the Empire State Building. This is a speculative observation with which in principle I strongly concur on the obvious grounds that *in dubio pro reo*. The point is that the observation is not very illuminating historically.

Rhetorical versus semantic complexity

Let us call the actual (surface-)grammatical and logical complexity of a sentence S the 'rhetorical complexity of S'.

Let us call the complexity of the conjunction of the grammatically and logically simplest possible sentences which jointly are equivalent to S the 'semantic complexity of S'.

Given these very rough definitions, which will suffice for our present purposes, we can then formulate simply our concrete problem concerning the passages from Seneca and Demosthenes more rigorously: Do our sentences quoted or mentioned from Seneca and from Demosthenes show any semantic complexity beyond the articulatory power of Classical or Literary Chinese grammar?

We are also in a position to formulate a more general and far-reaching question of comparative intellectual history: Can we find a significant set of Sanskrit, Classical Greek and Classical Latin sentences (especially in the scientific and philosophical literature) which show a semantic complexity beyond the articulatory power we have evidence of in the preserved Classical and Literary Chinese texts?

Translations of philosophical and logical texts into Literary Chinese

When it comes to the ability of the Chinese language to serve as a medium of logical discourse, one's general confidence in Classical Chinese is not an act of faith. We have the indigenous Chinese traditions of logical discourse as evidence. As for Literary Chinese, we have translations of highly technical Sanskrit Buddhist texts. R. H. Robinson (1967), pp. 73–88, discusses in commendable detail the achievements of Kumārajīva (+344 to +413) as a translator, and he concludes:

To summarise the virtues and faults of this translation: The Chinese is often more explicit than the Sanskrit. It relies less heavily on anaphora and so is clearer. It sometimes supplies explanatory phrases, such as one finds in the prose paraphrases of Sanskrit commentaries. . . . The Chinese copes successfully with syntactic features such as the locative absolute and statements of reason by means of ablative noun compounds. It possesses a device for handling the highly-important abstract-noun suffixes. . . .[1]

---

[1] Robinson (1967), p. 87. Robinson also points out some weak points, but none of these are of an inevitable kind: they just show that Kumārajīva's team did not always do as well as they might have done.

164 LANGUAGE AND LOGIC

Hsüan-Tsang's 玄奘 translations done in the +7th century mark a clear improvement over Kumārajīva's.[1] Indeed, I have argued that Hsüan-Tsang's Classical Chinese version of *Nyāyapraveśa* is *more* (not less) lucid and transparent than the Sanskrit original as we have it. Latin and Greek logical texts, for obvious historical reasons, have received less consistent and thorough attention from traditional Chinese translators, and the translations are correspondingly weaker.

'Aristoteles Sericus' and 'Platon Sericus'[2]

When Christianity was reintroduced[3] to China in Late Ming times, it too brought along its logic: the logical *Organon* of Aristotle. However, Christianity did not strike deep roots in Ming China. Correspondingly, Aristotelian logic (even more than Buddhist logic in the +7th century) remained a marginal phenomenon on the Chinese intellectual scene during the +17th century, when it was introduced. Aristotelian logic practically disappeared from the Chinese scene until Yen Fu 嚴復 (1853 to 1921) detached Western logic from Western religion and provided paraphrased translations of William Stanley Jevons's (1835 to 1882) primer *Logic* (Chinese translation published 1908) as well as John Stuart Mill's *System of Logic* (Chinese translation published 1905). Yen Fu's relative success is connected with his wise decision not to try to reproduce syntactic complexity but to go for a Literary Chinese paraphrase of what he (often rightly) thought was essential.

It is symbolically significant that Indian logic came to China because the Chinese monk Hsüan-Tsang 玄奘 took the initiative to go to India. He returned with a number of logical treatises which he felt were important in the context of Buddhism. On his return, Hsüan-Tsang translated a little primer of Buddhist logic into Chinese, and this primer became the point of reference for the tradition of Buddhist logic. By contrast, Aristotelian logic came to China because Father Francisco Furtado (+1587 to +1653)[4] took the initiative to go to China, and he took along a very bulky, detailed and theoretically advanced scholastic treatment of Aristotelian logic in the form of the detailed standard handbook *Commentarii Collegii Conimbricensis e Societate Jesu in Universam Dialecticam Aristotelis Stagiritæ*, published in +1616.

[1] I have compared the basic Sanskrit logical work *Nyāyapraveśa* with the +7th-century Chinese translation by Hsüan-Tsang. The results of this comparison are extremely flattering for Hsüan-Tsang's efforts and will be reported in some detail in my section on Buddhist logic. Hsüan-Tsang found no unsurmountable problems with syntactic complexity in the Sanskrit. Having compared Hsüan-Tsang's text with the Sanskrit original in considerable detail, I must report that the Chinese, if anything, is more – not less – transparent than the Sanskrit original. Surveying the general field of translations from Sanskrit into Literary Chinese, one does not find that semantic complexity in Sanskrit has proved to be an obvious impediment to the Chinese translators' success.

[2] Among Chinese translations of Plato we note Wu Hsien-Shu, *Li-hsiang Kuo*, Commercial Press, Shanghai 1929, reprinted Shanghai 1957, Yen Chhün-I, *Po-la-thu: Thai-a-thai-te*, Shanghai, Commercial Press 1961, Chang Shih-Chu, *Po-la-thu tuei-hua chi liu chung*, Shanghai, Commercial Press 1933, Chu Kuang-Chhien, *Po-la-thu wen-i tuei-hua chi*. Peking, People's Publishing Company, 1963, reprint 1980, Hu Hung-Shu, *Po-la-thu tuei-hua lu*. Taipei, Cheng Wen Book Company, 1966.

[3] Nestorian Christianity as well as Manicheism have a long history in China, going back to Thang times.

[4] Furtado had become a Jesuit in +1608, arrived in Macao with Trigault in +1619, joined Li Chih-Tsao 李之藻 in Hangchou from +1625 and in +1630 took charge of the Christian funeral for Li Chih-Tsao. He later became a leading missionary administrator.

This hefty work was partly translated into Chinese with the help of Li Chih-Tsao 李之藻 (died +1630), a crucial figure in the history of Late Ming Christianity in China.[1] Li Chih-Tsao was a thoroughly experienced translator when he began to work together with Furtado on the Chinese translation of the *Commentarii*. His translation was really an adaptation. He tried to make the material he translated more palatable by free adaptation, which involved the breaking down of long periods into shorter sentences. There is evidence in his translations that he had prepared himself for the task by looking at some texts on Buddhist logic. However, he did not take over from Hsüan-Tsang and his Buddhist colleagues the indispensable iron discipline of terminology which is as important in Aristotle as it is in *Nyāya* logic. The result was, from our point of view, a rather poor translation of Aristotle into Literary Chinese. But it is not at all clear to what extent the failure was due to insufficiencies of the Chinese language itself.

Let us, then, consider the much more recent Chinese translations of Plato, particularly a Literary Chinese version of Plato's *Republic*, by Wu Hsien-Shu (Shanghai 1929). Comparing this translation with the Greek original and English translations, one soon learns to predict where in the Greek or English text Wu Hsien-Shu's Literary Chinese version would show itself unable to cope with the syntactic and logical complexities of the original. On the basis of such translations of Plato into Literary Chinese, one might be tempted to conclude that Literary Chinese was indeed unable to articulate a predictable range of philosophically significant and syntactically complex passages in Plato's writings.[2]

There is nothing inherently impossible about such a conclusion, and there certainly is nothing prejudiced or biased about contemplating the possibility that Literary Chinese has shown itself to be logically less powerful than Plato's mother tongue.

The conclusion is not inconceivable, biased, or prejudiced. It simply needs careful strictly philological verification. What at first sight appear to be limitations of Classical Chinese, often turn out upon closer investigation to be very largely limitations of early translators like Wu Hsien-Shu 吳獻書. This becomes clear when one compares another translation from Plato's works done by Chang Shih-Chu 張師竹 and corrected by the well-known philosopher Chang Tung-Sun 張東蓀. These translations contain five important and representative works: the *Euthyphro*, the *Apology*, the *Crito*, the *Protagoras*, and the *Meno*. They present an interesting new

---

[1] Li was a close friend of Matteo Ricci, whom he met in +1601, with whom he studied extensively, by whom he was baptised 'Leo' in +1610, and whose funeral he administered in the same year. Li was impressed to see that Ricci had been able to draw a map of the world (first published in Nanking in +1600), and was determined to learn what he could about the approach that enabled Ricci to get so far in cartography. Ricci had used Li as his translator. A treatise on geometry and another one on arithmetic were both published in Li's Chinese translation in +1614.

In +1613, Li presented a now-famous memorial to the emperor, in which he listed fourteen discoveries of Western science that had never been discussed in the writings of ancient Chinese worthies. In +1628, Li published a translation of Aristotle's *De caelo*, to which he added an introduction of his own.

[2] Comparison of Hsiang Ta's 向達 translation of Aristotle's *Nicomachean Ethics* (ed. *WYWK*, 1933) with the Greek original seemed to confirm the conclusion reached on the basis of translations of Plato. For a while I was convinced that this would have to be the final conclusion on the translatability of Plato into Classical Chinese.

picture. Chang generally handles syntactic complexity with vastly greater intellectual care and syntactic skill than Wu Hsien-Shu or Hsiang Ta 向達.[1]

Having compared parts of the Greek original with the Literary Chinese translation, I have to report that where one might have expected the Literary Chinese translation to give up in the face of syntactic complexity, Chang Shih-Chu and Chang Tung-Sun usually manage reasonably well to address this complexity with the means available in Literary Chinese. One is led to the conclusion that as far as the above-mentioned five representative works of Plato are concerned, the structural and grammatical limitations of Classical and Literary Chinese are not such that they necessarily more than marginally affect the articulation of what Plato had to say in these pieces.

The point at hand is so important that I shall present to the reader a complex Greek passage *in extenso* with its English and Literary Chinese translation, so that he can make up his own mind. I have chosen *Protagoras* 313a, because it presents some interesting syntactic challenges. Let us consider the famous translation by Benjamin Jowett (1817 to 1893). This classic English version was first published in 1871; we shall use the third edition, 1892 (vol. 1, p. 134) and compare it with both the Greek and the Literary Chinese.

Hippocrates is about to become a follower of Protagoras the sophist. Socrates warns against this and says:

JOWETT'S TRANSLATION: . . . If you were going to commit your body to some one, who might do good or harm to it, would you not carefully consider and ask the opinion of your friends and kindred and deliberate many days as to whether you should give him the care of the body?

GREEK: ἢ εἰ μὲν τὸ σῶμα ἐπιτρέπειν σε ἔδι τῳ διακινδυνεύοντα ἢ χρηστὸν αὐτὸ γενέσθαι ἢ πονηρόν, πολλὰ ἂν περιεσκέψω εἴτ'ἐπιτρεπτέον εἴτε οὔ, καὶ εἰς συμβουλὴν τούς τε φίλους ἂν παρεκάλεις καὶ τοὺς οἰκείους σκοπούμενος ἡμέρας συχνάς.

GREEK LITERAL: Or if on the one hand that you entrust the body to someone was-necessary, risking it either to become good or bad, much you-would-have-examined whether it should-be-entrusted or not, and into counselling both friends you would probably call in, and the housemates, deliberating [for] many days . . .

CHINESE LITERAL: If you entrust your own body to a person and risk having the danger of either transforming to the good or changing to the bad, then you must first diligently enquire (whether) ultimately (you) should entrust or not. Furthermore it is proper to counsel with all relations and friends having gone through many days, only then to decide.

JOWETT: But when the soul is in question, which you hold to be of far more value than the body, and upon the good or evil of which depends the well-being of your all, –

GREEK: ὁ δὲ περὶ πλείονος τοῦ σώματος ἡγῇ, τὴν ψυχήν, καὶ ἐν ᾧ πάντ'ἐστὶν τὰ σὰ ἢ εὖ ἢ κακῶς πράττειν, χρηστοῦ ἢ πονηροῦ αὐτοῦ γενομένου,

GREEK LITERAL: . . . what, on the other hand, you consider as more valuable than the body, the soul, and in which resides either the being well or not being well of all (that is) yours, (according to) the thing turning out good or bad . . .

[1] Incidentally, Wu Hsien-Shu 吳獻書 seems to be a pen name. I have been unable to ascertain to whom the name refers.

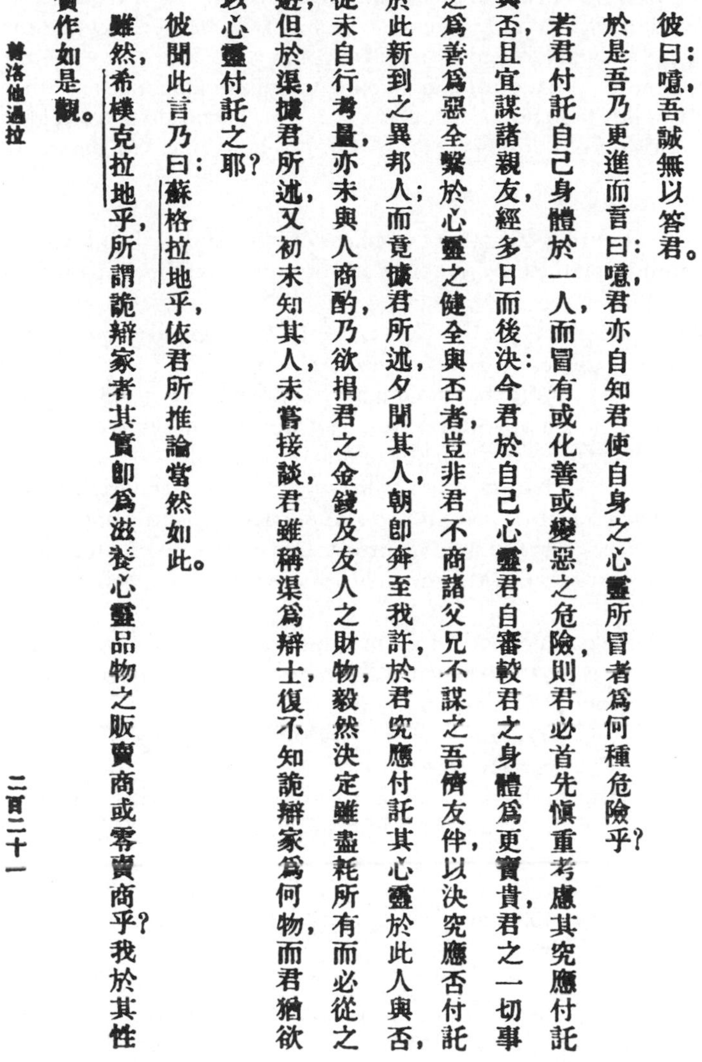

彼曰：噫吾誠無以答君。

於是吾乃更進而言曰：噫，君亦自知君使自身之心靈所冒者爲何種危險乎？

者君付託自己身體於一人，而冒有或化善或變惡之危險則君必首慎重考慮其究應付託

與否且宜謀諸親友，經多日而後決今君於自己心靈君自審較君之身體爲更寶貴君之一切事

之爲善爲惡全繫於心靈之健全與否者豈非君不商諸父兄不謀之吾儕友伴以決究應否付託

於此新到之異邦人而竟據君所述夕聞其人朝即奔至我許於君究應付託其心靈於此人與否

從未自行考量亦未與人商酌乃欲捐君之金錢及友人之財物毅然決定雖盡耗所有而必從之

遊但於渠據君所述又初未嘗接談君雖稱渠爲辯士復不知詭辯家爲何物，而君猶欲

以心靈付託之耶？

彼聞此言乃曰：蘇格拉地乎，依君所推論當然如此。

雖然，希模克拉地乎，所謂詭辯家者其實即爲滋養心靈品物之販賣商或零賣商乎？我於其性

質作如是觀。

普洛他過拉

二頁二十一

Fig. 1: *Protagoras* 313a–c in Chang Shih-Chu's translation

CHINESE LITERAL: Now as for your relation to your own heart-spirit, you yourself regard it as still more valuable than your body, and the goodness or badness of all your affairs depends on the well-being or otherwise of the heart-spirit,

JOWETT: . . . about this you have never consulted either with your father or with your brother or with any one of us who are your companions. But no sooner does this foreigner appear, than you instantly commit your soul to his keeping.

GREEK: περὶ δὲ τούτου οὔτε τῷ πατρὶ οὔτε τῷ ἀδελφῷ ἐπεκοινώσω οὔτε ἡμῶν τῶν ἑταίρων οὐδενί, εἴτε ἐπιτρεπτέον εἴτε καὶ οὐ τῷ ἀφικομένῳ τούτῳ ξένῳ τὴν σὴν ψυχήν,

GREEK LITERAL: . . . but about this neither with the father nor with the brother you have communicated, nor with nobody of us your friends, whether it should be entrusted or even not to this having-arrived foreigner the soul of yours.

CHINESE LITERAL: What do you do? You fail to discuss the matter with your father and your brother(s) and you do not take counsel with your friends in order to decide whether in fact you should or should not put your trust in this newly arrived person from a different country; . . .

JOWETT: In the evening, as you say, you hear of him, and in the morning you go to him, never deliberating or taking the opinion of any one as to whether you ought to entrust yourself to him or not; . . .

GREEK: ἀλλ᾽ ἑσπέρας ἀκούσας, ὡς φῄς, ὄρθριος ἥκων περὶ μὲν τούτον οὐδένα λόγον οὐδὲ συμβουλὴν ποιῇ, εἴτε χρὴ ἐπιτρέπειν σαυτὸν αὐτῷ εἴτε μή,

GREEK LITERAL: . . . but in the evening having-heard, as you-say, at dawn having-come about these things, on-the-one-hand, you pay no account nor counselling about this, whether it-is necessary to entrust yourself to him or not, . . .

CHINESE LITERAL: . . . but actually, according to what you report, hearing in the evening about the man, you come the next morning to me asking for my approval, and as regards the question whether you really should entrust your heart-spirit to this man or not, you do not at all proceed to consider your plans, neither do you discuss them with others,

JOWETT: — you have quite made up your mind that you will at all hazards be a pupil of Protagoras, and are prepared to expend all the property of yourself and of your friends in carrying out at any price this determination . . .

GREEK: ἕτοιμος δ᾽ εἶ ἀναλίσκειν τά τε σαυτοῦ καὶ τά τῶν φίλων χρήματα, ὡς ἤδη διεγνωκὼς ὅτι πάντως συνεστέον Πρωταγόρᾳ,

GREEK LITERAL: but you are willing, on the other hand, to spend your own and your friends' money, as if already having discerned that assuredly one should be with Protagoras, . . .

CHINESE LITERAL: on the contrary you want to spend your own money as well as your friends' wealth, having stubbornly determined that you will certainly go along with him, even if you use up everything,

JOWETT: . . . although, as you admit, you do not know him, and have never spoken with him: and you call him a Sophist, but are manifestly ignorant of what a Sophist is; and yet you are going to commit yourself to his keeping.

GREEK: ὅν οὔτε γιγνώσκεις, ὡς φῄς, οὔτε διείλεξαι οὐδεπώποτε, σοφιστὴν δ᾽ ὀνομάζεις, τὸν δὲ σοφιστὴν ὅτι ποτ᾽ ἔστιν φαίνῃ ἀγνοῶν, ᾧ μέλλεις σαυτὸν ἐπιτρέπειν;

GREEK LITERAL: . . . whom you neither know, as you say, nor even talked to ever, but a sophist you call him, but whatever the sophist is you appear to be ignorant of, to whom you are going to entrust yourself.

CHINESE LITERAL: . . . and yet when it comes to him, according to what you say, you do not even originally know him and have never talked to him; although you call him a disputer, you again do not know what kind of a thing a clever disputer is, and you are nonetheless determined to entrust your heart-spirit to him.

Jowett chooses to break down the Greek sentences by using full stops in his English version even when one would not punctuate the Greek in this way, and when in any case John Burnet's edition of 1903 does not punctuate this way. Friedrich Schleiermacher (+1768 to 1834), in his classic German version of Plato's complete works first published 1804–10 (second edition 1855–62), reproduces the Greek sentence without using a single full stop, following roughly a punctuation like Burnet's.[1]

There is always the possibility that other works by Plato (the *Theaetetus*, for example, or the *Republic*) do indeed pose significantly greater syntactic obstacles to the Literary Chinese translator from the point of view that we are concerned with. On past experience I am reluctant to predict that these obstacles will turn out insurmountable.

None the less, the clear impression, based on a wide range of translations, remains this: logical and grammatical complexity poses markedly greater syntactic and grammatical difficulties for a translator into Literary Chinese than for one into Ciceronian Latin[2] or into English.

Part of the reason for this is very simply that Literary Chinese writers as well as readers are not used to long and complex sentences. A given translator's mistakes may thus sometimes be arbitrary. On the other hand, they may also be symptomatic of the difficulties inherent in the specific task of translation in which he fails. There are structural reasons that make it likely that he should at least face severe difficulties.

In Modern Standard Chinese we have a quite unique work of detailed Plato scholarship, Chhên Khang's 陳康 *Pa-man-ni-tê-ssu phien* 巴曼尼得斯篇 (Chungking 1944).[3] The translation of Plato into Modern Chinese does raise its own fascinating problems regarding the profound Westernisation of Modern Standard Chinese, but our general conclusion is clear enough. Any disagreements I might have and do have with Chhên Khang's translation are very largely the sort of disagreements I might have and do have with French, English, or German translators. They are of no immediate consequence for our present enquiry, which is concerned with Literary Chinese.

[1] Cf. *Platon, Sämtliche Werke. In der Übersetzung von Friedrich Schleiermacher*, ed. Rowohlts Klassiker, Hamburg 1957, vol. 1, p. 55: Wie nun? Weißt du also, welcher Gefahr du gehst deine Seele preiszugeben? Oder würdest du, wenn du deinen Körper einem anvertrauen solltest auf die Gefahr, ob er gestärkt werden würde oder verdorben, dann wohl erst vielfach überlegen, ob du ihn ihm anvertrauen wolltest oder nicht, und zur Beratung deine Freunde herbeirufen und deine Verwandten, mehrere Tage lang der Sache nachdenkend: was du aber weit höher als deinen Körper achtest und dem gemäß alle deine Angelegenheiten gut oder schlecht gehen müssen, je nach dem es gestärkt oder verdorben, die Seele, hierüber hast du dich weder deinem Vater noch deinem Bruder mitgeteilt, noch irgendeinem von uns, deinen Freunden, ob du diesem eben angekommenen Fremdling deine Seele anvertrauen sollst oder nicht; sondern nachdem du gestern abend von ihm gehört, wie du sagst, kommst du heute mit dem frühesten Morgen, nicht etwa um noch darüber irgend Gespräch und Beratung zu pflegen, ob du dich selbst ihm hingeben sollst oder nicht, sondern ganz bereit schon, dein und deiner Freunde Vermögen daran zu wenden, also als wäre dieses schon fest beschlossen, daß du auf alle Weise dich mit dem Protagoras einlassen mußt, welchen du doch weder kennst, wie du sagst, noch auch jemals gesprochen hast; sondern du nennst ihn einen Sophisten, aber was ein solcher Sophist eigentlich ist, dem du dich selbst übergeben willst, darin zeigst du dich ganz unwissend.

[2] See Ax (1965), pp. 155b–187b, for a convenient bilingual Latin/Greek version of the *Timaevs*.

[3] Chu Kuang-Chhien's translation of the *Theaetetus* will provide particularly important additional material for the study of Modern Chinese translations of Plato, since Chu must undoubtedly rank as one of the China's leading philosophers in the 20th century. Unfortunately, I have not yet been able to procure this book.

Comparing standard translations of the Bible into Literary and Modern Chinese, one certainly gets the impression that the task is significantly easier in Modern Chinese.[1]

'Confucius Latinus': Séraphin Couvreur

Between +1654 and +1678, R. P. Ludovico Buglio, SJ (+1606 to +1682) began to publish four volumes of careful Literary Chinese translations from Thomas Aquinas, *Summa theologiae*,[2] which provide convincing evidence of the ability of the Literary Chinese language to articulate advanced systematic Western thinking. However, it is evidently a biased test of the articulatory power of a foreign language to ask how well it can be used to translate Plato, Aristotle, and Thomas Aquinas.

We need to ask the other way round: how well equipped is, say, Latin or Greek to translate the philosophical classics of China? Can one use Latin or Greek to make literal paraphrases of the syntactic complexities in the Chinese classics? This is an empirical question which needs detailed and open-minded study.

We have a wide range of translations of Chinese classics into ecclesiastic later Latin.[3] Stanislas Julien has provided the first attempt at a methodical and systematically literal paraphrase of a complete Chinese text in any European language. His *Meng Tseu, vel Mencium inter Sinenses philosophos ingenio, doctrina, nominisque claritate Confucio proximum, edidit, Latina interpretatione, ad interpretationem tartaricam utramque recensita, instruxit, et perpetuo commentario, e Sinicis deprompto, illustravit Stanislaus Julien, Societatis Asiaticae et comites de Lasteyrie impensis* (Lutetiæ Parisorum 1824) is perhaps the most richly annotated translation of the *Mêng Tzu* to date. Julien's Latin shows how an ancient Chinese text can be expounded word by word in comprehensible Latin, with only very few changes in the Chinese word order. His many mistakes are not due to any features of the Latin language. They are due to failures of Stanislas Julien to understand the text properly.

Séraphin Couvreur's literal Latin versions of the Four Books (*Lun Yü, Mêng Tzu, Ta Hsüeh, Chung Yung*), of *Shih Ching, Shu Ching*, and of *Li Chi* are less ambitiously systematic than those of the grammarian Stanislas Julien, but they too provide ample testing ground for the question whether Latin can in principle provide a comprehensible, very literal word-by-word paraphrase, and whether such literal paraphrases give a defensible literal interpretation of the Classical Chinese original.

[1] In the case of Literary Chinese Bible translations, as in the case of Literary Chinese Plato translations, it is important not to attribute weaknesses of the Chinese translations to weaknesses of the Literary Chinese language as such. Knotty passages such as the opening of Paul's Letter to the Colossians, for example, are opaque in early translations, but Wu Ching-Hsiung's 吳經熊 *Hsin Ching Chhüan Chi* 新經全集 published in 1948, has shown how the medium of Literary Chinese can be used to tackle Paul's convoluted prose both gracefully and efficiently. What is truly hard to reproduce in Literary Chinese is Paul's anacoluthically abrupt and sometimes syntactically incoherent intense style, and his parenthetic remarks. For a bibliographic survey of Bible translations into Chinese see Spillett (1975). For detailed analyses of five translations see Strandenaes (1987).

[2] Cf. the beautiful new edition (Shanghai 1930).

[3] Ecclesiastic later Latin differs substantially, of course, from Classical Latin. But not, I think, in ways that are very important for our present purposes.

One will not always agree with Couvreur's (and even less with Julien's) interpretations, but Couvreur has undoubtedly shown Latin to be an eminently viable medium in which to provide word-by-word paraphrases of the Chinese classics.[1]

In the area of syntactic complexity, Latin seems to cope quite easily and quite painlessly with just the kinds of problems that cause so much trouble for Classical or Literary Chinese literal paraphrases of Latin.

Classical or Literary Chinese could certainly not be used to provide literal paraphrases for the Latin Classics. R. P. Ludovico Buglio, SJ, did not and could not attempt a literal word-by-word paraphrase of Thomas Aquinas' text, although he provides remarkably careful Literary Chinese translations from the *Summa theologiae*. Indeed, he wisely chooses to provide only selective translations. Li Chih-Tsao, as we have noted above, encountered severe problems in translating Aristotle's categories.

Concluding remarks

It will be healthy to start out with some remarks A. C. Graham made a long time ago:

The present study does not encourage one to take it for granted that Chinese is either better or worse than English as an instrument of thought; each language has its own sources of confusion, some of which are exposed by translation into the other.[2]

All comparisons such as those I have summarised above raise serious doubts and queries: Are the reflections I have presented so far ultimately no more than a rather complex expression of basically simple cultural prejudice on the part of a biased Western observer? Such prejudice need not be malicious. It can simply be a reflection of the fact that the writer of these lines grew up with Latin and Greek, and did not grow up with Classical Chinese or Literary Chinese.

Such doubts and suspicions concerning the psychology of cross-cultural comparison deserve to be taken seriously. As we have seen in our survey of the history of sinology in the West, there is a long tradition of prejudice.

None the less I confidently conclude, firstly, that Classical Chinese writers show a very considerably lesser tendency to use rhetorical and semantic complexity (in the technical senses we have indicated above) than Classical Greek or Latin writers.

To the extent that the habit of reliably decoding highly complex articulated meanings constitutes a mental exercise that may be healthy for the conduct of

---

[1] That Latin (particularly ecclesiastic Latin with its Christian terminological overtones) faces tremendous terminological difficulties when used to translate Chinese texts, goes without saying and need not detain us here.

[2] Graham (1986a), p. 359. Graham continues, a little further on: 'Again, we know too little about Chinese grammar. We say that *ju* 如 has two senses, "if" and "like". In the former sense, it is obviously used quite differently from, for example *kou* 苟, also translated "if", and in the latter quite differently from *yu* 猶, the dictionary meaning of which is also "like". Has anyone ever clearly explained what these differences are? Until we can distinguish between the ordinary words with which classical Chinese deals with such basic ways of thinking as hypothesis and comparison, how can we tell whether it is a vague language or not? None of us yet knows classical Chinese. Even if the accusation of vagueness eventually proves to be true, it is a truth which it is unhealthy to keep too much in mind.' These are important sobering reflections which we must keep in mind in the context of our present investigation.

science insofar as science involves very complex interrelations between statements that need to be made explicit, aspiring scientists who were learning Classical Chinese were at a certain disadvantage when compared with speakers of Classical Greek or Latin. But, on the other hand, I suppose one might add that writers of Classical Chinese might have been led to acquire greater skills of simplification.

It seems profoundly significant that, when in logic classes one learns to translate sentences from natural languages into logical notation, one performs tasks that much of the time remind one strongly of the kinds of tasks one would perform if one had to translate that same sentence into Classical Chinese: one performs a process of logical 'factorisation' and reduction to the simplest possible form.

I conclude, secondly, that Classical Greek and Latin show a much greater systematic ability unambiguously and transparently to articulate logical and grammatical complexity than the comparable Classical or Literary Chinese.

To the extent that the rhetorically or semantically very complex sentences which cannot be reproduced in Classical or Literary Chinese should prove to be a significant aid in explicating, questioning and developing an overall view of a complicated constellation of scientific theories, the traditional Chinese scientists would also appear to be at a clear linguistic disadvantage.

Of special cultural interest is the grammatical possibility of a parenthesis which is present in Greek and Latin, since this allows for spontaneity of expression of current thoughts. This possibility is strongly discouraged by the structure of the Literary Chinese language. A parenthesis has to be expressed as an afterthought – when it may be too late. A truly parenthetic thinker is unthinkable in Classical Chinese. (Except that one feels Chuang Tzu comes pretty close to being a parenthetic thinker without using intra-sentential parentheses!)

Even on the technical grammatical point one must not generalise too freely. Here is a neat and unquestionable parenthesis from the great independent thinker Li Chih which was pointed out to me by Jacques Gernet:

Although he has never bowed to me as a master – he knows that I am unwilling to be the master of anybody – still he has already several times sent people over distances of thirty *li* to ask about the Law. Even if I wanted to not reply, could I?[1]

In spite of a striking example like this one, it remains significant that the parenthetic construction is very important in Indo-European languages and exceedingly rare in Chinese.[2]

Our results, even if accepted, raise a host of further questions: how *intellectually important* was it for the development of Socrates' thought that the Greek language, which he inherited from his forefathers, encouraged him to speak in sentences as complex as those in which he spoke, and how important is it that his language would have allowed him to speak much more complexly still without difficulty?

---

[1] Li Chih 李贄, *Fên Shu Hsü Fên Shu* 焚書續焚書, ed. Chung-hua-shu-chü, p. 183.
[2] Cf. Schwyzer (1939) and the brilliant thesis by Michael von Albrecht (1964). The study of the parenthesis in Modern Chinese is an important subject for future research.

How *ultimately essential* in the intellectual life of a person or a culture is the ability to speak in very complex sentences such as would be hard or impossible to translate into Literary Chinese?[1] To what extent is the extra logical and syntactic power of Latin and Greek culturally and intellectually operative and functional?

These are wide open questions which lie beyond the scope of our present enquiry.[2] They must be studied against the background of a wide range of philological facts regarding Latin, Greek and Chinese. They are not matters for abstract philosophical or linguistic speculation or dogmatism.

There is ample scope for a detailed and wide-ranging cultural study entitled something like *Platon Sericus*, which would take up questions concerning the general history of ideas: how much of the essence of Plato's thought can be rendered in sayable Classical Chinese or in readable Literary Chinese, and for that matter in Modern Standard Chinese? Where are the main losses in plausibility as Plato's views are transposed into the Chinese linguistic tonality or medium? Could it be that what does not travel across to such a very different language is in some philosophically significant way parochially Greek or Indo-European? Should not philosophy in principle aspire to rise above such parochialism? Should we not distrust what is parochial in Plato?

Since A. N. Whitehead could plausibly claim[3] that all of Western philosophy was mere footnotes to Plato, these might seem to be questions of some consequence.

(d) LOGICAL CONCEPTS

(1) PUNCTUATION AND THE CONCEPT OF A SENTENCE

In Section (c,2) we have seen that the ancient Chinese had sentence connectives. But did they also have the notion of a sentence as opposed to that of a text, a word or a character? And did they have the notion of a proposition as distinct from the sentence which can be used to express it? Let us look at the simpler notion of the sentence first.

The rambling sentence (to use Y. R. Chao's phrase), in which sentence boundaries between clause, phrase and sentence are vague, is common in ancient Chinese texts (as indeed it is in Classical Greek!) and it survives as a natural part of modern spoken Chinese. Only the rigid conventions of modern typography enforce strict sentence limits (full stops) where the language itself imposes no such hard-and-fast division.

There are many independent questions about the sentence which deserve a separate study.

[1] Similarly: how intellectually important was it that Classical Chinese had a concept of a class, but perhaps no concept of a property or characteristic?

[2] It is perhaps useful to say even at this point, though, that it does not seem from that present enquiry that the syntactic and grammatical limitations were the most important single factor shaping Chinese scientific thought through its history. Other social and intellectual factors were apparently much more important.

[3] Whitehead (1929), p. 53.

1. Could the ancient Chinese unambiguously mark off sentence boundaries when they wished to?
2. Did the ancient Chinese mark off sentence boundaries when they needed to?
3. Did the ancient Chinese have an unambiguous concept of the sentence as opposed to that of the word, passage or text?

The grammatical characterisation of the sentence

There is no doubt that the ancient Chinese could fix emphatic sentence boundaries when they needed or wanted to. A wide range of Classical Chinese particles (on one if not all of their distinct interpretations) clearly mark out sentence boundaries. The following are rules that allow us to recognise explicit sentence boundaries in ancient Chinese texts.

> Whenever one sees the extremely common particle *i* 矣, one may safely assume that a narrative or descriptive verbal sentence has come to an end. Already the dictionary *Shuo Wên Chieh Tzu* 説文解字 (postface +100) defines *i* as a sentence-final particle. Whatever exceptions there are to this rule are motivated by evident emphatic transpositions.
>
> Whenever one sees the common particle *hu* 乎 used as a question marker, one may safely assume that the interrogative sentence to which it belongs ends in *hu*.
>
> Whenever one sees the question markers *yü* 歟 or *yeh* 邪/耶, one may again assume that an interrogative sentence is coming to an end.
>
> Whenever one sees the (hard-to-define) final particle *i* 已 (literally: 'to finish'), one may again assume that a sentence is coming to an end.
>
> Whenever one sees the common combination *êrh i (i)* 而已矣 'and that is all' or one of its variants *êrh* 耳 and *êrh* 爾, one may safely assume that a generalising verbal sentence is coming to an end.
>
> Moreover, the ubiquitous particle *yeh* 也 provides a natural place to look for a sentence boundary even though that particle also has what one may be inclined to interpret as sentence-internal functions.[1]

These rules should not be taken to obscure the fact that Classical Chinese does not have a grammatical particle the exclusive function of which is to mark sentence boundaries. Sentence-final particles in Chinese always have a certain modal force. They are added at the end of sentences to indicate a certain modal nuance. Only incidentally do these particles mark sentence boundaries.

Moreover, the fact remains that many Chinese sentences do not make use of sentence-final particles, and there was in pre-Han times no known form of punctuation which unambiguously indicated the end of a sentence. Even more than in the case of Classical Greek, sentence boundaries could be left vague.

---

[1] For a remarkably detailed introduction to the techniques one may use to punctuate Classical Chinese sentences see Wu Tsung (*1988*), pp. 527–86.

Now, for example, to conclude that the Chinese had a poorly developed sense of historical time or contextual definiteness since they often do not make these categories explicit in their sentences would be plainly absurd. For example, few peoples are more time-conscious than the Chinese, and the elaborate tense system of the Sanskrit language has not prevented the Indians from paying considerably less attention to the dimensions of time and chronology than the Chinese.[1]

Similarly I believe we must be very careful indeed not to imply that the Chinese lacked a sense for the sentence simply because they often do not find it necessary or expedient to indicate explicit sentence boundaries. The fact that they *can* unambiguously indicate sentence boundaries proves conclusively that they do have an operative concept of a sentence. Without such an operative concept they would not be able to learn where to use their own sentence-final grammatical particles.

Having said this, we still can and must inquire whether the Chinese have in fact marked sentence boundaries when this was necessary. The crucial difficulty here is the inherent ambiguity of the word 'necessary': necessary to whom? We can soon agree that the Chinese have often failed to mark sentence boundaries when it would have been convenient for us to see them marked. The lack of explicit sentence boundaries made Classical Chinese more arcane than it might have been, but it clearly did not make the written language incomprehensible.

Punctuation

Classical Chinese and Classical Greek texts, like modern Chinese texts, were traditionally written in a kind of *scriptura continua* in which word-boundaries are not marked by empty spaces. The consistent printed punctuation of all texts in China is a phenomenon which belongs to the 20th century. The effect of *scriptura continua* on the reader is easily reproduced in English by writing out the beginning of Benjamin Jowett's translation of Plato's *Republic*:

IWENTDOWNYESTERDAYTOTHEPEIRAEUSWITHGLAUCONTHESONOFARISTONTHATIMI
GHTOFFERUPMYPRAYERSTOTHEGODDESSANDALSOBECAUSEIWANTEDTOSEEINWHATM
ANNERTHEYWOULDCELELBRATETHEFESTIVALWHICHWASANEWTHINGIWASDELIGHTE
DWITHTHEPROCESSIONOFTHEINHABITANTSBUTTHATOFTHETHRACIANSWASEQUALLY
IFNOTMOREBEAUTIFUL

When one uses *scriptura continua* in Chinese, the inconvenience is much slighter because the character boundaries (unlike the letters) conveniently mark out the boundaries of the morphemes – and to a large extent of the words. Thus as long as *scriptura continua* was used, the ancient Chinese were at a distinct advantage compared with the Europeans.

---

[1] The oracle bone inscriptions, which are more than 3,000 years old, already show a curious tendency to date events. The *Chhun Chhiu* annals dating to perhaps the −5th century are full of exact chronological detail. The *Tzŭ Chih Thung Chien* (+1084) provides a meticulously dated chronicle of 1,326 years of history.

When, on the other hand, the Europeans started early to mark word boundaries with gaps,[1] and clause boundaries as well as sentence boundaries with various marks, they developed a system which was clearly more convenient than the Chinese. Traditionally the Chinese have made up for part of this disadvantage by an extensive use of parallelism as an aid to the parsing of sentences.

Compare the case of Sanskrit. Word boundaries are not marked in the Devanāgarī script. But the earliest sizeable body of Sanskrit texts we have are the Aśoka inscriptions on rocks and columns in India. These inscriptions, as has long been noticed, have gaps in the text at certain predictable places, and these gaps can be viewed as a form of punctuation. The punctuation gaps occurred predictably in the following places: 1. after every sentence, 2. between two words loosely in apposition, 3. after every item in an enumeration or listing. There were also further gaps marking more idiosyncratic and scribally optional gaps in addition, and in this way the gaps served as an organising parsing system for the language.[2] From its beginnings, written Sanskrit has remained a fairly consistently punctuated language throughout its history, although it must be emphasised that the punctuation by strokes and double strokes of Sanskrit scribes is far from entirely regular and predictable.[3]

We have −5th-century Greek inscriptions in which phrases were sometimes separated by a vertical row of two or three points. In some of the oldest papyrus versions of literary texts we find a horizontal line called the *paragraphos* or *paragraphē* ('that which is written alongside') which was placed under the beginning of a line in which a new topic was introduced. Aristotle comments on this in *Rhetoric* 1420a19, emphasising that the end of a sentence should be marked not only *dia tēn paragraphēn* (by the paragraph-marker) but by rhythm.

In Greece, the discovery of the principles of punctuation is ascribed to the great Alexandrian lexicographer and philologist Aristophanes of Byzantium (*c*. −257 to *c*. −180). Aristophanes of Byzantium is said to have marked the end of a short section (called a *komma*) by a point after the middle of its last letter, that of a longer section (known as a *kōlon*)[4] by a point on the line, following the bottom of the letter, and that of the longest section (known as *periodos*) by a point above the line, following the letter. The *komma* and the *kolon* would seem to correspond to the traditional Chinese *tou* 讀 and *chü* 句, while the *periodos* corresponds naturally to the *chang* 章 in Chinese tradition.

The question mark first emerged in the +8th or +9th centuries in the West, when it was written as our semicolon (;). In the Chinese tradition, the question mark was

---

[1] Note that since word boundaries are notoriously unclear even in modern Chinese, the system of marking such boundaries by gaps would be quite difficult to introduce even today. Cf. Isaenko (1957), which includes an excellent essay on word boundaries in modern Chinese (pp. 241–318). The Chinese literature on the subject is vast.

[2] There actually is a monograph on the punctuation gaps in the Aśoka inscriptions: Klaus Ludwig Janert (1972). I owe this reference to Georg von Simson.

[3] Probably under Sanskrit influence, written Tibetan has from its very beginnings as we know them, had a consistent tradition of punctuation in addition to the wealth of Tibetan grammatical markers indicating clause and sentence ends.

[4] Cicero mentions these Greek terms in *Orator* 62, 211.

never introduced independently from the Western tradition. This is readily explained by the fact that the question-marking final particles combined with the explicit question words tend to suffice for making questions explicit.

Latin texts were often written in *scriptura continua*, but occasionally word boundaries were marked with a dot from the −1st to the +2nd centuries. The first consistently punctuated book we have is the Vulgate Bible by St Jerome (died +420). Jerome devised his own punctuation *per cola et commata* which was designed to facilitate the reading aloud of the text, especially in ceremonial contexts. Punctuation, for Jerome, served a primarily elocutionary rather than a grammatically clarificatory purpose.

In the +12th century intonational punctuation, based on the musical notation of neumes, became common, the *punctus elevatus*, the *punctus interrogativus* and the *punctus circumflexus*, the latter designating a rising inflection at the end of subordinate clauses. Even in Elizabethan times punctuation was almost entirely elocutionary rather than syntactic in nature. In +1556, Aldo Manuzio of Italy became the first to advocate the view that clarification of syntax is the main object of punctuation, in his book *Orthographiae ratio*. In England, Ben Jonson was the first to voice a similar view in his *English Grammar* (+1640).[1]

China

The case of China is markedly different.[2] In the oldest inscriptions we have from the oracle bones, punctuation, even in the form of gaps indicating pauses, is so erratic that no systematic pattern emerges.

From the Warring States period onwards we find marking of paragraphs,[3] but the various symbols we find on many bamboo strips do not coherently relate to the notion of a sentence in which we are interested. There is certainly no obligatory marking of sentence ends by gaps or otherwise. On the face of it, all the punctuation devices in the *Ma-wang-tui* manuscripts of the *Lao Tzu* taken together (−2nd century), for example, do not add up to any recognisable and coherent pattern. They constitute an epigraphic puzzle to be solved by future palaeographers. Certainly, there is no generally regular marking of clauses. Even less is there a regular marking of whole sentences. What we have of punctuation, as opposed to grammar, would not give us any set of reliable clues at all to sentence boundaries and clause boundaries in Classical Chinese.[4]

---

[1] My account here is based on the masterly survey of Brown (1984). For details on the Greek case see Turner (1971), pp. 9–16. For intonational aspects of punctuation see Peter Clemoes (1952). For punctuation in Elizabethan times see Partridge (1964), chs. 14, 15, and appendix viii. For a thorough study of the German case in comparative perspective see Alexander Bieling (1880).

[2] Lü Ssu-Mien (*1930*), written only ten years after the beginning of the current use of punctuation in China, is the classic contribution to the history of punctuation in China.

[3] A famous datable example is the *I Ching* text preserved on silk, and dating apparently from −168, where each comment first on the main meaning of a hexagram and then on each of the individual lines in the hexagram is systematically marked off by dots.

[4] Compare *Wu-wei Han-chien*, pp. 70f.

The *Han Fei Tzu* (−3rd century) preserves a story which became the ancestor of innumerable Chinese popular jokes about incorrect punctuation:

Duke Ai asked Confucius: 'I have heard it said that Khuei had one leg (*Khuei i tsu* 夔一足)[1], is that true?'
'Khuei is a man. How should he have one leg? That man has no other distinctions, he is only good at music. Yao said: "Khuei is good at one thing and that is enough."
Therefore the gentleman will say: "Khuei has one sufficient quality", not that he has one leg.'[2]

Here is an intriguing example from the Book of Ritual (*Li Chi*), where a certain Shih I approaches the Duke of Chou for approval of a certain course of action:

The Duke of Chou said (or meant to say): 'How? That is not acceptable! (*chhi pu kho* 豈不可)!'
Shih I went ahead with his plan.

Here the commentator Chêng Hsüan 鄭玄 (+127 to +200) explains that Shih I failed to understand that one must break the sentence after the *chhi* 豈 'how?' and took the Duke of Chou to give his approval. The commentator Chêng Hsüan writes this extraordinary comment:

The Duke of Chou says: 'How?' End of sentence (*chüeh chü* 絕句).
This means 'How should this be? End of sentence (*chüeh chü*).'
In relation to ritual this is unacceptable. End of sentence (*chüeh chü*).[3]

If punctuation had been a known standard procedure at all during his time, the great Chêng Hsüan would surely never have written in this cumbersome way.

The art of parsing texts correctly was recognised as a central educational achievement. In the same Book of Ritual (*Li Chi*) we read:

After the first year (of instruction) one examines whether (the children) parse the classics and distinguish meanings (*li ching pien chih* 離經辨指).[4]

Chêng Hsüan explains *li ching* 離經 'parsing the classics' as *tuan chü chüeh* 段句絕 'cutting off where the phrase ends'.[5] Khung Ying-Ta 孔穎達 explains *li ching* as 'they caused the paragraphs and sentences to be separated (*shih chang chü tuan chüeh yeh* 使章句段絕也)'.

Wang Chhung 王充 (+27 to +100) commends voracious readers for having acquired the skill of parsing (or punctuating) texts (*shen ting wên tou* 審定文讀).[6] Probably Wang Chhung just thought of a correct placing of pauses when reading out a text, possibly when declaiming it aloud. As we have seen, such declamatory practice was important in the history of punctuation in the West.[7]

---

[1] Confucius, in his answer, will punctuate: *Khuei i. Tsu* 夔一. 足, 'Khuei is specialised in one thing. That is enough.'
[2] *Han Fei Tzu* 33. 15.22; cf. Liao (1939), vol. 2, pp. 71f.
[3] *Li Chi*, ed. *Shih San Ching chu shu*, p. 1401, middle column.
[4] *Li Chi*, ed. Couvreur, vol. 2, p. 30.   [5] *Li Chi chi-chieh*, ed. Sun Hsi-Tan, *WYWK*, vol. 10, p. 4.
[6] *Lun Heng*, ed. Chung-hua shu-chü, p. 777; cf. Forke vol. 2, p. 295.
[7] Kao Yu (*floruit* around +205 to +212) tells us that he learnt to 'put commata and stops (*chü tou* 句讀) and to chant' when he was young. (Preface to *Huai Nan Tzû*, ed. Liu Wên-Tien, p. 2a.)

But the dictionary *Shuo Wên Chieh Tzu* 說文解字 (+100) already defines a written punctuation mark in the form of a dot as 'by this we know that there is a break (*yu so chüeh chih* 有所絕止)'.[1] The same book also defines a hook which was used as a punctuating device. The Records of the Grand Historian Ssuma Chhien 司馬遷 (−145 to *c.* −89) preserve a passage where the hook is even used verbally meaning 'to put in a hook'.[2]

The standard term for the parsing of texts into clauses during the Han dynasty was *chang chü* 章句.[3] Ho Hsiu 何休 (+129 to +182) speaks of the parsing or punctuating of sentences by *chü tou* 句讀 'stops and commas' in the preface to his edition of the *Kung Yang* 公羊 Commentary,[4] and he reports that the cases of people making fools of themselves by misplacing punctuation marks are innumerable. The *chü* 句, we are inclined to say, marks off a sentence, and the *tou* 讀 a clause, although we do not know how precise the distinction was in Ho Hsiu's mind any more than we know how precise the distinction was among traditional punctuators of European books.[5] Lü Ssu-Mien (*1930*), p. 13, makes a good case for maintaining that the terms *tou* and *chü* were synonymous rhyming terms during Han times, both being used simply to mark any break, which could also be called *chüeh* 絕. By Thang times the notion of the sentence is clear enough, but perhaps significantly it is first articulated by a Buddhist Thien-thai monk by the name of Chan Jan who explains:

In any sutra, where an utterance breaks off, we speak of a *chü* 句. Where the utterance is not yet broken off and we put a dot for convenient reading, this is called a *tou* 讀.[6]

Punctuation, then, was certainly known from Han times onwards. But the Tun Huang manuscripts dating from the +6th to +10th centuries still provide few consistent clues on punctuation. There often is evidence that such punctuation as we find in these manuscripts is later than the original writing itself.

Through the ages we do find punctuated books, especially in the imperial libraries. Yüeh Kho 岳珂 (+1183 to +1234) reports that imperial books, as opposed to other books, in his time had punctuation, and he notes that this is convenient for the reader.[7] Ku Chieh-Kang 顧頡剛 has examined a Song copy of one of Yüeh Kho's works and finds that in it there is a distinction between full stops, marked in place of a character, and what corresponds to our commas, marked by the side of the character after which they are to be construed. The mammoth work *Yung Lo Ta Tien* 永樂大典 written out in 11,095 volumes by +1408, for example, was profusely and consistently marked – at least in the case of those facsimile volumes I have seen – with red punctuation marks of elaborate kinds. From Ming times onwards, proper names were sometimes, marked off by a straight line along their left side in certain punctuated editions. Place names were, at this time, marked with a double straight line, also along the left side, in such editions. In early Chhing times we find a straight

[1] Cf. *Chung-wen Ta Tzû-tien*, vol. 1, p. 442.
[2] *Shih Chi* 126, ed. Takigawa, p. 15. This passage is not by Ssuma Chhien himself.
[3] See Lü Ssu-Mien (*1930*), pp. 1ff.   [4] Ed. *Shih-san-ching chu-shu*, p. 2191.
[5] Cf. Bieling (*1880*), already referred to above.
[6] See his *Fa Hua Wên Chü Chi* 法華文句集 as quoted in Wu Tsung (*1988*), p. 527.
[7] See Ku Chieh-Kang (*1983*), p. 66.

line across the breadth of a character used to indicate a unit smaller than a paragraph but often containing several sentences. Unfortunately, from mid-Chhing times onwards the increasing use of punctuation came to be felt to be a vulgar innovation, and the movement was rather abruptly discontinued. None the less, even the great philologist Wang Yin-Chih 王引之 (+1766 to +1834) made a point of privately printing his famous commentary on the *Kuang Ya* 廣雅 dictionary as well as his *Ching Chuan Shih Tzhu* 經傳釋詞 in punctuated editions.[1]

But, to take one example of infinitely many, the scholar Tuan Yü-Tshai 段玉裁 (+1735 to +1815) to whom we owe a fine commentary on the dictionary *Shuo Wên Chieh Tzu* 說文解字,[2] when he found occasion to insist on a certain punctuation of his outstanding work, resorted to writing the character *tou* 逗 at the relevant place. He, and many others before and after him, could not imagine imposing on his printers such a vulgar and unseemly thing as printing a punctuation mark would have been.[3] It must be emphasised none the less that punctuation of Chinese texts is useful enough even to the most learned of Chinese scholars.[4]

The art of traditional punctuation of Chinese texts (*chü tou* 句讀) has received detailed attention in a monograph by Wu I, published in +1789, and it remains a respectable discipline.[5] For example, Yang Shu-Ta's 楊樹達 classic *Ku Shu Chü Tou Shih Li* 古書句讀釋例 (second edn. 1954) contains careful analyses of 168 classified cases of controversies over punctuation of classical texts. One of these must suffice for our purposes:

Chi Khang Tzu asked: 'How can one inculcate in the common people the virtue of reverence, of doing their best and of enthusiasm?'
The Master said: 'Rule over them with dignity and they will be reverent; treat them with kindness and they will do their best. Raise the good and instruct those who are backward and they will be imbued with enthusiasm.'[6]

Here it turns out that the traditional punctuation wants us to read: 'Raise the good and instruct them. Then those without ability will feel encouraged', as many early quotations show. Modern interpreters have preferred the punctuation we have taken over above from D. C. Lau. Examples of this sort are not hard to find, but one must keep in mind that they are not hard to come by in Greek texts either.

The insistence on coherent and rigid printed punctuation of texts in China emerged along with the May-Fourth-Movement in the early 20th century, the campaign for the popularisation and democratisation of knowledge, and with the movement for writing in the Chinese vernacular. The only form of punctuation that was ever rigourously and systematically applied to a wide range of books and

[1] For information on these two dictionaries see our Section (*b*,5) on the history of dictionaries in China.
[2] See our Section (*b*,5) on the art of lexicography in traditional China.   [3] Cf. Tuan Yü-Tshai (*1981*), p. 717.
[4] Beginnings of paragraphs were often marked by circles in Sung editions. However, one might add that even the beginnings of new paragraphs often go unmarked in Classical Chinese books. As I write these lines, I look at Lu Wên-Chhao's 盧文弨 (+1717 to +1796) edition of Chia I's *Hsin Shu* 賈誼, 新書, ch. 8, p. 5b, where this scholar has to write the characters *i tuan* 'one paragraph' into the text in order to tell the reader what is happening.
[5] For a convenient account of this work see Chang Ti-Hua *et al.* (*1988*), p. 236.
[6] *Lun Yü* 2.20; cf. Lau (1983a), p. 15.

texts was a basically Western kind of punctuation,[1] and the movement for systematic punctuation, curiously enough, went along with the abandonment of Classical Chinese as the general medium of publication.

Even after the success of the modern vernacular as a medium of written communication there were those, like the great scholar Hsiung Shih-Li 熊十力 (1885–1965), who not only continued to write in Classical Chinese but also continued to write unpunctuated books.

It appears that the first emergence of punctuation was roughly contemporary in China and Greece. But while punctuation in Greece and in Europe was a comparatively trivial matter, the morphology of the language supplying ample cues for fairly reliable punctuation, the art of punctuating a Chinese text has remained a task that has had to be entrusted to men of considerable learning. A recent example of this being the series of the twenty-four dynastic histories punctuated by one of the greatest historians of his time, Ku Chieh-Kang 顧頡剛, during the 1960s.

By and large the vast majority of Literary Chinese texts were printed in unpunctuated editions until the 20th century. Punctuation was regarded as vulgar. The presumption was that those who deserved to understand these books would naturally be able to punctuate them properly. Literary Chinese was never a natural medium for popularisation, and the elitist nature of Chinese written culture has, in this instance, led to a refusal to make these texts more accessible through coherent punctuation. We must not fail to record an aesthetic reaction to punctuation, particularly typical of the middle and late Chhing, which still survives. To a seasoned Chinese scholar, an unpunctuated text has something virginal about it. Just as there is a special pleasure in that tiresome cutting open of a new French book, so there is a sublime quiet joy in punctuating a Classical Chinese edition as one reads it for the first time. After all, Classical Chinese was never meant for the impatient modern reader. Classical Chinese texts are meant to be savoured slowly by people who have the leisure to punctuate them.

The concept of a sentence or a statement

There are syntactic contexts or frames in Classical Chinese which must be filled in with a declarative sentence or a statement. Thus *ku (yüeh)* 故曰 . . . 'therefore (it is said) . . .', and *wu wên chih* 吾聞之 . . . 'I have heard it said that . . .'[2] are patterns where the empty places are expected to be filled with statements rather than any other string of words. In order to fill in the right sort of linguistic forms the ancient Chinese must have had an operative notion of a statement versus other strings. But is there a word or a technical term in Classical Chinese for the sentence or the statement versus other strings? The most common ancient Chinese word for a sentence would be the general *yen* 言 'speech':

---

[1] Traditionally, the Chinese used three kinds of marks: the full stop, indicated by a circle, and comma as well as indicated by a dash, a special dash to mark off items in lists or other co-ordinate structures.

[2] *Li Chi Chi Chieh, Than Kung*, ed. Sun Hsi-Tan, vol. 2, p. 59.

The Master Prefect has addressed to me nine sentences (*chiu yen* 九言): (1) 'Do not begin to cause trouble! (2) Do not trust wealth! (3) Do not rely on favour! (4) Do not go against common opinion! (5) Do not take ritual/politeness lightly! (6) Do not take pride in your abilities! (7) Do not get angry a second time (i.e., do not bear grudges)! (8) Do not plan what goes against your moral power! (9) Do not commit what goes against your duty.[1]

The ancient Greeks were in a similar condition and used their similarly vague *logos* 'speech, word' for a sentence. In Sanskrit we have the equally vague *vākya*.

However, Diogenes Laertius reports:

He (Antisthenes (−444 to −365)) was the first to define the sentence (*logos*) saying: 'The sentence is that which makes clear (*dēlōn*) that which is or which will be.'[2]

We can see that the only word Antisthenes could find for 'sentence' is still *logos*. And he defines the *logos* not in terms of the speaker's intended meaning but in terms of objective semantics.

Aristotle (−384 to −322) coined a technical term to express the precise idea of a declarative statement (*apophasis*), and he makes the logically crucial distinction between declarative and non-declarative sentences.

Every sentence (*logos*) signifies, but not every sentence is declarative (*apophantikos*): only those sentences in which one can be right or wrong are declarative. For example a prayer may be a sentence, but it is neither true nor false.[3]

In −3rd-century legal language *tzhu* 辭 was a technical term, apparently for a statement by the defendant, but we do not know how far back this legal term goes:

The person who makes a statement makes this statement in the court.[4]

Similarly, ancient Chinese philosophers used the technical term *tzhu* 'formulation, sentence' to designate what we are inclined to call a sentence (as opposed to the words (*yen* 言) and texts (*shu* 書)). But we find no clear explicit definition of the declarative versus non-declarative sentence like that just quoted from Aristotle. Let us look at some usages of the word *tzhu* where we suspect it must mean something like 'sentence'.

The sentence (*tzhu*) combines the names for different realities to bring out one intended meaning (*i* 意).[5]

In a large number of other contexts *tzhu* invites the translation 'sentence', but 'formulation' would do as well.

The sentence (*tzhu* 辭) is the external expression of an intended meaning (*i* 意). To concentrate on the external expression but to discard the intended meaning is contradictory.[6]

[1] *Tso Chuan*, Duke Ting, 4 fu, ed. Yang Po-Chün, p. 1542; cf. Couvreur (1951b), vol. 3, p. 506.
[2] Diogenes Laertius, *Vitae philosophorum* 6.3, ed. R. D. Hicks, Loeb Classical Library, vol. 2, p. 4.
[3] *De Interpretatione* 17a 1.   [4] Hulsewé (1985), D79.   [5] *Hsün Tzu* 22.39; cf. Köster (1967), p. 239.
[6] *Lü Shih Chhun Chhiu* 18.4, ed. Chhen Chhi-Yu, p. 1179; cf. Wilhelm (1928), p. 303.

In many cases sentences (*tzhu*) seem to be wrong but are right, or they seem to be right but are wrong. The guiding line of right and wrong (*shih fei chih ching* 是非之經) should be studied.[1]

The Later Mohists use *tzhu* as a technical term, as A. C. Graham has shown:

One uses names (*ming* 名) to refer to objects, and one uses sentences (*tzhu*) to bring out intended meanings (*i*).[2]

The section Names and Objects from the Mohist Dialectical Chapters deals with the problem of deciding whether apparently parallel sentences are really similar. We read:

The sentence (*tzhu*) is that which is engendered in accordance with a *ku* 故 'what is behind it, reason', becomes full-grown in accordance with a pattern, and proceeds in accordance with its kind.[3]

Moreover, when the same section makes a distinction between knowing and having an image, it would appear that it focusses on knowledge being knowledge of a sentence or statement being true, whereas 'having an image' implies no such judgement on the truth of a sentence.[4]

We shall discuss Later Mohist reflections on logic and language in Section (*f*,4). But it is worth noting that Wang Chhung, who was not particularly interested in the abstract science of logic, did have a clear concept of a sentence. Wang Chhung writes:

When written characters have a (complete) intended meaning (*i*), they make up a sentence (*chü* 句). When there are a certain number of sentences (*chü*), we string together paragraphs (*chang* 章) by means of them. When paragraphs have a certain coherent structure, we make chapters (*phien* 篇) of them. The chapter (*phien*) is the largest unit of sentences and paragraphs.[5]

The commentators of the +2nd and +3rd centuries liked to call their works *chang chü* 'NN in paragraphs and sentences', the most well-known of these being Chao Chhi's 趙岐 (died +200) *Mêng Tzu Chang Chü* 孟子章句, and in the subcommentary to this work we find a convenient listing of early examples of this genre.[6]

The literary critic Liu Hsieh 劉勰 (+465 to +562) makes a sequence *tzu* 字 'character, word', *chü* 'sentence', *chang* 'paragraph' and *phien* 'chapter'.

When men write literature, they make sentences (*chü*) on the basis of characters (*tzu*), they put together sentences to make paragraphs (*chang*), and they put together paragraphs to make chapters (*phien*). . . . A sentence (*chü*) deploys several characters, and these must link up before the sentence can be used.[7]

---

[1] *Lü Shih Chhun Chhiu* 22.5, ed. Chhen Chhi-Yu, p. 1527; cf. Wilhelm (1928), p. 401.
[2] Graham (1978), NO 10.   [3] *Ibid.*
[4] Graham (1978), NO 3. Hansen (1985) argues that the ancient Chinese did not have a concept of a sentence. He provides no new philological evidence to support his case.
[5] *Lun Heng*, ed. Chung-hua-shu-chü, p. 1589; cf. Forke (1911), p. 451.
[6] *Mêng Tzu Cheng I*, ed. Shen Wen-Cho, p. 32.
[7] *Wên Hsin Tiao Lung*, ch. 34, ed. Lu Khan-Ju and Mou Shih-Chin, vol. 2, p. 177. Cf. (1983), p. 186.

Yen Chih-Thuei 顏之推 (+531 to after +590) describes the function of the (mostly sentence-final) particle *yeh* as follows:

Yeh 也 is a word that finishes a sentence and aids punctuation (*chu chü* 助句).[1]

Yen Chih-Thuei goes on to note quite accurately that *yeh* is sometimes used after what we would call the subject, and is not always obligatory.

I am inclined to agree with Yen Chih-Thuei: *yeh* 也 tends to tell us where sentences end. That is indeed why *yeh* is such a useful and ubiquitous particle in Classical and Literary Chinese. Sentence-final grammatical particles like *yeh* carry much of the parsing burden that is carried by punctuation in Western languages.

One might suspect that the Chinese were unable to distinguish between the clause and the sentence, since they use the same punctuation mark for both. However, this is a dangerous assumption, because it is open to refutation by examples such as this one I came upon in the sayings attributed to Chu Hsi 朱熹 (+1130 to +1200) where he quotes *Hsün Tzu* 3.16: 'The gentleman, when he has large ambitions, will act according to Heaven and he will act according to the Way. When he has small ambitions, he will stand in awe of righteousness and he will be restrained.' There are many clauses here. But Chu Hsi is absolutely right when he takes it that there are exactly two sentences:

These two sentences are well said (*tzhu êrh chü shuo-ti hao* 此二句説得好).[2]

The sentence (*chü* 句), for Chu Hsi, seems closely connected to the concept of a point made (*tao li* 道理).[3] Discussing the correct way of reading, Chu Hsi comments:

One sentence (*chü*) makes the point (*tao li*) of one sentence. When you have exhaustively studied one sentence, you have got hold of the point made by this sentence. When you read books, then after you have understood the meanings of the characters, you should ponder what the sages were driving at.[4]

In Classical Chinese the correct parsing of sentences is aided in crucial ways by the pervasive use of parallelism. This is one of the subjects we have taken up above.

(2) THE CONCEPT OF MEANING

If we wish to understand Chinese conceptions of language and its relation to reality, we must begin by looking closely at the way in which the Chinese speak about the meanings of words, phrases, sentences and texts.[5] Did they ever use words which we are entitled to translate as 'meaning'? What did they attribute meaning to? To speakers using expressions? To the expressions as used in a given context? Or to

---

[1] *Yen Shih Chia Hsün*, ch. 17; cf. Teng Ssu-yü (1968), p. 161.

[2] *Chu Tzu Yü Lei*, ed. Wang Hsing-Hsien, p. 3253.

[3] The fact that Chu Hsi occasionally uses *chü* 句 loosely (much in the way we use the word 'sentence') to refer to subordinate sentences or even embedded sentences need not detain us here.

[4] *Chu Tzu Yü Lei*, ed. Wang Hsing-Hsien, p. 2978.   [5] Cf. in this connection Hess (1972).

expressions as such? These are important questions for an understanding of the philosophy of language which is either explicit or implicit in Chinese.

The meanings of characters are described in the traditional Chinese dictionaries.[1] Sentence meanings are often expounded in the vast traditional commentarial literature from the −3rd century onwards. The meanings of texts are systematically expounded as such in some commentaries. For example, the commentary to the *Mêng Tzu* by Chao Chhi 趙岐 (died +201) contains for each separate item (*chang* 章) a separate intended import of the item (*chang chih* 章旨). Taken together, the dictionaries and the commentaries give abundant evidence that the Chinese were much preoccupied with questions regarding the meanings of words, sentences and texts. Indeed, not the least important part of later Chinese philosophy consisted in attempts to explain and systematise earlier texts in a fashion that reminds us of the scholastic tradition in the West.

In the Chinese tradition, words are taken to have meanings primarily in so far as they function in concrete utterances on concrete occasions. A commentary gloss on a word will not tell you what that word as such means, but rather what the word is used to convey in the concrete context.

A dictionary entry will tell you not really what a word as such means but rather what it has been taken to mean in concrete contexts by various authoritative commentators or interpreters. Dictionaries originated as collections of glosses. Chao Chhi, whom we have just referred to, will not tell you what a sentence as such means but what in his opinion it was used to convey in one particular context. The notion of context-free meaning of a sentence as such is not current in traditional China.

Definitions, as we have seen in Section (*b*,4), tend to be of a thing, a value etc. Definitions of the meaning of a word tend to be definitions of the meaning of that word as used by a certain person on a certain occasion in a certain communicative context. The formal definitions of the Later Mohist logicians and the ethical definitions of Chia I provide neat exceptions to this generalisation.[2]

Logocentrism

We may imagine a logocentric civilisation which focuses on what is said or written down, on sentences, on what these sentences objectively mean, and on whether these sentences – so understood in accordance with their linguistic meaning – are actually true. Greek civilisation was, I think, in this sense predominantly logocentric. The dominant culture in India, certainly, was almost obsessively logocentric. The fixation on Vedic truth is symptomatic of this logocentrism.

China is not in this sense a predominantly logocentric civilisation. It is a civilisation which concentrates not on what words as such mean, but rather on what people really intended to convey by using these words. The Chinese see sentences as deeply embedded in personal and social reality. The meaning of sentences was not

---

[1] Cf. our Sub-section (*b*,5), on traditional Chinese lexicography.
[2] See Section (*b*,4) on definition for details.

for them a grammatical and lexical question. It was a historical question. Sentences are only messengers (*shih* 使) for meaning, they are not taken to articulate meaning literally.

In China, the crucial semantic concept is not that of sentence meaning but of speaker's meaning. The Chinese were preoccupied with the speaker's intended meaning and the intended import of words. That is why we have so little in the way of literal dogmatism in Confucianism. There was no inquisition which tested what people said against a set of *dogmata, Lehrsätze*, or a corpus of explicit doctrinal truths. The history of these explicit doctrinal truths is the history of logocentrism. The history of strictly scriptural fundamentalist theology in the West can be interpreted as the history of logocentric textual fetishism.

Christianity was able to become a heavily doctrinal religion because it emerged in a logocentric culture in Europe. It became an obsessively logocentric religion under the influence of Martin Luther's insistence on Scripture as the sole authority. Buddhism, originally not logocentric in the least, grew into a textual and doctrinal religion because it emerged in a fiercely logocentric Indian cultural context.

Confucianism, Taoism, Legalism, all the indigenous ancient Chinese 'schools' of thought were non-doctrinal in the sense that they did not define a *Credo* and a body of truths defining an orthodoxy. The form of Buddhism that became most intellectually significant in China, Chhan or Zen, was also non-doctrinal. We might say that Chhan Buddhism is defiantly and outrageously *anti*-logocentric. It would be interesting to investigate whether Buddhism in China generally remained less logocentric than it was in India.

The Mohists, on the other hand, did define their 'ten doctrines'.[1] These were more like ten topics or items on a party programme, but they do come the closest I know to a proper catechism of a creed in ancient China. They include religious topics (The will of Heaven, Elucidating the spirits), 'philosophical' issues (Rejecting destiny, Universal love), political aims (Elevating worth, Conforming to superiors, Rejecting aggression), and moral precepts (Economy in funerals, Economy in expenditure, Rejecting music).

Mo Tzu said: 'My words are worth using! Rejecting these words and engaging in new reflections is like omitting the harvest and collecting stray grains instead. To use one's words to refute my words is like throwing eggs against a rock. One can use up the world's eggs, but the stone is still right. It cannot be destroyed.'[2]

Mo Tzu certainly feels his philosophical position is unassailable, but still he emphasises the *use* rather than anything like the objective truth of his words. He may, in this sense, justly be called a pragmatist.

Of course, Confucius's sayings and Mo Tzu's sayings were collected because these sayings were felt to be important. Chinese culture was much preoccupied with the written word and was in *that* sense logocentric. But these sayings were important for what they were ultimately trying to get at, not for what they actually articulated.

---

[1] Cf. Graham (1978), pp. 1–3 and 10–15.   [2] *Mo Tzu* 47, end; cf. Mei (1929), p. 229.

The hermeneutics of these texts is personal (what did Confucius mean?), not scriptural or logocentric (what does the text mean?). It is only when one understands this fine distinction that one can truly appreciate the philosophical spirit of Chinese commentarial literature and Chinese concepts of meaning.

Asking for the meaning of something

It is hard to document or demonstrate within a brief space that the Mohists as well as the other philosophers were predominantly non-logocentric in their attitude towards their work. Chinese philosophers *could* ask what sentences and words meant. Let us look at the questions of this sort that they did ask.

Predominantly, Confucians were concerned with moral values: 'What is called "goodness" (*ho wei jên* 何謂仁)?' (*Meng Tzu* 7B25), 'What is called "good faith"?' (*Mêng Tzu* 7B25), 'What is called "to set one's mind on high principles"?' (*Mêng Tzu* 7A33). These questions are not ontological like Plato's 'What is the beautful (*ti estin to kalon*, τί ἐστιν τὸ καλόν)', they are linguistic and psychological: 'What does one call/do we call/is called goodness?'. Similar observations apply to questions like 'What is called "the Way"?' (*Chuang Tzu* 11.72), 'What is called a "true man" (*ho wei chên jên* 何謂真人)' (*Chuang Tzu* 6.9 and 6.20) and passages like 'What is called "sit and forget" (*ho wei tso wang* 何謂坐忘)? . . . This is called "to sit and forget" (*tzhu wei tso wang* 此謂坐忘).' (*Chuang Tzu* 6.92)[1] All these are formulated as linguistic rather than ontological questions as the use of the crucial word *wei* 謂 'call' indicates. The subject of *wei* 'call' is the person who uses the name in question. Thus we could have translated, and perhaps should have translated: 'What do we call "the Way"?', etc.

Of course, many of such questions are not concerned directly with values, but they still concern the intended meaning of a speaker rather than the inherent meaning or essence of a word.

Tzu Chang asked: 'What must a gentleman be like before he will win through?'
Confucius asked back: 'What on earth do you mean by "winning through" (*ho tsai êrh so wei ta chê* 何哉爾所謂達者)?'[2]
The Master said: 'Raise the straight and set them over the crooked. This can make the crooked straight.' What did he mean (*ho wei yeh* 何謂也)?'[3]

Again, the Mohists ask 'What do you call "the three basics" (*ho wei san pên* 何謂三本)?' (*Mo Tzu* 9.15) 'What do you call "the three standards"?' (*Mo Tzu* 35.7). These are requests for a specification of a certain list which the person asked is supposed to have in mind.

Han Fei Tzu 9.1 on *The Eight Villainies* asks no less than eight times for conceptual clarification within a context: 'What do we call/what is called X (*ho wei* 何謂 X)?' gives a general explanation of a villainy, and then concludes: 'This is what we call/is

---

[1] Cf. also *Chuang Tzu* 17.51f., 19.53, 20.56, 20.59, 20.60.
[2] *Lun Yü* 12.20; cf. Lau (1983a), p. 117.
[3] *Lun Yü* 12.22; cf. Lau (1983a), p. 117. For more questions about the meanings of expressions see *Lun Yü* 2.5, 3.8, 3.13, 4.15; 14.40, 20.2.

called *X*.' Again, the tenth chapter of the same book, *The Ten Faults*, asks ten times 'What do we call *X* (*hsi wei* 奚謂 *X*)?' and provides examples of ten faults. Note that all these questions are marked as linguistic by the verb *wei* 謂 'call', not an ontological verb like *esti* 'it is' or *wei* 為 'turn into, counts as', which would have been expected in Greek.

The Chinese were perfectly able and indeed sometimes inclined to ask what words mean, but even when a hostile Mohist by name of I Tzu asks Mêng Tzu 'what this phrase means', he is interested in the significance, the intended meaning, not the literal interpretation of the form of words as such.

According to the Way of the Confucians the rulers of old 'acted as if they were tending a new-born baby'. What do these words refer to (*tzhu yen ho wei yeh* 此言何謂也)? In my opinion it means that there should be no gradations in love, though the practice of it begins with one's parents.[1]

What one *wei* 謂 *X* 'calls "*X*"' or 'means by "*X*"' is recognised as a subjective matter:

What I call 'profit' is the root of righteousness. What the world calls 'righteousness' is the way of cruelty.[2]

Another important semantic verb is *yü* 喻 'to make one's meaning plain; to indicate by metaphor, a metaphor'. *Yü* is a psychological concept, as is demonstrated neatly in the following pair of examples:

Speech (*yen* 言) is the means by which one makes plain (*yü*) intended meaning (*i* 意).[3]
In general the purpose of speech (*yen*) is to make plain (*yü*) what is on one's mind (*fan yen chê i yü hsin yeh* 凡言者以喻心也).[4]

Yü, as far as we can ascertain, refers to the putting across by a person of his thoughts and more particularly his communicative intentions:

If names and objects are not unravelled, if the noble and the base is unclear, if the identical and the different are not distinguished, if things are like this, then we are sure to have trouble over the (communicative) intention (*chih* 志) not being made plain (*yü*).[5]

It is not a word or a sentence as such which *yü*s anything, it is a person using words or other means.[6]

---

[1] *Mêng Tzu* 3A5; cf. Lau (1983c), p. 111.
[2] *Shang Chün Shu*, ch. 7, ed. Kao Heng, p. 77; cf. Duyvendak (1928), p. 230.
[3] *Lü Shih Chhun Chhiu* 18.4, ed. Chhen Chhi-Yu, p. 1177; cf. Wilhelm (1928), p. 300.
[4] *Lü Shih Chhun Chhiu* 18.5, ed. Chhen Chhi-Yu, p. 1185; cf. Wilhelm (1928), p. 304.
[5] *Hsün Tzu* 22.13; cf. B. Watson (1963), p. 141.
[6] Cf. the following passage from *Mêng Tzu*:

That which a gentleman follows as his nature, that is to say, benevolence, rightness, the rites and wisdom, is rooted in his heart, and manifests itself in his face, giving it a sleek appearance. It also shows in his back and extends to his limbs, rendering their meaning plain without speaking (*pu yen êrh yü* 不言而喻). (*Mêng Tzu* 7A21; cf. Lau (1983c), p. 271)

It is only when things show in his facial expression or when they come out through his voice that his meaning becomes plain (*chêng yü sê fa yü shêng êrh hou yü* 徵於色發於聲而後喻). (*Mêng Tzu* 6B15; cf. Lau (1983c), p. 263)

Nominalised concepts of meaning

Let us turn to the concept *i* 意 which we often gloss as 'intended meaning'. The *Shuo Wên Chieh Tzu* 說文解字 (postface +100) defines:

I 'intended meaning' means intention (*chih* 志) . . . When examining words one understands intended meaning (*i*).[1]

The binome *chih i* 志意 'intention and intended meaning' is significant:

The Hu and Yüeh people do not understand each other's languages. They cannot communicate intentions and intended meanings (*chih i*),[2] but when mountainous waves arise about the boat they share, they rescue one another as though they were one group.[3]

Hsün Tzu commonly uses the compound *chih i* 志義:

Those 'words and sentences' are messengers of the intentions and purport (*chih i*).[4] When they are sufficient to communicate, one dismisses them (the words and sentences).[5]

The famous *thien i* 天意 of Mo Tzu are 'the intentions of Heaven' which man must follow. Quite frequently Mo Tzu also refers to *thien chih* 天志 'the intent of Heaven', and the distinction between these two concepts is not clear. In any case, the *thien i* are not just Heaven's impressions, or thoughts, or images in its heavenly mind. The *thien i* involves heavenly intentions/wishes regarding matters on Earth.

In the human sphere, Mo Tzu uses the term *i* in a similar way:

'Suppose there is a fire. One person is fetching water to extinguish it, and another is holding some fuel to make the fire worse. Neither of them has yet done anything. Which one would you value more highly?'
Wu Ma Tzu replied: 'I approve of the intentions (*i* 意) of the one who fetches water and disapprove of the intentions (*i*) of the one who holds fuel.'
Mo Tzu continued: 'And I, too, approve of my intentions and disapprove of yours.'[6]

I is used to refer to the intentionality of an action, e.g., when one moves of one's own accord as opposed to being moved by someone else.[7] When applied to speech, we understand the *i* to refer to just this sort of intentionality.

However, there is another side to the concept of *i*. The word can come to mean something close to 'thought' and then 'idea', 'mental image':

Therefore the gentleman must make his conscious mind sincere. 'Making one's thoughts (*i*) sincere' means 'to make sure one does not cheat oneself'.[8]

What one can explain with words is the crude side of things. What one can reach with one's thought (*i*) is the subtle side of things. But what words cannot explain and what the

---

[1] Tuan Yü-tshai (*1981*), p. 502.
[2] Cf. *Yen Tzu Chhun Chhiu*, ed. Wu Tse-Yü, pp. 182f., or another use of *chih i pu thung* 志意不通, this time to indicate that Confucius could not communicate his intentions and intended meaning (to the rulers whom he tried to convince).
[3] *Chan Kuo Tshe*, no. 450; cf. Crump (1970), p. 516.
[4] I take *chih i* 志意 to be the same in meaning as *chih i* 志義. Cf. Morohashi's dictionary, p. 4361.
[5] *Hsün Tzu* 22.49; cf. Watson (1963), p. 149.   [6] *Mo Tzu* 46.13; cf. Mei (1929), p. 214.
[7] *Kung Yang Chuan*, Duke Hsi, 1.3.   [8] *Ta Hsüeh*, ed. Legge, p. 366.

mind cannot reach – that is beyond the realm of the crude and the subtle aspects of things.[1] Therefore that by means of which people imagine (*i hsiang* 意想) things is called an image (*hsiang* 象).[2]

Hsün Tzu speaks of the senses producing 'images (*i*)' of things[3] by which we decide whether things belong to the same class or not. Next, *i* comes to mean 'what one intends to convey', 'what one intends to communicate' or 'intended meaning'. The purpose of speech was seen in terms of thought or intended meaning (*i*):

Hearing is the sensibility of the ear; to attend to what you hear so that you grasp the intended meaning (*i*) is discernment by the mind. Saying is the flow of the tongue; to make a case for what you say so that your intended meaning (*i*) can be seen is the subtlety of the mind.[4]

Fish-traps are the means by which one catches fish. Once one has the fish, one forgets the traps.

Hare-traps are the means by which one catches hares. Once one has caught the hare, one forgets the traps.

Words are the means by which to get to the intended meaning (*i*). Once one has the intended meaning (*i*), one forgets the words.[5]

Is this 'forgetting the words' not a characteristic sign of a non-logocentric tradition? In any case, the sentiment here expressed is by no means limited to the book *Chuang Tzu*:

In ancient times people would disregard words once they had got the intended meaning (*i*). When one listens to words, one uses the words to catch sight of the intended meaning (*i*).[6]

What we tentatively translate as 'intended meaning' primarily attaches to the person. It is a psychological concept describing not the meaning attached to speech or words as such but meaning as intended by the speaker.

One might of course simply translate *i* by 'thought', and this would make for smoother reading, but the thought involved is essentially linked with the communication of it, it is a thought or image as intended to be understood by some audience. Chuang Tzu even distinguishes between the meaning one is intending to convey, and the ultimate point one is trying to get at:

A book is no more than sayings. But there is something that is the most valuable in speech. That which is the most valuable thing in speech is the intended meaning (*i*). The intended meaning (*i*) has something it pursues. But what it pursues cannot be transmitted in words.[7]

Words do not exhaust intended meaning (*i*). Does this mean that we cannot get a glimpse of what the Sages intended to communicate?

The Master said: 'The Sages established the diagrams in order exhaustively to bring out their intended meanings (*i*).'[8]

---

[1] *Chuang Tzu* 17.24; cf. Watson (1964), p. 178.   [2] *Han Fei Tzu* 20.28.4; cf. Liao (1939), vol. 1, p. 193.
[3] *Hsün Tzu* 22.16; cf. Watson (1963), p. 142.   [4] Graham (1978), NO 9.
[5] *Chuang Tzu* 26.48, tr. Graham (1981), p. 190.
[6] *Lü Shih Chhun Chhiu* 18.4, ed. Chhen Chhi-Yu, p. 1179; cf. Wilhelm (1928), p. 303.
[7] *Chuang Tzu* 13.65; cf. Graham (1981), p. 139.
[8] *I Ching, Hsi Tzhu*, ed. Kao Heng, p. 541; cf. Sung (1969), p. 302.

Hsün Tzu, on the other hand, speaks as if intended meaning (*i*) is indeed conveyed by words, and his usage comes closest in ancient Chinese literature to that of sentence meaning as opposed to speaker's meaning:

In a sentence (*tzhu* 辭) we combine the names of different realities to explicate (*lun* 論) one intended meaning (*i*).[1]

One's general intended meaning is seen in contrast with factual instantiating details:

If you omit details of fact and put forward your intended meaning (*i*), you will be said to be weak-minded and less than thorough-going.[2]

When trying to persuade someone, Han Fei feels one does need to make plain one's intended meaning (*i*):

Difficulties in persuasion, generally speaking, are not difficulties relating to the knowledge with which to persuade someone. Neither are they difficulties relating to my rhetoric (*pien* 辯) being able to make plain my intended meaning (*i*)... Generally speaking, the difficulty in persuasion relates to one's understanding of the mind of the person one persuades, and to the ability to make one's persuasion fit his mind.[3]

Finally, it is important to remember that the notion of *i* 'intended meaning' is by no means limited to philosophical discourse. When King Hui of Chhin says:

'I want to order Wu-An Tzu-Chhi to go and convey this meaning/thought (*i* 意).[4]

The thought in question is not about something the king intends to do, it is something he intends to communicate, namely the thought that the various states are as likely to be united as two cocks are to share the same perch.

The technical term 'chih' 指

The dictionary *Shuo Wên Chieh Tzu* 說文解字 (postface +100) explains:

The first (type of character) is called 'referring to a matter (*chih shih* 指事)'. When one sees a graph of this type, it may be understood on seeing it. By looking at it one sees the intended meaning (*i*). The graphs *shang* 上 'above' and *hsia* 下 'below' are of this sort.[5]

For Hsü Shen 許慎 (died *c.* +149), a character points to (*chih*) a state of affairs or thing, and by looking at a character one may recognise its intended meaning (*i*).

---

[1] *Hsün Tzu* 22.16; cf. Watson (1963), p. 142.
[2] *Han Fei Tzu* 12.3.44; cf. Liao (1939), vol. 1, p. 108. The *ta i* 大意 is the overall intended purport of what one is saying. (Cf. *Han Fei Tzu* 12.4.32).
[3] *Han Fei Tzu* 12.1.1; cf. Liao (1939), vol. 1, p. 106.
[4] *Chan Kuo Tshe. Chhin* 1.3, ed. Chu Tsu-Keng, p. 139; cf. Crump (1970), p. 59.
[5] Preface to the *Shuo Wên*. Tuan Yü-Tshai (*1981*), p. 755; cf. Thern (1966), p. 9. Note that the term *i* 意 'intended meaning' could apply to the meaning of lexical items as such. For detailed comment see Hsiang Hsia (*1974*), pp. 38ff.

The original meaning of *chih* 指 was 'finger', then 'to point to with a finger', 'to indicate'.[1]

Therefore, when the name is sufficient to point out (*chih*) the object, and the sentence sufficient to display it to the utmost, one goes no further.[2]

The main use of *chih* 'to indicate, to point to' was nominal. The word came to mean something like 'intended import'.

Synonymity of words is explicitly described in terms of identity of intended import (*chih*) by Chuang Tzu:

The three words *chou* 周, *phien* 遍, and *hsien* 咸 are different terms for the same reality (*shih* 實); their intended import is one and the same (*chhi chih i yeh* 其指一也).[3]

This example is important because it clearly and explicitly attaches the notion of intended import (*chih*) not to a speaker who intends to convey it but rather to a word to which it is attached.[4]

The intent (*chih*) (like the intended meaning *i*) is something one can try to bring out or dredge out, as in the following passage which is uniquely rich in semantic vocabulary:

The disputant distinguishes separate kinds of things so that they do not interfere with each other, arranges in sequences different starting-points so that they do not confuse each other, brings out his intended meanings (*i* 意) and makes the intent (*chih*) intelligible, and clarifies what he has to say (*ming chhi so wei* 明其所謂). He shares his knowledge with others and makes it his business not to mislead them.[5]

Wang Chhung 王充 writes of men of literary accomplishments:

They dredge out the meaning and intent (of the books) (*shu chhi i chih* 抒其義指).[6]

The Chinese stylistic ideal is beautifully described – and exemplified – in a passage from the *Han Shu* 漢書:

His style is concise, but the intent (*chih* 志) is plain.[7]

The *Chhun Chhiu Fan Lu* 春秋繁露, a book of uncertain date which probably found something like its present shape some time around the +5th century, but which contains much earlier material, points out:

---

[1] There is a good account of the semantics of *chih* 指 in Graham (1979), pp. 547ff.

[2] *Hsün Tzu* 22.50; cf. Watson (1963), p. 149. Cf. also *Hsün Tzu* 22.14.

[3] *Chuang Tzu* 22.47. Graham (1978), p. 459, translates 'These three, *chou* 周, *pien* 遍 and *hsien* 咸, are different names for the same object.' However, it is not clear how these three words can be construed as names of an object, and of the same object. The Mohist definition of *chin* 盡 'all' as 'none not so' (Graham (1978), A43) shows that the Later Mohists in any case did not commit such a serious logical blunder as it would be to regard a quantifier as a proper name of an object. I do not think *shih* 實 can here have its technical meaning 'concrete object'.

[4] It is clear that this meaning does not fall under the definition 'the direction in which discourse points, its meaning or drift, the main point in contrast with details or side issues' as given in Graham (1978), p. 458.

[5] *Han Shih Wai Chuan* 6.6, ed. Hsü Wei-Yü, p. 208; cf. Hightower (1951), p. 196, and Graham (1978), p. 20.

[6] *Lun Hêng*, ed. Chung-hua shu-chü, p. 777; cf. Forke (1911), vol. 2, p. 295.

[7] *Han Shu*, ch. 53, ed. Chung-hua shu-chü, p. 2411.

Sentences (*tzhu* 辭) cannot reach. It all depends on the intent (*chih*) . . . When one has seen the intent (*chih*), one does not rely on the words. Only when someone does not rely on the formulations (or: sentences) (*tzhu*), can he come close to the Way.[1]

Conclusions

Our survey of nouns meaning 'meaning' is by no means complete. For example, sometimes verbs like *wei* 謂 'to call' are used in an *ad hoc* way as nouns, as when we are told:

Words are subservient to conveyed meaning (*yen chê wei chih shu yeh* 言者謂之屬也).[2]

We conclude that the ancient Chinese did not tend to attribute meanings to sentences as such. Rather, they tended to attribute intended meanings to users of sentences.[3] This tendency naturally connects with the elliptic nature of many Chinese sentences. The speaker's meaning contains much that is understood and not explicit in the overt sentences he uses. For the Chinese, then, a sentence would not naturally be regarded as a picture or representation of what it means. Instead, it is naturally taken as a contextually sufficient indicator to direct the audience to the meaning intended by the user, or to the user's thought. That thought, then, may or may not represent a fact. But that is a different question which will be addressed in our Section on the concept of truth.

(3) The Concept of Truth

We have seen that the ancient Chinese had logical sentence connectives and the notion of a sentence. But did they have the notion of the truth of sentences?

Note first that the ancient Chinese, unlike the ancient Greeks and their Western successors, distinguished carefully between problems of the all-important *tao* 道 'the right way of going about things', and the more ordinary questions of *truth* as 'being the case', for which they had a variety of words some of which will be introduced below.

The subjective moral truth one lives by (Sanskrit *dharma*) translates naturally into *tao* in Classical Chinese (although the Buddhist translators actually chose *fa* 法 'law' as the standard translation for the technical term), and this concept, though of no special concern for the logician, is certainly a crucial concept of Chinese philosophy. It is profoundly significant that Chinese preoccupations were not with the objective truth of factual or philosophical statements as such but rather with the right way (*tao*) to go about personal and political life. Donald Munro overstates his case, but he does have a significant point when he writes:

---

[1] *Chhun Chhiu Fan Lu*, ch. 3, ed. Ling Shu, pp. 33f.
[2] *Lü Shih Chhun Chhiu* 22.3, ed. Chhen Chhi-Yu, p. 1168; cf. Wilhelm (1928), p. 298.
[3] The truth predicates in ancient Chinese, on this interpretation, were applied to what sentences were intended to convey, not to what sentences mean as such. See our Section (*d*,3) on the concept of truth.

In China, truth and falsity in the Greek sense have rarely been important considerations in a philosopher's acceptance of a proposition; these are Western concerns.[1]

We shall want to ask in detail whether truth and falsity (in what D. Munro seems to think of as the Greek sense) really have not been Chinese concerns.

When it comes to the notion of factual truth (Sanskrit *satya*) we must distinguish between a range of questions of different orders of philosophical significance:

1. Did the ancient Chinese have an operative concept of truth, i.e., were they concerned to distinguish between true and untrue sentences or statements with predicates like our 'is true' or 'is not true'?
2. How exactly (if at all) did the ancient Chinese express or articulate the idea that a statement is true?
3. Did the ancient Chinese ascribe truth to such entities as sentences or statements or to truthful speakers only?
4. What were the connotations of the ancient Chinese ways of calling something true?
5. Did the ancient Chinese discuss abstract nominalised notions of truth?

Aristotle defines:

Thus a person has got it right (*alētheuei*) who considers that things which are separate are separate and that things which belong together belong together.[2]

The question is whether we have statements of this sort in ancient China. We do. Compare, for example:

We call it 'being straightforward' to declare something 'this (or: right)' if it is this (or: right), and to declare something 'not-this (or: wrong)' if it is not-this (or: wrong) (*shih wei shih, fei wei fei yüeh chih* 是謂是非謂非曰直).[3]

I want to stress two important contrasts here:

1. Aristotle is interested in a formal definition of what it is to be right, whereas Hsün Tzu defines what it is to *tell* the truth by *calling* what is right 'right'.
2. Aristotle thinks of the subject term and the predicate term being separate or belonging together whereas Hsün Tzu thinks of a predicate applying or not applying to a thing.

Both are profound differences.

For Aristotle the question is one of the theoretical definition of *alētheuein* 'being in accordance with the truth', whereas for Hsün Tzu the problem is the more social definition of *chih* 'being straight, telling the truth, getting things straight'. For Aristotle the paradigm of a statement is a general statement of the sort 'All philosophers are

---

[1] Munro (1939), p. 55. Smith (1980), p. 432, writes: 'Truth thus (i.e., in the Chinese way) conceived is a kind of performative: it is speech or deed aimed at effecting an intended consequence.' Hansen (1985), p. 492, draws the blunt conclusion: 'Chinese philosophy has no concept of truth'. For a discussion of Hansen's claims see Harbsmeier (1989).
[2] *Metaphysics* 1051b3.   [3] *Hsün Tzu* 2.12; cf. Köster (1967), p. 12.

humans' or 'All humans are mortal': his paradigm of a proposition relates two general terms. For Hsün Tzu, as for all other early Chinese philosophers of language, the main concern is not in this way essentially tied up with a relation between two terms. Chinese philosophers of language were mainly concerned with the relation between names and objects/things.

The question is simply this: did the ancient Chinese have the notion of the truth of statements? Wang Chhung declares:

The Confucians theorise saying: 'Heaven and Earth deliberately give birth to men.' This statement (*yen* 言) is mistaken.[1]

Yen must refer to a sentence or statement. Wang Chhung goes on to consider what he regards as false analogies between Heaven and Earth on the one hand, and a potter or founder. He rejects these and concludes with a general point which is crucial for our purposes:

When analogies do not correspond to the facts, then they cannot be said to bring out the facts. When written statements do not correspond to reality, they cannot be called right/true (*wên pu chhêng shih wei kho wei shih* 文不稱實未可謂是).[2]

This is a negative definition of the predicate 'is right/true (*shih* 是)', not of a noun 'truth'. The predicate is applied to *wên* 文 which in this context must be a written statement. In any case, what does the corresponding here is not a single predicate but a whole description. The correspondence is not with a physical concrete object but with reality in a wider sense.

There is no one standard Classical Chinese noun for truth. The Chinese have not tended to reify or hypostasise an abstract concept of truth as correspondence with facts. They have indeed shown a commendable reluctance, encouraged by the morphemic structure of their language, to hypostasise or reify *any* abstract concept.

On the other hand we shall see that ancient Chinese philosophers were in many ways concerned with semantic truth, with the truth of sentences. From Han times onwards, the Chinese have known the phrase 'Seek the truth on the basis of fact (*shih shih chhiu shih* 實事求是)',[3] which became an important slogan during the +17th and +18th centuries in China.[4] But note that what the common saying recommends one to do – and commends a certain king for doing – is to seek *facts of a matter* rather than *true propositions* or sentences. The Chinese have always tended to be interested in the practical aspect of truth. They have traditionally been inclined to ask of a statement not only whether it was true, but also what would happen if one held it to be true. An anecdote will illustrate our point:

Tzu Kung asked Confucius: 'Do the dead have knowledge or do they not have knowledge?'

[1] *Lun Heng*, ch. 15, ed. Chung-hua-shu-chü, p. 205; cf. Forke (1964), vol. 1, 103.
[2] *Lun Heng*, ch. 15, ed. Chung-hua-shu-chü, p. 206; cf. Forke (1964), vol. 1, p. 103.
[3] *Han Shu* 53, ed. Chung-hua-shu-chü, p. 2410. Yen Shih-Ku's 顏師古 (+581 to +645) comment on the saying is worth quoting: 'He made it his business to find out the facts (*shih shih* 實事). In each case he tried to find what was genuinely right (*chên shih* 真實).' What is genuinely right here are not facts but surely statements about them.
[4] Cf. Elman (1984).

The Master said: 'I might want to say that the dead have knowledge. But then I am afraid that filial sons and obedient grandsons will abandon their old folks and leave them unburied.

I might want to say that the dead have no knowledge, but I am afraid that unfilial sons would abandon their parents and leave them unburied. You want to understand whether the dead have knowledge or not. When you die you will find out in due course. There is no hurry now.'[1]

Note that even in this passage what 'will be found out in due course' is the factual truth concerning the knowledge of the dead. Confucius takes a pragmatic attitude *towards a concept of truth which he has.*

The Chinese pragmatism comes out in statements like these:

One's words hitting the mark a hundred times out of a hundred (*pai yen pai tang* 百言百當) is not as good as being fast to act and to act in a well-considered way.[2]

In view of such attitudes our question becomes even more urgent: did the ancient Chinese have an ordinary notion of factual truth, and of truth as applied to philosophical generalisations?

Han Fei (died −233), for one, did take the notion of truthfulness and of telling the truth seriously enough to declare the following (in what was to become the opening paragraph of his book):

Any minister, if not loyal, must be condemned to death. If what he says be not true (*tang* 當), he must be condemned to death, too.[3]

Han Fei is using the term *tang* 'fit the facts' to express an idea that I believe W. K. Liao's translation captures adequately. He speaks not of the appropriateness of saying something or the expediency of believing something. He speaks of the importance of good faith (standardly called *hsin* 信 in Classical Chinese) which is connected but not identical with the notion of the truth of sentences. Or could we take Han Fei to refer to the semantic concept of the objective truth of sentences? The context is not clear enough to allow us to decide.

The semantic concept of truth

It is one thing to say that a state of affairs exists. It is another thing to say that a sentence, statement or claim is true. We call the latter the semantic concept of truth. Consider the *Huai Nan Tzu* (−2nd century) which discusses the difficulties of accounting for such phenomena as the 'loving stone' (or magnet) and the sunflower's turning towards the sun and then continues:

---

[1] *Shuo Yüan* 18.31, ed. Chao Shan-I, p. 558, and *Khung Tzu Chia Yü*, ch. 2, no. 17; cf. Kramers (1950), p. 238. Compare the justly famous Buddhist-Confucian debate on the immortality of the soul during the +4th and +5th centuries. Precious material on this has been made available to me by Professor Yang Xianyi. (Cf. Liebenthal (1952)).

[2] *Huai Nan Tzu*, ed. Liu Wên-Tien, 18.12a.

[3] *Han Fei Tzu* 1.1.6; I quote the translation in Liao (1939), vol. 1, p. 1.

Therefore perception by the eyes and ears is not sufficient to sort out the attributes of things (*fên wu li* 分物理); discursive thinking of the mind is not sufficient to fix what is right and wrong (*pu tsu i ting shih fei* 不足以定是非).[1]

The context suggests that the right and wrong here is a scientific, not a moral right and wrong. But is truth ascribed to sentences?

One might imagine that the Chinese were able to wonder about states of affairs but that they did not speak or think of *sentences* or *statements* being true. For example, ancient logicians will speak of 'the facts of the case, literally: the be-so or not-be-so of a thing (*wu chih jan fou* 物之然否)',[2] a way of speaking which does not bring sentences or statements into play. Again, when the Chinese use the predicate *yu* 有 'to exist' to indicate that something is true, we might interpret them to claim not that a sentence is true but that a state of affairs exists. One might suspect that the ancient Chinese spoke and thought of facts of a case, not of the truth of a sentence.

This suspicion, however, is not borne out by the facts. The ancient Chinese did regularly apply predicates similar to our 'be true' to sentences, even in the context of abstract logical debate. Consider this example:

'Is the whiteness of white feathers the same as the whiteness of white snow, and the whiteness of white snow the same as the whiteness of white jade?'
'That is so (*jan* 然).'[3]

Is it our Western bias which misleads us into interpreting Mêng Tzu's question as a question regarding the truth of a sentence, and Kao Tzu's answer as confirming that that sentence is true? The question has to be decided against the background of a wide range of further evidence.

The *Lü Shih Chhun Chhiu* considers two alternative statements and calls one 'so (*jan*)' and the other 'not so (*pu jan* 不然)':

The path you follow is very difficult but leads to no results. If someone were to call you determined, that would be so (i.e., true enough) (*wei tzu yu chih tsê jan* 謂子有志則然), but if someone were to call you clever, that would not be so (i.e., true) (*wei tzu chih tsê pu jan* 謂子智則不然).[4]

Again we read in the same book:

Many statements look as if they were not-right (*fei* 非) but turn out right (*shih* 是). Many others look as if they are right (*shih*) but turn out not-right (*fei*). The guiding principle of right and not-right (*shih fei chih ching* 是非之經) must be distinctly understood. That is what the Sage is careful about. He goes along with the inherent essence of things (*wu chih chhing* 物之情) and keeps in touch with the inherent essence of man (*jên chih chhing* 人之情).[5]

---

[1] *Huai Nan Tzu*, ch. 6, ed. Liu Wên-Tien, p. 6a. One might be tempted to see the attributes/principles of things (*wu li* 物理) in the *Huai Nan Tzu* as an interesting approximation to the notion of a natural law.
[2] Graham (1978), p. 624.   [3] *Mêng Tzu* 6B3; cf. Lau (1983c), p. 223.
[4] *Lü Shih Chhun Chhiu* 20.1, ed. Chhen Chhi-Yu, p. 1322; cf. Wilhelm (1928), p. 348.
[5] *Lü Shih Chhun Chhiu* 22.6, ed. Chhen Chhi-Yu, p. 1527; cf. Wilhelm (1928), p. 401.

Shih and *fei* here apply to sentences rather than states of affairs. Moreover, the author of this passage clearly attached the greatest importance to whether words were *shih* 'this, right' or *fei* 'not-this, wrong'.

Han Fei explicitly rejects a philosophical thesis as untrue:

When the opponent says: 'One has to depend on a sage, then there will be good order', then that is not so (*pu jan* 不然).[1]

Mêng Tzu speaks of the truth of historical statements:

Wan Chang asked: 'It is said by some that "virtue declined with Yü who chose his own son to succeed him, instead of a good and wise man". Is there such a state of affairs (*yu chu* 有諸)?'
Mêng Tzu replied: 'No, it is not so (*fou, pu jan yeh* 否不然也).'[2]
Wan Chang asked: 'Some people say that I Yin tried to attract the attention of Thang by his culinary abilities. Is there such a state of affairs (*yu chu*)?'
Mêng Tzu replied: 'No. That is not so (*fou. pu jan* 否不然).'[3]

In all such contexts claims are quoted and the question is raised whether there are facts validating claims or statements. Sometimes it is even more explicit that the subject is a form of words. In *Mêng Tzu* a certain Hsien Chhiu-Mêng retails a story which he introduces by *yü yun* 語云 'the story goes', and then he goes on to complain:

I do not know whether this story (*tzhu yü* 此語) is genuinely so (*chhêng jan* 誠然).[4]

In instances like this it becomes almost ungrammatical to use the English 'be so' for *jan* 然. What we really need is 'be true'. Book 5A of *Mêng Tzu* contains an extended series of queries whether certain claims which are quoted are true or not.

Hsün Tzu addresses the philosophical question whether human nature is good as follows:

Mêng Tzu says: 'Human nature is good.'
(I) say: 'This (*shih* 是) is not so (*jan*).'[5]

'This (*shih*)' here clearly refers to Mêng Tzu's claim which has just been quoted. In this very chapter Mêng Tzu is quoted four times, and his claims are explicitly rejected four times.[6]

In the *Ta Tai Li Chi* (stabilised around +100) we read:

Tan Chü-Li asked Tseng Tzu: 'As for "Heaven is round and Earth is square", is there truly such a state of affairs (*thien yüan êrh ti fang chê, chhêng yu chih hu* 天圓而地方者誠有之乎)?'

---

[1] *Han Fei Tzu* 40.5.3; Liao (1939), vol. 2, p. 203.   [2] *Mêng Tzu* 5A6; cf. Lau (1983c), p. 191.
[3] *Mêng Tzu* 5A7; cf. Lau (1983c), p. 193.   [4] *Mêng Tzu* 5A4; cf. Lau (1983c), p. 185.
[5] *Hsün Tzu* 23.36; cf. Köster (1967), p. 307.
[6] *Hsün Tzu* 23.10; 23.14; 23.36; 23.45. Cf. also *Chan Kuo Tshe*, ed. Chu Tsu-Keng, p. 201; cf. Crump (1950), p. 63, *Yen Tzu Chhun Chhiu* 3, ed. Wu Tse-Yü, p. 239 and *Shuo Yüan* 7.37, ed. Chao Shan-I, p. 197, for instructive further instances where truth is explicitly ascribed to sentences.

Tseng Tzu replied: 'Li, have you heard that this is so?'

Tan Chü-Li said: 'I, your disciple, have not investigated the matter. That is why I presume to ask.'[1]

Thus this eminently Confucian compilation seems at this point urgently concerned with the concept of truth as applied to a (proto-) scientific statement.

A later Confucian compilation provides us with a splendid statement on the gentleman's attitude to truth versus rhetoric:

In his discourse (*lun* 論), the gentleman (*chün tzu* 君子) sets store by adequacy to the principles inherent in things. He does not set store by flowery sentences.[2]

We conclude that words or sentences were regularly called something like 'true' in Classical Chinese, as in Confucius's almost proverbial *chhêng tsai shih yen yeh* 誠哉 是言也 'how genuinely adequate are these words'.[3] There is no evidence to support the current claim that the Chinese thinkers could not or did not conceive of the truth of sentences, or that they were unconcerned with the truth of statements.

Some Chinese conceptual distinctions within the area of truth

Mêng Tzu distinguishes between factual truth which he confirms by *yu chih* 有之 'there is such a state of affairs', and moral acceptability (*kho* 可):

King Hsüan of Chhi asked: 'Is there such a state of affairs (*yu chu* 有諸) that Thang banished Chieh and King Wu marched against (the tyrant) Chou?'

Mêng Tzu replied: 'There is (such a state of affairs) in/according-to the records.'

'Is it (morally) acceptable (*kho*) that a minister kill his ruler?'

'He who trespasses against benevolence is a trespasser. He who trespasses against righteousness is a crippler. A man who is both a trespasser and a crippler is "an outcast". I have indeed heard of the punishment of "the outcast (tyrant) Chou". I have not heard of the assassination of a ruler.'[4]

Mo Tzu has a different distinction between *kho* 'be (subjectively) acceptable' and *jan* 然 'be (objectively) so':

Looking at it from the point of view of what you say, what everybody calls 'acceptable (*kho* 可)' need not necessarily be so (*jan*).[5]

Here a ruler is addressing Mo Tzu. The ruler attributes to Mo Tzu a distinction between subjective acceptability (*kho*) and objective truth (*jan*). Note that what

[1] *Ta Tai Li Chi* 58, ed. Kao Ming, p. 207; cf. Grynpas (1967), p. 115.

[2] *Khung Tshung Tzu*, ch. 12, ed. *SPTK*, p. 75a.

[3] *Lun Yü* 13.11. D. C. Lau (1983a), p. 125, translates: 'How true is the saying that after a state has been ruled for a hundred years by good men it is impossible to get the better of cruelty and to do away with killing.' Cf. also Yang Hsiung *Fa Yen* 10, ed. Wang Jung-Pao, p. 363, where *chhêng* 誠 is similarly applied to a generalisation. In both passages *chhêng* 'sincere, etc.' is glossed by *hsin* 信 'trustworthy, etc.' by the earliest commentators.

[4] *Mêng Tzu* 1B8; Lau (1983c), p. 39. This is a fine example of the Confucian 'correct use of names (*chêng ming* 正名)', which will be discussed in our Sub-section on the origins of logic in China.

[5] *Mo Tzu* 49.24; cf. Mei (1929), p. 246.

everybody calls acceptable is a form of words, a sentence, or a claim. And that same thing is said to be not necessarily *jan* 'objectively so'.

There is even some evidence that the ancient Chinese could distinguish between arguing for or maintaining (*chü* 舉) a proposition and a fact obtaining.

'I have heard it said that you argue that to be a sage is easy. Is there such a state of affairs (*yu chu* 有諸)?'
Thien Chhü replied: 'That is what I maintain (*chü*).'[1]

Quite properly, Thien Chhü replies 'yes, this is what I maintain', not 'yes, this is so'.

The verb *tang* 當 'fit the facts' is used in a most remarkable passage concerning the distinction between explanatory and factual truth:

Every thing's being so must necessarily have a reason (*ku* 故). And when one does not know the reason, then even though what one says fits the facts (*tang*), this is the same as ignorance.[2]

What one says can be true, and one can be right, but for the wrong reasons.

As we shall see in our Section (*f*,2), Kungsun Lung 公孫龍 used the concept *kho* to refer to the logical acceptability of a certain claim in a given logical context.

Finally we might mention in passing historical technical terms introduced by the Grand Historian Ssuma Chhien司馬遷 (−145 to −89), *khung yen* 空言 'empty words' and *khung wên* 空文 'empty writing', which refer to non-factual judgements and generalisations as opposed to judgements of historical fact.[3] The Grand Historian's terminological innovation here has not caught on in the Chinese tradition and was not taken up again for many centuries after his death. In general usage *khung yen* would be taken to mean 'empty, meaningless or untrue words', but Ssuma Chhien's technical usage is important. Ssuma Chhien was obsessed with exactly dated historical truth based on reliable sources, as his famous Chronological Tables[4] show with particular eloquence. One might say that the Chronological Tables are something of a positivistic *tour de force*. One thing that enabled Ssuma Chhien to indulge in this *tour de force* was his ability to distinguish historical factual truth from general comments.

Having established that the ancient Chinese do apply truth predicates to sentences, and having looked at the finer distinctions they made within the general semantic area around 'semantic truth', we shall now take a brief look at the expressions which may be used in Classical Chinese to express notions closely related to 'truth'.

Verbal notions like 'is true' in Classical Chinese

In translating from the ancient Chinese texts I have so far often avoided using the word 'true' in order not to beg the question at hand: how exactly do the ancient

---

[1] *Lü Shih Chhun Chhiu* 18.1, ed. Chhen Chhi-Yu, p. 1142; cf. Wilhelm (1928), p. 291.
[2] *Lü Shih Chhun Chhiu* 9.4, Chhen Chhi-Yu, p. 498; cf. Wilhelm (1928), p. 111.
[3] The evidence on this usage is discussed in B. Watson (1958), pp. 88f.
[4] *Shih Chi* 13–22. In the standard Takigawa edition these chronological tables cover exactly 756 pages.

Chinese treat the semantic area of truth? Instead of asking this question from a Western perspective, taking our notion of truth for granted and asking to what extent the Chinese expressed it, we aspire to start from the Chinese usages surrounding the notion of truth and ask ourselves just how these usages are to be understood.

What, then, were the words that the Chinese used to cover the semantic field of 'is true'? As a first orientation I shall simply list these words and try to explain their semantic nuances.

1. *Shih* 是 '(be) this, (be) it, (be) right'
2. *Fei* 非 'not be this, not be it, be wrong'
3. *Shih* 實 '(be) solid, (be) real, (be) based on fact'
4. *Hsü* 虛 'be tenuous, be unreal, not be based on fact'
5. *Jan* 然 'be so, be the case'
6. *Fou* 否 'be not so, not be the case'
7. *Yu* 有 'have, exist, there is such a state of affairs'
8. *Wu* 無 'lack, not exist, there is no such state of affairs'
9. *Chhêng* 誠 'be honest, be sincere, be genuinely so'
10. *Hsin* 信 'believe, be loyal, be trustworthy, be reliably so'
11. *Wei* 偽 'to fake, create artificially, be fake'
12. *Tang* 當 'to fit, fit the facts'
13. *Kuo* 過 'to go beyond, to not fit, to not fit the facts'
14. *Chên* 真 'be genuine, be genuinely so'
15. *Wang* 妄 'be misguided, be wrong'.

There are some other words which one might be tempted to include in this list: e.g., *chêng* 正 as 'be straight, be according to the rule, be morally correct (opposite of *hsieh* 邪), be factually accurate'. When Mêng Tzu has offended a king by explaining to him that he should be dismissed if he did not listen to remonstrations from ministers coming from his own family, the story continues:

> The King looked appalled.
> 'Your Majesty should not be surprised by my answer. Since you asked me, I dared not answer with anything but the truth (*pu kan i pu chêng tui* 不敢以不正對).'[1]

The concept of *chêng* here must be taken to involve the straightforwardness of a reply as well as the correctness of the point of view expressed in that reply. The concept of truth seems thus embedded in the concept of *chêng* as used on this occasion.

Another case in point is the word *kuo* 果 'fruit, reality, really, be borne out by the facts', as when Yang Hsiung 揚雄 (−58 to +18) meets the complaint:

> Remarkable, how the records and books are not borne out by the facts (*pu kuo* 不果)![2]

---

[1] *Mêng Tzu* 5B9; Lau (1983c), p. 219. Cf. also *Wei Liao Tzu*, ch. 12, ed. Chung Chao-Hua, p. 53.
[2] Yang Hsiung, *Fa Yen* 18, ed. Wang Jung-Pao, p. 749. Note the evidence on *kuo* 果 assembled in the commentary.

Nominal uses of truth predicates

I have demonstrated through examples that the ancient Chinese applied verbs meaning something like 'be true' to sentences, and that in this sense they had a concept of semantic truth. But did they focus on this concept as a subject of discussion? Did they have nominalised concepts of truth as opposed to expressions for 'be true'? The evidence is that they did.

When right and wrong (*shih fei* 是非) are in doubt, then assess (*tu* 度) the situation according to other independent facts (*yüan shih* 遠事) test (*yen* 驗) them on things at hand (*chin wu* 近物), check (*tshan* 參) them with a balanced mind.[1]

The assessing, the testing, and the checking are crucial concepts in the ancient Chinese attitude towards truth.

To ignore your own safety in the quest for wealth; to make light of danger and try to talk your way out of every difficulty; to rely on lucky escapes; to ignore the essential characteristics of right and wrong, of what is so and what is not so (*shih fei jan pu jan chih chhing* 是非然不然之情), and to make it one's purpose only to overpower others: such is inferior valour.[2]

Notice the place of the notion of *shi fei* 是非 in this passage from a book of the +4th century which contains much earlier material:

Yen Tzu asked Kuan Chung about 'tending life'. Kuan Chung answered: 'It is simply living without restraint; do not suppress, do not restrict.'
'Tell me the details.'
'Give yourself up to whatever your ears wish to listen to, your eyes to look on, your nostrils to turn to, your mouth to say, your body to find ease in, your will to achieve. What the ears wish to hear is music and song, and if these are denied them, I say that the sense of hearing is restricted. . . . *What the mouth wishes to discuss is truth and falsehood, and if this is denied it, I say that the intelligence is restricted.* What the body wishes to find ease in is fine clothes and good food, and if these are denied it, I say that its comfort is restricted. What the will wishes to achieve is freedom and leisure, and if it is denied these, I say that man's nature is restricted. All these restrictions are oppressive masters.'[3]

There are instances where the notions of truth and *reality* seem closely intertwined. The term *chhing shih* 情實 'the real facts' is unambiguously factual in the following passage:

The viscount Hsüan of Han said: 'My horses have had an abundance of madder and grain. Why are they so skinny? I am worried over this.'
Chou Shih replied: 'If the stableman feeds them with all the beans and grain, then they are bound to become fat. But suppose he gives them much in name but little in fact. Then they are bound to become skinny. If Your Highness does not investigate the real facts of the case (*chhing shih*) but sits there and worries about it, the horses will never become fat.'[4]

[1] *Hsün Tzu* 27.123; cf. Köster (1967), p. 362.
[2] *Hsün Tzu* 23.86; cf. Watson (1963), p. 170. Cf. also *Hsün Tzu* 8.28 for the same formula *shih fei jan pu jan chih chhing* 是非然不然之情.
[3] *Lieh Tzu* 7, ed. Yang Po-chün, pp. 140f.; tr. Graham (1960), p. 142.
[4] *Han Fei Tzu* 33.27; Liao (1939), vol. 2, p. 78.

The natural opposite of this *shih* 實 'reality, solid, real' is *ming* 名 'name, in name only', as in the passage I just quoted. This opposition between the solid facts and mere tenuous words or claims, is ubiquitous in ancient Chinese philosophical as well as historical literature. It plays a central part in the ancient Chinese intellectual construction of reality.

Another opposite to *shih* is *hsü* 虛 'tenuous, empty, unreal' as in

The way the ancients had it: 'if one bends one is preserved intact' is no empty saying (*hsü yü* 虛語).[1]

Hoshang Kung's commentary dating to at least the +2nd century, elaborates: 'These are correct words (*chêng yen* 正言) and not empty words (*hsü yen* 虛言).'[2]

Dialecticians or 'talkers' are accused of not being concerned with the real facts (*chhing* 情) but only with words. This notion of real facts must be carefully distinguished from that of verbal truth referred to in the passage just quoted from Lao Tzu:

When power is in single hands, then squabbling will stop. When squabbling stops, then dialecticians (*shuo chê* 説者) will stay away and the real facts (*chhing*) will become clear. When the real facts are unembellished, then solid reality (*shih shih* 實事) will become plain.[3]

Hsün Tzu complains that Têng Hsi and Hui Shih do not care for the 'essential characteristics of right and wrong or for what is so and what is not so (*shih fei jan pu jan chih chhing* 是非然不然之情)'.[4] The concept of *jan pu jan* 然不然 'what is so and what is not so' (or *jan fou* 然否 'what is so and what is not') is the closest the ancient Chinese came to an unambiguous concept of factual truth. Note that by using opposing terms in this way the Chinese avoid the sort of hypostasisation of an abstract entity suggested by such abstract forms as the English *truth*. The same technique is used in other relevant pairs of concepts:

The Sages established the emblems (of the Book of Changes) in order to give complete expression to their intended meaning. They established the diagrams to give full expression of the real-or-fake-character (*chhing wei* 情偽) of something.[5]

A recently discovered manuscript from the Ma-wang-tui tomb provides neat evidence of the place the Chinese accorded to truth in their scheme of things:

Only after understanding the correspondence between names and realities, only after exhaustively understanding what is real and what is fake (*chhing wei*) and being free of confusion, can one complete the Way of the emperors and kings.[6]

---

[1] *Lao Tzu* 22; cf. Lau (1983a), p. 35. Cf. also *Hsün Tzu* 18.71, Köster (1967), p. 233: 'When it comes to the sentence (*fu yüeh* 夫曰) "Yao abdicated to Shun", this is an empty saying (*hsü yen* 虛言).'

[2] Shima Kuniō (*1973*), p. 99. Cf. Eduard Erkes (1950), p. 49, and, for the dating of the commentary, *ibid.*, pp. 8–12.

[3] *Lü Shih Chhun Chhiu* 17.5, ed. Chhen Chhi-Yu, p. 1092; cf. Wilhelm (1928), p. 276.

[4] *Hsün Tzu* 8.28; cf. Köster (1967), p. 73. Cf. also *Hsün Tzu* 23.86.

[5] *I Ching, Hsi Tzhu*, ed. Kao Heng, p. 542. Khung Ying-Ta's 孔穎達 canonical commentary on the phrase *chhing wei* 情偽 confirms our interpretation.

[6] *Ching Fa*, ed. Wen-wu, p. 29.

Thus attaining Truth was seen as dependent upon obtaining truth. Importance was attached to truth, especially in the context of politics, because it ensured that the ruler was not confused about the real state of affairs, could act on the basis of fact:

If you do not check-and-compare (*tshan* 參) what you see and hear, then the genuine facts (*chhing* 情) will not reach your ear.[1]

This passage gives us the crucial connection of the notion of truth with that of comparative investigation (*tshan*). Han Fei is in no doubt about such comparative investigation:

A man who considers a matter as certain (*pi* 必) without checking-and-comparing it and testing the evidence (*tshan yen* 參驗) is stupid.[2]

Considering something as certain, for Han Fei, is considering something as certainly true. Throughout his book he emphasises the need to check the reliability or certainty of what one hears. He speaks of 'deciding the genuine facts of a matter (*chhêng* 誠) by checking and comparing'.[3] Han Fei talks about 'trying to find the genuine facts of a matter (*chhiu chhi chhêng* 求其誠)'[4] and recommends 'listening to others' words and trying to find what fits the facts (*chhiu chhi tang* 求其當)'[5] as well as 'comparing and checking words in order to understand the genuine facts of a matter (*chih chhi chhêng* 知其誠).'[6]

The nominal concept of *chhêng* 誠 is not always purely factual, for Han Fei can also ask:

Whom could one ask to decide the real facts of the case (*chhêng*) of the Mohists and Confucians?[7]

Here the question of *chhêng* clearly involves both factual and moral issues, just as it does in the case of Western concepts of truth.

Of course, Han Fei was by no means the only one to take such an attitude to the importance of finding out whether words are true or not.

When one gets hold of words one must check (*chha* 察) these. . . . Hearing something and investigating (*shen* 審) it brings good fortune. Hearing something without investigating it is even worse than not hearing it at all. . . . Whenever one hears words, one must carefully appraise the person (from whom they come), and one must test (*yen* 驗) them according to the attributes/principles of things (*li* 理).[8]

What more splendid evidence of the scientific spirit and interest in truth during pre-Han times can one look for than such programmatic statements?

[1] *Han Fei Tzu* 30.2.2; Liao (1939), vol. 1, p. 281.   [2] *Han Fei Tzu* 50.1.37; Liao (1939), vol. 2, p. 299.
[3] *Han Fei Tzu* 47.2.31; Liao (1939), vol. 2, p. 250.   [4] *Han Fei Tzu* 32.2.20; Liao (1939), vol. 2, p. 27.
[5] *Han Fei Tzu* 46.6.12; Liao (1939), vol. 2, p. 247.   [6] *Han Fei Tzu* 48.4.21; Liao (1939), vol. 2, p. 266.
[7] *Han Fei Tzu* 50.1.31; Liao (1939), vol. 2, p. 298.
[8] *Lü Shih Chhun Chhiu* 22.6, ed. Chhen Chhi-Yu, p. 1526; cf. Wilhelm (1928), p. 399.

Consider again the preface to the *Lü Shih Chhun Chhiu*:

Above I have explored Heaven, below I have tested (*yen*) the Earth, in between I have investigated (*shên*) men, and in this way what is right or wrong, what is acceptable or unacceptable (*shih fei kho pu kho* 是非可不可) has nowhere to hide.[1]

Consider again the case of historical truth. Here is what Pan Ku 班固 (+32 to +92) had to say about his predecessor the Grand Historian Ssuma Chhien 司馬遷 (−145 to *c.* −89):

Liu Hsiang 劉向 (−79 to −8), Yang Hsiung 揚雄 (−59 to +18), and other men of wide learning and copious letters all praise (Ssuma) Chhien as a man of excellent ability as a historian and testify to his skill in setting forth events and their causes. He discourses without sounding wordy; he is simple without being rustic. His writing is direct and his facts sound. He does not confer empty praise nor does he conceal what is evil. Therefore his may be termed a 'true record' (*shih lu* 實錄).[2]

Yang Hsiung, in his ardently Confucian work *Fa Yen* is asked:

'In five hundred years one sage arises', is there such a state of affairs (*wu pai sui êrh shêng jên chhu, yu chu* 五百歲而聖人出有諸)?[3]

Yang Hsiung is questioned about the truth of an honoured Confucian commonplace which has the unquestioned authority at least of Mencius. But after having given some perfectly plausible examples of sages who were near-contemporaries, like the legendary sage emperors Yao and Shun, he concludes that 'one cannot tell if there will be one even in a thousand years'. Li Kuei's 李軌 commentary (early +4th century) to the *Fa Yen* brings out the meaning perfectly:

Whether there is one person (i.e., sage) in a thousand years or a thousand persons (i.e., sages) in one year may not be known.[4]

Yang Hsiung also answers the following question in the affirmative:

'To use the methods of the sages of the past in order to govern the future, that is like glueing the tuning columns to the body of the *sê* 瑟-lute and then trying to tune the instrument', is there not such a state of affairs?
(Yang Hsiung) answered: 'There is.'[5]

In general, the *Fa Yen* shows a very considerable Confucian interest in the truth and consistency of Confucian doctrines. A wide range of un-Confucian objections are raised and answered in that book dating from around +5.

---

[1] *Lü Shih Chhun Chhiu, Hsü I*, ed. Chhen Chhi-Yu, p. 648; cf. Wilhelm (1928), p. 155.
[2] *Han Shu* 62, ed. Chung-hua-shu-chü, p. 2738; cf. B. Watson (1958), p. 68.
[3] Yang Hsiung, *Fa Yen* 11, ed. Wang Jung-Pao, p. 370. Cf. also the important, but unfortunately problematic, passage in *Fa Yen* 12, ed. Wang Jung-Pao, p. 450, where Yang Hsiung discusses the notion of *chên wei* 真偽 'genuine/fake, genuineness, truthfulness(?)'.
[4] *Ibid.* I see no need whatever to see a misprint here to the effect that Li Kuei was only saying there could be one sage in one year, as one commentator has suggested.
[5] Yang Hsiung, *Fa Yen* 12, ed. Wang Jung-Pao, p. 436.

In Han times *shih* 實 'solid, real' was commonly nominalised and came to mean something like 'the solid facts of a matter'.

Wang Chhung 王充 (+27 to +100) writes programmatically:

In (scientific) discourse (*lun* 論), what matters is what is right (*shih* 是).[1]

In his preface, Wang Chhung makes it quite explicit that his purpose in writing *Lun Hêng* is none other than to refute untruths and thereby to promote truth. Wang Chhung is indeed a living refutation of the view that Chinese philosophers did not place central emphasis on the notion of the scientific objective truth of doctrines. He says, in the same context:

> Explanatory discourse disputes about what is so and what is not so.[2]
> One discards what is fake (*wei* 偽) and retains what is genuine (*chên* 真).[3]

Aristotle defines philosophy as 'knowledge of truth/Truth (*epistēmē tēs alētheias ἐπιστήμη τῆς ἀληθείας*)'.[4] Wang Chhung might easily have voiced the same view as Aristotle – if he had had a concept of philosophy available to him which for interesting reasons not to be discussed here he did not have. In any case he was, in practice, a seeker of truth, asking such questions as whether man was deliberately created by Heaven and Earth, whether ghosts exist, etc., and demanding factually correct reasoned answers. His *opus magnum*, the *Lun Hêng* consists in weighing against each other conflicting answers to questions of this sort.

One might object that the *Lun Hêng* sees itself as exceptional, which of course it is. But consider now this authoritative main-stream definition of discourse (*lun* 論):

As a genre the *lun* performs the function of distinguishing and getting right what is true and what is not.[5]

Here the leading literary critic of traditional China declares the object of discourse (*lun*) to be the establishing of the truth.

Finally, let us note that the concept of truth is closely linked with that of evidence. Han Fei mentions the notion of evidence (*yen* 驗) establishing the substantiveness of a claim:

When arguments (*pien* 辯) are unsupported by evidence (*yen*), these are speeches for which one is blamed.[6]

Yang Hsiung insists:

As for the words of a gentleman, if they are obscure they have their evidence (*yen*) in something clear. If they are remote, they have their evidence (*yen*) in something close at hand. If

---

[1] *Lun Hêng* ch. 85, ed. Chung-hua-shu-chü, p. 1690; cf. A. Forke (1964), vol. 1, p. 73.
[2] *Ibid.*
[3] *Ibid.* The fact that *chên* 真 'genuine' and *wei* 偽 'fake' very often refer to persons rather than doctrines remains important.
[4] *Metaphysics* A, 993b20.
[5] Liu Xie (+465 to +522), *Wen Hsin Tiao Lung*, ed. and tr. Shih (1983), pp. 204f.
[6] *Han Fei Tzu* 18.3.16; Liao (1939), vol. 1, p. 152.

they are overwhelming, they have their evidence (*yen*) in something manageable. If they are subtle, they have their evidence (*yen*) in something plain. Speaking without evidence (*yen*) is mistaken. Is the gentleman mistaken? No he is not mistaken.[1]

Yang Hsiung's concern is, of course, with good government. But he takes his analogy from the science of arithmetic:

In running a state to hope for efficient results without following the law must be compared to the case of the calculating sticks.[2]

Yang Hsiung did not even have to explain that the calculating sticks serve to determine the correct result of arithmetic calculation. He wanted the operation of government to be governed by the sort of impartial concern for truth that is characteristic of arithmetic. Laws were a means to introduce such objectivity. – But this, of course, leads us astray from our subject, the concept of scientific truth, and towards Chinese social thought. The significant point is that the two realms are not as separate as one might imagine.

Conclusion

We conclude that far from finding the notion of truth inconceivable, ancient Chinese philosophers frequently asked themselves whether some statement was true or not, although they did not show the same degree of philosophical preoccupation with factual truth as Westerners might expect.

The Chinese regularly applied the predicate 'true' to words or statements. They often referred to the nominalised notion of truth. None the less, there is a crucial difference of attitude with Greek thinkers in ancient times which comes out in this quotation:

As for the gentleman's explanations (*shuo* 説) . . . when they are sufficient to understand the real state of things (*wu chih chhing* 物之情) and when having grasped the words men can live their lives, then one stops.[3]

This is where much of Greek philosophical analysis starts. Greek philosophy is predominantly not pragmatic but theoretical in ambition and approach.

The ancient Chinese may have taken a pragmatic approach to language and thinking. But as pragmatists should, they had plenty of use for the scientific notion of objective factual truth. Whereas Greek philosophers were very often preoccupied with the notions of factual and evaluative truth for its own sake, their Chinese counterparts looked upon language and thought as much more pragmatically embedded in social life. Their key concept was that of the Way (*tao*) of conducting human affairs, not of objective factual or doctrinal truth.

---

[1] Yang Hsiung, *Fa Yen* 8, ed. Wang Jung-Pao, p. 246. For a verbal use of the word *yen* 驗 meaning 'to provide evidence for, to prove' see *ibid.*, p. 391.

[2] *Fa Yen* 12, ed. Wang Jung-Pao, p. 459.

[3] *Lü Shih Chhun Chhiu* 16.8, ed. Chhen Chhi-Yu, p. 1019; cf. Wilhelm (1928), p. 258.

Moralist pragmatism is, of course, not an oriental wisdom alien to the West. There were plenty of people in ancient Greece who thought this way, and in more recent times Jean Jacques Rousseau's moralistic pragmatism had a profound political and educational influence:

Le Dogme n'est rien, la morale est tout.[1]

All this does not mean that Rousseau had no concept of truth. It means that he was particularly interested in certain kinds of things.

The objective truth of a statement was considered in China along with its political appropriateness, its practical usefulness for the organisation of society and for the cultivation of one's personal life. Those who, like the Later Mohists, to a certain extent tried to detach the theoretical considerations of objective truth from their political, social, and personal contexts, were doomed to a marginal place in Chinese intellectual history.

Collectively, Chinese intellectuals have tended towards a holistic view of meaning and truth which refuses to detach these concepts from their anthropological and historical context. The Neo-Confucian term *tao li* 道理 which we may be tempted to translate as 'truth' because this is what the word *tao-li* sometimes means in anglicised Modern Chinese, is in fact something much closer to our modern 'point' as in 'he has a point':

The poems of Su Tung-Pho's later years are certainly good. The only thing is that much of the wording is haphazard and not at all to the point (*chhüan pu khan tao li* 全不看道理).[2]

Thus we must be careful not to impose our Western scientific notions of objective truth upon Chinese texts where they really do not belong. There is excellent reason to warn against the uncritical assumption that the notion of scientific objective truth played the same kind of central rôle in China that it did in the West.

In China one tended to look upon someone who, we would say, is wrong, as someone who sees only part of the truth. The sage is thought to have the overarching vision reconciling all such partisan or partial views, seeing the point of each. Chu Hsi brings this out explicitly:

Ordinary people's studies are often partisan to one point (*li* 理), they are dominated by one theory (*shuo* 説). Therefore they are not catholic in their vision. As a result disputes arise. The Sage stands at the exact centre and is conciliatory, leaning to neither side.[3]

The Chinese quest for 'truth' tends to be the quest for this un-partisan catholic vision which is above controversy in the sense that it understandingly embraces rather than refutes seemingly opposing views.

The remarkable last chapter of the book *Liu Tzu* 劉子, which we must probably attribute to the literary critic Liu Hsieh 劉勰 (+465 to +522), lists and characterises

---

[1] 'Dogma is nothing, morals is everything.' Œuvres complètes, vol. 4 (Bibliothèque de la Pléiade), p. 1077.
[2] *Chu Tzu Yü Lei*, ed. Wang Hsing-Hsien, p. 3326.   [3] *Ibid.*, p. 130.

the nine schools of Chinese thought[1] and then gives the weaknesses of each of these schools. It does not ask which school is right. It asks what point each school is getting at and how far it gets.[2] This, it seems to me, remained a characteristic Chinese attitude even after the establishment of a Confucian orthodoxy during the Sung.

(4) THE CONCEPT OF NECESSITY

Aristotle was interested in the philosophical difference between contingent statements like 'It is raining' and necessary or analytic statements like 'Two and two is four'.

Ancient Chinese philosophers generally did not make this difference into a major topic of philosophical reflection, but the Mohists, in their opaque way, did try to grapple with the problem of exactly how to understand and define their concept *pi* 必 which we translate by 'necessarily'. The necessary logical relations obtaining between propositions, such as that of one proposition being a logical consequence of another, have not been widely discussed as such among Chinese thinkers. In this specific sense logical necessity was not an issue that preoccupied the ancient Chinese.

Suppose all white swans die of white swan plague tomorrow: then it will be quite generally true to say that 'All swans are black'. However, this statement, though quite generally true, would be contingently, not necessarily, true. It just so happens that all the white swans have died.

Factual inevitability in this world was defined as follows:

The necessary (*pi chê* 必者) is what is ordained by Heaven.[3]

But the problem I find important for the history of science is whether the Chinese had a notion of inevitability in all possible worlds, a logical kind of inevitability which goes beyond factual concomitance or invariance in the world as we have it.

To the extent that the ancient Chinese reasoned along logical lines we might say that they employed notions of logical consequence and thereby of various varieties of logical necessity. But my question is whether they explicitly focused on such relations of logical necessity which obtain not only in the world as we have it but in any world that can be conceived of.[4]

---

[1] Confucianism, Taoism, school of Yin and Yang, logicians, legalists, Mohists, diplomatists, syncretists, and agriculturalists.

[2] Cf. *Liu Tzu*, ed. Lin Chhi-Than and Chhen Feng-Chin, pp. 301ff. For the authorship and dating see *ibid.*, pp. 335–96.

[3] *Ching Fa*, ed. Wen-wu, p. 28.

[4] I am aware that the notions of logical necessity, of logical consequence, and of possible worlds are highly controversial philosophically, and that I am treading on treacherous ground. However, this will not prevent me from trying to chart ancient Chinese ways of using their own notion *pi* 'be necessary', and from investigating the extent to which we find that the Chinese employed this word or other devices to mark sentences as analytically true or necessarily true in all possible worlds. For it so happens that the notion of (necessary) truth in all possible worlds is important for the history of theoretical science.

In general *pi* marks invariable truth in this world in descriptive sentences:

A man of moral charisma will necessarily (*pi*) be the author of sayings. The author of say-ings does not necessarily (*pi*) have moral charisma. A benevolent man will necessarily (*pi*) possess courage, but a courageous man does not necessarily (*pi*) possess benevolence.[1]

Confucius is not at this point presenting an example of pure semantic analysis. He generalises about the world in which he lives. This kind of generalisation is marked by *pi*.

Contingent invariance and something more like a conceptual link seem to be expressed by the same grammatical form, in Chinese as in English.

People who are clever with their mouths are not necessarily (*pi*) trustworthy.[2]

Here we still have a general observation concerning the world in which Confu-cius lived. We do not have conceptual analysis.

The inevitability of a law of nature is again something which obtains in this world:

Everything that lives between Heaven and Earth must necessarily (*pi*) die. This is some-thing which cannot be avoided.[3]

Note that there is nothing logically impossible about a world in which things live eternally. It just so happens that things in this world invariably are not like that. Yang Hsiung 揚雄 (−59 to +18) explains this quite properly:

That which has life will necessarily (*pi*) have a death. That which has a beginning will neces-sarily (*pi*) have an end. Such is the Way of what is so of itself (*tzu jan chih tao yeh* 自然之道也).[4]

At times, these strict regularities of nature become the subject of explicit dis-course:

The sage knows the necessary patterns/attributes of things (*pi jan chih li* 必然之理).[5]

One does not have a very precise idea about what the necessary patterns of things were. In later times, the notion of *li* 理 'pattern, principle' became a crucial concept of Chinese philosophy. But note that these patterns or principles are always patterns or principles of this world as we have it. They are not logical principles or patterns.

At times, one does begin to perceive a purely abstract, conceptual interest, as when Tzu Lu asks:

Why is it necessary (*pi*) to read books before one can count as learned?[6]

[1] *Lun Yü* 14.4; cf. Lau (1983), p. 133.   [2] *Huai Nan Tzu*; ch. 10, ed. Liu Wen-Tien, p. 17b.
[3] *Lü Shih Chhun Chhiu* 10.2, ed. Chhen Chhi-Yu, p. 524; cf. Wilhelm (1928), p. 120.
[4] Yang Hsiung, *Fa Yen* 18, ed. Wang Jung-Pao, p. 768. Hsi Khang of the +3rd century (ed. Tai Ming-Yang, p. 262) resumes the phrase 'the Way of what is so of itself', which begins to remind one of the notion of a law of nature.
[5] *Shang Chün Shu* 18, ed. Kao Heng, p. 144; cf. Duyvendak (1928), p. 292.
[6] *Lun Yü* 11.25; cf. Lau (1983), p. 133.

One can read this question as asking whether the concept of learnedness necessarily involves the notion of bookishness. Note the neat logical contrast with the grammatically similar:

Does Hsü Tzu invariably (*pi*) weave a cloth before he wears it?[1]

Cases where one feels the ancient Chinese were moving towards conceptual analysis are not uncommon:

Now he who desires profit will necessarily (*pi*) abhor loss. . . . He who wants good government will necessarily (*pi*) abhor political chaos.[2]

This does sound like a conceptual reflection. I cannot imagine Han Fei seriously persuading anybody who desires profit that he should abhor loss, neither can I imagine prescribing such a procedure. The matter is too obvious. We do seem to have an analytic statement made for the purpose of conceptual clarification in the course of an argument.

When issuing a prohibition, one necessarily (*pi*) wishes something not to be done. When issuing an order, one necessarily (*pi*) wishes something to be done. When seeking something, one necessarily (*pi*) wishes to obtain it.[3]

These three sentences read like conceptual clarifications rather than descriptions of the state of the world as it is.

In the *Yin Wên Tzu* 尹文子 (+3rd century) we read:

That which has a form will necessarily (*pi*) have a name. But that which has a name does not necessarily (*wei pi* 未必) have a form.[4]

It turns out that the commonest form in which the notion of necessity occurs in ancient Chinese literature involves negation:

Chhü Tao was fond of water-chestnuts. King Wên preferred calamus pickles. Not that these were 'correct' tastes. The two talented men just set store by these things. What people have a taste for is not necessarily (*pu pi* 不必) of excellent taste. Duke Ling of Chin liked Shan Wu-Hsü. Kuai of Yen regarded Tzu Chih as the most talented. Not that these two were 'correct' gentlemen. The two rulers just honoured them. What one regards as a sage is not necessarily (*pu pi*) a sage.[5]

Han Fei seems to be answering not just the empirical question whether those who are regarded as sages in fact turn out to be sages in this real world. He seems at the same time to focus on a presumed logically necessary conceptual link between being regarded as a sage and being a sage. In any case, this type of discourse is common:

As for a great man, his words are not necessarily (*pu pi*) truthful, and his actions do not necessarily (*pu pi*) bear fruit. The point is in his sense of duty.[6]

---

[1] *Mêng Tzu* 3A4.4; Lau (1983c), p. 103.   [2] *Han Fei Tzu* 46.4.30 and 46.4.35; Liao (1939), vol. 2, pp. 242f.
[3] *Kuan Tzu*, ed. Tai Wang, vol. 1, p. 71; cf. Rickett (1985), vol. 1, p. 252.
[4] *Yin Wên Tzu*, ed. Lin Shih-Hsi, p. 4.   [5] *Han Fei Tzu* 39.12.2; Liao (1939), vol. 2, p. 197.
[6] *Mêng Tzu* 4B11; Lau (1983c), p. 163.

Chuang Tzu reflects:

Thus if you are enthroned as Son of Heaven, you are not necessarily (*pu pi*) honourable. If you stay stuck as a commoner, you are not necessarily (*pu pi*) base. Your share of honour or dishonour depends on how fair or foul your conduct is.[1]

It seems that Chuang Tzu shows an interest in the concept of honour as such. He is at pains to show that honour is not an external concept linked to positions of honour and the like. He denies that there is a conceptual or logical link between being an emperor and being honourable. His notion of necessity, at this point, is conceptual. Whether we should say that he speaks of strictly logical necessity here is a subtle question which must perhaps remain open, but passages like the ones here under discussion certainly raise such questions regarding early Chinese thought. Here is another representative instance:

A man must necessarily (*pi*) know ritual before he can show polite reverence.[2]

If one takes a very strict view of logical necessity, defining it in terms of truth in any conceivable or possible world, then I suppose there is no neat evidence that the ancient Chinese ever thought in such terms. They failed to make that explicit step from the consideration of the world as it appeared to them to the consideration of any possible world that anyone might conceive. On the other hand, they did habitually observe links between concepts, links which they considered to be either necessary (*pi*) or not necessary (*wei pi* 未必／*pu pi* 不必).

The Later Mohists maintain, quite generally:

Names and objects do not connect necessarily (*pi*).[3]

Finally, I need to consider the question whether the ancient Chinese ever problematised the notion of necessity and made it the subject of philosophical reflection. Not surprisingly, there is evidence that the Later Mohists did take this step. They certainly did try to define the concept *pi*. Unfortunately, their explanations remain textually and philosophically opaque to me.

Canon: The necessary (*pi*) is the unending.[4]

(5) THE CONCEPT OF CONTRADICTION

Shakespeare writes (*Othello* iv.4.14): 'Her honour is an essence that's not seen;/ They have it very oft that have it not.' On the face of it, Shakespeare is here asserting

---

[1] *Chuang Tzu* 29.62; cf. Graham (1981), p. 239.
[2] *Kuan Tzu*, ch. 5, ed. Tai Wang, vol. 1, p. 44; cf. Rickett (1985), vol. 1, p. 196.
[3] *Names and Objects* 1; Graham (1978), p. 470.
[4] Explanation: It applies to cases where complements are perfect. Such cases as 'elder brother or younger' and 'something so in one respect or not in one respect' are the necessary and the not necessary. Being this or not this is necessary. *Mo Ching* A51; cf. Graham (1978), p. 299. I quote the translation in Graham (1989), p. 143. A. C. Graham has changed his mind on the interpretation of this passage. I must confess that I still feel there is no satisfactory explanation or translation of the text.

incompatibles. But everyone realises at once that the contradiction is only an apparent one.

In China there are some thinkers (especially the Taoists) who were fond of paradoxes. Indeed we read in a recent paper that the Chinese were not bothered by contradiction at all:

Even the highest authorities on logic in China literally did not know what they were talking about, and frequently contradicted themselves *without being bothered by it!*[1]

This thought has a long and distinguished history:

Le principe de contradiction ne préoccupe pas plus les Chinois que le principe de causalité.[2]

Statements like those by Marcel Granet are useful because they force us to state clearly the reasons we might have for attributing to the ancient Chinese the concepts of incompatibility and of contradictories. Donald Leslie (1964) has tried to take up this challenge and to demonstrate that the notion of contradiction did play a rôle in ancient Chinese thinking.

Let us start out with some crucial conceptual distinctions. Firstly, we must distinguish between the view that two statements actually will never be true at the same time in the world as it is (factual incompatibility) and the view that contradictories can never be concurrently true under any conceivable circumstances (logical incompatibility). We are especially interested in the latter.

Secondly, we must distinguish between statements which cannot at the same time be true (incompatibles) and statements which cannot at the same time be true but one of which must necessarily be true (contradictories). Saying that something is a buffalo is incompatible with saying that it is a horse. 'This is a horse' and 'This is a buffalo' are incompatible predicates but not contradictory ones, because they may both at the same time fail to apply. As in the case of a dog.

Contradictory statements would be statements such as 'This is a horse' versus 'This is not a horse'. Ancient Chinese philosophers from Confucius onwards were concerned with incompatible predicates and statements, but they showed little theoretical interest in the notion of contradictories, i.e., predicates which are such that they cannot both apply or both not apply to a given object.

Inconsistency between actions

Very common in ancient China is the accusation of inconsistency in action, as here exemplified by Mêng Tzu:

Chhen Chen asked: 'The other day in Chhi the King presented you with a hundred i 鎰 of superior gold and you refused. In Sung you were presented with seventy i and you accepted.

---

[1] Becker (1986), p. 84. The italics are Becker's own.

[2] 'The principle of contradiction does not preoccupy the Chinese any more than the principle of causality.' Granet (1934), p. 385. Cf.: 'Neither the principle of causality nor the principle of contradiction can be said to take the rôle of guiding principles.' *Ibid.*, p. 329 (my translation).

In Hsüeh you likewise accepted fifty *i*. If your refusal in the first instance was right, then your acceptance on subsequent occasions must be wrong. If your acceptance was right, your refusal must be wrong. You cannot escape one or the other of these two alternatives.'[1]

The underlying thought here is that a morally responsible person should react similarly in similar situations. If he reacts differently to the same sort of situation there is a contradiction *in actu*. Mêng Tzu recognises this as a valid objection, but he defends himself by pointing out that the situations were not in fact similar in the relevant respect.[2]

Incompatibility between words and deeds

Donald Leslie has drawn our attention to a passage where Mo Tzu 墨子 (−5th century) speaks explicitly of the incompatibility of words and deeds:

That means that in speaking one rejects universality but in making (moral) choices (*tsê* 擇) one accepts universality. This is incompatibility between words and deeds.[3]

This notion remained current and perhaps even predominant in ancient China.

Incompatibility between psychological attitudes

Confucius, for one, was interested in incompatible desires:

Tzû Chang asked about . . . recognising confusion (*huo* 惑). The Master replied: '. . . If you love something you want it to live; if you hate something you want it to die. To wish something to live and (at the same time) to wish it to die is confused (*huo*).'[4]

Here Confucius focuses on the notion of being confused (*huo*) which seems to us closely related to incompatibility of desires.

Here is another explicit example formulated in terms of incompatibility of psychological attitudes:

Mohist arguments are self-contradictory (*tzû wei chhi shu* 自違其術). On the one hand the Mohists place light emphasis on burial, but on the other hand they honour the ghosts of the deceased.[5]

Incompatibility between desires and actions

Consider the realist philosopher Han Fei 韓非 (died −233):

Therefore, incompatible things do not coexist (*pu hsiang jung chih shih pu liang li* 不相容之事 不兩立). For instance, to reward those who kill their enemies in battle, and at the same time

---

[1] *Mêng Tzu* 2B4; Lau (1983c), p. 77.
[2] For contradiction in practice see e.g., Mêng Tzu 5B6 and Chhên Ta-Chhi (*1969*), pp. 55ff.
[3] *Mo Tzu* 16.33/45; cf. Mei (1929), pp. 90/92.
[4] *Lun Yü* 12.2; cf. Lau (1983a), p. 113, who does not recognise the logical edge of this passage.
[5] *Lun Heng*, ch. 67, ed. Chung-hua-shu-chü, p. 1321; cf. Forke (1964), vol. 2, p. 375.

to esteem deeds of mercy and generosity; to reward with ranks and bounties those who capture enemy cities and at the same time to believe in the theory of impartial love . . .[1]

Han Fei wants to point out that certain actions are incompatible with certain attitudes. We can speak here, in a loose sense, of real contradictions. So this is a notion which the modern Chinese by no means had to learn from Marxism.

Incompatibility between statements

Han Fei was also interested in the abstract concept of contradiction between statements (not only actions and attitudes) and constructed a famous joke about the matter. He was so fond of this joke that he repeated it twice in his book, and again referred back to it explicitly on two separate occasions. Janusz Chmielewski has drawn our attention to the logical significance of this passage:

In Chhu there was a man who sold shields and lances and who praised his merchandise as follows: 'My shields are so strong that nothing can pierce them.' He went on to praise his lances thus: 'My lances are so strong, they will pierce anything whatever.' Someone commented: 'What if you use one of your lances to pierce one of your shields?' The man from Chhu was lost for words.[2]

The man from Chhu was lost for words because he was caught in a contradiction. He was caught out making two incompatible statements. These statements, as Han Fei puts it, 'cannot at the same generation/time be established (*pu kho thung shih êrh li* 不可同世而立)'.[3] It turns out that Han Fei liked to point out that those he criticised held in this way incompatible positions (*mao tun chih shuo* 矛盾之説).[4]

Wang Chhung 王允 (+27 to +100) maintains that certain accounts of the Spring and Autumn period 'attacked each other (*hsiang fa* 相伐)', i.e., were incompatible and could not at the same time be true, in the sense that if one was right (*shih* 是) the other had to be wrong (*fei* 非) and *vice versa*.[5] By his time, this was a current intellectual ploy with a long history.

Confucius's disciples were well-known for attacking their master as being inconsistent in his comments:

The Master went to Wu Chheng. There he heard the sound of the the zithern and singing. The Master broke into a smile and said: 'Surely you don't need to use an ox-knife to kill a chicken.'[6]

---

[1] *Han Fei Tzu* 49.10.20; cf. Liao (1939), vol. 2, p. 287.

[2] *Han Fei Tzu* 36.4.20; cf. Liao (1939), vol. 2, p. 143. The same story is retold with slight variations in *Han Fei Tzu* 40.5.37.

[3] *Han Fei Tzu* 36.4.31; cf. Liao (1939), vol. 2, p. 143.

[4] Cf. *Han Fei Tzu* 36.4.33 and 40.5.44. For a collection of instances where Han Fei seems to use the concept of contradiction or incompatibility, cf. Chou Chung-Ling *Han Fei Tzu ti lochi*, Peking 1958, pp. 16–21. For the later history see the *Hsi Khang Chi*, ed. Tai Ming-Yang, pp. 298 and 277, with its commentaries.

[5] *Lun Hêng*, ed. Chung-hua-shu-chü, p. 1592; cf. Forke (1964), vol. 1, p. 452.

[6] Confucius's implication is that it is incongruous to play a noble instrument in such a base environment.

Tzû Yu 子游 commented: 'Some time ago I heard it from you, Master, that the gentleman instructed in the Way loves his fellow men[1] and that the small man instructed in the Way is easy to command.'[2]

Confucius seems to have recognised the inconsistency and simply pleaded that he had just been joking on the first occasion:

The Master said: 'My friends, what Tzû Yu said is right. My remark a moment ago was only made in jest.'[3]

In a text attributed to Teng Hsi but probably compiled in Han times we read:

One who speaks distinguishes different kinds and makes sure that they do not contradict each other (*hsiang hai* 相害), one keeps the theses in their proper order so that they do not get confused. In bringing out what one has on one's mind one should convey one's intended meaning and not to get involved in contradictions (*kuai* 乖).[4]

Incompatibility between terms

The Master said: 'I have never met anyone who is truly unbending.'
Someone objected: 'But what about Shen Chheng?' The Master replied: 'Chheng is full of desires. How can he (*yen tê* 焉得) be unbending?'[5]

Confucius plainly regards 'being unbending' and 'being full of desires' as logically incompatible predicates. It is on this basis that his argument works.
There is also incompatibility between nouns:

'Gentleman' is the opposite (*fan* 反) of 'little man'.[6]

The term *fan* refers to contraries rather than contradictories in such contexts.
Yang Hsiung 揚雄 (−53 to +18), when challenged, insists that subtlety (*wei* 微) and clarity (*ming* 明) are not incompatible. The case is logically complex:

As for the subtle (*wei*), in seeing it to be clear (*ming*), is that contradictory (*pei* 誖)?[7]

The notion of contradictories versus contraries

We have seen that the ancient Chinese thinkers were indeed concerned to avoid inconsistency of action, of intention or between statements. But did they develop the notion of contradictories as opposed to contraries?

---

[1] This is taken to imply that the gentleman spreads high culture among his (common) fellow men.
[2] *Lun Yü* 17.4; cf. Lau (1983a), p. 171.　[3] *Ibid.*
[4] *Têng Hsi Tzu* (dating to the Han dynasty?), ed. Wilhelm, p. 69.　[5] *Lun Yü* 5.11; cf. Lau (1983a), pp. 39f.
[6] *Hsün Tzu* 3.16; cf. Köster (1967), p. 23. *Hsin Shu* 8, ed. Chhi Yü-chang, pp. 927ff., contains dozens of examples of this kind of *fan* 反.
[7] Yang Hsiung, *Fa Yen* 9, ed. Wang Yung-Pao, p. 275. Wang Yung-Pao's generally very fine commentary does not grasp the logical force of *pei* 誖.

One may wonder whether the Chinese ever paid special theoretical attention to the notion of contradictories. In our survey of Later Mohist Logic we shall find that this is indeed the case.[1] At this point we give a brief preview of some essential evidence.

Canon: Logical analysis (*pien* 辯) is contending over claims which are the contradictories of each other (*chêng fan* 正反).
Explanation: One calling it 'an ox' and the other calling it 'not an ox', that amounts to contending over claims which are the contradictories of each other.[2]
Contradictories (*fan*) are what cannot be both regarded as unacceptable at the same time.[3]

If one maintains a sentence, one is thereby committed to denying the negation (or rejecting the denial) of what one says. Consequently, refusing to deny any sentences, as the Taoists are fond of doing, is simply inconsistent:

Canon: To reject denial is inconsistent.
Explained by: he does not reject it.
Explanation: If he does not reject the denial (of his own thesis that denial is to be rejected) then he does not reject denial. No matter whether the rejection is to be rejected or not, this amounts to not rejecting the denial.[4]

Suppose someone maintains that one should reject denial. Then, if he means what he says, the thesis that one should not reject denial should be denied. But if one admits that that thesis should indeed be denied, then one is not really maintaining that one should reject denial: In at least one instance (i.e., the thesis in which one rejects denial) one fails to reject denial. One's position is therefore inconsistent.

For a more detailed discussion of such technical Later Mohist arguments the patient reader is referred to our Section (*f*,4). But it is important to keep in mind that a keen awareness of contradiction did not disappear with the Mohists.

Concluding remarks

Even an 'anti-intellectual' Confucian attitude was expressed using the notion of logical contradiction:

An old saying goes like this: 'Not knowing is compatible with (*pu hai* 不害) being a gentleman. Knowledge does not make someone any less of a small man.'[5]

Indeed, one might argue that making contradictory claims is not incompatible with being a gentleman. Han Fei tells us that Confucius once defends himself against the charge that he is giving inconsistent explanations of good administration by pointing out that the different explanations were directed at different concrete political circumstances.[6] This suggests 1. that the charge of inconsistency was

---

[1] For a survey of Later Mohist Logic the reader must turn to our Section (*f*,4).
[2] Graham (1978), A40.   [3] *Ibid.*, A73.   [4] *Ibid.*, B79.   [5] *Yin Wên Tzu*, ed. Li Shih-Hsi, p. 14.
[6] Different versions of this story are conveniently assembled in *Shuo Yüan* 7, ed. Chao Shan-I, pp. 182ff.

taken seriously, and 2. that Confucius' (sometimes apparently incompatible) statements should perhaps be construed not as theoretical abstract explanations or definitions, but rather as concrete responses to historical situations.

There is nothing particularly Chinese about such a pragmatic view of public statements and contradiction. The public speaker Demades (−4th century) proudly claimed that he had often made claims that were contrary to his own, but never claims that were contrary to the interests of the city state (*hautō men auton t'anantia pollakis eirēkenai, tē de polei mēdepote*, αὐτῷ μὲν αὐτὸν τἀναντία πολλάκις εἰρηκέναι, τῇ δὲ πόλει μηδέποτε).[1]

(6) THE CONCEPT OF A CLASS

First order predicate logic can be considered as a theory of classes or sets. Chinese logical writings like the White Horse Dialogue by Kungsun Lung have in the past been interpreted as a part of Chinese reflections on classes or sets. But did the ancient Chinese actually have a clear concept of a class? Janusz Chmielewski (1969, part I on early Chin Logic I) argues that they did. A. C. Graham (1965) disagrees.[2] Neither of them offers a detailed conceptual history of the word *lei* 類 to back up their claims. We need to look carefully at the history of this concept in order to determine whether it designates a class of items, or whether it would seem to refer to a kind of stuff.

The notion of a category, a class or a set, is important in the history of science. It is through the categorisation and classification of his environment that man begins his long march towards science. Every word of a language represents, in a way, a classification or categorisation of the human environment, but it is not at all evident that every language must have an abstract notion of a class. Indeed the abstract notion of a class versus a natural kind is a fairly modern notion.

Few people can have been more preoccupied with classification and taxonomy than the Chinese. One striking and early example must suffice here to remind us of this fact: the chapter *Ku Chin Jên Piao* 古今人表 'List of Men Past and Present', which forms chapter 20 of the *Chhien Han Shu* 前漢書, completed by the great woman scholar Pan Chao 班昭 (died *c*. +120). This list ranks no less than 1,955 individuals according to nine grades of their moral and intellectual worth.[3]

In China, the development of the notion of a class or category started from that of an ancestral group, a racial group, an animal kind, a biological kind, a natural kind, to become, finally, a humanly defined set of things.

---

[1] Plutarch, *Parallel Lives*, ed. Loeb, vol. 7, p. 30.

[2] See Graham (1965), p. 143. 'Objects are conceived not as members of a *lei* 類 but as "of the same kind (*thung lei* 同類)" or "not of a kind (*pu lei* 不類)". One may say that oxen are "of the same kind", but one cannot use the term to relate white horses and horses.' We shall discuss the crucial notion of subsumption under a class in the next Sub-section.

[3] For a wonderful survey of this classificatory orgy see Bodde (1939), pp. 207ff.

In this Section I shall investigate the semantic history and the use of the central concept of *lei* 'kind', 'category' in ancient Chinese literature.[1]

The pre-history of the notion of a class

The word *lei* has not, as far as I know, been identified in the oracle inscriptions or in the bronze inscriptions.[2] Neither does it occur in the Book of Changes. In the Book of History, *lei* is the name of a solemn sacrifice performed only for the highest deity, the *shang ti* 上帝. This sacrifice is often mentioned in the ancient repository of ritual, the *Chou Li* 周禮, and occasionally in the Book of Songs as well as the *Kuo Yü* 國語. However, in general the sacrificial meaning of *lei* had all but disappeared during Warring States times.

Another early meaning of *lei* is 'good'. We find a clear definition relating the two meanings in the following passage:

As for '*lei*', it means 'not reducing to shame the former wise men (i.e., the ancestors)'.[3]

The link between the sacrificial and the moral meaning of *lei* is evident: both these concepts are linked to the notion of the ancestor. We note, in passing, that the Greek word *dikaios* 'just' originally meant 'observant of custom', and thus designated 'a good tribesman'.

As the meaning 'good' disappeared, a new meaning emerged, that of a racial group defined in terms of its common ancestor.[4]

I have heard it said that the spirits do not accept sacrifices from others than their own stock (*fei lei* 非類), and that the people do not sacrifice to others than their own clan (*fei tsu* 非族).[5]

Heaven and Earth are the bases of life. The former ancestors are the bases of races/clans (*lei*).[6]

We note that *fei lei* means nothing like 'dissimilar'. It will not do to take the notion of similarity as basic. Already in early sources we get the notions of *thung lei* 同類 'cognate' and *i lei* 異類 'non-cognate, agnate', and the old commentary to the *Kuo Yü* 國語 ascribed to Wei Chao 韋昭 comments simply: '*thung lei* means "of the same family name".'[7] What has the same family name is not a kind of similar stuff, but individuals. Moreover, they are defined in terms of kinship, not similarity.

---

[1] Note that the word *chung* 種 'seed', hence 'breed' (as in 'among the barbarians there are nine breeds (*chung*)' became current at a later stage although we find isolated instances like *tzhu ho chung yeh* 此何種也 'what sort of thing is this?' (*Han Fei Tzu* 32.31). When Buddhist philosophy was transmitted to China it was *chung* not *lei* which became the standard term for the characteristic Buddhist tradition of subclassification. *Chung* acquired a nuance of 'species', as opposed to *lei* 'genus'.

[2] See Wu Chien-Kuo (*1963*), p. 103.   [3] *Kuo Yü*, ed. World Book Company, p. 85.

[4] Compare, incidentally, the etymology of the Greek word *genos* 'genus, kind', which also originally meant a tribe of people.

[5] *Tso Chuan*, Duke Hsi 10, ed. Yang Bojun, p. 334; cf. Couvreur (1951b), vol. 1, p. 279. *Tso Chuan*, Duke Hsi 31, ed. Yang Po-Chün, p. 487; Couvreur (1951b), vol. 2, p. 423, uses the combination *tsu lei* 族類.

[6] *Hsün Tzu* 19.14; cf. Köster (1967), p. 243.   [7] *Kuo Yü, Lu Yü, apud* Wu Chien-kuo (*1963*), p. 108.

When Confucius declares that for him there are no *lei* 類 in education,[1] it is likely that he thereby meant he did not make any distinction between people according to which clan they came from.

Biological kinds

A natural development from this is the notion of *lei* as a biological kind in general.

Tigers and men belong to different kinds (*lei*).[2]

To some extent one may say the *lei* have a certain metaphysical force. Human and natural *lei* are seen in a profound analogy even by the Confucian Hsün Tzu.

Herbs and trees spring from their own kind (*lei*). Birds and beasts live in flocks. That is because every thing follows its kind (*lei*).[3]

Good fortune and bad fortune are then defined relative to the category a thing belongs to.

That which is in harmony with a (biological) kind (*shun chhi lei chê* 順其類者) is called good fortune. That which goes against a kind (*ni chhi lei chê* 逆其類者) is called ill fortune. Such is the government of Heaven.[4]

Dictionaries, from the *Shuo Wên Chieh Tzu* 説文解字 (postface +100) onwards, will describe animals in terms of the *lei* to which they belong. Typically, for the Chinese, this *lei* is not necessarily defined in terms of a superordinate more general genus under which a species is subsumed, but quite often by example.[5]

The next step is to take *lei* not as a specifically biological kind but as a generally natural kind. The *lei* to which a thing belongs determines its behaviour in the world.

The ten thousand things are of different kinds (*shu lei* 殊類) and they have different characteristic shapes. They all have distinct functions and cannot replace each other.[6]

The practical usefulness of an awareness of natural kinds was obvious enough to the ancient Chinese:

Whenever you undertake anything, you should not go against the Way of Heaven. You must follow the seasons of things and you must adapt to the relevant natural kinds (*lei*).[7]

This notion of a natural kind is importantly linked, still, to the notion of a common origin, as the Book of Changes insists:

Things that draw their root from heaven tend towards what is above. Things that draw their root from the earth cleave to what is below. So every thing follows its kind (*ko tshung chhi lei yeh* 各從其類也).[8]

---

[1] *Lun Yü* 15.39; cf. Lau (1983a), p. 159.   [2] *Chuang Tzu* 4.62; cf. Graham (1981), p. 72.
[3] *Hsün Tzu* 1.15; cf. Dubs (1928), p. 34.   [4] *Hsün Tzu* 17.12; cf. Dubs (1928), p. 176.
[5] See Tuan Yü-tshai (*1981*), p. 458, for one of many examples of the practice of defining a biological kind in terms of similarity to a member.
[6] *Lü Shih Chhun Chhiu* 3.5, ed. Chhen Chhi-Yu, p. 172; cf. Wilhelm (1928), p. 38.
[7] *Lü Shih Chhun Chhiu* 8.1, ed. Chhen Chhi-Yu, p. 422; cf. Wilhelm (1928), p. 94.
[8] *I Ching* 2a; cf. Sung (1969), p. 9.

'Metaphysical' categories

The notion of a kind must not be identified in a facile way with biological or physical classification. We often have a more 'metaphysical' kind of classification under the five elements or five phases (*wu hsing* 五行):

The mare belongs to the category (*lei*) of earth.[1]

The setting up of categories the manipulation of which will give one control over the things belonging to them is a central concern in the Book of Changes. The Great Appendix to this book explains:

In ancient times Fu Hsi ruled over the world. . . . and he devised the eight trigrams to show fully the Power of demonic understanding and to categorise (*lei*) the essences of the ten thousand things.[2]

Very interestingly, the *I Ching* verbalises the notion of *lei* to make it mean 'categorise, divide into categories', and it is clear that the categories (*lei*) here involved are of a profound metaphysical nature. It is often hard to find out how much of this metaphysical dynamism of *lei* survives in less metaphysically laden contexts than the Book of Changes.

Since things in some sense follow their kind, one must understand the kinds in order to understand the things. This, in any case, is the ambition of the Sage.

The ten thousand things in the world are the body of one man. That is called the Great One (*ta thung* 大同) . . . Heaven fills all things (with life). The Sage surveys them to ascertain their kinds (*lei*).[3]

The reasons why the sage must observe the kinds or categories of things become clear in the following passage:

It is inherent in things of one kind that they call each other up. If the vital energies (*chhi* 氣) are the same, they act together, when a sound correlates, it responds. When you strum the note *shang* 商 or *chüeh* 角 on one instrument, the same string is stirred on another. If you pour water on flat ground it flows towards the damp. If you put fire to wood of equal quality it tends towards the dry.[4]

The same in kind (*thung lei* 同類) go along with each other, the same in sound respond to each other.[5]

Things affect each other according to their kinds (*wu lei hsiang tung* 物類相通).[6]

The idea of cosmic resonance has been studied in detail in Charles Le Blanc (1986).

One Taoist attitude towards this cosmic resonance was to regard it as inaccessible to discursive reasoning.

---

[1] *I Ching* 2; cf. Sung (1969), p. 16.   [2] *I Ching Hsi Tzhu* 2; Sung (1969), p. 309.
[3] *Lü Shih Chhun Chhiu* 13.1, ed. Chhen Chhi-Yu, p. 659; cf. Wilhelm (1928), p. 160.
[4] *Lü Shih Chhun Chhiu* 13.2; cf. Wilhelm (1928), p. 161, and Graham (1978), p. 373. Compare also *Lü Shih Chhun Chhiu* 20.4, tr. Wilhelm, p. 355, for a similar passage.
[5] *Chuang Tzu* 31.12; cf. Graham (1981), p. 249.   [6] *Huai Nan Tzu* 3, ed. *SPTK* 3, p. 1b.

As for the correspondance of things (*lei chih hsiang ying* 類之相應), this is something mysterious and profound. Knowledge cannot sort it out (*lun* 論), logical analysis (*pien* 辯) cannot unravel it.[1]

This notion of a metaphysical resonance of like to like is of the utmost importance in Chinese intellectual history. The pervasive speculations on *yin* 陰 and *yang* 陽 moving the universe in a sort of cosmic music of resonance are but one instance of the popularity of this way of thinking in China. The enthusiastic acceptance of the Buddhist doctrine of karma is another instance.

In ancient Chinese medicine, the concept of *lei* 類 comes to play a significant part, as when the famous *Huang Ti Nei Ching Su Wên* 黃帝內經素問 systematically subsumes things cosmologically under the five elements or phases (*wu hsing* 五行) by describing things as *lei huo* 類火 'of the category of fire', etc.[2] Within Chinese theories of art the idea of cosmic resonance also had its impact.[3]

Similarity groups

The next step was that from a natural kind or metaphysical category to that of similarity in general, as in the dramatic story where criminals enter the bedroom of Duke Hsiang of Chhi, kill the man lying on the Duke's bed, but then they find they have made a mistake:

'It's not the ruler. He's not like him (*pu lei* 不類)!'[4]

Mencius makes an important categorial statement:

All things of the same category (*lei*) are similar (*ssu* 似). Why should we have doubts on this just in the case of man?[5]

Belonging to the same category, according to Mencius, is not simply the same thing as being similar. But things which belong to the same category will be similar (in some relevant way). Often the question whether things are of the same *lei* is thought to be controversial:

Things often are alike but are not the same (*hsiang lei êrh fei* 相類而非): the dark coloured grasses that grow between wheat sprouts resemble wheat. A brindled yearling(?) is like (*ssu*) a tiger. Bleached bone might be suspected of being ivory. The *wu-fu* 武夫 precious stone resembles (*lei*) jade. These are all cases of being like a thing but not being the thing.[6]

---

[1] *Huai Nan Tzu*, ed. Liu Wên-Tien, 6.3a; cf. Le Blanc (1986), pp. 116ff.
[2] *Huang Ti Nei Ching Su Wên*, ch. 1, p. 14. Compare also *Lü Shih Chhun Chhiu* 20.4; tr. Wilhelm (1928), p. 355, for a similar passage. On the early history of the concepts of Yin and Yang as well as the Five Phases see Graham (1986b), pp. 70–94.
[3] See Munakata (1983), pp. 105–31.
[4] *Tso Chuan*, Duke Chuang 8, ed. Yang Po-Chün, p. 176; cf. Couvreur (1951b), vol. 1, p. 144.
[5] *Mêng Tzu* 6A7; Lau (1983c), p. 229.
[6] *Chan Kuo Tshe*, no. 295, ed. Chu Tsu-Keng, p. 1139; cf. Grump (1970), p. 372.

This notion of a similarity group can then be used in a nominal way:

He can raise to office good people. Only those who are good can raise to office those who are like them (*chü chhi lei* 舉其類).[1]

By the −4th century the notion of *lei* as a concept referring to a similarity group was firmly established. Categories were no longer entirely traditional or given by nature. They were also conceived as set up by man. The difference between natural and artificial kinds (*lei*) comes out beautifully in this passage:

Therefore laws are not established by themselves. Categorisations cannot apply generally of themselves (*lei pu nêng tzu hsing* 類不能自行).[2] If one finds the right person (to administrate them), they will exist. If one fails to find the right man, they will cease to exist.[3]

Evidently, the concept of *lei* played its part in administration theory where it referred to administrative categories of people and actions. On the other hand, the concept was also to play a crucial part in Chinese conceptions of reasoning and explanation.

Explanatory categories

For Mo Tzu the concept of *lei* was a central one. When accused of incoherence in his opposition to offensive warfare, Mo Tzu replies:

You have not properly investigated the categories to which my words belong (*wu yen chih lei* 吾言之類) and you have not quite understood my reasons (*ku*).[4]

Mo Tzu mentions the *lei* 'categories' and the *ku* 故 'reasons' on a par. The name of such a category can specify the *lei* involved:

To enter other people's states and to associate with those who rob others, these are not things of the category of dutifulness (*fei i chih lei yeh* 非義之類也).[5]

Understanding categories (*chih lei* 知類) was evidently a matter of profound importance for Mo Tzu. Not understanding categories was in fact tantamount to not understanding him at all. Mo Tzu's conflict and dialogue with Kungshu Phan provides an important clue on the concept of understanding categories. Mo Tzu begins by asking Kungshu Phan to kill one of his (Mo Tzu's) enemies, which Kungshu Phan refuses on the grounds that it is in contravention of what he perceives as his duty (*i* 義). Mo Tzu gets up, bows twice, and proceeds to a moralistic harangue against Kungsun Phan which culminates and ends as follows:

To consider it as one's duty not to kill relatively few and go on to kill (relatively) many, that cannot be called knowing categories (*pu kho wei chih lei yeh* 不可謂知類也).[6]

---

[1] *Tso Chuan*, Duke Hsiang 3, ed. Yang Po-Chün, p. 927; cf. Couvreur (1951b), vol. 2, p. 192.
[2] Compare the technical use of the term *hsing* 行 'proceed' in connection with *lei* 類 among the later Mohists see Graham (1978), pp. 177f.
[3] *Hsün Tzu* 12.2; cf. Köster (1967), p. 152.   [4] *Mo Tzu* 19.32; cf. Watson (1967), p. 59.
[5] *Mo Tzu* 39.41; cf. Mei (1929), p. 207.   [6] *Mo Tzu* 50.6; cf. Mei (1929), p. 258.

Mêng Tzu seems to have taken over Mo Tzu's strategy on this point:

Mencius said: 'Now if one's third finger is bent and cannot stretch straight, though this neither causes any pain nor impairs the use of the hand, one would think nothing of the distance between Chhin and Chhu if someone able to straighten it could be found. This is because one's finger is inferior to other people's. When one's finger is inferior to other people's, one has sense enough to resent it, but not when one's heart is inferior. This is what is called "not knowing (the relevant) categories (*lei*)".'[1]

Knowing explanatory categories, according to Mo Tzu and Mêng Tzu, seems to be the ability to extrapolate from one similar case to the other, and the relevantly similar cases are then said to be of one *lei*. The process by which one extrapolates came to be called that of *thui* 推 'pushing' by the Later Mohist logicians.[2]

Kungshu Phan's ruler wants to attack Sung, but Mo Tzu opposes this move with a number of interesting analogies which he summarises by referring to the notion of *lei*.

Mo Tzu saw the ruler and said: 'Suppose there is a man who, putting aside his elegant carriage, desires to steal his neighbour's shattered sedan chair; or someone who putting aside his embroidery and finery desires to steal his neighbour's short jacket; or someone who putting aside his meat and grains desires to steal his neighbour's husks. What kind of a man would this be?'

The lord said: 'He must suffer from the disease of kleptomania.'

Mo Tzu continued: 'The land of Ching amounts to five thousand square *li* 里, while that of Sung is of only five hundred. This is similar to the contrast between the elegant carriage and the shattered sedan. Ching possesses Yün Mêng which is full of rhinoceroses and deer. The fish, tortoises and crocodiles in the Yangtse and the Han rivers are the most abundant in the empire. While Sung is said to possess not even pheasant, rabbits, or foxes. This is similar to the contrast between meat and grains on the one hand and husks on the other. In Ching there are tall pines, spruces, cedars and camphor trees, while Sung has no tall trees at all. This is similar to the contrast between embroidery and finery on the one hand and a short jacket on the other. When your ministers and generals set out to attack Sung, it seems to me they belong to the same category as this (*yü tzhu thung lei* 與此同類). As I see it, you, my ruler, are bound to violate the demands of duty without getting anything in return.'[3]

For Mo Tzu in this dialogue a *lei* 類 is not just a fixed natural kind, it is a relevant similarity group, a set of things that are similar in a relevant respect. 'Knowing categories', then, amounts to recognizing similarities that are relevant to the argument at hand.

Mencius made use of the term in this connection in a justly celebrated passage:

'What is the difference in form between refusal to act and inability to act?'

'If you say to someone "I am unable to do it" when the task is one of striding over the Northern Sea with Mount Thai under your arm, then this is a genuine case of inability to act. But if you say "I am unable to do it" when it is one of making an obeisance to your

---

[1] *Mêng Tzu* 6A12; Lau (1983c), p. 237.

[2] The concept of knowing categories recurs, for example, in *Lü Shih Chhun Chhiu* 10.3, ed. Chhen Chhi-Yu, p. 537, *Lü Shih Chhun Chhiu* 13.4, ed. Chhen Chhi-Yu, p. 698, and *Lü Shih Chhun Chhiu* 20.5, ed. Chhen Chhi-Yu, p. 1374.

[3] *Mo Tzu* 40.12; cf. Mei (1983c), p. 258.

elders, then this is a case of refusal to act, not of inability. Hence your failure to become a true King is not of the same category (*lei*) as "striding over the North Sea with Mount Thai under your arm" but of the same category (*lei*) as "making an obeisance to your elders".'[1]

The unicorn and other animals, the phoenix and other birds, Mount Thai and small mounds, rivers and seas and running brooks, these are all of a (i.e., the same) kind (*lei yeh* 類也). The sage and a commoner are also of a kind (*lei yeh*).[2]

Mencius called the pushing of categorial thinking to extremes '*chhung lei* 充類 "filling up the category"':

To maintain that taking anything that does not belong to one is robbery is 'filling up a category' (*chhung lei*) and carrying the concept of dutifulness (*i* 義) to its limit.[3]

For Mencius, *chhung* 充 means the taking of a category to its logical limits, to take it as far as it will go. The notion of *lei* can be quite abstract, as in the following puzzling passage:

Mêng Tzu said: 'For every man there are things he cannot bear. To extend this to what he can bear is benevolence. For every man there are things he cannot get himself to do. To extend this to what he can get himself to do is righteousness. If a man can extend to the full (*chhung*) his natural aversion to harming others, then there will be an overabundance of benevolence. If a man can extend to the full (*chhung*) his dislike for boring holes and climbing over walls, then there will be an overabundance of rightness . . .

To speak to a gentleman who must not be spoken to is to use speech as a bait. On the other hand, not to speak to one who could be spoken to is to use silence as a bait. In either case, the action is of the same category (*lei*) as that of boring holes and climbing over walls.'[4]

Han Fei, explaining the difficulties of speaking, knows that making such analogies between things can have an effect contrary to that intended:

. . . If I use many words and complicated expressions, if I string together categories and compare things (*lien lei pi wu* 連類比物), then this will be looked upon as empty and useless . . .[5]

However, Han Fei too is convinced of the importance of understanding categories, as the following reflection of his shows:

The possessor of a thousand *li* 里 of rocky land cannot be called rich. The possessor of a million puppets cannot be called strong. Not that the rocks are not big or that the puppets are not numerous. The possessors cannot be called rich and strong simply because great rocks do not produce grain, and because puppets cannot be used to resist enemies.

Now men who get office through purchase and who play tricks, they eat without cultivating the land. They are as unproductive as uncultivated land, in the same category as rocks. In the same way the literati and the knights render no meritorious army service but are celebrated and prosperous. They are useless people, the same thing as puppets. Those who know what disasters rocky lands and mere puppets are, but who fail to understand that the office purchasers, the literati and the knights are as harmful as uncultivated land and unemployable people, they are the sort of people who do not understand the categories (*lei*) of things.[6]

---

[1] *Mêng Tzu* 1A7; Lau (1983c), p. 17.   [2] *Mêng Tzu* 2A2; Lau (1983c), p. 61.
[3] *Mêng Tzu* 5B4; Lau (1983c), p. 211.   [4] *Mêng Tzu* 7B31; Lau (1983c), p. 298.
[5] *Han Fei Tzu* 3.1.10; cf. Liao (1939), vol. 1, p. 23.   [6] *Han Fei Tzu* 50.6; cf. Liao (1939), vol. 2, p. 305.

In one passage Hsün Tzu mentions *lei* 'categories' and *li* 禮 'ritual' side by side. According to him, the aim of speech is to capture the natural regularities which are inherent in the natural categories as well as in the artificial categories set up by man:

. . . This is the sign of a great Confucian (*ju* 儒). His words are in accordance with (the relevant) categories (*chhi yen yu lei* 其言有類), his deeds are in accordance with ritual (*chhi hsing yu li* 其行有禮).[1]

For Hsün Tzu the word *lei* becomes a term of high intellectual praise. The categories which he takes *lei* to refer to are of course those set up by the Confucian tradition.

If someone speaks a lot and keeps in accordance with (the relevant) categories (*to yen êrh lei* 多言而類) he is a sage.[2]

The Taoist philosopher Chuang Tzu lifts himself above the sort of reasoning that involves categorisation. In his essay entitled 'On regarding all things as equal' he sets up just one highest category in which everything is equal.

Now suppose that I speak of something, and do not know whether it is of the same category (*lei*) as the thing at hand or whether it is not of the same category (*lei*) as the thing at hand. If what is of the same category (*lei*) and what is not of the same category (*lei*) are all considered to be of the same category (*lei*), there is no difference between the thing at hand and the other thing.[3]

If Chinese civilisation was able to become the cradle of great scientific discoveries, this was because Chuang Tzu's attitude was not in fact typical of Chinese intellectual practice. More typical is the attitude displayed in the chapter entitled 'Distinguishing categories' in the encyclopaedia *Lü Shih Chhun Chhiu* 呂氏春秋. Here the difficulty of setting up categories is clearly realised:

Many things look as if they are of a certain category but are not. As a result states are lost and peoples come to grief in an endless series. Among the plants there are the *hsin* 莘 and the *lei* 藟. When one eats one of these separately, they kill one; if people eat them together, they give them a longer life . . . One cannot with certainty infer from one member of a category to another of the same category (*lei ku pu pi kho thui chih yeh* 類固不必可推知也).

A set of four small horses is of the same category (*lei*) as a set of four large horses. A small horse is of the same category (*lei*) as a large horse. But small knowledge is not of the same category (*lei*) as large knowledge.[4]

Here the author reflects on the potential of the notion of a *lei* for formal deduction. You might have thought that if two plants are equally poisonous, the two of them together are doubly poisonous. Not so. You might have thought that small horses being of the same category as large horses, by analogy, by a process of 'pushing categories' (*thui lei* 推類), you could conclude that small (i.e., petty) knowledge is

---

[1] *Hsün Tzu* 8.86; cf. Dubs (1928), p. 109.   [2] *Hsün Tzu* 27.126; cf. Köster (1967), p. 363.
[3] *Chuang Tzu* 2.48; cf. Graham (1981), p. 55.
[4] *Lü Shih Chhun Chhiu* 25.2, ed. Chhen Chhi-Yu, p. 1642; cf. Wilhelm (1928), p. 435.

of the same category as large (i.e., superior) knowledge. Not so. According to the writer, such conclusions are unwarranted, and he even obliges with a joke to illustrate his point:

> In Lu there was a man by the name Kungsun Cho. He told someone: 'You know, I can bring the dead back to life again!'
> 'How so?'
> 'You see, I can certainly cure hemiplegia, and if I take twice the amount of medicine with which I cure hemiplegia, I should be able to raise the dead to life again.'[1]

What better illustration would one want for the close relation between Chinese logic and Chinese jokes? Like so many others, this joke is based on a false analogy, and a false analogy is essentially a case of putting into the same category what is not in the relevant respect similar.

In the same chapter on 'Distinguishing categories' we find another entertaining and philosophically more subtle story:

> A connoisseur of swords said: 'White (metal) makes a sword hard, brown (metal) makes it elastic. If you combine the white and the brown, the sword will be hard and elastic and you will get the best sword.'
> Someone raised an objection: 'The white (metal) will make the sword not elastic. The brown (metal) will make it not hard. If you mix the white and the brown, you will make the sword neither hard nor elastic. Moreover, if a sword is elastic, it gets easily bent. If it is hard, it is easily broken. A sword that is easily bent and easily broken, how can you call that a sharp (i.e., good) sword?'
> The real nature (chhing 情) of the sword was unchanged but one considered it as good, the other as bad. Their explanations (shuo 説) bring about this (divergence).[2]

Here the author focuses his attention on the fact that by categorising something under a certain lei like 'hard' you are at the same time committed to categorising under the lei 'not elastic'.

The Later Mohists made extensive abstract use of the term lei as a category of things. We shall investigate their theories in the Sub-section on Mohist logic. But even a philosopher like Han Fei, in speaking of 'the category of things with shapes' (yu hsing chih lei 有形之類),[3] was clearly capable of a quite abstract notion of a category insofar as these shapes were going to be sharply different and dissimilar.

The very intriguing text Liu Tzu 劉子, which we may very tentatively attribute to the literary critic Liu Hsieh 劉勰 (+465 to +522), argues logically that a similarity group does not establish a lei. We might say without distorting the issue that the Liu Tzu also argues against the transitivity of the similarity relation.

When you generalise uniquely on the basis of lei, because this is like that, then if you maintain that a dog is similar to a chüe 玃-ape, that a chüe-ape is similar to a chü 狙-ape, and that the chü-ape is similar to man, then a dog will be similar to a man. If you maintain that

---

[1] Ibid.   [2] Lü Shih Chhun Chhiu 25.2, ed. Chhen Chhi-Yu, p. 1642; cf. Wilhelm (1928), p. 435.
[3] Han Fei Tzu 21.10.1.

white is similar to light yellow, light yellow is similar to brown, brown is similar to red, red is similar to purple, purple is similar to dark purple, and dark purple is similar to black, then white turns out to be black.[1]

This amounts to advanced abstract theorising about categories (*lei*). Ludwig Wittgenstein would have been delighted to find this piece of reasoning, I am sure.

There is a deep contrast between the current Classical Chinese notions like *lei* and the purely logical notion of class which is central to our concern. The notion of an ordered hierarchy of categories, or of a *lei* of different *leis*, a category of categories, a class of classes, does not seem to have been focused on by the ancient Chinese as it has been by the ancient Greeks, and particularly by Plato in his theory of *diairesis* διαίρεσις. However, this did not in any way prevent the Chinese, when the time came, from becoming consummate taxonomers of biology, as previous volumes of *SCC* have shown.

The old Classical Chinese notion of *lei* corresponds not to the logical notion of a mathematical set or a class, but evolved more along the lines of what we might call 'a natural kind' and of explanatory categories.[2] Natural kinds are classes of things that we regard as of explanatory importance; classes whose normal distinguishing characteristics are 'held together' or explained by deep-lying mechanisms. Categories are kinds we are conscious of having set up ourselves.

Semantic categories, categories and subcategories

However, we do have the concept of a semantic class in traditional Chinese lexicography, as when Liu Hsi 劉熙 writes in the introduction to his dictionary *Shih Ming* 釋名 written around +200:

As for the relation between names and objects (*shih* 實), there are classes of meaning (*i lei* 義類).[3]

Chêng Hsüan 鄭玄 (+127 to +200) seems already to employ something like a notion of a subcategory *chung* 種 when he defines *chiu I* 九夷 'the Nine I-Barbarians' of *Lun Yü* 9.14 as *tung fang chih I, yu chiu chung* 東方之夷有九種 'the I-barbarians of the East of which there are nine subgroups'.[4]

The *Chhun Chhiu Fan Lu* 春秋繁露, a work which contains a fair amount of early Han material but was perhaps compiled no earlier than the +5th century, introduces a notion of *fan hao* 凡號 'comprehensive appellation' to refer to a general term or genus and *san ming* 散名 'spread names' for its subcategories or species.

Things all have a general appellation (*fan hao*). And appellations all have their subcategories (*san ming*).[5]

---

[1] *Liu Tzu*, ch. 16, ed. Lin Chhi-Than and Chhen Feng-Chin, p. 94. Chapters 16 and 17 are concerned with names (*ming* 名). I mention this because the book is not mentioned in the excellent bibliography to Graham (1978).
[2] See Putnam (1975).   [3] Cf. Chhien Chien-Fu (*1986*), p. 151.
[4] *Lun Yü*, ed. Liu Pao-Nan, vol. 2, p. 97.   [5] *Chhun Chhiu Fan Lu*, ch. 34, ed. Ling Shu, p. 234.

Examples will bring out the nature of this division

The general (*fan chê* 凡者) only brings up the main matter. Entertaining the ghosts and spirits has one (general) appellation: sacrifice (*chi* 祭). The subcategories (*san ming*) of 'sacrifice (*chi*)' are as follows: . . .

Hunting birds and beasts has one appellation: hunt (*thien* 田). The subcategories (*san ming*) of 'hunt (*thien*)' are . . .[1]

B. C. Cohen (1982) argues that understanding natural kind terms requires extensive shared common-sense beliefs about the characteristic properties of a kind. The task of the historian of Chinese intellectual history is indeed to reconstruct these extensive networks of common-sense beliefs which – for the ancient Chinese – defined their categories and their changing conceptual schemes. What we have been studying in this Sub-section, then, is not the abstract mathematical concept of a class or set. It is the absolutely fundamental cultural concept of the cognitive (as well as administrative) category in ancient China.

(7) ABSTRACTION AND THE CONCEPT OF A PROPERTY

The concept of a kind or category plays an important part in Chinese intellectual history. We shall also see that the notion of subsumption under a category was known to the Chinese. But the abstract concept of the quality, property, or characteristics as such is an idea which needs separate investigation in its own right. After all, it is conceivable that the ancient Chinese were nominalists in the sense that they spoke only of things, of kinds of things, of distinctions between things, and of names applying to things, but never of abstract properties or qualities as such as attaching to things.

The Classical Chinese language lacks such morphemes as the English '-ness'. Antisthenes would have had trouble telling the ancient Chinese in their own language what he told Plato: 'I can see horses, but I cannot see horsehood.' Plato, the story goes, replied: 'That, Antisthenes, is because while you have what it takes to see horses, you do not have what it takes to discern horsehood, that is: intelligence.'[2] Do we have to conclude that the ancient Chinese 'did not have what it takes to discern horsehood, that is a grammatical morpheme like *-otēs* or '*-ness*'? Does this mean that the Chinese had no way of speaking of abstract concepts or properties? Or does this only mean they had a more limited rôle for abstraction in their scheme of things? We must look at the philological evidence.

Abstraction in ancient Chinese thought

The ancient Book of Songs has the enigmatic lines:

Heaven gave birth to the multitude of people. There are creatures, and there are rules (for these creatures) (*yu wu yu tsê* 有物有則).[3]

---

[1] *Ibid.*, p. 233.   [2] Cf. Simplikios (1882), p. 208, lines 30ff.

[3] *Shih Ching* 260.1; Karlgren (1950), p. 228. Note Couvreur (1934), p. 399, who translates: *sunt res (quibus constant homines), est lex moralis.*

We seem to have a dualism here between *wu* 物 'thing, creature' and *tsê* 則 'rule, pattern'. One is inclined to find *tsê* distinctly more abstract than *wu*.

It is possible to speak of a property without actually mentioning the abstract concept 'property' as such. We can have abstraction without suffixes indicating abstraction.

The being a circle of a small circle is identical with the being a circle of a large circle (*hsiao yüan chih yüan yü ta yüan chih yüan thung* 小圓之圓與大圓之圓同 lit.: SMALL CIRCLE'S CIRCLE WITH LARGE CIRCLE'S CIRCLE IDENTICAL).[1]

What is identical here is not the circle. It is the circularity of the circle. The Later Mohist is concerned with the abstract property of being a circle. There is no natural entification or hypostatisation of 'being a circle' in Chinese. Unlike Plato, the Mohist had no obvious morpheme available to mark off which occurrences of the word CIRCLE refer to the physical round object, and which occurrences refer to the property of circularity as such. However, if his contemporaries understood this sentence at all, they must also have understood a distinction between a circle and circularity.

One might object that a parlance of this sort was limited to the very special kind of scientific literature of which the Mohist Dialectical Chapters from which we have just quoted are remnants. But the *Mêng Tzu* surely belongs to the mainstream of Chinese philosophical literature:

Mêng Tzu asked: 'Is the being white of white feathers the same as the being white of white snow; and is the being white of snow the same as the being white of jade?'
(Kao Tzu replied:) 'That is so.'
'In that case, is the nature of a dog the same as the nature of an ox; and is the nature of an ox the same as the nature of man?'[2]

There is a Socratic flavour to this passage, which is essentially an attempt at a *reduction ad absurdum*. There is an explicit concern with 'being white' as opposed to 'white things'.

Moreover, we have philosophically quite innocent common ways of referring to abstract properties, as in the opening of the Taoist book *Chuang Tzu*:

As for the size of the Khun (*khun chih ta* 鯤之大), I do not know how many thousand *li* 里 it is.[3]

However, here one might object that the many thousand *li* apply to the Khun as a proper subject, not to its size. One cannot reject this objection out of hand.

The *Hsün Tzu* is a central Confucian text, and in it we find disquisitions like this:

That by means of which man is man (*jên chih so i wei jên chê* 人之所以為人者) is not just that he has two feet and no hair. It is that he has the capacity to set up distinctions.[4]

---

[1] *Mo Tzu, Ta Chhü* 墨子, 大取, cf. J. Chmielewski (1969), Pt. VII, p. 125, and Graham (1978), pp. 474ff.
[2] *Mêng Tzu* 6B3; Lau (1983c), pp. 223f.   [3] *Chuang Tzu* 1.1; Graham (1981), p. 43.
[4] *Hsün Tzu* 5.24; Dubs (1929), p. 71.

Hsün Tzu is interested in what we could call certain properties that make man into man, although he uses no abstract term for 'property'.[1]

Even without the use of lexical, syntactic or morphological special devices the Chinese were able to refer to such abstract concepts as that of 'fatherhood' also in non-philosophical contexts like the following gruesome story:

Yüeh Yang was Wei's general in the attack against Chung Shan. His son was hostage in Chung Shan at the time and the ruler of that state had him boiled alive and made into a soup which was sent to Yüeh Yang. Yüeh Yang ate it and was praised by one and all.

Yüeh Yang ate his son to show his trustiness. He knew he committed an offence against fatherhood (hai fu 害父), but he sought to conform to the law (which required his loyalty).[2]

The earliest commentary glosses fu 父 'father' as fu tao 父道 'the way of a father'. On the other hand it is most instructive to observe the ingenuity with which later commentators have attempted a highly speculative emendation of the text (involving the addition of one character, and the explanation of two characters as miswritings for a personal name) because they failed to recognise or did not find natural the abstract use of the word fu to mean fu tao 'the proper way of being a father, fatherhood'.[3] This shows how abstraction in Chinese, being often not clearly marked, is less palpable, more elusive, as long as it is not attached to clear and explicit syntactic devices.

In philosophical contexts there often is no doubt about the abstract nature of Chinese concerns. When we read a Classical Chinese sentence like

Ssuma Niu wên chün tzu 司馬牛問君子 SSUMA NIU ASK GENTLEMAN 'Ssuma Niu asked about the gentleman',[4]

we have exactly the same ambiguity in the English translation as we have in the Chinese. Ssuma Niu may either be taken to inquire about a certain individual whom he identifies as a gentleman, or he may be concerned with the abstract notion of a gentleman. It so happens that we have not the slightest doubt in this particular passage that what is being asked about is the gentleman in general. In Classical Chinese no morphological device such as 'gentlemanhood' is available which might encourage one to hypostasise a corresponding abstract 'Platonic' entity. Similarly in English we happily discuss the abstract notion of the gentleman without normally resorting to such grammatical variants as gentility or gentlemanhood or Trollope's softer sounding gentlehood.

But how often do the Chinese ask questions of this 'abstract' sort? Are not such general concerns unnatural to the Chinese bent of mind? Let us take the crucial text of the Confucian tradition, the Lun Yü. Compare:

The governor of She asked Tzu Lu about Confucius.[5] Chhen Ssu-Pai asked whether Duke Chao knew ritual.[6]

---

[1] As we shall see in a moment, Hsün Tzu did have a technical term for 'perceptible property', namely chuang 狀.

[2] Chan Kuo Tshe, no. 489, ed. Chu Tsu-Keng, p. 1733; Crump (1970), p. 574.

[3] Two such attempts are quoted in extenso in Chu Tsu-Keng's edition, p. 1734.

[4] Lun Yü 12.4; Lau (1983a), p. 109.   [5] Lun Yü 7.19; Lau (1983a), p. 61.   [6] Lun Yü 7.31; Lau (1983a), p. 65.

These are concrete questions. We also have:

Fan Chhih asked about knowing/being knowledgeable (*chih* 知) . . . He asked about being good/goodness (*jên* 仁).[1]

These are abstract questions insofar as they are concerned with *chih* 'knowing' and *jên* 'being good' as such and not as attaching to any particular person. Surveying the vast amount of Confucian discussion of the term *jên*, one may well feel that this particular concept *was* hypostasised by Chinese thinkers. In idiomatic English we are tempted to translate *jên* as 'goodness' rather than as 'being good'. The gloss 'good person, good persons' does not always seem to work. The problem is to what extent this natural tendency of translators is justified and based on grammatical facts of Classical Chinese.

Note that in English we discuss good and evil without needing to write explicit morphemes to indicate the nominality of what we speak about. 'Good' and 'evil' are not less abstract than 'goodness' or 'evilness' would have been. Plato could have formulated his theory of the ideas of 'good' and 'evil' without recourse to such morphemes as '-ness'. Indeed Plato himself tends to ask not at all about the perfectly possible neologism *kalotēs* 'beautifulness' but about *to kalon* 'the beautiful'.[2] Something analogous happens pervasively in Chinese. Thus we have a word *ming* 明 'be clear about', but this word, when used as the object of *wên* 問 'ask about', comes to mean the abstract property of 'being perspicacious' rather than any concrete perspicacious person.[3] *Chiao* 交 means 'to be friends with', but when used as the object of *wên* 'ask about', it comes to mean 'being friends with someone', and it does not come to be synonymous with *yu* 友 'friend' which in the same position can come to mean 'treating someone as a friend'.[4]

The ancient Chinese had no problems speaking about such things as the concrete case of somebody's death and about the dead. But could they and did they ask about death as such? Let us see:

Chi Lu asked about how the spirits of the dead and the gods should be served.
The Masters said: 'You are not able even to serve man. How can you serve the spirits?'
'May I ask about death (*kan wên ssu* 敢問死)?'
'You do not understand even life. How can you understand death?'[5]

A question about the dead, in Classical Chinese, would be a question about *ssu chê* 死者 'those who have died'.[6] This question is not about dead individuals, but about the more abstract matter of dying.[7]

---

[1] *Lun Yü* 6.22; Lau (1983a), p. 53.
[2] In point of fact, *eidos*, *idea*, etc., are not very frequent words even in those dialogues that happen to experiment with a 'theory of forms'.
[3] *Lun Yü* 12.6; Lau (1983a), p. 111.   [4] *Lun Yü* 19.3; Lau (1983a), p. 191, versus *Lun Yü* 12.23; Lau (1983a), p. 119.
[5] *Lun Yü* 11.12; Lau (1983a), p. 99. The old commentary, ed. Liu Pao-Nan, 2.33, explains: 'The ghosts and the matter of dying (*ssu shih* 死事) are difficult to illuminate. To talk about them will bring no profit. Therefore he does not reply.'
[6] For *ssu chê* 死者 'the dead' see, e.g., *Lü Shih Chhun Chhiu* 16.3: If the dead have consciousness, how am I to face up to Kuan Chung?
[7] The old commentary (ed. Liu Pao-Nan, *WYWK*, vol. 3, p. 31) glosses *ssu* 死 'die' as *ssu shih* 死事 'the matters of death'.

In the *Lun Yü* alone, a text which fills no more than forty-two printed pages in the Harvard Yenching Index, I find thirty-two such abstract or conceptual questions like '*X* inquired about *Y*'.

The history of the art of definition in China is the history of semantic answers to some such abstract questions.[1] Confucius was notable for refusing to come up with abstract definitions, just like many of the interlocutors of Socrates. Hsün Tzu, a later Confucian, on the other hand, paid great attention to the art of abstract definition.

Some ancient Chinese concepts seem fairly specialised in carrying an abstract meaning. *I* 義 'righteousness' is a case in point.[2] It is not clear how *i* 'righteousness' in its nominal uses is radically less abstract than the Greek *dikaiosynē* 'justice, righteousness'. Consider Mencius's discussion with Kao Tzu:

Kao Tzu said: 'Human nature (*hsing* 性) is like the *chhi* 杞 willow. Righteousness (*i*) is like cups and bowls.'[3]

Here *i* 'righteousness' is considered as a thing, hypostasised if you like, and as such compared to cups and bowls. It will not do to translate *i* here as 'being righteous', or as 'a righteous person', because we have the parallelism with *hsing* 'nature' which is not a concrete noun referring to a natural individual, but an abstract noun 'nature, inborn properties'.

I 'righteousness' is something that the Taoists could *chhi* 棄 'reject'. But let us look at the way in which such rejection, which we tend to see as the rejection of an abstract value, is expressed:

Get rid of sageliness (*shêng* 聖), discard wisdom (*chih* 智),
And the people will benefit a hundredfold;
Get rid of benevolence (*jên* 仁), discard righteousness (*i*),
And the people will again be filial   [4]

We could avoid attributing all abstraction to the Chinese in this passage: instead of 'sageliness' we could translate 'sages'; instead of 'wisdom', 'the wise'; instead of 'benevolence', 'the benevolent'; instead of 'righteousness', 'the righteous'. And the crucial question is this: do these alternative translations represent a true ambiguity in Chinese or are they only a complication introduced into the text by our English translating idiom? Is the ambiguity only in the mind of the translator or is it in the Chinese text?

There is a clear distinction in Classical Chinese between *shêng jên* 聖人 'a sage' and *shêng tao* 聖道 'the way of a sage'. It makes sense to ask within the framework of Classical Chinese which of these is rejected in the above passage from the *Lao Tzu*. The word *shêng* 聖 is ambiguous not only in terms of English translations but also in terms of the available paraphrases in Classical Chinese.

---

[1] Hansen (1983), p. 77, writes: 'Classical Chinese philosophical theories had no roles for abstractions.' Certainly it has to be said that Confucius's answers to the questions we have just discussed do not involve a 'Platonic' attempt to account for the world in terms of an ontology of a higher world of 'ideas' or 'forms' corresponding to the concepts he discusses.

[2] However, *i* does occasionally function verbally 'be just'.

[3] *Mêng Tzu* 6A1; Lau (1983c), p. 223.   [4] *Lao Tzu* 19. cf. Lau (1983), p. 27.

In translating ancient Chinese texts it is a sound principle to attribute to them as little abstraction as is compatible and seems philologically unavoidable to attribute to them. It is all too easy to read our own habitual abstractions into ancient Chinese texts. Take an example:

When one sees what is right (*i* 義) but does not do it, that is lack of courage (*chien i pu wei wu yung yeh* 見義不為無勇也).[1]

Here, I suggest, one can and therefore probably should resist any temptation to take *i* as an abstract noun, righteousness, along the lines 'when one is aware of right-eousness but does not act according to it . . .'.

Tao 道 'the Way', *tê* 德 'charismatic power, power' as well as *yin* 陰 'the female/dark force Yin' and *yang* 陽 'the male/bright force Yang', on the other hand, do strike us in various ways as 'abstract' hypostasised concepts which are sometimes spoken about as if they were referring to a higher realm of ultimate reality to which Chinese mystical thinkers gain direct access through meditation not totally unlike that cultivated by some Neo-Platonists.[2]

There are thus inherently abstract concepts in Classical Chinese. One is often quite confident that ancient Chinese thinkers were in fact concerned with an abstract concept as such. But it is as if there is in Chinese – as there is indeed in the best of modern philosophy – a natural and strong gravitational force towards the non-abstract down-to-earth use of words. There is no tendency towards, indeed not much room for, the inessential or fatuous abstract use of words which could easily be replaced by an equivalent less abstract way of speaking. The abstract use of words tends to be introduced only where it is essential to convey a message. It is never – as sometimes in Latin and Greek, and often in modern academic prose – a preferred stylistic device. But could the Chinese focus on the non-concrete concep-tually by an abstract notion of a property as opposed to a thing? We must investigate this question in an open-minded way.

It is facile to move from the observation that abstraction is morphologically unmarked to the judgement that it is anthropologically unimportant in ancient China. Such a judgement would be like someone observing that in spoken Modern Standard Chinese the pronouns for 'he' and 'she' are identical, and that there are no masculine or feminine genders, jumping to the conclusion that the Chinese do not distinguish between the female and male rôles.

The concept of a property

It is one thing to distinguish abstract concepts from others. It is another to have an abstract concept of a property. It is quite conceivable that the ancient Chinese indulged in abstraction without having the benefit of a word like 'property'. Note,

---

[1] *Lun Yü* 2.24; Lau (1983a), p. 17.
[2] Cf. Vol. 2, pp. 36ff. and 433 *et passim*, and, particularly for *yin* and *yang*, pp. 268f. For the book *Chuang Tzu* see particularly Chang (1982), pp. 61–130.

however, that the concept of a property does not have to be seen in a philosophical context. In a narrative context we find passages like this:

The Shang King was most unruly. He was immersed in 'the virtue of wine (*chiu tê* 酒德)'.[1]

Presumably 'the virtue of wine' was to do with the properties of wine.

Let us consider the early history of some Chinese explicit abstract words related to that of a property:

1. *hsing* 性 'nature, natural characteristics, property(?)'
2. *chhing* 情 'real state of affairs, essence, essential properties(?)'
3. *mao* 貌 'external shape, shape, property(?)'
4. *chuang* 狀 'form, visual characteristics, property(?)'
5. *li* 理 'pattern, attribute(?), property(?)'

Hsing

The basic meaning of *hsing* 'nature (from birth)' is attested already for Confucius.

By nature (*hsing*) men are close to each other. Through their practice they differ from each other.[2]

It is the goodness or otherwise of this nature of man that Mêng Tzu and Hsün Tzu discussed at great length. By analogy, the notion *hsing* is also applied to animals,[3] and then to inanimate things such as mountains or water[4] or to wood, as when Hsün Tzu speaks of wood being by nature (*hsing*) straight or not straight.[5] Chuang Tzu speaks of the genuine nature of the horse, of the nature (*hsing* 性) of water, and of things generally losing their nature (*hsing*).[6] In a text from the −2nd century we read:

It is the nature (*hsing*) of a cocoon to make silk thread, but if it is not first heated in boiling water by the woman whose job it is, and then unravelled and put into order, it will not become silk thread. It is the nature (*hsing*) of an egg to make a chick; but without a good hen to cover it and protect it for days on end, it will not become a chick.[7]

In *Huai Nan Tzu* we read that 'it is the nature (*hsing*) of bamboo that it floats'.[8] The *Po Hu Thung* speaks of humaneness, wisdom, ritual propriety, good faith, and dutifulness as of natural characteristics (*hsing*).[9]

---

[1] *Lü Shih Chhun Chhiu* 16.1, ed. Chhen Chhi-Yu, p. 976; cf. Wilhelm (1928), p. 234.

[2] *Lun Yü* 17.2; Lau (1983a), p. 171.

[3] For example, we have passages like this:

The nature (*hsing*) of an ox is not like that of a sheep, and the nature (*hsing*) of a sheep is not like that of a pig. (*Lü Shih Chhun Chhiu* 23.5, ed. Chhen Chhi-Yu, p. 1569)

It is interesting to note that the ancient commentator glosses *hsing* 性 'nature' as *thi* 體 'body'.

[4] *Mêng Tzu* 6A3, 6A8.   [5] *Hsün Tzu* 23.48ff.; cf. also *Mêng Tzu* 6A1 and *Kuo Yü* 13, ed. *SPTK*, p. 2b.

[6] *Chuang Tzu* 9.1, 15.16, and 16.5.   [7] *Han Shih Wai Chuan* 5.17; cf. Hightower (1951), p. 176.

[8] *Huai Nan Tzu*, ch. 11, ed. Liu Wen-Tien, p. 6b.   [9] *Po Hu Thung* 12.6; Tjan (1952), vol. 2, p. 469.

Considering the usages of *hsing* in the ancient literature, it seems clear that a property like 'being bamboo' as such is not what is primarily intended by *hsing*, while 'floating on water' belongs to the *hsing* 'nature' of bamboo. It has been current to translate *hsing* by 'nature', and this captures well the fact that every thing only has one *hsing* 'nature', whereas we would say it has many properties.

Chhing 情[1]

While *hsing* refers to natural endowments or characteristics of a thing in general, *chhing*, which is often used in very similar contexts, refers more specifically to *essential* characteristics or properties of a thing. A. C. Graham was tempted to risk the translation 'essence' while warning the reader not to take this term in any technical Aristotelian sense. Again, every thing will only have one *chhing*, not many. A frequent topos in ancient Chinese literature is the following:

It is the essence of man (*jên chih chhing* 人之情) that men all are concerned for their own person.[2]

Han Fei speaks of the essence (*chhing* 情) of the Way,[3] and Hsün Tzu of the *chhing* of the sense of duty.[4] Even Chuang Tzu speaks of the *chhing* of benevolence and duty.[5] Not only positive qualities have *chhing*. Han Fei speaks of the *chhing* of good or bad government[6] and even of the *chhing* of wickedness.[7] Sometimes no moral qualities whatever are involved, as when we hear of 'being blind and wanting to understand the *chhing* of black and white'.[8]

Chhing as a semi-technical term refers to the qualities that entitle a thing to its name. What is in this sense essential is not necessarily accessible to sensory perception:

The visible to sight is shape and colour. The audible to hearing is name and sound. How sad it is then that worldly people think shape and colour, name and sound, sufficient means to grasp the *chhing* of objects! If shape, colour, name, sound, are really inadequate means to grasp their *chhing*, then, since knowers do not speak and speakers do not know, how can the world understand them?[9]

The aim of Chuang Tzu was in some way to grasp this profound 'essence of the ten thousand things (*wan wu chih chhing* 萬物之情)'.[10] The ten thousand things, of course, do have characteristics that can be perceived by the senses. The Way (*tao* 道) is said to have only *chhing* essence' but no *hsing* 形 'manifest external form, shape, *Gestalt*'.[11] Here we enter a higher metaphysical realm.

---

[1] For an exemplary attempt at the reconstruction of the conceptual history of *chhing* 情, see Graham (1986a), pp. 59–65.

[2] *Han Fei Tzu* 10.9.36; Liao (1939), vol. I, p. 90.   [3] *Han Fei Tzu* 20.27.51.

[4] *Hsün Tzu* 16.79.   [5] *Chuang Tzu* 13, cf. Graham (1966), p. 61.   [6] *Han Fei Tzu* 14.5.2 and 14.5.18.

[7] *Han Fei Tzu* 50.10.10.   [8] *Han Fei Tzu* 14.2.25; Liao (1939), vol. I, p. 118.

[9] *Chuang Tzu* 13.66; Graham (1981), p. 139.   [10] Cf. *Chuang Tzu* 17.38 and 20.8.   [11] *Chuang Tzu* 2.16.

Hsün Tzu makes logical use of the notion of *chhing*:

Things which are of the same kind (*lei* 類) and have the same essence (*chhing* 情) will be apprehended by the senses as being one and the same thing. Therefore by comparing such things with other similar things, one will settle upon a common conventional name to establish a common convention.[1]

The *chhing* is not necessarily accessible to sensory perception. But without it a thing would not be called what it is called. It can be – and is frequently said to be – lost (*shih* 失) so that a thing loses the right to its original name.

Mao 貌

Mao 'external characteristics, aspect' can always be perceived by the senses, and it is subject to change in time. At any given time a thing has only one *mao* 'aspect' although we should say it may have many visual properties. *Mao* is often contrasted as external (*wai* 外)[2] to the quintessential *chhing*:

Ritual must count as that which is the external shape of the essence (of man) (*li wei chhing mao chê yeh* 禮為情貌者也).[3]
The gentleman opts for the essence (*chhing*) and disregards the (external) shape (*mao*).[4]

In the Later Mohist texts, A. C. Graham has identified *mao* as a technical term. Judging from the many examples he mentions, the primary technical meaning is 'visible properties'.[5] In commentarial glosses the term *mao* is constantly used in describing the semantics of descriptive adjectives.

In *Shuo Wên Chieh Tzu* 説文解字 this *mao* is a technical term for descriptive adjectives written with the same graph *mao* that is also often found in the Dialectical Chapters of the *Mo Tzu*.

Chuang 狀

Hsün Tzu uses the term *chuang* 'form' as a technical term to refer generally to the perceptible aspect of an object:

There are things with the same form (*chuang*) but in different places and things with different form (*chuang*) but in the same place; they are to be distinguished. Things with the same form (*chuang*) but judged to be in different places, although they may be put together, are said to be two objects; a thing which alters in its form (*chuang*) without objects dividing and being judged different is said to be transformed, and what is transformed without division is said to be one object.[6]

The technical usage of *chuang* is special to the *Hsün Tzu*, except that something like it recurs in a later work:

---

[1] *Hsün Tzu* 22.17; cf. Watson (1963), p. 142.   [2] *Han Fei Tzu* 21.17.4.
[3] *Han Fei Tzu* 20.7.1; Liao (1939), vol. 1, p. 173.   [4] *Han Fei Tzu* 20.7.3; *ibid.*
[5] Graham (1978), pp. 194–6 and 223.   [6] *Hsün Tzu* 22.27; cf. Graham (1978), p. 197.

Hui Shih said: 'Let us suppose we have a man who does not know what a *tan* 彈 is. If he says: "What is the form (*chuang*) of a *tan* like" and you answer "Like a *tan*", will it be conveyed to him?'

'It will not.'

'If you proceed to answer instead: "A *tan* in its form (*chuang*) is like a bow, but with a string made of bamboo", will he know?'

'He will know.'[1]

It is hard to be sure of any clear semantic distinction between *mao* 貌 and *chuang*.

Li 理[2]

Li, standardly glossed as 'principle' in many English writings, is a concept which came to play an increasingly central part in the history of Chinese philosophy, the early history of which still awaits detailed study. The concept played a particularly important part in Neo-Confucian thought. Chhên Chhun 陳淳 (+1155 to +1223), a disciple of Chu Hsi, writes in his justly famous *dictionnaire philosophique*:

Li 理 'principle' and *tao* 道 'Way' are by and large the same thing but since there is a division into two characters, there must be a distinction.[3]

The concept *li* has been discussed in Vol. 2 and in the large literature concerned with the history of Chinese philosophy.[4] The specific question that concerns us at this point is whether *li* was used by ancient Chinese philosophers to refer to anything like the properties of things.

In one of the most remarkable documents of early Taoist theorising, the chapter *Chieh Lao* 解老 'Explaining Lao Tzu' from the book *Han Fei Tzu* 韓非子 (−3rd century), the concept of *li* is extensively used and discussed as a technical term,[5] but, as it often happens, the key term turns out to be impossible to translate satisfactorily. I shall risk by way of an experiment the translation 'attribute' for *li*, which is an extension of the meaning 'visible pattern', well attested in the early literature:

Being short or being long, being large or being small, being square or being round, being strong or being brittle, being light or being heavy, being bright or being dark, these are called attributes (*tuan chhang, ta hsiao, fang yüan, chien tshui, chhing chung, pai hei chih wei li* 短長, 大小, 方圓, 堅脆, 輕重, 白黑之謂理).[6]

Generally, attributes (*li*) are an apportioning of being square or being round, of being short or being long, of being coarse or being fine, of being strong or being brittle.[7]

---

[1] *Shuo Yüan* 11.8, ed. Chao Shan-I, p. 307. Cf. Graham (1978), p. 445.

[2] My treatment of *li* 理 is inspired by a fascinating essay by D. C. Lau entitled 'Taoist Metaphysics in the *Chieh Lao* and Plato's Theory of Forms', of which I have only seen the proofs but not the printed version. Although my account differs considerably from that of D. C. Lau, I am indebted to him in many ways.

[3] *Pei Hsi Tzu I* 北溪字義 ed. Chung-hua shu-chü, p. 41. I note in passing that the word *li* is glossed no less than 6 times as *tao* in the old commentary to the *Huai Nan Tzu*.

[4] Cf. Vol. 2, pp. 411ff., 438ff., 472ff., 557ff. *et passim*. Graham (1958), pp. 8–22, provides a singularly lucid account of Neo-Confucian conceptions of *li* 理.

[5] I count no less than 22 occurrences of *li*.   [6] *Han Fei Tzu* 20.34.16; Liao (1939), vol. 1, p. 200.

[7] *Han Fei Tzu* 20.29.2; cf. Graham (1978), p. 429.

We note that *tuan chhang* 短長 'literally: being short or being long, or: length' cannot here refer to short and long things. And most importantly, it cannot be taken to refer to things being *either* short *or* long. These pairs of concepts are not just arbitrary illustrations or examples of *li* 理. They seem deliberately lined up to indicate range of dimensions or parameters along which physical objects may be said to have attributes: one-dimensional length, three-dimensional size, shape, elasticity/strength, granulation, weight, pigmentation. It is interesting that smell and taste, for example, are omitted. On the other hand, Han Fei provides some further examples of attributes (*li*) of a different type:

Among the fixed attributes (*ting li* 定理) there are (those of) being persistent or perishing, being dead or living, flourishing or declining.[1]

These attributes (*li*) of things combine to form one *Gestalt*:

Out of the flow and flux things were brought forth. When the things came about, they brought forth attributes (*li*), and these were called the *Gestalt* (*hsing* 形).[2]

From the *Lü Shih Chhun Chhiu* we have an interesting definition of *li* which in the light of *Chieh Lao* 解老 we might try to translate as follows:

Therefore, if distinctions do not fit the attributes (*li*), then they are false. If knowledge does not fit the attributes (*li*), it is misleading. Misleading or false people were executed by the former kings. Attributes (*li*) of things are the ultimate source of right and wrong.[3]

In other words: we are right in calling a horse white on the ultimate basis that the attribute 'being white' is present in the horse. Knowledge as well as disputation must be measured against this ultimate standard.

Here is how Han Fei explains *li*:

As for the attributes (*li*), they bring about the patterning of things.[4]

Han Fei states the obvious in the theoretical manner of a logician:

Things have attributes (*li*).[5]

Again he insists on making explicit what might seem obvious in the manner of a logician:

The ten thousand things each have their different attributes (*wan wu ko i li* 萬物各異理).[6]

---

[1] *Han Fei Tzu* 20.29.4. For our interpretation see Liang Chhi-Hsiung (*1982*), vol. 1, p. 159, and also Lau (1989), p. 7.

[2] *Chuang Tzu* 12.39; cf. Watson (1964), p. 131.

[3] *Lü Shih Chhun Chhiu* 18.4, ed. Chhen Chhi-Yu, p. 1178; cf. Wilhelm (1928), p. 301, and Graham (1978), p. 192, for alternative interpretations of this important definition.

[4] *Han Fei Tzu* 20.27.4; Liao (1939), vol. 1, p. 191. The phrase *chhêng wu chih wên* 成物之文 is grammatically ambiguous. We could also translate: 'is the pattern which brings things about'. This difference is inconsequential for our present purposes.

[5] *Han Fei Tzu* 20.27.11; Liao (1939), vol. 1, p. 191.

[6] *Han Fei Tzu* 20.27.15; Liao (1939), vol. 1, p. 192. Cf. *Chuang Tzu* 25.63.

We have seen that Han Fei speaks of the 'fixed attributes (*ting li* 定理)'. We now need to inquire what he means by *ting* 定 'fixed' in such contexts. Han Fei says:

When the attributes (*li*) are fixed/distinct, it is easy to divide things up (into classes (*lei* 類)).[1]

It is only when attributes (*li*) are fixed that we can distinguish between distinct things, since we must use the attributes (*li*) of things as our criteria.

If this interpretation is right, it would help us explain Han Fei's generalisation 'Therefore only when attributes (*li*) are fixed can one attain the Way'.[2] The Way, after all, is what governs things having or not having their attributes (*li*). Any proper understanding of the Way presupposes an understanding of these fixed attributes (*li*) of things.

As our rough approximation we have translated *li* as 'attribute'. It seems that *li* is ambiguous between a phenomenological sense ('overt attribute') and a metaphysical sense ('disposition to show an overt attribute'). There is nothing strange in this ambiguity. The English word 'attribute' happens to have the same ambiguity. An overt feature like baldness is as much of a human attribute as is a (hidden) disposition like irascibility.

One basic meaning of *li* 理 is 'pattern, arrangement'. What Han Fei did in his technical use of the term was to identify some dimensions or parameters along which one physical object makes patterns. The patterns thus specified then come to resemble physical attributes, on the one hand simple primary qualities of things such as shape and weight, and on the other hand more complex attributes like being alive or being dead. This is how I reconstruct Han Fei's account of *li*.

To the extent that Han Fei speaks of *li* in an abstract way, and of things 'having' attributes, he speaks like a realist, not like a nominalist. But the issue of nominalism versus realism is a Western one which does not arise in ancient China. It is not an issue which needs to detain us at this point where we are attempting to reconstruct Chinese conceptual schemes.

Conclusions

We have set out with the question whether the ancient Chinese used abstract concepts and whether they had a concept of a property. We have ended up by looking into the semantics of some relevant Classical Chinese words, none of which, of course, are exact equivalents for 'property'. Let us now summarise our results. In non-technical Classical Chinese contexts, *li* 理 tends to refer to the pattern of things, the (often hidden) dispositions of a thing, its potential for showing certain qualities at given times. The *Chieh Lao* 解老 chapter of the *Han Fei Tzu* introduces a technical usage according to which *li* refers to any inherent patterns or attributes of things. The attributes as such are fixed (*ting* 定) (by definition, on the one hand, and by the

---

[1] *Han Fei Tzu* 20.34.17; Liao (1939), vol. 1, p. 200.   [2] *Han Fei Tzu* 20.29.3; Liao (1939), vol. 1, p. 194.

constancy of features of things on the other), but the presence of attributes in things changes (in accordance with the Way).[1]

The *chhing* 情 'essential characteristics' are those dispositions which qualify what is called X for the name 'X'. The *hsing* 性 'natural characteristics' are those dispositions with which a thing comes into existence. *Mao* 貌 'form' is used as a technical term for those perceptible properties which a thing has. *Chuang* 狀 'form' is used synonymously with *mao* 'shape' as a technical term to refer to the perceptible properties of things.[2] The crucial thing about all these latter notions is that there is nothing plural-like about them in Chinese. To the ancient Chinese mind things have only one essence, one nature, one shape, and one form.

(8) The Concept of Subsumption

The concepts of the subsumption of one item under a class, and of the inclusion of one class under another class are basic for the structural organisation of a conceptual scheme. In some sense we first understand a thing when we are able to subsume it under a relevant category of things to which it belongs. Setting up a conceptual scheme can be viewed as a systematic strategy of subsumption.

It would be quite easy to imagine a culture with the practice of subsumption but without a formal concept of subsumption, a culture in which the thought that an item belongs to a class or that one class is included in another is expressed by some form of words roughly comparable to our English phrases like *is a*, as in (*a*) *man is a creature*. We do find sentences like *jên wu yeh* 人物也 MAN CREATURE YEH in ancient Chinese literature.[3] It remains systematically unclear in such cases whether we have class inclusion or general subsumption in Chinese.

Theoretically, it would also be quite thinkable that a culture might conceive of subsumption as a part/whole relation where the subsumed is conceived of as a part of an abstract mass-like object consisting of what most other cultures consider as the set of objects belonging to that set. Such theoretical possibilities must be tested against the evidence from the ancient Chinese sources.

In this Section we shall ask whether the ancient Chinese had an explicit concept of subsumption, and we shall try to get a sense of how they conceived of the notion of subsumption. The word we shall concentrate on is *shu* 屬, which originally meant 'to be subordinate to, attached to, to pertain to' and then came to mean 'to belong to (a class)'. Used nominally, *shu* can refer to those who are under somebody's

---

[1] Adapting to the given (dispositions for) properties (*li*) of things is a method of attaining the Way which figures prominently in many early texts.

[2] Hansen (1983), p. 31, contrasts Chinese and Western conceptual schemes concerning objects as follows: 'The mind is not regarded as an internal picturing mechanism which represents the individual objects in the world, but as a faculty that discriminates the boundaries of substances or stuffs referred to by names. This "cutting up things" view contrasts strongly with the traditional Platonic philosophical picture of objects which are understood as individuals or particulars which instantiate or "have" properties (universals).' Neither the crucial book *Han Fei Tzu* nor any of the crucial concepts *hsing, chhing, mao, chuang* or *li* 性, 情, 貌, 狀, 理 figure in the index of that book.

[3] E.g., Wang Chhung, *Lun Heng*, ch. 43, ed. Chung-hua-shu-chü, p. 1392.

domination, subordinate in rank, those who belong to someone's retinue, and finally to things that belong to a given class or kind.

Losing one's retinue (*shu*) and one's army is in itself a serious matter.[1]

Adherents of a philosophical school could be called *shu*:

The disciples and adherents of Confucius and Mo Tzu (*Khung Mo chih ti tzu thu shu* 孔墨之弟子徒屬) fill the whole world.[2]

Shu may sometimes be understood to mean '(belong to) someone's legal jurisdiction':

Chao Meng asked who the officers of the district were, and it turned out that the man belonged to his jurisdiction (*chhi shu yeh* 其屬也). He called the old man and apologised to him . . .[3]

From the notion of personal political or judicial dependence, where someone's function is defined by whatever he belongs to (*shu*) the notion of cosmological dominance is naturally derived:

Chhen is dominated by the element of water (*Chhên shui shu yeh* 陳水屬也).[4]

This dominance can be of a fairly abstract kind:

The human body is composed of three hundred and sixty joints with four limbs and nine passages as its important equipment. Four limbs plus nine passages make thirteen. The functioning of all these thirteen things is subsumed under/dominated by (*shu*) life.[5]

The *Huai Nan Tzu* speaks of 'the five intestines being able to subordinate themselves under the heart and not to go against it'.[6]

In non-scientific, ritual contexts, too, the notion of *shu* plays a part that seems connected with dominance:

The *wang* 望 sacrifice belongs under (*shu*) the *chiao* 郊 sacrifice. If one does not perform the *chiao* sacrifice, one need not perform the *wang* sacrifice either.[7]

In moral contexts, the verb *shu* is used to indicate a person's moral choice or ability to be dominated by a moral value:

A man may be the descendant of a king, duke, prefect or officer, if he is unable to be dominated by ritual and duty (*pu nêng shu yü li i* 不能屬於禮義) he must be relegated to the common ranks. Even if someone is the descendant of a commoner, as long as he has accumulated

[1] *Tso Chuan*, Duke Hsüan 12.3, ed. Yang Po-Chün, p. 727; cf. Couvreur (1951b), vol. 1, p. 618.
[2] *Lü Shih Chhun Chhiu* 25.3, ed. Chhen Chhi-Yu, p. 1651; cf. Wilhelm (1928), p. 437.
[3] *Tso Chuan*, Duke Hsiang 30, ed. Yang Po-Chün, p. 1172; Couvreur (1951b), vol. 2, p. 545; cf. Legge (1872), p. 556.
[4] *Tso Chuan*, Duke Chao 9.3, ed. Yang Po-Chün, p. 1310; cf. Couvreur (1951b), vol. 3, p. 166.
[5] *Han Fei Tzu* 20.30.6; Liao (1939), vol. 1, p. 196.
[6] *Huai Nan Tzu* 7, ed. Liu Wen-Tien, p. 3b. A little earlier in the same passage (p. 3a) the heart is described as the ruler (*chu* 主) of the body.
[7] *Tso Chuan*, Duke Hsüan 3, ed. Yang Po-Chün, p. 668; cf. Legge (1872), p. 293; Couvreur (1951b), vol. 3, p. 575.

learning, has corrected his personal conduct, and is able to be dominated by ritual and duty, he should be elevated as minister, prime minister, officer or prefect.[1]

Moreover, whoever keeps his nature dominated by/subordinate to (*shu*) benevolence and duty, though as intelligent as Tseng and Shih, is not what I would call a fine man.[2]

Those who are thus dominated by a value or a feature *X* are called the *X chih shu* 之屬:

You must be courteous and temperate! Pick out and promote those who are loyal and public-spirited (*kung chung chih shu* 公忠之屬), allow no flattery or favouritism, and then who of your people will venture to be unruly?[3]

Perhaps in this passage we must translate 'people who are loyal and public-spirited and their likes', for the word *shu* 屬 does occur after referring expressions as in *fan tzhu chih shu* 凡此之屬 'in general, things of this sort'.[4]

The notion of dominance is, as we have seen, largely absent in the concept of *lei* 類. None the less, *lei* and *shu* are clearly two closely related concepts.[5]

Hairy and feathered creatures are of the walking or flying kinds (*fei hsing chih lei yeh* 飛行之類也), therefore they belong to (the male principle) Yang (*shu yü yang* 屬於陽).[6]
The kinds of things that have vital spirits and blood (*yu hsüeh chhi chih shu* 有血氣之屬) are bound to have knowledge. All creatures endowed with knowledge (*yu chih chih shu* 有知之屬) know to love their own kind (*lei*) . . . Of the kinds of things that have blood and vital energy (*yu hsüeh chhi chih shu*) none is more knowledgeable than man.[7]

In the *Shuo Wên Chieh Tzu* 説文解字 the idea of subsumption under a genus is regularly expressed by *shu*, as in *shih hao shu* 菩蒿屬 'the *shih*-plant is a species of *hao*-plant'.[8]

This usage is not at all restricted to moral, to formal logical or to scientific discourse. We find the combination in highly informal contexts that are quite unconcerned with logical problems:

When carpenter Shih came home, the sacred oak appeared in a dream and said to him: 'With what do you propose to compare me? Would it be with the fine-grained woods? As for the sort (*shu*) that bear fruit or berries, the cherry-apple, pear, orange, pumelo, when the fruit ripens they are stripped and in being stripped they are disgracefully abused, their branches broken, their twigs snapped off. These are trees which by their own abilities make life miserable for themselves; and so they die in mid-path without lasting out the years assigned to them by Heaven, trees which have let themselves be made victims of worldly vulgarity. Such are the consequences with all things.[9]

The *Kuo Yü* 國語 speaks of *yü po chih lei* 玉帛之類 'things belonging to the category of jade and silk'[10] and of *ping chhê chih shu* 兵車之屬 'things belonging to the category of armoured carts'.[11]

---

[1] *Hsün Tzu* 9.2; cf. Dubs (1928), p. 121.
[2] *Chuang Tzu* 8.36; cf. Graham (1981), p. 202. There are three more relevant examples in the context.
[3] *Chuang Tzu* 12.47; cf. Watson (1964), p. 132.   [4] E.g., *Huai Nan Tzu*, ch. 13, ed. Liu Wen-Tien, p. 27b.
[5] Indeed, the *Kuang-yün* 廣韻 dictionary (cf. our Section on dictionaries [*b*,5]) defines *shu lei yeh* 屬類也.
[6] *Huai Nan Tzu*, ch. 3, ed. Liu Wen-Tien, p. 2a.   [7] *Hsün Tzu* 19.98; cf. Köster (1967), p. 256.
[8] *Shuo Wen Chieh Tzu*, ed. Tuan Yü-Tshai (*WYWK*), vol. 1, p. 54.
[9] *Chuang Tzu* 4.69; cf. Graham (1981), p. 73.   [10] Ed. *SPTK*, p. 18.1b.   [11] Ed. *SPTK*, p. 6.10b.

In biological classification the notion of dominance has already largely given way to that of simple class membership. This becomes even clearer when a set is defined not in terms of a general term but of examples:

The disciples of Hsiangli Chhin, the followers of Wu Hou, and the Mohists of the South, Khu Huo, Chi Chhih, Teng-ling Tzu and their likes (*shu* 屬), all recited the Mohist canons but diverged and disagreed. They called each other heretical Mohists, abused each other in disputation about 'the hard and the white' and 'the same and the different', answered each other with propositions at odds and evens which do not match.[1]

By now, *shu* is purely classificatory:

... If you reject those who respect their superiors and follow the law, and cultivate roving knights and private swordsmen and their likes (*yu hsia ssu chien chih shu* 遊俠私劍之屬), if you act like that, it is impossible that your government should be firm.[2]

The army of Chhi invaded our state (i.e., Lu) and the duke was about to fight when a certain Tshao Kuei requested to be introduced to him. One of Kuei's fellow-villagers said to him: 'The meat-eaters are planning for the occasion. You have no business to interfere.'

Tshao Kuei replied: 'The meat-eaters are vulgar people. They are unable to make far-reaching plans.'

He went in to the duke and asked him on what basis he wanted to go to battle.

The Duke replied: 'Clothes and food are things that I enjoy, but I do not monopolise them. I make sure I give others their share.'

'That is just petty generosity and is not quite all-embracing. The people will not follow you.'

The Duke said: 'In sacrificing animals and offering jade and gems I do not go beyond what is required, so the prognostications are bound to be trustworthy.'

'That is just petty trustworthiness, it does not cover everything. The spirits will not send you their blessings.'

'In all legal matters, whether small or great, I may not be able to investigate them (completely), but I make a point to decide according to the real circumstances.'

'*That* must be subsumed under loyalty (i.e., is a matter of true loyalty to the people, doing one's best for the people) (*chung chih shu ye* 忠之屬也).'[3]

Here it seems that *shu* 屬 is used as a particle or noun of emphatic or contrastive subsumption. There is no question of subordination of one individual under another. *Shu* seems to have taken on a purely categorising function.

In the *Hsün Tzu*, *shu* is even used as a transitive verb 'to subsume as, categorise as':

Why is it that people are brawling? I would want to categorise (*shu*) such people as mad and sick, but that would not be right. The sage king will go on to punish them. I would want to categorise (*shu*) them as birds, rats, wild birds and beasts, but that would not be acceptable. Their physical bodies are those of men.[4]

---

[1] *Chuang Tzu* 33.29; cf. Graham (1981), p. 277.   [2] *Han Fei Tzu* 49.10.32; Liao (1939), vol. 2, p. 287.
[3] *Tso Chuan*, Duke Chuang 10.1, ed. Yang Po-Chün, p. 183; cf. Couvreur (1951b), vol. 1, p. 149: '*c'est de la loyauté!*'; Legge (1872), p. 86.
[4] *Hsün Tzu* 4.14; cf. Köster (1967), p. 31.

(9) The Concepts of Knowledge and Belief

The distinction between knowledge (*epistēmē*, ἐπιστήμη) and belief (*doxa*, δόξα) was basic for the development of Greek philosophy, and for science. For the task of philosophy (a concept which in ancient Greece included that of science) can simply be defined as the art of distinguishing things we know from those which we merely believe.

It is not a foregone conclusion that the Chinese had anything like our concept of knowledge which is at the heart of Western notions of science and the philosophy of science.[1] In any case, the ancient Chinese did not have a noun that corresponds to belief as opposed to knowledge.[2] On the other hand, there seems to be a clear distinction between believing and knowing in ancient texts. The notion of belief as opposed to knowledge comes out clearly in the following passage:

South of the mouth of the Hsia River there was a man called Chüan Chu-Liang. In disposition he was stupid and very fearful. When the moon was bright, he was walking in the dark. He looked down, saw his shadow and thought (*i wei* 以為) it was a ghost following him. He looked up, saw his hair and thought (*i wei*) it was a standing ogre. He turned round and ran. When he got to his house he lost his breath and died.[3]

The Later Mohists take a technical interest in the distinction between belief and knowledge:

Dreaming is supposing something to be so while one is asleep (*mêng wo êrh i wei jan yeh* 夢臥而以為然也).[4]

In this Section I wish to consider ancient Chinese ways of speaking about knowledge in order to reconstruct and understand ancient Chinese ways of thinking. Is the paradigm of what is learned in ancient China really only 'the Confucian virtues',[5] or is there a clear factual and scientific paradigm as well? Is there only knowledge how to treat things, and acquaintance or familiarity with facts in ancient China, or is there also knowledge that certain statements are true? How exactly are we to understand Hsün Tzu's formal definition of knowing when he says:

Considering this (or: what is right (*shih* 是)) as this (or: right), and considering not-this (or: what is wrong (*fei* 非)) as not-this (or: wrong) constitutes knowing. Considering this as

---

[1] Hansen (1982) and (1985) argues that they had no concept of propositional knowledge. His claim is stimulating in the sense that it forces us to test his conclusion against the philological evidence, something he conspicuously fails to do.

[2] When Chinese writers used *hsin* 信 '1. be faithful, be in good faith, 2. believe' nominally in Chinese, the meaning is unequivocally 'good faith, truthfulness', never, as far as I can determine, 'belief' conceived nominally. By contrast, when the character *chih* 知 'to know, to be aware of' is understood nominally (usually read in the fourth, falling, tone, and sometimes written with a different character *chih* 智), the meaning is something like 'wisdom' or 'knowledgeableness'.

[3] *Hsün Tzu* 21.74; cf. Dubs (1928), p. 275.

[4] Graham (1978), A24. Cf. also B10: 'Is it knowing? Or is it supposing the already ended to be so?'

[5] It has been claimed about the concept of knowledge in ancient China: 'Knowledge is a product of learning in the sense of training, not in the sense of the acquisition of data items called concepts and facts. The paradigm of what is learned is the traditional Confucian virtues.' (Hansen (1982), p. 66)

'not-this' and considering what is not-this as 'this' is making a fool of oneself (*shih shih, fei fei, wei chih chih. Fei shih, shih fei wei chih yü* 是是非非謂之知。 非是是非非謂之愚).[1]

When the pronoun *shih* 'this, right' is used as a transitive verb, it is customary to translate it either psychologically as I have done 'to consider as right' or more pragmatically as 'to treat as right'. The profound question is whether in ancient China 'considering something as right' was simply the same as 'treating it as right'. Whether belief and knowledge should be viewed in Classical Chinese as pragmatic concepts.

The present Sub-section is a philological inquiry into the anthropology of believing and knowing in ancient China.

The trouble with the person who makes a mistake is that he does not know but believes he does (*kuo chê chih huan pu chih êrh tzu i wei chih* 過者之患不知而自以為知).[2]

I thought Your Majesty already knew this (*chhên i wang wei i chih chih* 臣以王為已知之).[3]

The notion of believing or supposing something to be the case is translated by the Classical Chinese *i wei*. The slightly different notion of 'believing in' something that has been said is expressed by *i wei jan* 以為然 'believe to be so', or by *hsin* 信 'believe to be trustworthy, believe to be true'.

Han Fei took the distinction between knowledge and belief very seriously indeed. He made a distinction between the factual truth of words on the one hand and psychological attitudes (e.g., of belief) on the other.

It is in the nature of words that they are taken to be trustworthy (*hsin*) when many people advocate them. Take a thing that is not so (*pu jan chih wu* 不然之物). When ten people maintain it, one has one's doubts. When one hundred people maintain it, one thinks it is probably so (*jan* 然). When one thousand people maintain it, it is incontrovertible.[4]

What is this 'thing (*wu* 物)'? Since it is something which one advocates, it appears to be something like a sentence. This sort of story is of central importance to the issue at hand. *Hsin* 'be trustworthy' here refers quite specifically to what is trustworthy because it is true. Good faith is plainly not at issue. It is worth looking at a variant of this story:

(Phang Kung) said: 'Suppose one person maintains there is a tiger in the market. Would you believe it?'
(The King) replied: 'No.'
'Suppose two people maintained that there is a tiger in the market. Would you believe it?'
'No!'
'Suppose three people maintained there is a tiger in the market. Would you believe it?'
'Yes, I would.'
Phang Kung continued: 'It is perfectly clear that there is no tiger in the market, but when three maintain that there is one, that makes a tiger!'[5]

[1] *Hsün Tzu* 2.12; cf. Köster (1967), p. 13.
[2] *Lü Shih Chhun Chhiu* 15.2, ed. Chhen Chhi-Yu, p. 1642; cf. Wilhelm (1928), p. 434.
[3] *Lü Shih Chhun Chhiu* 9.4, ed. Chhen Chhi-Yu, p. 499; cf. Wilhelm (1928), p. 113.
[4] *Han Fei Tzu* 48.6.6; Liao (1939), vol. 2, p. 269.   [5] *Han Fei Tzu* 30.17; Liao (1939), vol. 1, p. 291.

Han Fei took a central interest in the notion of objective truth (which may or may not be known) and which he clearly distinguished from the psychological notion of belief. And in case anyone should think that this clear distinction was limited to philosophers like Han Fei, we can point, for example, to the fact that the story is retailed again in the historical compilation *Chan Kuo Tshê* 戰國策.[1]

The realms of knowledge in ancient China

There are several realms of knowledge that must be distinguished for our purposes: 1. knowing things; 2. knowing how to do things; 3. knowing whether a sentence or statement is true. In addition there are three important distinct ways of reflecting on knowing: 4. reflecting on the nature and structure of knowing; 5. reflecting on the social usefulness of knowing; 6. reflecting on the objective reliability of knowledge.

Before we turn to the more detailed questions, recall the following saying:

It is not the knowing that is difficult. It is the acting that is difficult. The Master knew it, but I was not up to it (i.e., I was unable to follow his advice).[2]

There is little room in traditional Chinese culture for knowledge for its own sake. There was little enthusiasm for 'academic knowledge' as cultivated by philosophers such as Plato and Aristotle, who continued the heritage of Socrates. For the ancient Chinese what mattered was action, i.e., personal action and political action. Insight was valued insofar as it led to successful action.

'Knowing things'

We turn first to the cases when the object of the verb *chih* 知 'know' is a noun, which we call cases of 'knowing things'. For example, the ancient Chinese commonly spoke of the importance of 'knowing men (*chih jên* 知人)'. KNOW SHUN (*chih Shun* 知舜) meant 'know Shun', while KNOW HIS SHUN (*chih chhi Shun yeh* 知其舜也) meant 'know that he was Shun'.[3] The first kind of knowledge we call familiarity, the second we call discursive knowledge.

No knowledge is more important than (the knowing) of people.[4] This 'knowing people' includes knowing how to handle people. Again, ancient Chinese thinkers frequently commend someone for 'knowing ritual (*chih li* 知禮)', and by this they certainly mean that the person in question is properly educated, ritually well-trained. 'Academic' or theoretical knowledge of ritual by itself would not in ancient China have qualified one as *chih li* in every sense of the word. Knowing ritual in ancient China is usually taken not as a purely cerebral awareness of the truth of propositions. One might plausibly argue that it is an acquired skill.

When it comes to 'knowing the Way (*chih tao* 知道)' (*Mêng Tzu* 2A4, 6A6 *et passim*), this is not cerebral knowledge that something is the case. It is mainly understood as a moral and prudential skill.

---

[1] No. 302, ed. Chu Tsu-Keng, p. 1232; see Crump (1970), p. 377.
[2] *Tso Chuan*, Duke Chao 10.5, ed. Yang Po-Chün, p. 1319; cf. Couvreur (1951b) vol. 3, p. 178.
[3] *Lü Shih Chhun Chhiu* 18.1, ed. Chhen Chhi-Yu, p. 1142.   [4] *Huai Nan Tzu* 20, ed. Liu Wên-Tien, p. 27a.

The ruler knows the Way, the minister knows the tasks (*chün chih chhi tao yeh, chhên chih chhi shih yeh* 君知其道也。臣知其事也).[1]

Knowing the Way in ancient China is knowing how to go about things. Consider the following philosophically fascinating passage:

People all use their lives and live, but they do not know that whereby they live; they use their knowledge and know (*i chhi chih* 以其知知), but they do not know that whereby they know. Knowing that whereby one knows is called 'knowing the way (*chih tao* 知道)'. Not knowing that by which one knows is called 'throwing away the treasure'.[2]

What exactly is this thing called 'that whereby one knows'? Is it a knack of some sort, or is it some discursive knowledge? Compare:

There are many people who are not aware what they practise, who do not inquire into what they repeatedly do, who follow a way all their lives but do not understand (*chih*) it.[3]

Notice here the insistence that the common people, the *hoi polloi*, in practice do follow (and in that sense know) their way. In practical terms they do know what they are doing and where they are going. The crucial point made by *chih* is that they do not 'understand'. Confucianism is not just about training people in certain ways of treating things and in moral skills, it is crucially about making them 'understand'. This involves lifting things up into consciousness. The best of Confucian philosophy consists in articulating these things in 'philosophical statements'. Without understanding this point one does not begin to understand the intellectual edge of Confucian thinking.

Interestingly, the early Confucians do not, in fact, speak of 'knowing benevolence (*chih jên* 知仁)' in this way very often at all.[4] On the contrary, Confucius keeps complaining that he does not know whether such-and-such is benevolent or not (*pu chih chhi jên* 不知其仁):

As for Yu, he may be given the responsibility of managing the military levies in a state of a thousand chariots, but whether he is benevolent or not, I do not know (*chih*).[5]

This is a way of saying that he did not know (*chih*) whether the statement 'he is humane' would be true. Thus, contrary to appearances, we seem to have a case of knowing that something is the case.

'Knowing how to do things': competence

Consider the following two kinds of knowledge:

He knew how to harm others but he did not know that others (would) harm him (*chih hai jên êrh pu chih jên hai chi yeh* 知害人而不知人害己也).[6]

[1] Fragments of Shen Pu-Hai 1.7; see Creel (1974), p. 350.
[2] *Lü Shih Chhun Chhiu* 5.3, ed. Chhen Chhi-Yu; p. 265; cf. Wilhelm (1928), p. 58.
[3] *Mêng Tzu* 5A5; Lau (1983c), p. 265.
[4] See *Lun Yü* 4.7, which according to Lau (1983a), p. 30, does not involve 'knowing benevolence'.
[5] *Lun Yü* 5.8; cf. Lau (1983a), p. 39, where there are three comparable examples.
[6] *Lü Shih Chhun Chhiu* 22.1, ed. Chhen Chhi-Yu, p. 1482; cf. Wilhelm (1928), p. 389.

Knowing how to harm others is 'knowing how', whereas knowing that others might harm one is 'knowing that', discursive knowledge. Knowing how to do things is expressed in the pattern KNOW + VERB PHRASE. Knowledge that something is the case is normally expressed in the pattern KNOW + NOMINALISED SENTENCE, although the sentence that is known may also be unnominalised, as we shall see presently. The point to notice here is that 'knowing how' and 'knowing that' are syntactically distinct in Classical Chinese.

'Knowing that': discursive knowledge

The question arises whether the ancient Chinese had the notion of propositional knowledge at all. In recent times this has been summarily denied. The suggestion is that when the Chinese believe X to be a Y, they really have no mental attitude to a proposition at all. They only have an attitude towards X, namely that attitude which is appropriate to things of the kind Y. Believing X is Y is just treating X as Y. Again, when the Chinese know that X is Y – according to this account – that does not involve the notion of a proposition at all, it only involves *successfully* treating X as Y.

A typical Greek scientist would look at a statement, consider what is intended by it, its content, the proposition, and then believe the statement to be true, or know that it is true. His Chinese counterpart might never entertain any belief concerning a statement. He might never claim to know anything concerning a statement. He might only know about things. The only thing he does is to learn to treat X as Y, and if he does so successfully, he will use the word *chih* 'know' to indicate this success:

References to knowing (*chih*) in Chinese philosophical texts are most naturally treated as either knowing how or knowledge by acquaintance. Knowing the virtues (e.g., knowledge of benevolence), can be read as either knowing (how) to be benevolent or knowing of (being acquainted with) benevolence.[1]

Let us consider, then, one rare case where 'knowing benevolence' is discussed in ancient Chinese literature. Mo Tzu has an important passage on the nature of 'knowing benevolence' which has no parallel in Confucian literature so far as I know:

The blind will say that that which is bright is white, and that that which is dark is black. Even the clear-sighted cannot improve on this. But if we should mix up the black and white objects and let the blind pick out (the black or the white) among them, they could not do it. Hence the reason that I say the blind do not know black and white is not terminological. It is in the picking out. The reason why the gentlemen of the world do not know benevolence is not terminological. It is in the picking out (of the benevolent from the non-benevolent).[2]

The Mohist recognises purely terminological (*i chhi ming* 以其名) knowledge concerning definitions or meanings as independent from *a posteriori* knowledge of the world beyond language. He carefully distinguishes this from the more important

---

[1] Hansen (1982), p. 66.   [2] *Mo Tzu* 47.23f.; cf. Mei (1929), p. 225.

'knowing how' to pick out (*chhü* 取) things according to definitions. The blind man does *not* know how to successfully treat bright things as white, but he *does* know as a matter of terminology that bright things are white.

We may safely conclude that the Mohists spoke of purely discursive knowledge. According to A. C. Graham, the Later Mohists drew a distinction between knowledge gained by experience and knowledge which does not require experience other than that concerning the language one is using. With good reason, A. C. Graham is tempted to use the Western term 'a priori' knowledge to describe what the Later Mohists were getting at. The Later Mohists thought that the logical interdependence of predicates creates a network of necessary 'a priori' relations between things.

From the things that follow from each other or exclude each other, it is admissible that we know 'a priori' what it is (*hsien chih shih kho* 先知是可).[1]

The Later Mohists' use of the concept of *a priori* knowledge (*hsien chih* 先知) may be conveniently studied in Graham (1975).

A detailed survey of the uses of the word *chih* 知 'know' in the non-logical Classical Chinese literature yields plenty of cases of *chih* 'knowing' that are neither 'knowing how' nor 'knowing by acquaintance'. On the other hand, a preliminary survey of the verb *ming* 明 'be clear about' suggests that the ancient Chinese notion of *ming* tended to be one of knowledge by intellectual familiarity. Han Fei recommends:

Compare words and know>ascertain whether they are true (*tshan yen êrh chih chhi chhêng* 參言 而知其誠).[2]

For 'know>ascertain whether they are true' we normally read in Classical Chinese 'know>ascertain their truth'. Similarly, we normally say something that looks like 'He knows its being so' for 'He knows that it is so'. Now in English we can say 'He knows it is so', and in Classical Chinese we find cases where the sentence which forms the object of *chih* 知 'know' is not – or in any case not overtly – nominalised:

This is how I know (*chih*) the knights and gentlemen of the world know petty things but do not know Heaven.[3]

When the ancient Chinese said they did not know whether something was the case, such wondering certainly does not presuppose that there is a fact to wonder about.

I do not know (*pu shih* 不識) whether Shun failed to realise that Hsiang intended to kill him.[4]

The object of knowledge may also be an overt question:

I do not know (*pu shih*) how you knew (*chih*) this.[5]

---

[1] Graham (1978), p. 342.   [2] *Han Fei Tzu* 48.4.21; Liao (1939), vol. 2, p. 266.
[3] *Mo Tzu* 26.8f.; cf. *ibid.* 26.14, 27.42. and 28.5off., which has no less than four cases in point. Cf. also *Han Fei Tzu* 10.11.75, 31.28.25 and 31, 32.12.7, 35.6.115, 50.10.11, *Tso Chuan*, Duke Chao 30.4, *Hsün Tzu* 28.42.
[4] *Mêng Tzu* 5A2; Lau (1983c), p. 181.   [5] *Hsün Tzu* 31.41; cf. Köster (1967), p. 385.

In counterfactual cases[1] no event or fact is envisaged:

Suppose that Chieh and Chou had known (*chih*) that their state would be bound to be ruined and that they personally would be bound to die and would be left without offspring, then I am not so sure (*wei chih* 未知) that their cruelty and their wayward actions would have gone this far.[2]

The interesting case is the *wei chih* 未知. Chieh and Chou are explicitly presupposed not to have known the fate they would suffer. There is no fact concerning their cruelty or waywardness under those hypothetical conditions to be acquainted or unfamiliar with. The issue in this sort of sentence is one of what propositions one would know to be true on certain assumptions which are known to be untrue.

From this point of view I am not so sure (*wei chih*) that the ruler of a ruined state cannot count as a talented ruler (*wei chih wang kuo chih chu pu kho i wei hsien chu yeh* 未知亡國之主不可以為賢主也).[3]

If I understand the idiom *wei chih* correctly, the truth of the object of ignorance here is not at all presupposed. On the contrary, the suggestion is that rulers of a ruined state can sometimes count as talented.

Again, Confucius expresses uncertainty about a possible fact rather than unfamiliarity with an actual fact when he advises us, sensibly:

It is fitting that we should hold the young in awe. How is one to know (*chih*) that/whether coming (generations) will be inferior to the present one?[4]

Our conclusion at this point is that discursive knowledge in ancient China (as in ancient Greece and in the modern West) was not just familiarity with things and knowing how to apply names to things. The ancient Chinese had notions of '*a priori*' terminological knowledge and of 'knowing that/whether' statements were true. And yet it remains a most interesting fact that the sentential object of the verb *chih* 'to know' in Classical Chinese never seems to take a syntactically complex form involving conditional and other sentence connectives. Thus in Classical Chinese we never find sentences like this one:

I know (or: believe) that if Shakespeare, although he knew no Chinese, had read some translations of Yüan drama, or if he had had a chance to see a performance, he would have been most excited.

Explicit objects of knowledge and of belief tended to be syntactically simple, even more simple than Classical Chinese sentences generally tended to be.[5]

The typical attitudes of the ancient Chinese towards knowledge, their assessment of the social importance and the intrinsic nature of such knowledge, and the question

---

[1] Cf. our Section (*c*,2).
[2] *Lü Shih Chhun Chhiu* 7.4, ed. Chhen Chhi-Yu, p. 402; cf. Wilhelm (1928), p. 88. There is another exactly parallel example in the context.
[3] *Lü Shih Chhun Chhiu* 4.5, ed. Chhen Chhi-Yu, p. 232; cf. Wilhelm (1928), p. 53.
[4] *Lun Yü* 9.23; Lau (1983a), p. 83.   [5] Cf. our Section (*c*,6) on grammatical and logical complexity.

whether they regarded such knowledge as ultimately reliable will be discussed separately in what follows.

'Explaining knowledge': views and theories about knowledge

For the historian of science traditional Chinese attitudes to knowledge are of primary interest insofar as science is concerned with the accumulation and advancement of learning and knowledge. Such accumulation of knowledge was not popular among Confucian philosophers:

The point in knowledge (*chih*) is not quantity, it is in carefully examining what one knows.[1]

Wide knowledge is of no special concern to Confucius:

The Master said: 'Do I have knowledge (*chih*)? I don't have knowledge! There was a vulgar person who asked me something; his mind may be all empty, but I will get cracking[2] from both ends and will do my best.'[3]

After listing some issues prominent among the sophists that he regards as futile, Hsün Tzu concludes:

Ignorance (in matters of their disputation) is not inconsistent with being a gentleman.[4]

Confucius's priorities for *chih* 'knowledge/wisdom' are clear:

Fan Chhih asked about knowledge/wisdom (*chih*). The Master replied: 'To work for the things the common people have a right to and to keep one's distance from the gods and spirits while showing them reverence may be called knowledge/wisdom (*chih*).'[5]

Confucius attempts an interesting 'definition' of knowledge:

Yu, shall I teach you to know (*chih*) things properly? When you know something, consider that you know it. When you do not know something, consider that you do not know it. That constitutes knowing.[6]

This is as close as Confucius comes to defining *chih* 'knowing'.

Hsün Tzu has a carefully balanced view on the need for knowledge:

Knowledge (*chih*) without benevolence (*jên* 仁) (in a minister) is unacceptable. Benevolence (*jên*) without knowledge (*chih*) is unacceptable. If someone is both knowing and benevolent, then he is a treasure for a ruler of men.[7]

---

[1] *Hsün Tzu* 31.10; cf. Köster (1967), p. 381.

[2] This translation may sound excessively colloquial. But the original, unless it contains textual corruption, seems to be of a similarly colloquial nature.

[3] *Lun Yü* 9.8; Lau (1983a), p. 79.   [4] *Hsün Tzu* 8.35; cf. Köster (1967), pp. 74f.

[5] *Lun Yü* 6.22; Lau (1983a), p. 52.   [6] *Lun Yü* 2.17; Lau (1983a), p. 15.

[7] *Hsün Tzu* 12.67; cf. Köster (1967), p. 162. The *Lü Shih Chhun Chhiu* 1.4, ed. Chhen Chhi-Yu, p. 45, summarises feelings current in ancient China concerning being ('academically') knowledgeable:

> If one occupies a high position, one does not want to be inquisitive in a small way. One does not want to be knowledgeable in a small way. . . .
> Being knowledgeable and pursuing private aims is not as good as being stupid and pursuing unselfish ends.

Hsün Tzu also had a rather subtle distinction between two meanings of *chih* 'know(ing)':

The means of knowing that is within man is called '*chih* (intelligence)'; his intelligence tallying with something is (also) called '*chih* 智 (knowing)'.[1]

It is not improbable that Hsün Tzu derived inspiration for this distinction from the Later Mohists.

The Mohist theoretical account of knowledge

It is the considerable merit of the Mohists that they recognised the central importance of the concept of knowledge in their intellectual scheme of things. They took a conceptual interest in knowledge which is alien to earlier Chinese thinking. To start with, the Later Mohists defined a series of concepts in the semantic field of knowledge as follows:

'Intelligence (*chih*)' is the capability.
Explanation: It being the means by which one knows, one necessarily does know. (Like (the case of) eyesight.)

'Thinking (*lü* 慮)' is the seeking.
Explanation: By means of one's intelligence one seeks something, but does not necessarily find it. (Like peering.)

Lü 'thinking' is interpreted here as trying to achieve knowledge. This trying is not necessarily successful.

'Knowing (*chih*)' is the connecting.
Explanation: By means of one's intelligence, having passed the thing one is able to describe it. (Like seeing.)

This definition of knowledge is curiously close to the etymology of the Greek word for to know, *oida* which literally means 'I have seen'.[2]

'Understanding (*chih*)' is the illumination.
Explanation: By means of one's intelligence, in discourse about the thing one's knowledge of it is apparent. (Like clearness of sight.)[3]

The Mohists proceeded to a threefold classification of knowledge in terms of methods or sources of knowledge and objects of knowledge:

Knowing (*chih*) is by hearsay, by explanation, or by personal experience.
Explanation: (How one knows:) Having received it at second hand is 'knowing by hearsay'.
Knowing that something square will not rotate is by 'explanation'.
Having been a witness oneself is 'knowing by personal experience'.

---

[1] *Hsün Tzu* 22.5; cf. Graham (1978), p. 269.
[2] Another curious instance of such coincidence of Western etymology and Eastern definition is the Mohist definition of *pi* 必 'necessary', where the Latin etymology of *necessarius* is 'not ceasing, not ending'. This interesting observation was pointed out to me by William B. Boltz.
[3] For the text and an excellent interpretation of this sequence of definitions see A. C. Graham (1978), p. 267.

(What one knows:) What something is called by is its 'name'.
What is so called is the 'object'. The mating of name and object is 'relating'.
To intend and to perform are to 'act'.[1]

Thus the Later Mohists recognised and practised 1. a science of names, 2. a science of objects, 3. a science of how names apply to objects, and 4. a science of human action. Thus they had an explicit scientific scheme of knowledge according to which they proceeded.[2]

Having thus defined a range of concepts, the Mohists proceeded to conceptual analysis:

When one knows, it is not by means of the 'five roads.'[3] Explained by: duration.
Explanation: The knower sees by means of the eye, and the eye sees by means of fire, but fire does not see. If the only means were the 'five roads', knowing as it endures would not fit the fact. Seeing by means of the eye is like seeing by means of fire.[4]

Finally, the Later Mohists explained an apparent paradox of knowledge: that we can be said to know what we do not know. This, we are told, is because we are able to choose between what we know and what we do not know.[5] It seems likely that this extremely sensible explanation is a reaction to Taoist sceptical rejections of knowledge.

We must now turn to a closer examination of the traditions of anti-intellectualism and of scepticism. Anti-intellectualism and scepticism must be carefully separated: I call anti-intellectualism the negative views on the social and pragmatic importance of (proto-)scientific knowledge, and I call scepticism the pervasive doubt concerning the objective reliability of human knowledge. Ancient Chinese anti-intellectualism is concerned with the social function of knowledge, scepticism belongs firmly to the realm of epistemology, the general theory of the relation between knowledge and reality.

'Dismissing knowledge as useless': anti-intellectualism

It is one of the interesting paradoxes of the history of Chinese science that some of the Taoists, who contributed most to the progress of science in China through the ages, have also gone on record as 'rejecting knowledge(ableness)':

Cut off sageliness, reject intellectual excellence (*chih*), and the people will benefit a hundredfold.[6]

The obvious question is what exactly the ancient Taoists rejected when they rejected *chih* 'knowledgeableness, intellectual excellence'.[7] How exactly are we to interpret the noun *chih* 'wisdom, knowledgeableness, shrewdness, wiliness' in the relevant contexts?

---

[1] Cf. Graham (1978), p. 327. Contrast Vol. II, p. 177.
[2] See Graham (1978), pp. 231–4.   [3] I.e., the five senses.
[4] See Graham (1978), p. 415. Contrast Vol. II, p. 178, 'Knowledge of duration'.
[5] Cf. Vol. II, p. 179 and Graham (1978), p. 417.   [6] *Lao Tzu* 19; cf. Vol. II, p. 87.   [7] Cf. Vol. II, pp. 86ff.

The Taoists certainly did not reject intuitive or practical aptitude and skill. On the contrary, they cultivated it under the provocative name of non-action (*wu wei* 無為). The widely celebrated 'knack-passages' in *Chuang Tzu*[1] bear eloquent witness to this. Like everybody else the Taoists rejected *chih* understood as the negative quality of 'wiliness', but such trivial rejection of an intrinsically unattractive quality would be too facile to deserve so much attention and emphasis.

A detailed survey of the uses of *chih* 知 and *chih* 知/智 (falling tone) in the indexed literature has led me to the conclusion that these negative attitudes to *chih* constitute a rejection of what we today might call 'intellectual excellence' or even 'academic excellence' and of the sort of discursive 'academic' knowledge that defines 'academic excellence'. Included in this concept of intellectual excellence is knowledgeableness in matters of lofty Confucian moralising philosophy. Our ancient texts often mention *chih* 'knowledgeableness' together with *pien* 辯 'sophistry', which after all was a (derogatively used) term for what was the pursuit of scientific knowledge for its own sake. The kind of knowledgeableness attacked as *chih* is the rhetorical knowledgeableness of the sophists and the theoretical knowledgeableness of scientists like the Later Mohists, as well as the traditionalist moralistic knowledgeableness of traditional Confucian learning.

To the extent that scientific knowledge and intellectual excellence became instrumental in the pursuit of the good Taoist life (good health, long life as well as immortality), such intellectual excellence was encouraged and cultivated by later Taoists. The preceding volumes of *SCC* bear witness to the large scale on which this has happened throughout Chinese history.

Confucians, for their own reasons, rejected *chih* 'intellectual excellence' when it was not constructively instrumental in the good conduct of the moral life and of political life. Confucius and his followers thought that intellectual excellence was secondary to moral excellence, and that the effects of intellectual excellence upon the moral qualities of the individual as well as the political qualities of the state were predominantly negative. Intellectual excellence was therefore not to be especially cultivated except as a handmaid to moral edification or to political administration.

When ancient Confucian and Legalist texts address the problem of *chih* 'knowledgeableness, intellectual excellence', they do not address a problem of epistemology at all. Often they address a problem of public administration comparable to the recurrent question of how much weight a president in the United States of America should put on academic expertise in his government, and how many distinguished Harvard professors he should employ.

The Legalist Han Fei was very much preoccupied with the thought that intellectual excellence easily combines with insubordination and thus gets in the way of the strict discipline which is necessary for the well-regulated running of a state. When Han Fei used the word *chih* 智 (now read in the falling tone), he mostly

---

[1] Cf. Vol. II, pp. 121ff., and Graham (1981), pp. 6ff. Note that there are no such passages in *Lao Tzu*, while there are many in *Lieh Tzu*.

thought of it as 'academic knowledgeableness' together with such words as *pien* 辯 'rhetoric' and the like:

So-called knowledgeableness (*so wei chih chê* 所謂智者) consists in subtle and mysterious talk.[1]

There are literally hundreds of passages attacking such politically unproductive ('academic') knowledgeableness and its rhetorical as well as philosophical derivatives in *Han Fei Tzu*. One less polemical and philosophically more interesting remark concerns the limitations of knowledge:

Knowledge (*chih*) is like eye(sight): (the eye) can see further than a hundred paces, but it is unable to see its own eyelids.[2]

This thought, one might suspect, belongs to the Taoist strain in Han Fei's work.

'Doubting the reliability of knowledge': scepticism

Consider a piece of knowledge like 'two plus two equals four'. Conceivably, one might dismiss this statement as inconsequential and morally irrelevant. That would be one way of dismissing such a piece of knowledge. But a hard-headed mathematician might also do something entirely different. He might doubt that we can be absolutely certain that two plus two actually equals four. That would be a case of scientific epistemological scepticism.

The question I now propose to discuss is this: did the ancient Chinese develop epistemological scepticism as clearly distinct from what we called 'anti-intellectualism' in our last Sub-section? Did the ancient Chinese cast general doubt on the ultimate reliability of human knowledge?

Consider a famous saying attributed to Lao Tzu:

Knowing to not-know is superior (*chih pu chih shang i* 知不知尚已).[3]

Wang Pi 王弼 (+226 to +249) comments:

If you do not know that knowing cannot be relied upon (*pu tsu jên* 不足任), you are at fault (*ping* 病).[4]

Chuang Tzu asks:

Is it when not knowing that one knows? Is it that when one knows one does not know? Who knows the knowing which is not-knowing?[5]

---

[1] *Han Fei Tzu* 49.11.3; Liao (1939), vol. 2, p. 288.   [2] *Han Fei Tzu* 21.20.9; Liao (1939), vol. 1, p. 226.

[3] *Lao Tzu* 71; cf. Lau (1982), p. 105: 'To know yet to think that one does not know is best.' The Chinese text does not support this interpretation.

[4] See Rump and Chan (1979), p. 194.

[5] *Chuang Tzu* 22.61; cf. Graham (1981), p. 163. In his usual, pithy style Hsün Tzu points out a profound paradox of epistemology which would seem to be Taoist-inspired:

Great efficiency consists in what one does not achieve deliberately. Great knowledge (*ta chih* 大知) consists in what is beyond (deliberate) cogitation (*tsai so pu lü* 在所不慮). (*Hsün Tzu* 17.16; cf. Köster (1967), p. 217)

This not-knowing is quite possibly Lao Tzu's 'knowing to not-know'. The theme, in any case, is a common one in the *Chuang Tzu*:

Men all set store by what wit knows, but none knows how to know by depending on what his wits do not know; may that not be called the supreme uncertainty?[1]

The book *Chuang Tzu*, particularly chapter 2, asks unflinchingly and with crystalline clarity of thought the all-important question:

How do I know that what I call knowing is not not-knowing?
How do I know what I call not-knowing is not knowing?[2]

Chuang Tzu remains uncertain. For every level of knowledge one may have achieved in one's life there is a higher level of uncertainty concerning the reliability of that knowledge one has achieved. Knowledge is thus inevitably built on an uncertain basis, on what we do not know, on not-knowing. And a recognition of this latter state of affairs constitutes part of the true wisdom of the Taoist sage. It is a wisdom which consists in not-knowing. A wisdom which deserves to be closely compared with the famous *oida ouk eidōs* (οἶδα οὐκ εἰδώς) 'I am aware that I do not know' attributed to Socrates.

For example: how do we know that we are not dreaming as we think we know something? The justly celebrated story of Chuang Tzu and the butterfly tries to illustrate that we cannot be sure.[3]

The Taoist sceptical attitude is that all knowing of theorems is never quite certain. There may be delusion. We may 'wake up' to higher insight which might invalidate whatever we think we know. Chuang Tzu also gives us an example from the philosophy of life:

How do I know that to take pleasure in life is not a delusion?
How do I know that we who hate death are not exiles since childhood who have forgotten the way home? . . .
Who banquets in a dream, at dawn wails and weeps. Who wails and weeps in a dream, at dawn goes out to hunt. While we dream we do not know that we are dreaming, and in the middle of a dream interpret a dream within it. Not until we wake up do we know that we are dreaming. Only at the Great Awakening shall we know that this was a Great Dream. Yet fools think they are awake, so confident that they know what they are, princes, herdsmen, incorrigible! You and Confucius are both dreams, and my calling you a dream is also a dream.[4]

Through this poetic speech in a fictitious dialogue Chuang Tzu suggests that our knowledge is uncertain because we may wake up to find that it was an illusory dream. As the history of Taoism shows, a theoretical conviction of this sort and a commitment to the book *Lao Tzu* as an authoritative source is perfectly compatible with the conduct of empirical science.

---

[1] *Chuang Tzu* 25.52; cf. Graham (1981), p. 102.   [2] *Chuang Tzu* 2.66; cf. Graham (1981), p. 58.
[3] *Chuang Tzu* 2.94; tr. Graham (1981), p. 61.
[4] *Chuang Tzu* 2.78/9; cf. Graham (1981), p. 59 and Vol. II, p. 87.

Apart from the argument from delusion, Chuang Tzu appeals to arguments from the inevitable subjectivity of human viewpoints. Chuang Tzu observes that deictic terms like 'this' and 'that' are inevitably subjective, and he speculates whether all our assessments of things might not be of the same kind:

'It's acceptable!' Then it is acceptable. 'It's unacceptable!' Then it is unacceptable. The Way comes about as we walk it. Things become 'so' because we call them 'so'.[1]

Chuang Tzu delights in the thought that, names being arbitrary, anything can become anything else by a mere change in nomenclature. If I call oxen horses, then all oxen have thereby become horses. Thus, he concludes, all knowledge is subjective and relative to the knower. Taoist enlightened not-knowing consists in an awareness of this ultimate subjectivity and relativity of nomenclature which leads one not to know whether or not 'ultimately' to call things by one name or another.

Chuang Tzu's relativism is remarkably close in mood to Heraclitus. You may ask whether a non-level road 'really' goes up or down. There is no objective answer to this, as Heraclitus points out, and as Chuang Tzu no doubt would have delighted in pointing out if he had thought of it:

A road is, upwards and downwards, the same.[2]
The sea is purest and most unclean water: for fish, drinkable and life-giving; for men, undrinkable and deadly.[3]

For Chuang Tzu subjective distinctions and oppositions are indistinguishable in the Way. Similarly we have in Heraclitus:

God is day night, winter summer, war peace, surfeit famine; but he is modified, just as fire, when incense is added to it, takes its name from the particular scent of each different spice.[4]

The connection between knowing and not-knowing is discussed in chapter 12 of the *Huai Nan Tzu*, which tells of a fictitious person Translucence asking the equally fictitious Infinitude whether he knew the Way. Infinitude replied he did not know. Translucence went on to ask Non-Action the same question, and Non-Action replied that he did know. Confused, Translucence went on to ask No-Beginning which was right, the not-knowing of Infinitude or the knowing of Non-Action.

No-Beginning replied: 'The not-knowing is profound. The knowing is superficial. The not-knowing is internal, the knowing is external. The not-knowing is subtle, the knowing is crude.'
Translucence threw back his head and sighed: 'So not-knowing is knowing. Knowing is not-knowing. Ah, who knows that knowing is not-knowing and that not-knowing is knowing?'[5]

Not-knowing, then, is not at all ignorance, it is an advanced Taoist version of *docta ignorantia*. It is not an anti-intellectual rejection of scientific knowledge. It is

---

[1] *Chuang Tzu* 2.33; cf. Graham (1981), p. 53.   [2] H. Diels (1954), fr. 60.
[3] *Ibid.*, fr. 61.   [4] Heraclitus, fr. 67; cf. Hussey (1972), p. 46.
[5] *Huai Nan Tzu* 12, ed. Liu Wên-Tien, pp. 1aff. Cf. the probably earlier *Chuang Tzu*, ch. 22; Graham (1981), pp. 162f.

itself the product of an advanced piece of scientific theorising. Chuang Tzu thought that our knowledge, however well-founded empirically and theoretically, is ultimately uncertain.

Chuang Tzu nowhere directly and dogmatically states that we cannot know. He only persists in asking 'How do we know?'. He is not an adherent of the dogma that we cannot know anything (which would make him into what in Greek philosophy would be called a dogmatic Academic). Chuang Tzu simply cannot see how we can avoid uncertainty (and must count as a true sceptic in the way that Sextus Empiricus (*fl.* +180 to +200) was famous for in the West). The dialogue between the fictitious characters Gaptooth and Wang Ni brings this out:

> Gaptooth asked Wang Ni: 'Do you know what all things agree in calling right?'
> 'How would I know that?'
> 'Do you know that you do not know that?'
> 'How would I know that?'
> 'Then do (we) creatures know nothing?'
> 'How would I know that? But suppose I try saying something. What way do I have of knowing that if I say I know something I don't really not know it? Or what way do I have of knowing that if I say I don't know something I don't really know it? Now let me ask you some questions . . .'[1]

Compare the redoubtable German mathematician and theologian Nicolas of Cusa (+1401 to +1464). When he published his great work of philosophy, *De docta ignorantia* (1440), he quite properly entitled the first chapter *Quomodo scire est ignorare* (How knowing is not-knowing) and then proceded to explaining *Quod praecisa veritas incomprehensibilis sit* in his chapter 2. Nicolas of Cusa finishes this beautiful third chapter with the paradox: *Et quanto in hac ignorantia profundius docti fuerimus, tanto magis ipsam accedimus veritatem* (And the more profoundly learned we become in this ignorance, the closer we come to the truth.)[2] Nicolas of Cusa was, in a way, more dogmatic than Chuang Tzu. He maintained a dogma on the uncertainty of theological knowledge, a dogma which on our interpretation Chuang Tzu would have found questionable.

It is significant that Nicolas supports his sceptical doctrine with a wealth of geometrical arguments. He was an accomplished mathematician and interested in medicine and biology, as well as in many branches of applied science. Through experimental methods he established such important insights as that air has weight and that plants absorb nourishment through the air. Being ultimately a sceptic and a Neo-Platonist mystic did not prevent him from being a proficient practitioner of science.[3]

I dwell on Nicolas of Cusa because his way of thinking about the ultimate statements, about theology, was close to that of Chuang Tzu: *Et ex his manifestum est,*

---

[1] *Chuang Tzu* 2.64; cf. Watson (1964), p. 45.   [2] Nicolas of Cusa (1964), p. 14.
[3] Or, incidentally, from discovering a dozen unknown comedies by the Roman dramatist Plautus.

quomodo negationes sunt verae et affirmationes insufficientes in theologicis (From this we see clearly how in matters theological negations are true and affirmations are insufficient).[1]

The doubting attitude to knowledge, the insistence on the justification for claiming the objective validity of apparently self-evident or commonly accepted knowledge, is a central part of rationality and a crucial factor in early Chinese intellectual history.[2] Chuang Tzu's attitude of pervasive uncertainty concerning the question of the reliability of our knowledge is the result of such rational doubt, as expounded in chapter 2 of the *Chuang Tzu*. This chapter, therefore, must count as a central document in the history of Chinese science and epistemology.

Pyrrho of Elis (*c.* −360 to −270), the father figure of Greek scepticism, was roughly contemporary with Chuang Tzu. Pyrrho's sceptical tradition became important in Europe from the −3rd century onwards. It appears that Pyrrho visited India, and it is said that he mixed there with the 'naked sophists' or fakirs (*gymnosophistai*) and the magicians (*magoi*).[3] We are told:

(Pyrrho) used to say that nothing was beautiful or ugly, just or unjust, that all human action was invariably by custom and by habit, that nothing was more this or more that.[4]

Thus the sceptical tradition in the West, which was so important for the development of self-critical science, may have benefited from Indian inspirations. Like his contemporary Chuang Tzu, Pyrrho seems to have been a colourful and poetic soul. There was an anti-authoritarian and non-conformist streak in both men. But then it is important to remember that there is nothing particularly unscientific about being anti-authoritarian, non-conformist and doubtful about everything.[5]

The early Socrates (as depicted by Plato) was a case in point. He refused to be certain about anything. He was a thorough sceptic. The Norwegian philosopher (and mountaineer) Arne Naess has argued eloquently for the philosophical and scientific feasibility of scepticism in his philological reconstruction and philosophical analysis of Pyrrhonism as expounded in the work of Sextus Empiricus (*fl.* +180 to +200), which constitutes the earliest coherent philosophical exposition of scepticism that has come down to us from European antiquity.[6] Indeed, many philosophers of science have recognised the sceptic Sextus Empiricus as a crucial figure in the history of Western scientific methodology. And yet, Sextus Empiricus' purpose in cultivating a sceptical non-committal attitude to all dogma was not immediately scientific. Like Chuang Tzu, he – and Pyrrho many centuries before him – cultivated a sceptical stance as a means to obtain serenity, peace of mind, *ataraxia* 'unruffled balance of thought'. Through this quest he made a significant contribution to the history of science.

---

[1] *Ibid.*, p. 112.   [2] Cf. Vol. II, p. 365.   [3] See Diogenes Laertius, *Vitae Philosophorum* 9.61.
[4] *Ibid.*   [5] A good historical survey on Pyrrho and early Greek scepticism is that of L. Robin (1944).
[6] Naess (1968).

(e) LOGICAL PRACTICE

(1) ARGUMENTATION AND RATIONALITY IN ANCIENT CHINA

We have seen that the ancient Chinese had the linguistic and logical tools which are necessary to construct arguments. But what argumentative use, if any, did they make of these tools? On the one hand it is perfectly possible for a person – or a culture – to live a happy life without ever explicitly arguing logically for any conclusions or deliberately proving any points. On the other hand it is far from certain whether any people could ever get by without in their speech and conduct showing evidence that they – at least implicitly – reason logically. One might well be tempted to argue that such a people is unthinkable.

In this Section, then, I shall be concerned to show some examples of Chinese explicit reasoning. I also want to show how logical reasoning is clearly implicit in certain ancient Chinese ways of speaking and writing. The first point is far from trivial. In the second case I think it is worth while to illustrate through examples what I believe might well seem self-evident to a logician or to a philosopher.

Rationality and argumentation arise when a thinker seriously contemplates the pervasiveness of the possibility that he may be wrong, that he needs reasons and arguments to support the validity of his views. This rationality is not something universal, but it seems to me to be not unconnected with the notion of a philosophical culture.

The question of whether rationality developed among the Chinese is entirely separate from the question whether the ways of speaking of the Chinese do or do not involve what we can recognise as logical trains of thought. Recognising a train of thought as logical, in turn, is in no way the same as recognising it as plausible or acceptable. A person or a culture can be desperately wrong in our eyes, but at the same time they may be entirely rational and logical in elaborating what we may perceive as a wrong-headed way of conceiving of things. Thus rationality and acceptability have to be kept carefully separated.

There are, then, several separate questions:

1. Did the ancient Chinese try to justify claims which they were making about the world? (The answer to this one is an emphatic 'Yes, of course they did. They did it all the time.')
2. Did the ancient Chinese construct explicit logical arguments? (The answer is 'Yes, they certainly did, but the rôle of these arguments in their discourse was significantly smaller than for example in Greece, and the logical structure of such arguments was somewhat simpler.')
3. Can we reconstruct arguments as plainly implicit in ancient Chinese texts without violating the spirit of these texts or reading alien logical elements into them? (The answer is 'Yes we can, and in order to understand Chinese texts properly traditional Chinese commentators as well as modern Western translators have to do this all the time.')

4. Are some of the Chinese patterns of explicit argumentation (if indeed there are any) recognisable by us as valid?[1] (The answer is 'Yes, there are many such valid explicit and implicit arguments.')

I am aware that relativists will argue that in arriving at the answers outlined above I impose our categories on the Chinese texts. On the other hand that is, in principle, also what we are reduced to doing when we understand the logical structure of what members of our own culture say, or even what we ourselves are saying. In each of these cases we have to translate something into our own idiom, and the fact that the translation sometimes is homophonous with the original does not, philosophically, matter at all. At this point I can only remind the reader of the remark by W. V. O. Quine:

Wanton translation can make natives sound as queer as one pleases. Better translation imposes our logic upon them, and would beg the question of prelogicality, if there were a question to beg.[2]

To all philosophical intents and purposes we are natives ourselves to ourselves when we try to understand what we are saying. Relativism applies also in that case. There is nothing special about the case of the Chinese in this respect.

In any case, the above – quite separate – issues are basic to the history of science and civilisation in China, but unfortunately they are not at all easily studied in any conclusive way.[3]

Even if we lined up some instances where the ancient Chinese would appear to argue or think along familiar logical lines, we would still be open to the objection that the examples we have chosen to concentrate on are collected from a biased point of view and not at all representative of the essence and the natural bent of the Chinese mode of thinking. One might argue that the predominant intellectual mode in ancient China was not that of argument or proof but of classification or correlation.[4] Such correlation was naturally expressed in sets of parallel sentences, hence the tendency towards parallelism. The Chinese, one might feel, were not so much interested in proving general theses as they were in correlating concrete phenomena with supernatural agents, *shen* 神, and classifying them, e.g., under such broad categories as *yin* 陰 and *yang* 陽, 'hot' and 'cold' in medicine and natural philosophy, or the *wu hsing* 五行 'five phases'.

In many cases these different explanatory and argumentative strategies were in competition with each other, as in the case of the sickness of the Duke of Chin in

---

[1]   Note that by now remarkably few of Plato's arguments, for example, would strike philosophers as acceptable as they stand. The point is that we seem to perceive Plato as being in pursuit of what we would recognise as rational and convincing arguments.

[2]   Quine (1969), p. 58.

[3]   One of the most detailed studies of argumentation in early China is Leslie (1964b)'Processes of Reasoning in Confucian Texts down to the +1st century.' Ms at Needham Research Institute, n.d., (1964?). Liu Chhi (1966) provides a wealth of examples, and Garrett (1983) attempts to interpret ancient Chinese reasoning in the light of Piaget's developmental psychology. Bodde (1938), ch. 11 'Types of Reasoning in Li Ssu', was the pioneering study of ancient Chinese argumentative rhetoric.

[4]   For a stimulating analysis of correlative thinking see Graham (1986d).

−539, on which our historical records have preserved extensive discussion. We are told that the divination specialists attributed the matter to interference from two spirits unknown to all the scribes. Tzu Chhan arrives on the scene, explains who the two spirits are, but insists that they are the sort who have nothing to do with the Duke's illness.

The ruler's health depends on his conduct, on his drinking and eating, on his griefs and joys. The spirits of the mountains, rivers and heavenly constellations have nothing to do with it.[1]

Tzu Chhan argues that it must be the Duke's irregular and improper daily routines that have caused the sickness. Moreover, he points out, there are four concubines in the Duke's harem who bear the same surname as the Duke: couldn't that breach of the taboo against endogamy have been a reason for the sickness? Tzu Chhan recommends that the four concubines be dismissed. The Duke of Chin is impressed by these arguments but requests more medical advice from Chhin, and a certain Dr Ho from Chhin comes up with a remarkable psychological explanation:

The sickness cannot be treated. This is a case of being sick to the point of bewitchment as soon as one gets into the female chambers. It is neither a matter of the ghosts or of food. It is a matter of getting confused and losing ones mental control (*sang chih* 喪志).[2]

The Duke is worried:

'Must I, then, keep away from women?'
'You must show moderation. . . . Heaven has the Six Ethers, which descending generate the Five Tastes, issue as the Five Colours, are evidenced by the Five Sounds, and in excess generate the Six Diseases. The Six Ethers are *yin* 陰 and *yang* 陽, wind and rain, dark and light. They divide to make the Four Seasons, in sequence make the Five Rhythms, and in excess bring about calamity. From *yin* in excess cold diseases, from *yang* hot; from wind in excess diseases of the extremities, from rain of the stomach; from dark in excess delusions, from light diseases of the heart. Woman being a thing of *yang* but from a dark season, in excess she generates the diseases of inward heat and deluding poisons.[3]

Compare the reasoning in −486 by Chao Yang of Chin. He hesitates whether or not to go to the rescue of the state of Cheng which is being attacked by the state of Sung. By divination it is determined that the event in question is characterised by the juncture of fire being extinguished by water. The divination specialists decide that this signifies that a war is possible, and that since a war (like water) belongs to the *yin* or female dynamism, this is dominant. The specialists further explain that if Chao Yang wishes to go to war, then, being of a *yin*-clan himself, he should not go to war with another *yin*-clan, that of the princes of Sung. On the contrary, the specialists argue, he could attack the prince of Chhi whose lineage descends from an officer of the fire and therefore must be said to be a *yang*-clan. We are told that Chao

---

[1] *Tso Chuan*, Duke Chao 1, ed. Yang Po-Chün, p. 1220; cf. Couvreur (1951b), vol. 3, p. 33.
[2] *Tso Chuan*, Duke Chao 1, ed. Yang Po-Chün, p. 1221; cf. Couvreur (1951b), vol. 3, p. 36.
[3] *Tso Chuan*, Duke Chao 1, ed. Yang Po-Chün, p. 1222; cf. Couvreur (1951b), vol. 3, p. 37. For the intellectual context of this passage, cf. Graham (1986d), particularly p. 71.

Yang avoided attacking Sung in support of Cheng but instead lead a successful campaign against Chhi.[1] The kind of practical reasoning here summarised was of great importance in Chinese intellectual history. In one of the texts attached to the *Ma-wang-tui* 馬王堆 manuscript of the *Lao Tzu* we read:

In general, in theoretical discourse one must use *yin* and *yang* to make plain the grand scheme (*ta i* 大義).[2]

Chinese arguments of a correlative sort may be seen to be rational even when they seem quite implausible to us:

In the fourth month there was a fire in Chhen. Pi Tsao of Cheng said: 'In five years the state of Chhen will be re-established, and after 52 years it will finally perish.' Tzu Chhan asked for a reason (*ku* 故).

Pi Tsao replied: 'Chhen belongs to the element of water. Fire is the element antagonistic to water and is under the supervision of Chhu. Now the fire-star has come out and brought fire to Chhen. This indicates that Chhu is ousted and Chhen is established. Antagonistic elements are ruled by the number five. That is why I say in five years. The year-star must five times come to the configuration *kuo hou* before Chhen perishes. That Chhu will gain control over it is in the nature of Heaven. That is why I say 52 years.[3]

For all we are told, Tzu Chhan was satisfied with this piece of reasoning, or at least he was expected to find it satisfactory and plausible. Today, we badly need the ancient commentary to understand what is going on. The commentator Tu Yü 杜預 (+222 to +284) explains that after five years 'water' will gain the ascendance in the sky and that as a result of that Chhen (representing water) will be re-established. From then on four years will pass before the configuration *kuo huo* will appear in the sky, and this configuration appears every twelve years. Thus it will have appeared five times after 52 years $(4 + (4 \times 12))$. When properly seen against the background of the beliefs apparently current at the time, this argument was a rational one. The difficulty in many cases is how to reconstruct the rational force of apparently wild Chinese arguments.

One might argue that the ancient Chinese were more inclined to argue 'analogically', by analogy or comparison, rather than logically by demonstration or proof.[4]

---

[1] *Tso Chuan*, Duke Ai 9, ed. Yang Po-Chün, p. 1652f.; cf. Couvreur (1951b), vol. 3, pp. 657–60; Legge (1872), p. 819. Vandermeersch (1988), p. 66, summarises the affair.

[2] *Ching Fa* 經法, pp. 94f. The text continues to provide a list of 22 *yang/yin* 陽陰 pairs: Heaven/Earth; spring/autumn; summer/winter; day/night; big states/small states; important states/unimportant states; action/inaction; stretching/contracting; ruler/minister; above/below; man/woman; father/child; elder brother/younger brother; older/younger; noble/base; getting on in the world/being stuck where one is; taking a wife or begetting a child/having a funeral; controlling others/being controlled by others; guest/host; soldiers/labourers; speech/silence; giving/receiving. (Cf. Graham (1986d), p. 27.)

[3] *Tso Chuan*, Duke Chao 9.3, ed. Yang Po-Chün, p. 1310; cf. Couvreur (1951b), vol. 3, pp. 166f., and Legge (1872), p. 626.

[4] Cf. J. S. Cikoski (1975) as well as J. Chmielewski (1979), which offers critical comments on Cikoski. For detailed analysis of the method of analogy in the *Mêng Tzu* see Lau (1983c), pp. 334–56, and more recently Reding (1986).

There is no doubt that the ancient Chinese, like the Greeks, were fond of arguments by analogy, but it still makes good sense to study the non-analogical arguments in China and Greece. E. R. Dodds, in his book *The Greeks and the Irrational* (Berkeley, 1951), has shown that the ancient Greeks generally were very far indeed from thinking predominantly in ways that we would call rational. But Dodds adds an important sequel to his observation:

The evidence which is here brought together illustrates an important, and relatively unfamiliar, aspect of the mental world of ancient Greece. But an aspect must not be taken for the whole.[1]

Logical argumentation (as opposed to correlative explanation) is only one aspect of ancient Chinese intellectual culture and must not be taken for the whole. None the less it makes good sense to study the evolution and the patterns of non-correlative rational thinking and argumentation in ancient China.

I wish to demonstrate that there was room in ancient Chinese culture for this kind of logical reasoning. I shall try to discover some of the typical strategies of argumentation and inference explicitly or implicitly present in the ancient Chinese texts that have come down to us. In the following Sub-section I shall then go on to look a little more closely at some specific forms of argument such as the syllogism, *modus ponens*, *modus tollens*, the sorites. We are interested in the presence or absence of these forms of (explicit or implicit) reasoning in China because we are interested to know whether the ancient Chinese mind moved along entirely different logical avenues from ours.

Arguments in non-technical Classical Chinese texts are in general designed not to *prove* a proposition but to *convince* a reader with plausible reasons of a proposition which the philosopher, most often on the independent basis of his superior wisdom, holds to be true. Such reasoned argumentation must be carefully distinguished from persuasion (*shui* 説), which falls into the realm of rhetoric or psychological tactics and is not of immediate concern in the context of the history of science in China, although it was viewed as important by the Chinese themselves at the time.[2] Indeed, most reasoning in ancient China was designed to make someone of superior rank not just *believe* but *use* (*yung* 用) certain words, i.e., *act on them*.[3]

In a scientific context the crucial notion is that of proof and of the art of plausible reasoning. The art of formal proof was little developed in China by comparison with Greece. The art of plausible reasoning on the other hand, formed an integral part of intellectual life in China from the time of Confucius's disciples onwards, and especially since the Confucian teachings were challenged by rival schools of thought such as that of Mo Tzu.

---

[1] Dodds (1951), p. viii.
[2] The *Chan-Kuo Tshê* 戰國策 or 'Intrigues of the Warring States' (see Crump (1970)) and the *Shuo Yüan* 説苑 'Garden of Persuasions' contain particularly fascinating material on the art of persuasion in ancient China, but the theme recurs throughout the philosophical and historical literature.
[3] See, e.g., *Kuo Yü* 國語 21 no.1, ed. Shanghai Ku-chi, p. 643.

One unique feature of intellectual life in Greece was precisely the demand for formal proof in informal contexts.[1] Plato's tendency towards strict argumentative rationalism amounted essentially to just this: to demand rigid clarity of meaning in philosophy, and to demand clear deductive evidence for any philosophical thesis maintained, preferably in the form of a proof from true or plausible premises.

In China, such rigid rationalism remained a marginal phenomenon. A typical, almost programmatic, statement of Chinese rationalism and empiricism would be this:

When one hears words, one must investigate them (i.e., their truth). Better not to hear about something at all than to hear about something and not investigate the matter.[2]

No Chinese philosopher would ever say about an academy[3] anything like *mēdeis ageōmetrētos eisitō*, μηδεὶς ἀγεωμέτρητος εἰσίτω, 'let no one ignorant of geometry enter'. In point of fact, the ascription of this saying to Plato is probably apocryphal, too, but it adequately expresses the thought that access to wisdom depended on mathematical training.[4] This thought was not entirely alien to the traditional Chinese thinkers, but it must be fair to say that among those whom we traditionally regard as 'philosophers' such an emphasis is rare to say the least.

In any case Cicero, in an interesting comparison of the achievements of Greek and Roman culture, notes around −45:

In summo apud illos (i.e., the Greeks) honore geometria fuit, itaque nihil mathematicis inlustrius; at nos metiendi ratiocinandique utilitate huius artis terminavimus modum.[5]

Most Chinese philosophers might well have sympathised with Cicero's general attitude, and they might more naturally have assigned to history the place which Plato and many other Greeks gave to geometry.

Compare Confucius with Plato. Confucius made many claims. To what extent he was prepared to argue for his claims, or how he did argue for them, is difficult for

---

[1] I recall that in 1965, when I was interviewed for a scholarship to study at Oxford, I was asked by the learned board why I had shown such inordinate interest in philosophy. I remember clearly that I replied that Aristotle had asked himself the question why one should philosophise, in a dialogue which unfortunately was lost, but of which fragments remain. I said Aristotle's answer was of unsurpassable elegance and ran somewhat like this: 'Either we must philosophise or we need not philosophise. This much is clear. Now if we must philosophise, then we must philosophise. This much is evident. If, on the other hand, we need not philosophise, then we need to give philosophical reason why we need not philosophise. Therefore, quite generally, we must philosophise.' It was rightly felt to be comical to use formal proof in answer to an everyday question. And yet, what I was doing was quite in the spirit of Greek rationalism. I must apologise for this autobiographical footnote, but I do feel it illustrates the point at hand rather well.

[2] *Lü Shih Chhun Chhiu* 22.6, ed. Chhen Chhi-Yu, p. 1526; cf. Wilhelm (1928), p. 399.

[3] And we do find the equivalent of an academy in −4th-century China, notably the Chi-hsia academy in the state of Chhi in present-day Shantung Province, which attracted large numbers of intellectuals during the −4th and −3rd centuries.

[4] *Der kleine Pauly, Lexikon der Antike in 5 Bänden*, vol. 4, p. 901. It should be pointed out, of course, that the emphasis articulated in this description does not describe the intellectual or even the philosophical climate of ancient Greece in general. For an important general survey of the differences between the intellectual climate in ancient Greece and ancient China see Nakayama (1984).

[5] 'Geometry held the highest place of honour among them, and nothing was more illustrious than mathematicians. But we have circumscribed our concern for these by their usefulness for the art of measurement and of calculation.' *Tusculanae disputationes* 1.11.5.

us to know. His disciple Tzu Lu showed a more combative and argumentative disposition, pointing out inconsistencies.

In the *Mo Tzu*[1] three standard tests for the acceptability of a claim were introduced: 1. the opinion of the authorities of the past, 2. observation by the people, and 3. the beneficial or harmful effect on society. The *Mo Tzu* not only sets up these standards but defends its tenets on the basis of them. Many Chinese thinkers from Mo Tzu onwards tried to justify whatever claims they wished to make. Logical argumentation played a certain limited part in this. But we must emphasise that logical argumentation was not the preferred way of justifying or supporting one's thesis in ancient China.

What then *was* the preferred way of supporting one's claims? It was through what we may call paradigmatic arguments from historical examples. This was noticed by G. B. Bilfinger in 1724:

The Chinese find agreeable a mode of demonstration which to our ears, at least, is not very usual: it consists in not substantiating their instructions by arguments but by examples from kings and emperors.[2]

Arguments from historical examples

Ancient Chinese philosophy (or in any case what is preserved of it) was quite predominantly *social* philosophy. It was concerned with the conduct of human affairs. This conduct was discussed on the basis of recorded historical episodes and events. Such episodes and events were used to illustrate and justify more general or abstract 'philosophical' statements in much the same way as *exempla* or *bispel* 'edifying illustrative stories' were used in medieval sermons.[3] And just as we have collections of *exempla* and *bispel* from medieval times onwards in Europe, so we have plenty of collections of illustrative anecdotes to be used in support of philosophical, ethical, social or political points in China.[4] Historical annals and other historical sources were used as sources for such *exempla* or *topoi* to be adduced, marshalled (or, some would say: trotted out) in support of a given point of view. At the same time such illustrative historical anecdotes were designed to demonstrate the practical application of the point of view supported by them.

These points of view did not have to be Confucian. For example, chapter 21 of the *Han Fei Tzu* 韓非子 (−3rd century) and chapter 12 of the *Huai Nan Tzu* 淮南子 (−2nd century) consist predominantly of a series of anecdotes designed to substantiate the truth and practical application of a series of statements from the Lao Tzu. One brief example will suffice to indicate the genre:

[1] *Mo Tzu*, chapters 35, 35, 37; see Mei (1929), pp. 183, 189, and 194, as well as Watson (1967), p. 118.
[2] Placet (i.e., to the Chinese) inusitatum demonstrandi modus, nostris quidem auribus non nimium consuetus, quo præcepta sæpenumero non argumentis firmantur, sed Regum atque Imperatorum exemplis. Bilfinger (1724), p. 30.
[3] Cf. Moser-Rath (1964), for a survey of the later development of the genre and a bibliographic orientation. For Aristotle's treatment of arguments by historical examples, see Lloyd (1966), pp. 405ff.
[4] See *Han Fei Tzu*, chs. 30 to 35; *Huai Nan Tzu* chs. 16 and 17.

The ruler of Yü desired the chariots from Chhü Chhan and the jade from Chhui Chi, and he did not listen to Kung Chih-Chhi's advice. As a result his country was laid waste and he himself died.

Therefore it is said:[1] 'No fault is more painful than the urge to get hold of things.'[2]

Chapter 18 of the *Huai Nan Tzu* lays out sixteen theses which are substantiated in each case by a paradigmatic illustrative anecdote. The interesting thing about these theses is that in each case they are in a certain sense *proved* by the examples provided.

The *Shuo Yüan* 説苑 and the *Hsin Hsü* 新序 (both compiled during the −1st century on the basis of earlier material) are extensive collections of edifying historical evidence for the use of Confucians who needed to substantiate and argue for their points of view.

The points of view thus supported did not necessarily have to be philosophical in our sense. The *Han Shih Wai Chuan* 韓詩外傳 (−2nd century) is full of illustrative historical anecdotes to show the practical application, truth and relevance of certain lines from the ancient Book of Songs.[3]

A modern reader of the surviving ancient Chinese literature might be forgiven for suspecting that a thesis or point of view counted as plausible to the ancient Chinese to the extent that it could be 'exemplified' by historical anecdotes and episodes. If a point was unsusceptible to such exemplification, this raised a suspicion that it was either irrelevant to the conduct of human affairs or simply untrue. No wonder scientific or abstract theoretical claims had a hard time in such an intellectual climate.

The *Lü Shih Chhun Chhiu* 呂氏春秋 occasionally speaks of something being 'expounded on (*chieh tsai* 解在) . . .'.[4] What is intended is not theoretical justification but historical illustration. Demanding an explanation (*shuo* 説) in ancient China was often tantamount to demanding a historical *exemplum* or some other kind of illustrative anecdote. 'Philosophising' in ancient China tended to be an intensely and pervasively historical exercise. Reasoning tended to consist in an appeal to historical example and traditional authority.

But this is not the whole story. Some people, like the Later Mohists and a number of others, did think of explanation (*shuo*) for a thesis very much in terms of definitions and theoretical logical reasoning. It is this rather specialised – and intellectually important – kind of reasoning in its implicit and explicit forms which holds our special interest in the context of the history of science.

Implicit logical arguments

Arguments do not have to be explicit to be clearly present. An ancient little story about a child will illustrate the point:

When Wang Jung 王戎 (+234 to +305) was seven years old, he was once roaming about with a group of children and they saw a pear tree by the wayside. The tree had so much fruit that

[1] *Lao Tzu*, ch. 46.　　[2] *Han Fei Tzu* 21.4; Liao (1939), vol. 1, p. 208.

[3] For another aspect of the purpose of the *Han Shih Wai Chuan* see Hightower (1951), p. 3.

[4] See *Lü Shih Chhun Chhiu* 13.1 to 13.7. Each of these chapters contains the phrase *chieh tsai* 解在 'the explanation is in'.

its branches were breaking under the weight. All the children rushed forward to get the fruit. Only Jung did not move. Someone asked him why. He replied: 'If the tree by the wayside has much fruit, that must be because the pears are bitter.' They picked the pears and he turned out to be right.[1]

This kind of story is common in Classical Chinese literature, and it crucially involves logical reasoning. Wang Jung reasons as follows: if the pears were good there would not be so many of them left, since they would have been eaten by others. But there are many pears left. *Ergo*: they can't be any good.

One may object that this is just plain common sense, but the point is precisely that in applying plain common sense the Chinese habitually proceed along the lines of plain formal logic. Even when this does not show at all in the surface structure of their arguments. Implicit ways of speaking often show a keen sense of logical structure, as when Yang Hsiung 揚雄 (−58 to +18) is attacked with the following question:

Someone said: 'You despise philosophers (*chu tzu* 諸子). Isn't Mêng Tzu a philosopher (*Mêng tzu fei chu tzu hu* 孟子非諸子乎)?'[2]

Yang Hsiung's answer shows that he understood the implicit argument which runs something like this: You despise all philosophers. Mêng Tzu is a philosopher. Therefore you must despise Mêng Tzu. But you don't despise Mêng Tzu. Therefore your position is inconsistent. Yang Hsiung defends himself by claiming that Mêng Tzu is not a *chu tzu* 'philosopher' in his sense of the word. He would not have needed to defend himself in this way if he had not understood the argument along the lines we have outlined.

Reasoning and scientific explanation

In general, the early Chinese, and particularly the Confucians, were not much given to deductive reasoning. And in particular it was not common for such reasoning to be applied to questions of scientific explanation. But that does not mean such reasoning was not common among the ancient Chinese people. The book *Lieh Tzu* 列子, compiled probably around the +4th century but containing much earlier traditional material,[3] pokes fun at Confucius's refusal to engage in such reasoning concerning natural phenomena:

Confucius was travelling in the east when he came upon two small boys arguing with each other. One boy said: 'I think the sun is closer at sunrise and further away from us at noon.' The other boy said: 'The sun is further away at sunrise and closer at noon.'

The one boy said: 'At sunrise the sun is as large as the cover of a carriage. At noon it is as small as the cover for a plate. The sun is closer at sunrise and further away from us at noon, for surely that which is further away is smaller than that which is closer.'

---

[1] *Shih Shuo Hsin Yü* 世説新語 6.4; Mather (1976), p. 181.
[2] Yang Hsiung, *Fa Yen* 18, ed. Wang Jung-Pao, p. 724.
[3] See Graham (1986), pp. 216–82, for a brilliant discussion of the authenticity of this important text.

The other boy said: 'At sunrise the sun is refreshingly cool. At midday it feels as hot as putting your finger into hot soup. The sun is further away at sunrise, for surely (the same object) which feels hotter is closer to us than when it feels cooler.'

Confucius was unable to solve the puzzle.

The two small boys laughed at him and said: 'Who would think you knew anything?'[1]

The first boy – who for all we know may represent a widely held popular attitude towards Confucian aloofness throughout the ages – is thinking and arguing along the following lines:

1. The sun is larger at sunrise than at noon.
2. Everything that appears larger to us gets closer to us.
3. *Ergo*: The sun gets closer to us at sunrise than at noon.

The *Lieh Tzu* uses this argument in a humorous spirit, and the traditional Chinese response to this sort of satirical attack has become proverbial:

An old saying goes like this: Not knowing is compatible with being a gentleman. Knowledge does not make someone any less of a small man.[2]

We must balance this example with another one where Confucius is said to apply strict scientific logic to a case of ethics:

When Confucius was on his way to Chhu a fisherman insisted on presenting him with a fish. Confucius refused to accept, but the fisherman insisted: 'It's awfully hot. If I try to sell this fish in the market place, which is a long way from here, I won't be able to get rid of it. I was thinking of throwing it away. But then I thought it would be better to present it to a gentleman.'

Confucius bowed twice, accepted the gift and told his disciples to broom the sacrificial area in order to make a sacrifice of the fish.

'That fellow was about to throw it away. How can you use it as a sacrificial animal?' asked the disciples. 'I have heard it said that he who distributes leftovers and does not let them rot away is a sage. Now I have received a gift from a sage. How could I fail to sacrifice it?'[3]

Note the Master's logic:

1. He who distributes leftovers and does not let them rot away is a sage.
2. That fisherman from Chhu distributed his leftovers and did not let them rot away.
3. *Ergo*: That fisherman of Chhu is a sage.

A semi-barbarian fisherman is deduced to be a sage! And Confucius seems to be delighted to make this claim and thereby to offend current notions on who was and who was not a candidate for sagehood.

---

[1] *Lieh Tzu*, ch. 5, ed. Yang Po-Chün, p. 105; cf. Graham (1960), p. 104. This story goes back to Huan Than (died +28). For the genre of stories in which children poke fun at Confucius see Soymié (1954) and Graham (1986a), p. 218.
[2] *Yin Wên Tzu*, ed. Li Shih-Hsi, p. 14.   [3] *Shuo Yüan* 5.20, ed. Chao Shan-I, p. 120.

Cultural logic and the logic of the joke

We leave it to the reader to reconstruct the unfilial son's underlying reasoning in the following story in detail. It is enough to emphasise that in order to get the joke in this passage one must understand the apparent logical force of the unfilial son's argument.

A learned man from Sung returned home after three years of study and called his mother by her personal given name. The mother said: 'You have been away on study for three years and now you come back and call me by my given name. Why do you behave like this?'

The son replied: 'There is no one I regard as more talented than Yao and Shun, and still I call them by their given names. There is nothing I regard as larger than Heaven and Earth. But I call them by their given names. Now a mother is no more talented than Yao and Shun, and she is no grander than Heaven and Earth. Therefore I call my mother by her given name.'[1]

This is perfect cultural logic, wonderfully outrageous. If one wants to understand the way logic entered the everyday lives of the ancient Chinese, this sort of evidence is invaluable.

A man from Wen is arrested by the Chou border guards but declares himself to be a subject of the Chou.

'You are not a Chou citizen. How can you insist that you are not a foreigner?'
'From my youth I have recited the Book of Songs, and one Song says:

> Every place in the world
> Is the King's land.
> Anyone within the circling sea
> Is the King's servant.[2]

Now Chou rules the world and consequently I am the subject of Chou. How could I count as a foreigner?'[3]

The argument can be expounded as follows:

1. I am within the circling sea.
2. Everyone within the circling sea is a Chou subject.
3. *Ergo*: I am a Chou subject.

Even in the case of early Chinese logicians like Kungsun Lung and Hui Shih we have good reason to believe that their logical argumentation was conducted partly as comic logical entertainment. In any case, the tradition for argumentative entertainment of this sort provided the social setting within which Chinese 'sophistry' flourished.

---

[1] *Chan Kuo Tshe*, no. 359, ed. Chu Tsu-Keng, p. 1255; cf. Crump (1970), p. 431.
[2] Cf. *Shih Ching*, no. 205.   [3] *Chan Kuo Tshe*, no. 42, ed. Chu Tsu-Keng, p. 29; cf. Crump (1970), p. 51.

The logic of insult

Logic can function well in the service of rudeness and insult. Here is a translation of the ancient insult involving logic:

> Yen Tzu 晏子 went as an ambassador to Chhu . . .
> The King asked: 'Are there no people in Chhi?'
> Yen Tzu: '(The capital of Chhi) Lin Tzu contains as many as 300 boroughs. . . . How should there be no people?'
> 'How come then that you act as an ambassador?'[1]

When Yen Tzu acts as an ambassador to Chhu, the king of that country asks whether there aren't any people in Chhi. For if there were any other people there, he suggests, Chhi would surely have chosen someone 'bigger' than Yen Tzu as ambassador. The logic of the king's rudeness may be summarised as follows:

1. Chhi would only send something too insignificant to count as a man, like Yen Tzu, if there were no people in Chhi.
2. Chhi has sent someone as insignificant as Yen Tzu.
3. *Ergo*: There cannot be people in Chhi.

If one fails to understand the underlying logic as we (perhaps too laboriously) reconstruct it, one cannot fathom the depth of rudeness in the king's opening remark. I am not saying that the king of Chhu is here presenting a logical argument. He is not about to chop logic. He is trying to insult someone. But in order to insult effectively, he relies on a string of plain logical reasoning which he expects his audience to grasp in order to appreciate the insult. Without this implicit natural logical train of thought the insult simply would not work.

Of course, the king is not being serious. And neither is Yen Tzu 晏子, when he replies along similarly insulting but strictly logical lines, expecting the ruler to grasp something like the following argument:

1. Chhi sends to each country an ambassador who corresponds to the quality of that country's king.
2. Chhu has the most untalented of kings.
3. Chhu deserves the most untalented of ambassadors.
4. Yen Tzu is the most untalented of ambassadors.
5. *Ergo*: Chhi sends Yen Tzu as an ambassador to Chhu.

Now Yen Tzu does not, of course, formally go through the motions of this argument. Here is how he very effectively delivers his logically exquisite piece of rudeness:

When the state of Chhi assigns ambassadors, it is in each case governed by certain considerations. The talented men are sent to serve as ambassadors to talented kings. The untalented men are sent to serve as ambassadors to untalented kings. I am the least talented. That is why they sent me straight to Chhu.[2]

---

[1] *Yen Tzu Chhun Chhiu*, ch. 6, ed. Wu Tse-Yü, p. 389.   [2] *Ibid.*

Again: in order to understand this insult, the King of Chhu would certainly have to appreciate something like the logical structure outlined above. And the insult becomes even more effective by the fact that it is coupled with the obligatory ritual humility, as when Yen Tzu insists that he is the most untalented man in Chhi. The humility, here as so often elsewhere, is plainly ironical.

However, there is something profoundly unsatisfactory in this method we have pursued of scouring Chinese literature for traces of *Western* forms of reasoning, for what we are interested in is not just to what extent the Chinese argued and thought like Westerners. We are much more interested in their autochthonous tendencies and habits of reasoning and discourse in their own right, in their own terms, and for their own sake. We simply want to know: how did the Chinese tend to justify claims which they felt needed justification. What was the anthropology of argumentation in ancient China? Not argumentation in our sense, on our terms. Argumentation in terms of the Chinese tradition itself. The arguments from historical example which we have discussed above fall into this category. But what else is there?

The concept of 'ku' 故[1]

What are the objective criteria, based on the language itself, that justify our interpreting a given passage as an argument? An argument, after all, must be intended as such in order to count as an argument.

The key word in the history of Chinese reasoning is *ku*, which has the meanings 'old, that which comes earlier, cause, reason, (used adjectivally:) former, (used as sentence-initial particle:) therefore'. Arguments in ancient China may be construed as answers to (explicit or implicit) questions like *ho ku* 何故 'why' or its many equivalents.[2] Answers to such questions may be designed to provide either a motive for an action, or the cause for an event, or a plausible reason for a point of view.

Looking through the oldest indexed literature, one finds that in the vast majority of cases ancient Chinese answers to questions like 'why' are in terms of motives for action or causes for events. In the context of the history of science, on the other hand, we are particularly interested in explicit arguments of another kind.

To what extent, then, were the ancient Chinese in the habit of asking for plausible reasons for a given claim or point of view? If we consider the standard request for a reason, there is striking and highly significant contrast between Confucian texts and texts from other sources. In the *Lun Yü*, the *Mêng Tzu* and the *Hsün Tzu* there is not one single explicit request for a *ku* 'reason'.[3] Similarly, there are hardly any cognitive questions like *ho i chih chhi jan* 何以知其然 'how do we know that this is so?' or *ho i ming chih* 何以明之 'how do we explain this?' in the early Confucian texts. They first make their occasional appearance in the *Hsün Tzu*. In the more

---

[1] Weidmann (*c.* 1970), pp. 54–92, deals in detail with the uses of *ku* 故 in *Mêng Tzu* and in *Mo Tzu*.

[2] *Ho ye* 何也 'why is that' occurs 45 times in *Han Fei Tzu* alone. *Shih* (or: *tzhu*) *chhi ku ho* (*yeh*) 是 (此) 其故何 (也) is extremely common in *Mo Tzu*.

[3] There are, however, seven cases of *ho yeh* 'what about that?' in the *Mêng Tzu* and many more cases of this sort in the *Hsün Tzu*.

scientifically orientated book *Mo Tzu*, on the other hand, we find over fifty explicit requests for a *ku* 'reason', and over thirty overtly epistemological questions like 'how do we know/explain that this is so?'.

This does *not* mean that early Confucians were incapable of arguing or of giving reasons. It *does* mean that they were less inclined than others to demand such reasoning from their teachers. Confucians were used to appealing to ancient authorities and they liked to presume that they spoke with the natural authority of wise men themselves. Thus their moral excellence and sageliness constituted the main argument in favour of their views.

Confucius's intellectual achievement was not in supplying cogent arguments for his views, it was in the coherence of his moral interpretation of the past, his attempt at finding the one moral thread (*i i kuan chih* 以一貫之).

The other schools of thought – such as the Mohists – had to compete with a pre-existing Confucian tradition from the start, and for them argumentation became an important constitutive part of their legitimacy. They needed the better arguments to legitimise their claims. It is therefore not by chance that logic developed so strongly within the Mohist school.

Within Confucianism, argumentation developed to the extent that it developed from two sources: firstly from the necessity to refute well-argued attacks from the Mohists, legalists and others; secondly Confucian argumentation developed from the need to argue out internal differences that had arisen within the Confucian movements (such as the different views on the goodness or wickedness of human nature). A. C. Graham (1964), p. 45, describes the matter with his usual masterly elegance: 'The pressure of competition with other schools, which had forced Mencius into public debate, drove Hsün-tzu (*c.* −298 to −238) to a much more thorough rationalization of the Confucian position.'

However, the victory of Confucianism as state doctrine in the −2nd century also meant a battle won for the habit of reasoning from authority. Contrast the constant demand for reasons and explanations in the *Mo Tzu*. Chapter 28 of that book provides a neat example:

> Mo Tzu said: 'Be obedient. Be careful. Be sure to do what Heaven desires and avoid what Heaven abominates. Now, what does Heaven desire and what does Heaven abominate?
> Heaven desires righteousness and abominates unrighteousness.
> How do we know this?
> Because righteousness is the standard.
> How do we know that righteousness is the standard?
> Because with righteousness the world will be orderly. Without it the world will be disorderly. So this is how I know that righteousness is the standard.'[1]

One may not find these arguments impressive, but one cannot fail to notice the sturdy insistence on giving reasons for the views maintained. It is important to watch Mo Tzu's method closely if one wants to get a closer picture of rationality in ancient China.

---

[1] *Mo Tzu* 28.8; cf. Mei (1929), pp. 151f.

The gentlemen of the world who desire to be righteous must therefore obey the will of Heaven.

What is the will of Heaven that is to be obeyed?

It is to love all the people in the world universally.

How do we know it is to love all the people in the world universally?

Because Heaven accepts sacrifices from all.

How do we know Heaven accepts sacrifices from all?

Because from antiquity to the present day there is no distant or isolated country but that feeds oxen and sheep, dogs and pigs with grass and grains and prepares clean cakes and wine to worship the gods, hills, rivers, and the spirits. So therefore we know Heaven accepts sacrifices from all. Given that Heaven accepts sacrifices from all, it must love them all . . .

That Heaven loves all the people of the world is proved not only by this. In all the countries of the world and among all the people who live on grains, the murder of one innocent individual brings down calamity. Now who is it that murders the innocent? It is man. Who is it that sends down the calamity? It is Heaven. If Heaven really did not love (all) the people, why should Heaven send down calamities for the murder of an innocent person?[1]

It would be unrealistic, though, not to pay attention to the fact that even some people within the Mohist circles do reason through a traditionalist appeal to precedent and ancient authority.[2]

Thus Mo Tzu said: 'To have music is wrong.'

How do we know that this is right? It is found in the Code of Punishment of Thang among the books of the ancient kings.

This book proclaims: 'To have constant dancing in the palace is called the witches' pleasure.'[3]

We need not illustrate in detail at this point that the characteristic Confucian arguments from historical precedents as well as scriptural authority are not alien to the Mohists. They are an integral part of the Chinese intellectual heritage.

On the other hand, even the traditionalist Confucian Mêng Tzu recognised the importance of objective causes or reasons (ku 故). He was evidently impressed by the astronomic achievements of his time and his famous comment on these is worth quoting in full.

In talking about human nature people in the world merely appeal to (clever) reasons (ku). Those who appeal to (clever) reasons (ku) take self-interest as their basic consideration. What I dislike in clever men is that they bore their way through. If clever men could act as Yü did in guiding the flood waters, then there would be nothing to dislike in them. Yü guided the water by imposing nothing on it that was against its natural tendency. If clever men can also do this, then great indeed will their cleverness be. In spite of the height of the heavens and the distance of the heavenly bodies, if one seeks out the (real) reasons (ku), one can calculate the solstices of a thousand years hence without stirring from one's seat.[4]

---

[1] *Mo Tzu* 28.26ff.; cf. Mei (1929), pp. 154ff.

[2] According to Graham (1985) the discrepancy can be explained on the basis of a division of the Mohist school into three wings. Graham's paper is a major contribution to the reconstruction of Mohist philosophy.

[3] *Mo Tzu* 32.43; cf. Mei (1929), p. 180.

[4] *Mêng Tzu* 4B26; tr. Lau (1983c), p. 169. For a very important detailed discussion of this passage see Graham (1986a), pp. 52f.

It is remarkable to see how Mêng Tzu argues from the scientific to the philosophical paradigm. He recommends a scientific observation of realities also in non-'scientific' contexts. His attack on the logicians (the 'clever men who bore through') is not only that of a moralist, but that of a sensitive observer of contemporary science. The logicians should not keep arguing abstractly but should base their theories of human nature on precise observation of fact.

In ordinary discourse the question *ho ku* 何故 'why is that?' is a demand for a reason, and questions of this sort are the beginnings of reasoning and arguments. If we want to understand what indigenous Chinese arguments are we must study the replies to questions like *ho ku*.

'The army of Chhin will come to grief.'
'Why is that?'
'Their army is slight and arrogant. If an army is slight, it has few options. If it is arrogant, it does not follow the rules of propriety. If it does not follow the rules of propriety, the soldiers leave the ranks. If they have few options, they will trap themselves and when they are in trouble they will leave the ranks. How could they fail to be defeated?'[1]

Here the argument starts from the assumption that the Chhin army is slight and that it is arrogant. It is then shown how each of these assumptions will lead to its soldiers leaving the ranks and the army therefore getting defeated. Instead of saying *quod erat demonstrandum* the ancient Chinese, here as so often elsewhere, ends with a rhetorical question. If one needed formal proof that Chinese arguments are designed to convince rather than to prove, this would be it.

Ancient Chinese attitudes to argumentation

Argumentation was used to restrain customary traditionalist credulity, as the following story will demonstrate:

The King of Wei said: 'I have heard that the Taoists who ascend Mount Hua become immortal. I wonder whether you have also heard this?'
'In ancient times there was no such practice. I want nothing to do with it.'
'But I have heard it is true.'
'I have not yet investigated (*shen* 審) what you have heard. Have you personally heard it from an immortal, or have you heard it second hand? If you have heard it second hand, then the person whom you heard it from was mistaken. If you have heard it from an immortal, where is he now? If he is still around, you should learn from him and have no doubts. If he is not around, you should not learn from him and have no doubts.'[2]

The sanity and down-to-earthness of this piece of reasoning is characteristic of the best tradition within Confucianism. Another characteristic feature is the pragmatic attitude to philosophical questions as displayed in the following episode[3]

---

[1] *Kuo Yü*, ed. *SPPY*, ch. 2, p. 6b.   [2] *Khung Tshung Tzu* 孔叢子, ed. *SPTK*, ch. 15.
[3] We have already had occasion to discuss this passage in Section (*d*,3).

which constitutes a fitting conclusion to our general deliberations on reasoning in ancient China:

Tzu Kung asked Confucius: 'Do the dead have knowledge or do they not have knowledge?'

The Master said: 'I might want to say that the dead have knowledge. But then I am afraid that filial sons and obedient grandsons will harm the living and send them to accompany the dead. I might want to say that the dead have no knowledge, but I am afraid that unfilial sons would abandon their parents and leave them unburied. You are not to understand whether the dead have knowledge or not. There is no hurry now. You will naturally come to know soon enough.'[1]

When all is said and done, epistemological pragmatism, the endemic nature of the pragmatic suspense of judgement on tricky questions of the sort Confucius here demonstrates, coupled with the Confucian variety of scholastic traditionalism, are the things which put the greatest constraint on the evolution of formal logical method in China.[2] The place of intellectual endeavour in the Confucian scheme of things is well summarised by Hsün Tzu:

Hearing of something is better than not hearing about it. Seeing a thing for oneself is better than hearing about it. Understanding something is better than seeing it. Performing something is better than understanding it. The learning stops when performance is reached. Performing (a desirable act) is enlightenment (ming 明). Being enlightened about a thing amounts to being a sage (shêng jên 聖人).[3]

Logical argument is considered as a handmaid of moral orthodoxy:

If one's words are outside the area of humaneness, then it would be better to be silent than to speak, and it would be better to be fumbling than to be well-argued.[4] All talk that is not in accordance with the former kings and with propriety and righteousness we call cantankerous. Even if the words are well-argued, the gentleman will not listen to it.[5]

Arguing well for an *unorthodox* thesis or for a thesis which was felt to be morally and politically irrelevant was usually felt to be either flippant or wicked in traditional China. This will not surprise anyone familiar with Western intellectual history: *logica ancilla theologiae*. Logic as well as philosophy remained ancillary to theology in medieval and Renaissance Europe, and far beyond . . .

---

[1] *Shuo Yüan* 18.31, ed. Chao Shan-I, p. 558, and *Khung Tzu Chia Yü* 孔子家語, ch. 2, no. 17; cf. Kramers (1950), p. 238.

[2] Of course, such suspense of judgement is common in Greece as well as in China, but it was not endemic in Greece. There was a significant part of Greek intellectual culture in which the pursuit of tricky questions of this sort was cultivated. Moreover, this pursuit gained a certain general respectability and certainly some considerable popularity in Greece which it never won in China.

[3] *Hsün Tzu* 8.103; Dubs (1928), p. 113.   [4] *Hsün Tzu* 5.55; cf. Köster (1967), p. 51.

[5] *Hsün Tzu* 5.40; cf. Köster (1967), p. 48.

(2) Some Forms of Argument in Ancient China

The syllogism

Consider the following passage by Wang Chhung (+27 to +100):

The books of the literati relate that the Prince of Huai Nan in his study of Taoism assembled all the Taoists of the empire and humbled the grandeur of a princedom before the expositors of Taoist lore. Consequently, Taoist scholars flocked to Huai Nan and vied with each other in exhibiting strange tricks and all kinds of miracles. Then the prince attained the Way and rose to Heaven with his whole household. His domestic animals became genii too. His dogs barked up in the sky, and the cocks crowed in the clouds. That means that there were drugs of immortality in such plenty that dogs and cocks could eat of it, and follow the prince to Heaven. All who have a fad for Taoism and would learn the art of immortality believe in this story, but it is an untrue story (*hsü yen*).[1]

Man is a creature. He may be as noble as a king or feudal lord, but his nature is no different from that of a creature. All creatures without exception are mortal.

How can man become immortal?

Birds having feathers and plumes can fly, but they cannot rise to Heaven. How should man without feathers and plumes be able to fly and rise?

Were he feathered and winged, he would only be equal to birds, but he is not. How then should he ascend to Heaven?[2]

Wang Chhung is concerned to establish the truth or otherwise of a well-known story. He provides what effectively is a neat Aristotelian syllogism in support of his thesis that the story must be untrue:

All men are creatures.
All creatures are mortal.
Ergo: All men are mortal.[3]

Paradoxically, Wang Chhung's syllogism is more purely and more precisely 'Aristotelian' than the standard medieval Western form that has become current in the history of Western logic since medieval times:

All humans are mortal.
Socrates is human.
Ergo: Socrates is mortal.

Or, as Origen has it in a serious theological context:

If every man is made a fool by knowledge, and Paul is a man, then Paul was made a fool by knowledge.[4]

---

[1] Cf. our Sub-section (*d*,3) on the notion of truth.
[2] Wang Chhung, *Lun Hêng*, ed. Chung-hua-shu-chü, p. 410; Forke (1911), vol. 1, p. 335.
[3] Cf. the following excuse put forward by someone who refuses to commit suicide in the service of his master:

All service of another is because one considers it beneficial. To die is not beneficial. Therefore I do not die. (*Lü Shih Chhun Chhiu* 18.4, ed. Chhen Chhi-Yu, p. 1178, cf. R. Wilhelm (1928), p. 303.)

[4] *Patrologia Graeca*, vol. 13, pp. 343–4cd.

Jan Łukasiewicz (1957) has argued that the standard Western example of a syllogism does not in fact conform to Aristotle's rules because it has a non-generic subject term 'Socrates'. No such objection can be raised to Wang Chhung's syllogism. On the other hand Wang Chhung's formulation of the argument is not deliberately formalistic. But then Wang Chhung was not in the business of formal logic. What he demonstrates abundantly is that the notion of a valid argument in support of a conclusion, and the use of a syllogism form to prove a conclusion, are in no way completely alien to ancient Chinese thought.

Even in the earlier literature there is no shortage of strictly logical trains of quasi-syllogistic thought. Especially in jokes:

In Chhi there was a servant who refused to commit suicide for the sake of his master who got into trouble. In the street he met an old acquaintance who said: 'How come you're still alive?'

'Well, everyone serves others to gain an advantage. Dying is not an advantage. Therefore I refused to die.'

'But will you be able to face other people (having failed to commit the obligatory suicide)?'

'Well, do you imagine I could face them if I were dead?'[1]

One argument or logical train of thought in this joke can be set out as follows:

1. Everyone who serves others aims at profit for himself.
2. If I commit suicide that does not aim at profit for myself.
3. *Ergo*: I did not commit suicide.

Such logic-chopping in ancient Chinese is not always intended as funny:

Liang Chhiu-Wei told Yen Tzu: 'I will never reach your level of attainments until I die.'

Yen Tzu replied: 'I have heard it said that "In action those who persevere succeed. In walking those who persevere arrive." I am no different from other men. I persevere in my actions and do not let up. I persevere on my path and do not rest. There is nothing hard about it.'[2]

If the ancient Chinese understood Yen Tzu here, they must also have understood an elementary pattern of logic.

1. Everyone who perseveres succeeds.
2. I persevere.
3. *Ergo*: I succeed.

This is not a well-formed Aristotelian syllogism because it contains a singular term, 'I'. But it comes close enough to have been traditionally confused with a syllogism in Europe. In any case, the crucial point is that it represents sound recognisable logic.

---

[1] *Lü Shih Chhun Chhiu* 18.4, ed. Chhen Chhi-Yu, p. 1178; cf. Wilhelm (1928), p. 303.
[2] *Yen Tzu Chhun Chhiu*, ch. 6, ed. Wu Tse-Yü, p. 425.

In the *Mêng Tzu* we find another argument close to the traditional medieval (and non-Aristotelian) quasi-syllogism:

Tzu Kung once asked Confucius: 'Are you, Master, a sage?' Confucius replied: 'I have not succeeded in becoming a sage. I simply never tire of learning nor become weary of teaching.'
Tzu Kung said: 'Not to tire of learning is wisdom. Not to weary of teaching is benevolence. Since you are both benevolent and wise you are, by that token, a sage.[1]

Confucius admits that he does not tire of learning or teaching. From this premise Tzu Kung derives the desired conclusion that Confucius is a sage through two intermediate steps as follows:

1. Confucius does not tire of learning or teaching.
2. Everyone who does not tire of learning or teaching is wise and benevolent.
3. Everyone who is wise and benevolent is a sage.
4. *Ergo*: Confucius is a sage.

Confucius's disciple Tzu Kung argues according to a pattern which is more complicated than a syllogism. The pattern involves two intermediary universal propositions, each of which in turn are complex.

Having pointed out one example of a Chinese syllogism and having introduced a sampling of quasi-syllogisms from earlier literature, we must hasten to emphasise that from a modern logical point of view the syllogism is of no special systematic logical significance. There is nothing logically fundamental about the syllogism. It just so happens that historically this particular form has played an important part, since it was much favoured by Aristotle's attentions. The traditional preoccupation with the question whether the Chinese did or did not have a syllogism is theoretically quite misplaced. The fact that Wang Chhung used a syllogistic form is no more than an entertaining logical curiosity which we use as an introduction to our enquiries because it shows that, of course, the Chinese *could* use such forms of argument. But it would not make a big difference if the Chinee had not used it. In fact, formal syllogisms are exeedingly rare beyond texts specifically discussing abstract logic even in the West.

Wang Chhung aims to support everything he claims with good arguments just as any Western philosopher aims to do. We may not always find his arguments convincing, but we must not fail to acknowledge his pervasive attempt at coherent argumentative rationality throughout his great work, the *Lun Hêng* 論衡.

Sorites

The ancient Chinese were fond of arguing in sequences of implications that have been called 'sorites'.[2] These arguments take something like the following general form:

$$S_1 \text{ implies } S_2, S_2 \text{ implies } S_3, \ldots S_{n-1} \text{ implies } S_n,$$

[1] *Mêng Tzu* 2A2; cf. Lau (1983c), p. 59.
[2] In the O.E.D.'s second or transferred usage of the term. Cf. Leslie (1964b).

where the conclusion to the effect that S_1 *implies* S_n is typically left unexpressed but may or may not be relevant to the argument. Let us look at two formally imperfect examples by Han Fei in which he sets out to justify Lao Tzu's statement that misfortune lurks in good fortune.

When a man has good fortune, wealth and honour come to him. When wealth and honour come to him, he dresses and eats well.

When he eats and dresses well, arrogance will arise.

When arrogance arises, his behaviour will be wicked and his actions will be contrary to principle (*li* 理).

When his behaviour is wicked, he will die an untimely death.

When his actions are contrary to principle (*li*), he will not be successful.

On the one hand he will have the misfortune of dying an untimely death. On the other hand he will be known as unsuccessful. That is misfortune.

Therefore it is said: 'Misfortune lurks in good fortune.'[1]

Remarkably, Han Fei ends up considering something like the definition of misfortune: he considers whether the consequences of being very lucky do not in fact add up to misfortune. The sequence does indeed look like a logical argument. There is no doubt that Han Fei here focuses on the transitivity of the relation he denotes by the Chinese word *tsê* 則.

However, Han Fei is not, in fact, setting up a sequence of logical implications by using the particle *tsê*. By using *p tsê q* Han Fei only claims a regular concomitance in this world. He is not concerned with the logical connection between *p* and *q* in any possible world.

When a person suffers misfortune, his mind grows fearful.

When his mind grows fearful, his actions will be straight.

When his actions are straight, his deliberations are mature.

When his deliberations are mature, he gets to the inner principles of things.

When his actions are straight, he will suffer no harm. When he suffers no harm, he will live out his natural span of life.

When he gets to the inner principles of things, his mind achieves its proper end.

When he lives out his natural span of life, he stays intact until old age.

If his mind achieves its proper end, he gets rich and noble.

To stay intact until old age and be rich and noble, that is called good fortune.

Therefore it is said: 'Misfortune is the basis of good fortune.'[2]

Let us now turn to the most famous example of the 'sorites', or chain of reasoning, which is indeed formally impeccable, except that the conclusion is not stated explicitly:

When names are not correct, what is said will not sound reasonable.

When what is said does not sound reasonable, affairs will not culminate in success.

When affairs do not culminate in success, rites and music will not flourish.

When rites and music do not flourish, punishments will not fit the crimes.

[1] See *Lao Tzu*, ch. 58, for this dictum. This passage will be found in *Han Fei Tzu*, 20.11.1ff.; Liao (1939), vol. 1, p. 176.

[2] *Han Fei Tzu* 20.10.1; Liao (1939), vol. 1, p. 176.

When punishments do not fit the crimes, the common people will not know where to put hand or foot.

Thus when the gentleman says something, the name is sure to be usable in speech, and when he says something, this is sure to be practicable. The thing about the gentleman is that he is anything but casual where speech is concerned.[1]

The sorites at the beginning of this quotation consists of five clauses. The crucial logical question is whether Confucius intended to demonstrate the conclusion that 'if names are not correct, then the common people will not know where to put hand and foot.'[2]

By writing and thinking in the form of a sorites, the ancient Chinese established not logical laws, but sequences or chains of regular concomitance in their concrete world. They were engaged not in formal logic but in 'real' logic, in the regularity of the concomitance of certain features of the world. The theoretical importance of the sorites was that it allowed the Chinese thinkers to establish *indirect* links between different features of the world. A passage from the unjustly neglected book *Hsin Shu* 新書 by Chia I 賈誼 (−200 to −168) conveniently illustrates our point:

Education (*chiao* 教) is the basis of government. The Way is the basis of education (*chiao*). Only when there is a Way, can there be education. Only when there is education, can there be proper government. Only when there is proper government, do the people give it its moral support. Only when the people give the government its moral support, can the state get rich.[3]

Chia I shows that the Confucian Way, *indirectly*, is the source of the wealth of the state. His reasoning here is entirely typical of early Chinese thinking, and it does have a logical base, although it in no way involves purely formal logical reasoning.

The habit of constructing chains of regular concomitance was pervasive in Chinese intellectual history. It was a dominant strategy of patterning or structuring the world among the Chinese. An exquisite instance of such patterning is almost the whole of *Ta Hsüeh* 大學 or 'Great Learning', one of the influential Four Books of Confucianism.[4]

Modus tollens

An argument of the form:

(A) p implies q
     not−q
     Therefore: not−p

---

[1] *Lun Yü* 13.3; cf. Lau (1983a), p. 121.

[2] It is far from clear that we should attribute this passage to Confucius at all. The question has long been discussed. Creel (1974), pp. 116ff., summarises the discussion and casts considerable doubt on the 'authenticity' of this passage.

[3] Chia I, *Hsin Shu*, ch. 9, ed. Chhi Yü-Chang, p. 1015.

[4] Compare incidentally:

Because of a nail a shoe was lost. Because of a shoe a horse was lost. Because of a horse a rider was lost. Because of a rider a battle was lost, and all for the loss of a horse-shoe nail.

is traditionally called an argument by *modus tollens*. This form of argument turns out to be much more popular in ancient Chinese texts than its seemingly more straightforward counterpart, the *modus ponens* which we shall discuss below.

Consider a typical Chinese example of *modus tollens*:

Jan Chhiu said: 'It is not that I am not pleased with your way, but rather that my strength gives out.'

The Master replied: 'A man whose strength gives out gives up mid-way. Now in your case you set the limits beforehand.'[1]

Confucius makes the assumption here that anyone whose strength is insufficient to achieve something will give up mid-way. He assumes as evident that Jan Chhiu has not given up mid-way (he has failed to try properly). The conclusion is that the reason for Jan Chhiu's behaviour was not that his strength was insufficient. A tentative formalisation of this argument might run like this:

(B) (for all *x*) (*fx* implies *gx*)
  not–*ga*
  Therefore: not–*fa*

Here *f* stands for having insufficient strength to achieve something. *g* stands for giving up mid-way, and *a* stands for Jan Chhiu. We have an imperfect case of *modus ponens* because quantification is involved in the first premise. My point is not that this is a very sophisticated argument. However, the argument is clearly a formally valid one. Given the premises, the conclusion follows logically. And moreover, though one of the premises is not mentioned explicitly, what Confucius says is quite sufficient to enable us to reconstruct the logical structure of his thought with confidence. In the *Mêng Tzu* we find a more straightforward case of *modus tollens*:

For this reason there are no talented men. If there were, I would be bound to know about them.[2]

The underlying argument is clearly as follows: if there were men of talent I would know about them. But I do not know about any talented men. Therefore there are no talented men. This plainly does not amount to a formal proof for the non-existence of talented men, but the train of thought seems logically very clearly structured. Here is another example from the same source:

Between father and son there should be no reproving admonitions with respect to what is good. Such reproofs lead to estrangement, and there is nothing more inauspicious than estrangement.[3]

A logical paraphrase of this argument might run like this: If there are reproving admonitions between father and son, there will be strife. Strife is inauspicious (and must be avoided as such). Therefore one must avoid admonitions between father and son.

---

[1] *Lun Yü* 6.12; cf. Lau (1983a), p. 49.   [2] *Mêng Tzu* 6B6(5); cf. Lau (1983c), p. 251.
[3] *Mêng Tzu* 4A18; cf. Lau (1983c), p. 151.

Mencius said: 'Fish is what I want; bear's paws is also what I want. If I cannot have both, I would rather take bear's paws than fish. Life is what I want; dutifulness is also what I want. If I cannot have both, I would rather take dutifulness than life. On the one hand, though life is what I want, there is something I want more than life. That is why I do not cling to life at all costs. On the other hand, though death is what I loathe, there is something I loathe more than death. That is why there are troubles I do not avoid. If there were nothing a man wants more than life, then why should he have scruples about any means, so long as it will serve to keep him alive? If there were nothing a man loathes more than death, then why should he have scruples about any means, so long as it helps him to avoid trouble? However, there are ways of remaining alive and ways of avoiding death to which a man will not resort. In other words, there are things a man wants more than life and there are also things he loathes more than death. This is an attitude not confined to the moral man but common to all men. The moral man simply never loses it.'[1]

One strand of this argument runs like this:

1. If people's highest values were the preservation of life and the avoidance of death, they would do everything to preserve life and avoid death.
2. Under certain conditions, people will choose to do things that lead to their death or fail to do things that save their lives.
3. *Ergo*: There are things that people value higher than the preservation of life and the avoidance of death.

Finally, we may just note the current ancient Chinese practice of refuting a generalisation by a counterexample which is related to the *modus tollens*, though of course quite distinct, since it involves the use of the concept of necessity or quantification:

Thang and Wu succeeded in attaining supremacy without following the ancient practices, and as for the downfall of the dynasties Yin and Hsia, they were ruined without rites having been altered. Consequently, those who acted counter to antiquity do not necessarily deserve blame, nor do those who followed established rites merit much praise.[2]

Modus ponens

An argument by *modus ponens* is traditionally taken to be of the form

(C) p implies q
　　p
　　Ergo: q

Note the characteristic fashion in which the levels of thought and action are intertwined in this passage:

Pi Hsi summoned the Master, and the Master wanted to go. Tzu Lu said: 'Some time ago I heard it from you, Master, that the gentleman does not enter the domain of one who in his

---

[1] *Mêng Tzu* 6A10; cf. Lau (1983c), p. 235.
[2] *Shang Chün Shu*, ch. 1, ed. Kao Heng, p. 17; cf. Duyvendak (1981), p. 173.

person does what is not good. Now Pi Hsi is using Chung Mou as a stronghold to stage a revolt. How can you justify going there?'[1]

We may summarise the essential argument of Tzu Lu as follows:

1. The gentleman should not enter the domain of a person who does what is not good.
2. Pi Hsi is a person who does what is not good.
3. You, Confucius are a gentleman.
4. Ergo: You, Confucius, should not enter Pi Hsi's domain.

As our analysis shows, this is far from being a neat case of an argument by *modus ponens*, but the connection with *modus ponens* is plain enough. Confucius shows a distinct lack of interest in the plain logical side of Tzu Lu's query when he replies, disarmingly:

'It is true, I did say that. But has it not been said that "hard indeed is that which can withstand grinding"? Has it not been said that "white indeed is that which can withstand black dye"? Moreover, how can I allow myself to be treated like a gourd which, instead of being eaten, hangs from the end of a string?'[2]

Compare Mo Tzu's response to the same sort of complaint:

Master Mo Tzu was ill. Tieh Pi entered his room and asked: 'Master, you said that ghosts and spirits were omniscient, that they have the power to send blessing and misfortune, and that they reward the good and punish the evil. However, you, being a sage, have fallen ill. How is this possible? Can it be that your teaching was not entirely correct, that the ghosts and spirits are after all not omniscient?'[3]

Characteristically, Mo Tzu takes the charge of inconsistency seriously in his reply:

Master Mo Tzu replied 'Even if I am ill, how could the spirits' lack of omniscience be held responsible for that? Diseases are caused by excessive work or hardship. If there are one hundred gates and you shut only one of them, how can you prevent the burglar from getting in?'[4]

A fortiori

Arguments *a fortiori* or *par plus forte raison*, were already familiar to Aristotle, who called them arguments *kata poiotēta*.[5] Aristotle promises in several places that he

---

[1] *Lun Yü* 17.7; cf. Lau (1983a), p. 173. Argumentation along such lines was apparently a characteristic of Tzu Lu:

> Tzu Lu went in and asked: ' I have heard it said that Heaven rewards those who do good with good fortune, and that Heaven rewards those who do evil with misfortune. Now for a long time you, Master, have piled virtue upon righteousness, and good action has been close to your heart. How come that you live in obscurity?' (*Hsün Tzu* 28.33; cf. Köster (1967), p. 370).

Confucius answers along similar logical lines:

> I suppose you think that intelligent men inevitably find employment, but in that case Pi Kan should not have had his heart cut out. (*Ibid.*)

[2] *Lun Yü* 17.7; cf. Lau (1983a), p. 173.   [3] *Mo Tzu* 48.70; cf. Mei (1929), p. 240.
[4] *Ibid.*   [5] Cf. W. and M. Kneale (1962), p. 105.

would write a treatise on the matter. Unfortunately, no such treatise has come down to us. Let us consider this form of argument in ancient China:

Even the Duke of Chou left something to be desired in the way of benevolence and wisdom. How much more (*khuang*) must this be so in the case of Your Majesty![1]

Here the Duke of Chou is assumed to be a paragon of benevolence and wisdom. The comparison in an *a fortiori* argument can be between objects:

King Kou Chien of Yüeh once saw an angry frog and saluted it. The coachman asked: 'Why does Your Majesty salute it?'
'A frog that shows a fighting spirit like this! How could I fail to salute it?'
When the knights heard this they said: 'Even a frog, if it shows spirit, our king will salute. How much more (*khuang* 況) will he salute a knight who has courage!'[2]

Han Fei uses arguments *a fortiori* quite systematically to argue from the immediately plausible to the less immediately plausible.

Moreover, parents' relation to their children is such that when they get a son, they congratulate each other. When they get a daughter, they kill her. Now both boys and girls come from their parent's womb, and yet the parents congratulate each other when it is a boy and kill the child off if it is a girl. This is because they are concerned with later convenience and because they calculate long-term profit. Thus even parents dealing with their own children still use their calculating minds. How much more should one expect this when there is no such bond of affection between those involved![3]

Chinese arguments *a fortiori* can also turn on a comparison of verbs:

If the ruler of a state with a thousand chariots cannot even hope to *become acquainted* with him, how much less (*khuang*) can he hope to *summon* such a man![4]

There often is something poetic and emphatic about *a fortiori* arguments, in China as elsewhere:

If mere water clarifies when it is still, how much more the stillness of the quintessential-and-demonic, the heart of the sage![5]

(f) LOGICAL THEORY

(1) TÊNG HSI 鄧析 AND HUI SHIH 惠施[6]

Têng Hsi (died −501) and Hui Shih (*c.* −370 to −310) were both closely involved in the legal and political conflicts of their times. Têng Hsi is justly regarded as the founding father of the Chinese logical tradition. It is not a coincidence that he was earning a living as a specialist in law: his significance in the history of logic is closely

---

[1] *Mêng Tzu* 2B9; cf. Lau (1983c), p. 85.   [2] *Han Fei Tzu* 30.34; Liao (1939), vol. 1, p. 302.
[3] *Han Fei Tzu* 46.2.24; Liao (1939), vol. 2, p. 239.   [4] *Mêng Tzu* 5B7; D. C. Lau (1983c), p. 217.
[5] *Chuang Tzu* 13.4; cf. Graham (1981), p. 259.
[6] Sections (*f*,1) to (*f*,4) should be compared with Vol. 2, pp. 171–203.

connected with his legal interests. It was on this legal ground that Chinese logic grew. First in close connection with legal rhetoric, and then more independently as an autonomous preoccupation.

The creation of Chinese law codes must be placed perhaps as early as the −8th century, and even in the state of Chhin it must have occurred significantly earlier than the −4th century.[1] By the −3rd century the complexity of the legal system (penal as well as administrative) was considerable. A. F. P. Hulsewé (1985) presents crucial new archeological evidence on this matter in an exemplary way. We shall illustrate the logical significance of this legal material by considering the invention of one crucial device of abstract thinking, the invention of variables to refer to arbitrary items.

Variables in Chhin law

Variables like X, Y, Z are singularly useful devices for the articulation of any generalisations, and particularly of scientific generalisations. Aristotle's systematic deployment of variables to make clear the general and abstract nature of his concerns was a crucial part of his success as a logician. W. Kneale and M. Kneale (1962), p. 61, comment on Aristotle's invention of term-variables as follows:

This is a new and epoch-making device in logical technique. It is used for the first time without explanation in the second chapter of the *Prior Analytics*, which deals with conversion, and it seems to be Aristotle's invention. In earlier works generality is indicated by a rather clumsy use of pronouns or by examples in which it is left to the reader to see the irrelevance of the special material. Both methods are used by Plato and Aristotle. An example of the former from the *Republic* is: 'When things are of such a nature as to be relative to something, then those that are of a certain sort are relative to something of a certain sort.'[2]

It is clear that Plato would have been able to express himself more clearly if he had had variables available. His Greek sentence without his illustrations would have remained opaque. Aristotle solves this problem of logical articulation.

But was Aristotle's invention of the variables repeated anywhere else? When (if ever) did the ancient Indians or the ancient Chinese develop symbols that function like variables, symbols that have no material or epideictic function but serve purely to abstractly identify arbitrary items or person?

In Sanskrit we find the phrase *evaṃnāmā* 'of such-and-such a name, such-and-such' which corresponds roughly to the Chinese *mou* 某 'such-and-such'. Let us call this sort of expression a pseudo-variable. When more than one such-and-such are involved, we find that these are all referred to by the same term *evaṃnāmā*, so that we have X, X, and X all referring to different people as in the following generalised rule for the conduct of Buddhist communities:

This Monk such-and-such is able to instruct such-and-such and such-and-such as an *upādhyāya* in secret.[3]

---

[1] See Hulsewé (1985), p. 1.   [2] *Republic* 438b.   [3] Härtel (1956), p. 77.

This use of pseudo-variables seems to have been current in Sanskrit at least from the −3rd or the −4th century.

Consider now the following Chinese legal text from the −3rd century:

X (*chia* 甲) and Y (*i* 乙) do not originally know each other. X goes on to rob Z (*ping* 丙). As he arrives, Y also goes along to rob Z (*ping*), speaks with X, and they then each proceed to the robbery. The value of the goods was four hundred (cash) each. Having left the place they all got (profit). Had they made plans beforehand, then one would have to try them on the combined amount. Since they did not make plans, they must each be held responsible for their amounts.[1]

When X (*chia*) and Y (*i*) each have intercourse with the woman Z (*nü tzu ping* 女子丙) and X and Y for this reason come to injure each other, then if Z does not know about this, how is she to be accused? She is not to be accused.[2]

I found one case in which the number of people involved was four, with an additional pseudo-variable *mou* 某:

The commoners W and X of a certain (*mou*) village bound and brought in the men Y and Z...[3]

X (*chia*) steals cash in order to buy silk. He sends this to Y (*i*). Y receives it but does not know that it is stolen. What is Y (*i*) to be accused of? He is not to be accused.[4]

The symbols which the ancient Chinese used for variables could not be letters of the alphabet since they did not know of any alphabet. Instead they used the so-called ten stems, a series of characters used from ancient times in the Chinese system of dating.[5]

The variable in Chinese legal texts may be used as a modifier as in *tai fu chia* 大夫甲 'the dignitary X'[6] and *nü tzu chia* 女子甲 'the woman X'.[7] It is a standard part of the linguistic repertory of the practitioners of law at least since the −3rd century, and there is no particular reason for assuming that this use of variables began at that time, considering the long stretch of legal history before that time.

Some early practitioners of the law, it turns out, were particularly fond of paradoxes. Têng Hsi and Hui Shih were cases in point.

Têng Hsi 鄧析

We can get a realistic picture of Têng Hsi's activities from the following passage:

When people were involved in lawsuits, Têng Hsi would enter agreements with them: for a large case he could take one overcoat, for a small case he would take a jacket with trousers.

---

[1] *Shui Hu Ti Chhin Mu Chu Chien* 睡虎地秦墓竹簡, p. 156; cf. Hulsewé (1985), D11.

[2] *Shui Hu Ti Chhin Mu Chu Chien*, p. 225; cf. Hulsewé (1985), D152.

[3] *Shui Hu Ti Chhin Mu Chu Chien*, p. 252; cf. Hulsewé (1985), E9.

[4] *Shui Hu Ti Chhin Mu Chu Chien*, p. 165. There are two similar cases immediately preceding the ones I have translated. Further relevant cases are pp. 158, 163, 165, 166, 167, 168, 169, 171, 180, 193, 210, 218, 222f., 225, 230, 249, 251f., 254ff. It is interesting to note that the legal texts sometimes use a variable even in a case where only one individual is spoken about, i.e., where there is only an X but no Y, as *ibid.*, p. 180 and p. 204. But when there is no Y, the normal technique is to use the term *huo* 或 'someone'.

[5] For a concise tabulation of the ten stems and their use in dating see Giles (1892), pp. 29f.

[6] *Shui Hu Ti Chhin Mu Chu Chien*, p. 206.   [7] *Ibid.*, p. 222.

Innumerable people handed up their coats and jackets with trousers to take his legal advice. Têng Hsi turned wrong into right and right into wrong. There was no standard of right and wrong. What was admissible and what was not changed from day to day. If Têng Hsi wanted someone to win a case that man would win as a result. If Têng Hsi wanted someone to be found guilty, that man would be found guilty as a result. There was great confusion in Cheng, and the people kept arguing with each other.[1]

There is ample evidence that Têng Hsi 鄧析 developed his debating skills in the legal courts of the economically very advanced state of Cheng. The *Tso Chuan* 左傳 commentary credits him with the authorship of a penal code written on bamboo (*chu hsing* 竹刑)[2] which was in deliberate opposition to the official bronze code of Têng Hsi's apparent political enemy, the formidable Tzu Chhan 子產, who was the effective ruler of Cheng from −542 to −522.[3]

Têng Hsi's legal and political activities clearly brought him into conflict with Tzu Chhan and his successors. They also made him famous throughout the land for his extraordinary rhetorical skills. We can still get a glimpse of the spirit and subtlety of the sophistry attributed to Têng Hsi by looking at the following story:

There was a flood around the river Wei. A rich man from Cheng drowned. Somebody fished out the dead body and the rich man's family wanted to buy back the body. However, the man who had found the body demanded a lot of money. The rich family complained to Têng Hsi, who replied: 'Leave him, no one else will buy the body anyway.' The man who had found the body got angry and he too complained to Têng Hsi, who replied: 'Leave them! They have nowhere else to buy the body from anyway.'[4]

This is one of the many examples of jokes from early Chinese literature, and like many jokes it tells us a great deal about the perceptions and sensibilities of its time. Têng Hsi was felt to be a menial rhetorician who would provide convincing arguments for whatever his clients wanted to hear. Significantly, both his statements are perfectly true. The outrageous thing is that Têng Hsi 鄧析 seems unconcerned with the fact that his advice to the two parties is inconsistent. Within each situation he argues from his client's point of view. The inconsistency of the overall stance he is taking is of no concern to him. He is an adherent of what was later called *liang kho chih shuo* 兩可之説, the doctrine that both sides are right.[5] As a legal adviser, of course, one might feel that he was obliged to take such a stance, but the development of sophistry consisted in the detachment of the legal argumentation from its social context, the development of arguments for their own sake, and particularly the attempt to justify apparent paradoxes by irrefutable arguments. We shall see this development in the thought of Hui Shih 惠施.

---

[1] *Lü Shih Chhun Chhiu* 18.4, ed. Chhen Chhi-Yu, p. 1178; cf. Wilhelm (1928), p. 302.
[2] *Tso Chuan*, Duke Ting 9, ed. Yang Po-Chün, p. 1571; cf. Couvreur (1951b), vol. 3, p. 550.
[3] Cf. Rubin (1965).
[4] *Lü Shih Chhun Chhiu* 18.4, ed. Chhen Chhi-Yu, p. 1178; cf. Wilhelm (1928), p. 301.
[5] This term was first used, as far as I know, in Liu Hsin's preface to the (lost) original *Têng Hsi Tzu* 鄧析子: 'He (Têng Hsi) practised the doctrine that both sides are right and established arguments (*tzhu* 辭) that were unassailable.' (Wu Fei-Po (*1984*), p. 863).

Hui Shih[1]

Hui Shih (−370 to −310) is celebrated in our earliest sources as a man of many talents – and of many books. In point of fact he is perhaps the most versatile of all pre-Chhin philosophers we know of, being an eminent politician, an important legal specialist, a prominent theoretician of science, a distinguished practical scientist, and a noted conversationalist and friend of the eccentric Taoist Chuang Tzu 莊子. No wonder the ancient Chinese books are so full of stories on Hui Shih.[2] There is abundant evidence that he was a very active politician and it appears that his unlikely friendship with Chuang Tzu flourished after his political demise in Wei and his flight to Chuang Tzu's native Sung 宋.[3] His political and administrative career provides the indispensible background for our understanding of his logical and scientific views.[4]

We have, of course, no way of judging the quality of Hui Shih's legislation, but we may safely conclude that it was widely perceived to be unnecessarily subtle and complex, perhaps pedantic, just as legislation is often perceived in our own time. As a legislator Hui Shih was ahead of his time, and by all accounts he paid the sort of formal attention to the 'letter' of the law which struck his contemporaries as misplaced. This was the very same formal attention to words which, as we shall see, Hui Shih applied to philosophy, where it was felt to be no less unnatural and pedantic.

Again, Hui Shih's scientific interest was not limited to mundane politics and mundane legal practice. He was also accused in the Taoist book *Chuang Tzu* of 'chasing after things and not turning back'.[5] He proudly announced:

What nobler theme is there than Heaven and Earth? I keep my male rôle. I depend on no tradition.[6]

Defiantly, Hui Shih declared an interest in mundane questions of the science of his time.

Weak in the Power within, strong on the things outside, his was a crooked path. If we consider Hui Shih's abilities from the viewpoint of the Way of Heaven and Earth, yes, his were no more than the labours of a mosquito or a gnat. Even within the realm of things, what use were they?[7]

[1] Cf. Vol. 2, pp. 189–97, for a detailed survey of Hui Shih's paradoxes. Our present, briefer, account tries to take account of the the last 35 years' research.

[2] Apart from the book *Chuang Tzu* in which Hui Shih figures prominently as the Taoist philosopher's personal friend, Hui Shih is mentioned very frequently in *Lü Shih Chhun Chhiu, Han Fei Tzu, Huai Nan Tzu, Chan Kuo Tshe, Shuo Yüan* and in no less than six chapters of the *Hsün Tzu*. Cf. Moritz (1974) and Reding (1985).

[3] Cf. R. Moritz (1974), pp. 74–91.

[4] Cf. particularly *Lü Shih Chhun Chhiu* 18.5; tr. Wilhelm (1928), p. 306. By far the best treatment of the early sources on Hui Shih is Reding (1985). Reding has the advantage of being quite as well at home in the Greek texts as he evidently is in his Chinese sources. He suggests (pp. 321ff.) that five stories about Hui Shih in the *Lü Shih Chhun Chhiu* may go back to one original book on the sophist. The currency of such a book would indeed explain why Hui Shih is so often mentioned in pre-Chhin literature.

[5] *Chuang Tzu* 33.86; tr. Graham (1981), p. 285.   [6] *Chuang Tzu* 33.81; tr. Graham (1981), p. 284.

[7] *Chuang Tzu* 33.84; tr. Graham (1981), p. 285.

All this reminds us of the tradition attributing an agricultural invention to Hui Shih's predecessor Têng Hsi.[1] It does seem that the 'logicians' stood in some deliberate and defiant opposition to traditional highbrow interests among the more ethically inclined philosophers of their time. They are thus key figures in the history of 'proto-science' in China.

In the south there was a queer man named Huang Liao who asked why the sky did not fall and the earth did not sink; also about the causes of wind, rain and the rolling thunder. Hui Shih answered without hesitation, and without taking time for reflection. He discussed all things continuously and at great length, imagining that his words were but few, and still adding to them strange statements.[2]

Hui Shih's debating style is preserved for us in a long series of anecdotes of varying degrees of likely authenticity.

Khuang Chuang spoke as follows to Hui Shih in the presence of King Hui of Wei: 'A peasant will kill a locust if he has caught one. Why does he do this? Because they harm the grain. Now you travel around often with several hundred carts and several hundred followers, and at the very least with some dozen carts and some dozen followers. That means that you do not work but still eat. The harm to the grain is greater than that caused by the locusts.'
King Hui said: 'Hui Shih will have trouble answering you. Still, let him speak.'
Hui Shih said: 'If for example one builds a city wall, then some people will stand on top and tread down the earth. Others will carry along the baskets with earth down below. Still others look at the drawings and supervise the building activities. I am one of the supervisors with the drawings in their hands.
If you transform a spinner into silk, she won't be able to go on to work silk. If you transform a carpenter into wood, he won't be able to continue to work wood. If you transform a sage into a peasant, he won't be able to work the peasants. Why do you insist on comparing me to a locust?'[3]

Hui Shih's fondness for analogies was proverbial in ancient China, and in one story he defends his constant use of this method:

A client said to the King of Wei: 'When Hui Shih talks about anything he is prone to use illustrative comparisons. If you forbid him illustrative comparisons he won't be able to speak.'
The King agreed. At the audience next day he said to Hui Shih: 'When you speak about something I wish you would simply speak directly, without illustrative comparisons.'
'Let's suppose we have a man who does not know what a *tan* 彈 is', Hui Shih said. 'If he says "What are the characteristics of a *tan* like?" and you answer "like a *tan*", will it be conveyed to him?'
'It will not.'
'If you proceed to answer instead "A *tan* in its characteristics is like a bow, but with a string made of bamboo", will he know?'
'It will be known.'

---

[1] Cf. Wu Fei-Po (*1984*), p. 871.
[2] *Chuang Tzu* 33.82; cf. Graham (1981), p. 284. We follow Derk Bodde's translation in Fêng Yu-Lan (1953), vol. 1. p. 196.
[3] *Lü Shih Chhun Chhiu* 18.6, ed. Chhen Chhi-Yu, p. 1197; cf. Wilhelm (1928), p. 308.

'It is inherent in explanation that by using something he does know to convey what he does not know one causes the other man to know. To give up illustrative comparisons as you are telling me to do is inadmissible.'

'Well said!', said the King.[1]

The dialogues between the mystic Chuang Tzu and the logician Hui Shih are among the more memorable pieces of ancient Chinese philosophical prose that have come down to us. Here is an example which exemplifies Hui Shih's spirit of logical analysis:

Chuang Tzu and Hui Shih were strolling on the bridge above the Hao river.
'Out swim the minnows, so free and easy!' said Chuang Tzu. 'That's how fish are happy.'
'You are not a fish. Whence do you know that the fish are happy?'
'You aren't me. Whence do you know that I don't know the fish are happy?'
'We'll grant that not being you I don't know about you.
You'll grant that you are not a fish, and that completes the case that you don't know the fish are happy.'
'Let's go back to where we started. When you said "*Whence* do you know that the fish are happy?", you asked me the question already knowing that I knew. I knew it from up above the Hao.'[2]

When Hui Shih asks 'You are not a fish. How do you know that the fish are happy?', he really raises a profound epistemological question which will not come as a surprise to any Western philosopher. To Chuang Tzu, on the other hand, and to most of his contemporaries, Hui Tzu's question will have sounded quite hopelessly pedantic. Hui Shih's interest in politics and in the practical affairs of this world was below the spiritual dignity of Chuang Tzu.

When Hui Shih was chief minister of Liang, Chuang Tzu went to visit him. Someone told Hui Shih: 'Chuang Tzu is coming. He wants your place as chief minister.'
At this Hui Shih was frightened and searched for him throughout the state for three days and three nights.
Chuang Tzu did go to visit him. 'In the South there is a bird,' he said 'its name is the phoenix. Do you know of it? The phoenix came up from the South Sea to fly to the North Sea; it would rest on no tree but the sterculia, would eat nothing but the seeds of the bamboo, would drink only from the sweetest springs. Just then an owl had found a rotting mouse. As the phoenix flew over, it looked up and glared at it: "Shoo!". Now am I to take it that for the sake of that Liang country of yours you want to shoo at me?'[3]

Hui Shih's petty logic-chopping as well as his 'petty political interests' were like an interest in an intellectual rotting mouse to Chuang Tzu.

An interpretation of Hui Shih's paradoxes

J. P. Reding (1985), pp. 274–385, takes up the crucial question whether Hui Shih's *dicta* are properly understood as scientific paradoxes in the first place. He provides a

---

[1] *Shuo Yüan* 11.8, ed. Chao Shan-I, p. 307; cf. Graham (1978), pp. 444f.
[2] *Chuang Tzu* 17.85; tr. Graham (1981), p. 123.   [3] *Chuang Tzu* 17.84; tr. Graham (1981), pp. 122–3.

detailed alternative interpretation intending to show that they are not scientific paradoxes at all. A. C. Graham (1970), p. 140, on the other hand, writes: '. . . although Hui Shih's explanations no longer survive, the whole list can be read, like Zeno's paradoxes, as a series of proofs that it is impossible to divide space and time without contradiction.'

Unfortunately, all we have are Hui Shih's theorems and paradoxes as preserved in the last chapter of the book of his friend Chuang Tzu, to which we shall now turn.[1] The cases of Zeno and of Hui Shih differ profoundly in that we do have a fairly precise idea of the stringent logical arguments Zeno used to support his theses,[2] whereas we know so little of the intellectual context of Hui Shih's paradoxes that one can evidently raise doubt that they are scientific paradoxes in the first place, as Reding does.

The first 'thesis' is in fact not a thesis at all but a set of two crucial definitions. It runs like this:

Thesis 1
Definitions:
The perfectly large thing, having nothing outside it, is called the macro-one.
The perfectly small, having nothing inside it, is called the micro-one.[3]

The concept Hui Shih is concerned with is 'that than which nothing is bigger' and 'that than which nothing is smaller'. In order to appreciate the intellectual achievement represented by this sort of definition, it is healthy to compare it with the following dialogue:

Duke Ching 景公 asked Yen Tzu 晏子: 'Is there an infinitely large thing in the world?'
Yen Tzu replied: 'There is. In the Northern Lake there is the Pheng. Its feet float on the clouds, its back reaches the azure sky. The tail wafts between the heavens, and it hops and takes a sip from the Northern Sea. Its tail fills out the space between Heaven and Earth. So vast is it that it is not even aware where its wings are.'
'Is there a smallest thing in the world?'
Yen Tzu replied: 'Yes there is. In the Eastern Sea there is an animal which builds its nest on the eyelids of a mosquito. Again and again it lays its eggs and flies off but the mosquito does not get disturbed by it. I do not know its name, but the fishermen at the Eastern Sea call it the Chiao Ming.[4]

For Yen Tzu 'the largest thing' is just a very, very big thing. The distinct notion of infinity, the logical definition which Hui Shih is aiming for and achieving, lies beyond his intellectual horizon. And the tragedy for Hui Shih 惠施 as well as so many other logicians was that, quite generally, what they were concerned with was so far beyond the intellectual horizons of their contemporaries. Logical opinions were perceived as bizarre intellectual curiosities, as 'strange words', as grotesque

---

[1] All Hui Shih's theorems that follow are taken from *Chuang Tzu* 33; cf. Graham (1981), pp. 283ff.
[2] The evidence on Zeno is assembled in Diels (1954), pp. 247–58.
[3] As Solomon (1969), p. 5, points out, the term *ta i* 大一 recurs elsewhere in contemporary literature. (Cf. *Chuang Tzu* 24.106f.) However, its definition here is quite unfamiliar from other sources.
[4] *Yen Tzu Chhun Chhiu* 晏子春秋, ed. Wu Tse-Yü, p. 514.

paradoxes (*yin tzhu* 淫辭). As in the case of Kungsun Lung's 公孫龍 White Horse Dialogue, which we shall discuss later, they were recorded for their entertainment value rather than their intellectual content.

The macro-one and the micro-one are Hui Shih's abstract logical constructs. Paradoxical theses arise when these abstract constructs are applied to the concrete things of this world.

> *Thesis II*
> Definition:
> That which has no dimension (*wu hou* 無厚) cannot be accumulated, but its size is a thousand *li* 里.

The micro-one, since it has nothing inside it, has no dimension (*wu hou*). Yet things of size must presumably be thought of as composed of micro-ones. Thus something which does not have size, when added to other things which also have no size, adds up to something that has size. This is a paradoxical thought. However, it does seem to follow from the assumption that the smallest unit has no parts or size, and that sized things are compounds of these smallest units.

Thesis II follows naturally from the first part of 'thesis' I, given the extra assumption that sized things are made up of 'micro-ones'. Again, thesis III can be construed to follow naturally from the second part of 'thesis' I.

> *Thesis III*
> The sky is on the same low level as the earth. Mountains are on the same level as the marshes.

If the macro-one is infinite, then when compared with this infinitude the difference in height between the sky and the earth, between mountains and marshes, should be negligible, and this would mean that the sky is on the same low level as the earth and the mountains are on the same level as the marshes. The theses II and III are given a coherent explanation in

> *Thesis V*[1]
> Things which have more in common being different from things which have less in common, that is called micro-differentiation.
> All things having something in common, and all things differing from each other in some respect, that is called macro-differentiation.

One can obviously set up a difference between any two distinct things, but making this sort of difference is called micro-differentiation, while that which Hui Shih is concerned with for the purpose of his (sophistical) argument is macro-differentiation. By saying that a mountain is on the same level with a marsh he says something which he knows people will find absurd. Just as the later sophist Kungsun Lung, knowing that people would find it absurd for him to claim that a

---

[1] For our exposition we deviate from the order in which the theses are presented in the last chapter of the *Chuang Tzu*.

white horse is not a horse, insists on showing that he can defend this thesis on a given interpretation, so Hui Shih, knowing full well that there is a difference in height between a mountain and a marsh, insists on showing that there is a 'macro'-interpretation under which the two are on one level.

Chuang Tzu's 莊子 chapter 'Seeing All Things as Equal'[1] consists in a coherent attempt to take such a 'levelling' perspective on all things:

Nothing in the world is bigger than the tip of an autumn hair, and Mount Thai is small. No one lives longer than a doomed child, and Pheng Tsu died young. Heaven and Earth were born together with me, and the myriad things and I are one.[2]

One can imagine how Chuang Tzu and Hui Shih could become close intellectual companions on the basis of such profound common perceptions. However, what for Hui Shih was a result of some logical definitions was for Chuang Tzu the starting point of a philosophy of life.

In the book *Kuan Tzu* 管子 we find a passage that might throw light on the connection between Hui Shih's alleged egalitarianism and his talk of mountains and marshes:

How can a small rise make a flat plain high?
(Commentary:) The four major failings of men are for the ruler to be predatory, the father and mother to be cruel, the minister and subordinate official to be disloyal, and the son and his wife to be unfilial. Hence if any one of these major failings is present in a person, it is impossible for him to become a worthy, even though he may have some minor good points. 'A flat plain' refers to low marshland. Even though there may be a slight rise on it, it cannot be considered high.

How can a small depression make a lofty mountain low?
(Commentary:) The four highest forms of conduct for men are for the ruler to be kind, the father and mother to be compassionate, the minister and subordinate official to be loyal, and the son and his wife filial. If any one of these highest forms of conduct is present in a person, he cannot be considered unworthy, even though he may commit minor errors. 'A lofty mountain' refers to a mountain that is high. Even though it may have a small depression, it cannot be considered low.[3]

Though we have no way of telling, there may be, behind Hui Shih's paradox of mountains and plains, a social paradox as well. This, of course, does not mean that the thesis does not have a logical/scientific dimension as well.

Some of Hui Shih's paradoxes sound like intellectual pranks, but they can still be connected to his general definitions:

Thesis VIII
Interlocking rings can be separated.

We have, as usual, no record of the reasoning which lay behind this paradox. But suppose that a ring is defined as something round. Then the perfectly round thing

---

[1] See Graham (1981), pp. 48–61.   [2] *Chuang Tzu* 2.52; cf. Graham (1981), p. 56.
[3] *Kuan Tzu* 管子, ed. Tai Wang, vol. 1, p. 5 and vol. 3, p. 33; tr. Rickett (1985), p. 72. Cf. Rickett's earlier translation of the *Kuan Tzu* (1965), p. 133, note 63.

would be a circle of no thickness (a circle which lacks internal dimension (*wu hou* 無厚)). Two circles of *this* kind – one might tease an audience by claiming – could indeed be separated.

At the macro-level of space there are also paradoxes.

Thesis IX
I know the centre of the world. North of Yen up in the north, and south of Yüeh down in the south is the place.

This paradox is phrased almost like a *bon-mot*. What Hui Shih is surely intending to convey is the fact that in a universe that is as infinite as he defines it, the centre will be absolutely everywhere, even in the most 'marginal' places.[1]

Another spatial paradox is that about the limitless South, the scene of seemingly endless Chinese cultural and political expansion throughout the ages.

Thesis VI
The South is inexhaustible and yet it is exhaustible.

Here again, we do not have the arguments which Hui Shih used to defend this or any of his other paradoxes. However, we can attempt to reconstruct what may have been his reasoning.

Suppose that the South is inexhaustible. Then there is no other inexhaustible outside this inexhaustible, i.e., the inexhaustible is, after all, delimited, excluding other inexhaustibles outside itself.

Such an interpretation (in which we deliberately avoid modernizing precision of mathematical terminology) may still sound hopelessly modern and unnatural to impose on an ancient Chinese thinker. Can we really attribute to the ancient Chinese theories about infinitudes? It is hard to be sure.

The notion of infinity as opposed to a very large number was, in any case, familiar to the authors of the *Chuang Tzu* 莊子 who deliberately contrast the literal meaning of *wan* 萬 'ten thousand' with the metaphoric meaning 'indefinitely many'.

Suppose you were counting the number of things. Then you would not stop at ten thousand, and yet by convention we speak of them as 'the ten thousand things'. This is because we use the largest number to call and designate them by.[2]

Consider next the following most remarkable story from a part of the book *Lieh Tzu* which we have some reason to suspect quotes from a lost chapter of *Chuang Tzu*.

Emperor Thang of Yin asked Chi of Hsia: 'Were there things in ancient times, at the beginning of the world?'
Chi of Hsia said: 'If there had been no things then, how come there are things now? Suppose that later men maintain there were no things at this time, would that be acceptable?'

---

[1] This, I think, would be the view of modern physicists.   [2] *Chuang Tzu* 25.66; cf. Graham (1981), p. 151.

Thang of Yin said: 'Does that mean that there is no "before" or "after" to the realm of things[1]?'

Chi of Hsia replied:

> 'The endings and startings of things
> Have no limit from which they began.
> The start of one is the end of another,
> The end of one is the start of another.
> Who knows which came first?

But beginning from the point where the realm of things ends and which is before events, then I know nothing.'

Thang of Yin said: 'Does this mean that above and below and in the eight directions there is a limit, and that things are finite?'

Chi said: 'I do not know.'

When Thang pressed the question, Chi continued: 'Nothingness is without limit. Existence is finite. How am I to know the answer to your question? However, outside the limitless there is no further limitless. Inside the infinite there is no further infinite. On this basis I know their limitlessness and their infinitude, but I do not know their limit and their finiteness.'[2]

This argument may well echo, as A. C. Graham has suggested to me in conversation, the original arguments of Hui Tzu in favour of his sophism. The basic thought is that beyond the infinitude of space there is a higher order of infinitude which we do not know.

Perhaps this way of thinking was what lay behind Hui Shih's paradox. If it was, this would do more to explain the affinity and friendship which the Taoist Chuang Tzu felt for the sophist. But, whereas for Hui Shih his result is an abstract result of reasoning,[3] for Chuang Tzu it is a poetic expression for the human condition at the centre of a whole hierarchy of unknowable infinitudes.

The theses we have hitherto considered were all concerned with the notion of space. When we turn to the notion of time, his paradoxes are again of a kind that will naturally have amused a Taoist.

Thesis IV

When the sun is at the zenith, the sun at the same time declines. A thing is alive and at the same time dying.

Hui Shih seems to be thinking of the infinite divided into indivisibles in time: at the indivisible moment when the sun reaches the zenith it must also count as beginning its decline. The moment of being in the zenith is part of its decline if decline is to stretch over more than one indivisible moment. Similarly, the moment of death must be at the same time a part of a person's life. For Hui Shih these are theoretical consequences of his definitions of the infinitely great and the infinitely small. For

---

[1] Thang of Yin asks whether this means that there is no 'before' or 'after' the totality of things. I.e., 'before' and 'after' apply only within the realm of things.

[2] *Lieh Tzu*, ch. 5, ed. Yang Po-Chün, p. 92; cf. Graham (1960), p. 94.

[3] The Later Mohists carry on the debate in what presumably was Hui Shih's spirit. Cf. Graham (1978), B73.

the Taoist Chuang Tzu these results have a very different mystical appeal: for him it appears that the sophist by his formal methods independently arrives at results that coincide with his deep philosophical conclusions.

Another paradox deliberately combines the spatial and the temporal dimension:

Thesis VII
I go to Yüeh today and arrived yesterday.

Suppose I cross the border to Yüeh exactly at the moment where today and yesterday meet. Then I will have gone to Yüeh today and will all the same have arrived today, all depending upon under what day you count the moment I crossed the border. The moment, like that of death, belongs to both sides. It belongs both to yesterday and to today.[1]

Aristotle was of the opinion that the origin of philosophy was in the sense of wonder/amazement (*thaumazesthai*, θαυμάζεσθαι). I find it not at all implausible that it was the amazement at paradoxes like those we have just surveyed which sparked off logical speculation in China, as indeed it probably did in Greece. Diogenes Laertius (+3rd century), the loquacious historiographer of philosophy, observed that the *dialektikon*, τὸ διαλεκτικόν or 'dialectic' part of philosophy derived from Zeno the Eleatic who was well known for his paradoxes.[2] In China, as in Greece, the interest in *pien* 辯 'disputation' was inspired by the discovery of paradoxes.

(2) KUNGSUN LUNG AND THE WHITE HORSE DIALOGUE[3]

Of the book *Kungsun Lung Tzu* 公孫龍子 only two chapters may plausibly be attributed to the sophist Kungsun Lung 公孫龍 (*c.* −320 to *c.* −250): the essay on Meanings and Things (*Chih Wu Lun* 指物論) and the White Horse Dialogue (*Pai Ma Lun* 白馬論).[4] The strong arguments A. C. Graham has lined up for considering the other chapters as dating from the +4th to the +6th century has, in the West, led to a neglect of these later parts of the *Kungsun Lung Tzu* as forgeries ever since. However, even as forgeries of the +4th to +6th centuries, also the remaining chapters deserve our close attention.

We have an old commentary on the *Kungsun Lung Tzu* which is usually attributed to Hsieh Hsi-Shen 謝希深 (+995 to +1039), but is very probably of Thang date.[5] A

---

[1] For an alternative recent account of these paradoxes cf. Reding (1985) which represents an interesting effort to relate the paradoxes less to logical reflections than to the social and political realities of the time.

[2] *Lives of Eminent Philosophers* 1.18.

[3] Cf. Vol. 2, pp. 185–9. The present section tries to take account of the last 35 years of research on Kungsun Lung. Particularly the works of A. C. Graham mentioned below have completely changed our views on the authenticity of the *Kungsun Lung Tzu* and the nature of Kungsun Lung's thought.

[4] Graham (1986c) has shown this very convincingly and in admirable philological detail. His other very crucial contributions to the study of Kungsun Lung include Graham (1965), Graham (1955) and Graham (1978), pp. 457–69. It would obviously be tempting in the present Sub-section simply to summarise the latest findings of A. C. Graham who has been working on this material for 23 years. But for such an authoritative summary the reader is referred to Graham (1986b).

[5] Graham (Review of Kou Pao-koh, *Deux sophistes chinois*, in *Journal of Oriental Studies*, **2**, no. 2, July 1955). For this commentary see Kandel (1974).

postface to the *Kungsun Lung Tzu* by Hsieh Hsi-Shen survives. There is another interesting commentary by Fu Shan 傅山 (+1607 to +1684), written in almost colloquial Chinese.[1]

The modern Chinese literature on Kungsun Lung is considerable.[2] Since A. Forke (1902) drew attention to Kungsun Lung, there has also been a fair amount of work done in Western languages.[3]

Of the genuine works of Kungsun Lung, the potentially crucial essay on Meanings and Things has received a number of radically different interpretations which have one striking feature in common: they have by and large failed to convince other students of the text. The leading scholar of Kungsun Lung, A. C. Graham, is alone responsible for two radically distinct interpretations, and he remarks: 'There are more rival interpretations of the *Chih wu lun* 指物論 than of any other document in early Chinese philosophical literature.'[4] It is not my intention to add another interpretation at this point. Instead, I propose to turn to the White Horse Dialogue which I do believe we can interpret with some confidence, although the interpretation of this dialogue is, of course, also far from uncontroversial.

The text of the White Horse Dialogue

The White Horse Dialogue is our most important source on Kungsun Lung's thought. And the first problem one is faced with is of a textual nature. Recent interpretations of the White Horse Dialogue[5] have had this in common that they interpret not the *textus receptus* as it has come down to us but a version ingeniously reconstructed in Graham (1965) which involves, among other things, several major transpositions.

Chinese scholars on the other hand have generally tried to understand the *textus receptus* without such major textual surgery.[6] One is tempted to follow current Western practice and interpret the text as rearranged by A. C. Graham, but I shall aspire to present a plausible interpretation of the dialogue which resists the temptation to rearrange the text as it has come down to us.

A. C. Graham has commented: 'It does not seem too presumptuous to say that no one has yet proposed a reading of the dialogue as a consecutive demonstration

[1] See Luan Hsing (*1982*), p. 156, for bibliographic detail.

[2] Chhên Chu (*1947*), Than Chieh-Fu (*1957*), Hsü Fu-Kuan (*1966*), Phang Phu (*1979*), Luan Hsing (*1982*), Wu Fei-Po (*1984*), pp. 500–628, and Ho Chhi-Min (*1967*) are useful annotated editions. Hu Tao-Ching (*1934*) remains an invaluable collection of material.

[3] E.g., Perleberg (1952), Mei (1953), Kou (1953), Kao and Obenchain (1975). Hansen (1983) and the decisive works reprinted (with some additions) in Graham (1986a), pp. 125–215.

[4] Graham (1986a), p. 210.   [5] Including Reding (1985) and Hansen (1983).

[6] See Feng Yao-Ming (*1985*) for an attempted formalisation of the dialogue without rearranging it. The most recent interpretations include Phang Phu (*1974*), Phang Phu (*1979*) and Chhü Chih-Chhing (*1981*). I have found especially useful Wu Fei-Po (*1984*), pp. 500–628, and the most thorough interpretations, Luan Hsing (*1982*) and Ho Chhi-Min (*1967*). Taken together, these books provide exhaustive bibliographic and biographic information on Kungsun Lung, and they very conveniently lay out the sources on which an interpretation of the *Kungsun Lung Tzu* will have to be based. However, none of these books take into account A. C. Graham's crucial conclusions on the authenticity of the various parts of the *Kungsun Lung Tzu*.

which does not turn it into an improbable medley of gross fallacies and logical subtleties.'[1] Certainly, on our interpretation Kungsun Lung will not emerge as a brilliant and always superbly coherent theoretician of logic advancing one consecutive
logical demonstration. Indeed, I do not find it historically plausible that Kungsun
Lung was such a theoretician.

We must ask: What kind of a person was Kungsun Lung and what kind of
discourse does the White Horse Dialogue belong to? What is its proper cultural
context?

The social and historical context of the White Horse Dialogue

Some scholars, notably Fêng Yu-lan and Janusz Chmielewski, have read the dialogue as sustained serious logical discourse by a theoretician. Others, from earliest
times onwards, have considered it a facetious piece of sophistry.[2] I believe that the
Dialogue is both these things. I believe that Kungsun Lung (c. −320 to c. −250), as
our earliest sources describe him, was indeed an intellectual entertainer at the court
of the Lord of Phing-yüan.

There were many entertainers at ancient Chinese courts displaying different
skills or tricks. Kungsun Lung was one of these. His trick, his piece of entertainment, was of an intellectual kind. His showpiece or standard ploy was to declare
that he could prove that a white horse was not a horse. He would declare that he
could meet any objections anyone could raise against this thesis. By an extraordinary case of serendipity,[3] we still have an account that gives us quite a good idea of the
arguments Kungsun Lung was in the habit of using in defense of his thesis. I take the
text we have today in the White Horse Dialogue[4] as a demonstration of the sort of
sophists' dialogue that Kungsun Lung would engage in.[5] In the Dialogue, Kungsun
Lung is supposed to successfully defend his plainly outrageous thesis against an
opponent.

Kungsun Lung's repertory was not limited to the thesis that a white horse is not a
horse. Here is one of many other examples taken from a source roughly contemporary with Kungsun Lung:

Khung Chhuan and Kungsun Lung disputed at the palace of the Lord of Phing-yüan. They
spoke with profundity and rhetorical skill (*pien* 辯). When they came to 'Tsang has three

---

[1] Graham (1986a), p. 193. By the way, there is nothing historically improbable about a mixture of logical subtlety and elementary logical blunders. The history of the modern logical interpretation and criticism of Plato's
dialogues proves this point beyond all doubt.

[2] Cf. Chou Yün-Chih (*1981*).

[3] Most direct evidence about early Chinese 'sophistry' comes from sources hostile to this 'sophistry'. It just so
happened that during the +3rd and +4th centuries, when a certain intellectual flippancy was the fashion, someone still found access to some genuine material from the −3rd century.

[4] We also have the Dialogue on Meanings and Things, but this seems to me to be a document for which there
are many widely different interpretations, none of which carry conviction: like Janusz Chmielewski, I simply
cannot pretend that I can make sense of the text.

[5] The suggestion that a philosophically important text may ultimately involve a *flippant* use of logic is by
no means unheard of. Plato's dialogue *Parmenides* has been interpreted as a logical game rather than a serious
dialogue.

ears', Kungsun Lung propounded the thesis with great rhetorical skill. Khung Chhuan failed to come up with an answer and left.

The next day, when Khung Chhuan came to court, the Lord of Phing-yüan said to him: 'What Kungsun Lung said yesterday was presented with great rhetorical skill.'

Khung Chhuan replied: 'Yes. He almost managed to make Tsang have three ears. He may have been skilful, but his case was a difficult one. May I ask you a question: That Tsang has three ears is very hard to maintain, and it is that which is the wrong thesis. That Tsang has two ears is very easy to maintain; it is that which is the right thesis. Would you prefer to be right by taking the easy alternative or would you prefer to be wrong by taking the difficult alternative?'[1]

Saying that the White Horse Dialogue belongs to a tradition of light-hearted entertainment is not to imply that it is devoid of serious logical interest. For in the Dialogue Kungsun Lung shows a considerable interest in problems which we are inclined to say belong properly to the philosophy of language. The Later Mohists argued dead seriously about logic. Kungsun Lung was not necessarily as seriously interested in logical problems as they were, although he delighted in using logic to prove unlikely conclusions. Let us look at the logical structure of his famous white-horse arguments in some detail.

The grammatical background

In our Sub-section on definition and quotation marks we have noticed that the ancient Chinese had several syntactic devices available to them that come close to our quotation marks. These devices are an integral part of the Chinese grammatical system, and like so many other grammatical markers they are optional: they can be omitted when the context permits without this affecting the meaning.

Again, in our Sub-section on negation we have seen that the particle *fei* 非 in front of nominal predicates has two very distinct uses as exemplified in the following two sentences:

A. *tzu fei wo* 子非我 'you are not (the same as) me'.
B. *tzu fei yü* 子非魚 'you are not (a case of) a fish'.

Against this background we can translate a sentence like *pai ma fei ma* 白馬非馬 along two lines:

A. 'White horse' is not (the same as) 'horse'.
B. A white horse is not (a case of) a horse.

We note that the English 'White horses are not horses' and 'A white horse is not a horse' are similarly open to two such opposing interpretations.

Version B is, of course, the ordinary way in which one would understand *pai ma fei ma* 白馬非馬. But from the late first chapter and preface of the *Kungsun Lung Tzu* it

---

[1] *Lü Shih Chhun Chhiu* 18.5, ed. Chhen Chhi-Yu, p. 1186; cf. Wilhelm (1928), p. 305.

would appear that interpretation A was the one assumed by the authors quoted in the preface to that book:

Kungsun Lung and Khung Chhuan once met in Chao at the palace of the Lord of Phing-yüan. Khung Chhuan said: 'I have long heard about your noble conduct, and I have wanted to become your disciple. However, I cannot accept your thesis that a white horse is not a horse. Please reject this thesis, and I will beg to become your disciple.'

Kungsun Lung replied: 'Your words are self-contradictory. The reason I have made a name for myself is none other than this discourse on the white horse. If you now make me abandon this discourse, I have nothing to teach. Moreover, if one wishes to make someone one's master, this is because one's wisdom and learning is not up to his. If now you make me abandon my thesis, this would be first teaching me and then making me your master. First teaching someone and then making him into one's master is self-contradictory. Moreover, that "white horse" is not "horse" is precisely a thesis which Confucius advocated. I have heard it said that the King of Chhu was once wielding his bow Fan-jo with his (famous) Wang-kui arrows. He was shooting at crocodiles and rhinoceroses in his (legendary) Yün-meng Park, when he lost an arrow. His ministers were scrambling to collect the arrow, but the King replied: "That is not worth while. The king of Chhu has lost his arrow, and a man from Chhu (*Chhu jên* 楚人) will find it. Why go and look for it?"

When Confucius heard about this he commented: "The King of Chhu may be humane and just, but he is not perfect yet. Surely he should simply have said: 'A man (*jên* 人) has lost the arrow and a man (*jên*) will find it.' What reason is there to bring in the 'Chhu'? In this way Confucius made a distinction between *Chhu jên* 'a man of Chhu' and what is called *jên* 'a man'."

Now to approve Confucius's distinguishing between *Chhu jên* and *jên* but to disapprove of my distinguishing *pai ma* 白馬 "white horse" and *ma* 馬 "horse" is self-contradictory.'[1]

We note that the story about Confucius and the arrow is also preserved in a pre-Han text.[2] Confucius suggests in this story that the King should have said *jên* instead of *Chhu jên*. We note that the state of Chhu may be the place of origin of the Chhu man, but that piece of land is in no way a proper part of the Chhu man. There is nothing to suggest that the ancient Chinese ever thought of anything as strange as the idea of a Chhu man consisting of a piece of land and a man. On the other hand, there is clear evidence in the preface that the construction *Chhu jên* is construed similarly to *pai ma*.

What the preface to our text focuses on is the distinction between '*Chhu jên*' and '*so wei jên* 所謂人'. '*Chhu jên*' and '*jên*' are not the same (just as *pai ma* and *ma* are not the same). There can be occasions where one should use one expression rather than the other. That is the issue as it is understood in the preface to our text. And there is good reason to think that our passages concerning Kungsun Lung and Khung Chhuan are in the main of pre-Han origin.[3]

---

[1] *Kungsun Lung Tzu*, ch. 1, ed. Luan Hsing, p. 4.

[2] *Lü Shih Chhun Chhiu* 1.4, ed. Chhen Chhi-Yu, p. 44. A slightly later version is in *Shuo Yüan* 14.7, ed. Chao Shan-I, p. 390.

[3] See Graham (1986a), p. 179. Khung Chhuan was a descendant of Confucius. We may suspect that the story is of Confucian origin.

Khung Chhuan's answer concerning the difference between *jên* 人 and *Chhu jên* 楚人 is elaborated in the book *Khung Tshung Tzu* 孔叢子 (*c*. +3rd century):

Whoever uses the word 'man (*jên*)', refers generally to men, just like anyone who uses the word 'horse (*ma* 馬)' refers generally to horses. Chhu is basically a state, and white is basically a colour. If you want to refer more generally to men you must remove the 'Chhu'. If you want to make sure of correctly naming the colour you must not remove the 'white (*pai* 白)'.[1]

This story and its elaboration in the *Khung Tshung Tzu* take it to be Kungsun Lung's argument that there was a difference between *pai ma* 白馬 and *ma*. The reason why *pai ma* and *ma* are different is that *pai ma* involves being white plus being a horse, whereas *ma* does not involve the combination of two such concepts.[2] This, of course, is and was a trivial point, but what makes it entertaining is that in stating it in Chinese as '*pai ma fei ma* 白馬非馬' one says something that will naturally be understood along the lines of the absurd statement 'a white horse is not a case of a horse'. *Pai ma fei ma* sounds as obviously untrue as Zeno's provocative claim that the flying arrow never moves. Kungsun Lung – like Zeno – became famous for being able to give a plausible interpretation of an outrageously implausible statement. In this Kungsun Lung continues the line of Hui Shih before him which we have discussed in the last Section. By demonstrating his trivial truth Kungsun Lung can manage to appear to demonstrate an outrageously untrue sentence, a paradox, to be true.[3]

Kungsun Lung may, of course, have been misunderstood by those who compiled his book, by those who are quoted in its preface, and by the sources on which the *Khung Tshung Tzu* was based. However, I find no strong evidence that there was such a misunderstanding in the dialogue or in other early sources.

Kungsun Lung was, of course, widely criticised and misinterpreted as having said that a white horse does not count as a horse. Consider the following joke preserved in many versions,[4] some of which go back to the −3rd century. I quote a version by Huan Than 桓譚 (−43 to +28):

Kungsun Lung kept arguing that a white horse was not a horse, but people would not agree with him. After this he rode a white horse and wanted to pass a customs point without a licence (enabling him to take a horse out). The customs official was unable to understand him (i.e., his subtle arguments that 'a white horse' is not 'a horse'). This goes to show that empty words have a hard time removing facts.[5]

---

[1] *Khung Tshung Tzu* 12, ed. *SPTK*, 1, 75b. In this context we find the thought-provoking phrase *i Chhu jên yü so wei jên* 異楚人與所謂人 'distinguish "man of Chhu" from what is called "man"'.

[2] The evidence assembled in Graham (1986a), pp. 157–60 and 172f., documents the ancient Chinese view that *pai ma* consists of two semantically distinct parts whereas *ma* consists only of one part. Being white plus being a horse is not equal to being a horse. Similar observations apply to *Chhu jên* and *jên*. Being from Chhu plus being a man is not equal to being a man.

[3] Note that it is equally trivial to say that the flying arrow at no point in time moves: the arrow has obviously not got a chance to move as long as we only look at one point in time. Zeno – like Kungsun Lung – is guilty of playing on tedious trivialities. But like Kungsun Lung he is justly famous.

[4] Many versions of the story of Kungsun Lung or someone else crossing a border riding a white horse are collected in the excellent survey by Hu Tao-Ching (*1934*), pp. 16–18.

[5] *Hsin Lun* 新論, ch. 9, in Wu Fei-Po (*1984*), p. 597.

Here it is plain that the common-sense way in which to take *pai ma fei ma* 白馬非馬 was along the lines 'a white horse does not count as a horse'. This is obviously not what Kungsun Lung had in mind.

I assume that Kungsun Lung, like everybody else, knew that the ordinary way one would take a sentence like *pai ma fei ma* is along the lines of the customs official, or B. The *skandalon* was that Kungsun Lung insisted on taking the sentence – unnaturally – along the lines of A. On this more technical interpretation the sentence turns out to be plainly true. Thus a situation arises where Kungsun Lung can defend as plainly true a sentence which sounds plainly untrue.

We must and can imagine that such an intellectual stunt, such a piece of argumentative acrobatics, had a certain entertainment value.[1] Together with riddles and puzzles it fits well into the court life of musicians, dwarfs and other entertainers of which we have much indirect evidence. At the same time Kungsun Lung's dialogue shows a very considerable interest in the techniques of formal argument as well as in what we would call the philosophy of language.

Unfortunately, what we have of Kungsun Lung's dialogue on the white horse is not a transcript of an actual discussion. Rather it seems to be a paradigmatic summary of the ways in which Kungsun Lung was prepared to argue his outrageous thesis. None the less, this summary is a unique piece of direct evidence on the argumentative practice of the Chinese sophists.

Interpretation of the White Horse Dialogue

I

A. Is it admissible that 'a white horse is not a horse (*pai ma fei ma* 白馬非馬)'?

B. It is admissible.

A. Why should this be so (*ho tsai* 何哉)?[2]

B. 'Horse' is that by which we name the shape. 'White' is that by which we name the colour. Naming the colour is not (the same as) naming the shape. Therefore I say ' "white horse" is not (the same as) "horse" (*pai ma fei ma* 白馬非馬)'.

Kungsun Lung is asked whether a white horse is a horse and deliberately misconstrues this question as asking whether 'white horse' and 'horse' are the same. Since 'white horse' also names the colour (besides the substance (*chih* 質)),[3] it cannot be the same as 'horse' which does not name the colour. The argument at this point is

---

[1] I hasten to add that – seen with modern eyes – Kungsun Lung's sophism does turn out to be a bit of a lame joke (to put it mildly), a fairly straightforward case of an ambiguity exploited to create a puzzle. 2,000 years after the event (or shall we say *post festum*, after the feast?) it has therefore naturally become tempting – especially for sophisticated thinkers – to try to read into it more sophisticated thoughts that would make Kungsun Lung a most remarkable sophisticated logician treating the kings of his time to arguments of a sustained precision reminiscent of current analytical philosophy.

[2] Note that *ho tsai* 何哉 'why on earth?' is not synonymous with the *ho yeh* 何也 'why?' later in the dialogue, an important point which Graham (1986a) does not notice. Compare the difference between the simple question particle *hu* 乎 versus the combination *hu tsai* 乎哉 that makes rhetorical questions. Such distinctions are important in our context. The objector is not only asking the cool logical question 'why?', as Graham takes it. He is expressing surprise.

[3] Cf. *Khung Tshung Tzu*, ch. 12, ed. *SPTK*, p. 75a.

simple. The empirical question whether in point of fact all horses happen to be white does not affect the issue.[1]

Kungsun Lung's argument would be thoroughly boring if it did not have the effect of proving as clearly true something that appears to be clearly untrue. Note that similar observations can be made concerning the tricks of magicians less intellectual than Kungsun Lung. If you explain their techniques, there is nothing very exciting about what is happening. Similarly for a joke which loses its force when elaborately explained. The sophism, like its relatives, the puzzle, the joke, and even the magician's 白 trick, lives and thrives on a surprise effect. Once that element of surprise is gone, they lose their living force. This, however, does not mean they never really had such force.

II

A. But if one has a white horse one cannot be said not to have a horse; and if one cannot be said not to have a horse, how can (what one has) not be a horse? (In other words) if having a white horse counts as having a horse, why should the white one not be a horse?

The objector A comes up with a common-sense objection: supposing that a white horse was not a horse, then if you have a white horse, one would have to say you don't have a horse. But clearly you *do* have a horse when you have a white horse. Therefore, by *modus tollens*, ' a white horse is not a horse' cannot be true, i.e., a white horse must be a horse. We shall see that the technique of *modus tollens*, or if one prefers of *reductio ad absurdum*, is a standard device used in our dialogue.

B. If someone is looking for a horse, you can offer him a brown or a black horse. If (on the other hand) he is looking for a white horse, you cannot offer him a brown or a black horse. Supposing (now) that 'white horse' was nothing other[2] than 'horse'. Then what he would be looking for would be the same thing. And if what he is looking for was the same thing, then the white one would not be different from 'a horse'. But if what he is looking for (in those two cases of 'white horse' versus 'horse') was the same, why should it be that the brown and black horses were acceptable in one case but not in the other? It is evident that the admissible and the inadmissible are not the same as each other. Therefore, since a black or brown horse remains the same, but since one can answer that there is 'horse' and cannot answer that there is 'white horse', this means that the thesis that a white horse is not a horse is conclusively demonstrated true.

The reply, again, is plain enough in outline: looking for a white horse is not the same as looking for a horse *tout simple*. However, if 'white horse' was nothing other than 'horse', then looking for a white horse would be the same as looking for a horse. Therefore, by *modus tollens* or *reductio ad absurdum* he concludes that it cannot be true that 'white horse' is the same as 'horse'. If it *was* true, unacceptable consequences would follow.

---

[1] Note that *pai* 白 'white' and *ma* 馬 'horse' are not the same sort of thing. *Ma* 'fixes' a thing, *pai* does not, to use Kungsun Lung's phrase.
[2] Graham (1986b) translates *nai* 乃 with quite uncharacteristic vagueness as 'after all'.

The sophist goes on to use yet another argument by *modus tollens*. Supposing now, he argues, that what the two are looking for were one and the same thing, then it should not make a difference whether you come up with a brown or a black horse in either case. But it does make a difference. Therefore, by *modus tollens*, it cannot be true that they are looking for the same thing.

There is another logical technique in the sophist's arguments which is of special interest to the historian of logic: he seems to apply something like Leibniz's law of identity which says that if X and Y are identical, one must be able to substitute them for each other in any sentence, *salva veritate*, without this affecting the truth or otherwise of the sentence involved. Thus the sophist reasons that if 'white horse' was the same as 'horse', then 'I am looking for a white horse' should be true under exactly the same circumstances as 'I am looking for a horse'. But this is not so. Therefore, 'white horse' and 'horse' cannot be the same.[1]

Why does the sophist change the example from 'having a white horse' to 'seeking a white horse'? Here again, it looks as if something logically quite subtle may be going on. For 'have an X' counts as what analytical philosophers call an extensional context, whereas 'seek an X' creates an intensional context in the relevant sense. The relevant basic idea behind this distinction between extensional and intensional contexts is not hard to explain. If I have a certain horse, and without my knowing about it this horse is deaf, then in every sense of the word I do 'have a deaf horse'. If, on the other hand, I am looking for a certain horse, and without my knowing it this horse happens to be deaf, then – in one psychological sense of 'look for' – I am not looking for a deaf horse, i.e., my mental attitude is not towards a horse under the description 'deaf horse'. After all, I do not know that the horse is deaf.

'Looking for white horses' is no more 'looking for horses' than 'looking for black swans' is 'looking for swans'. On the other hand it is perfectly true that 'having a white swan' is indeed 'having a swan'.

When the sophist changes the example from 'having a white horse' to 'seeking a white horse', I cannot help feeling that he had a perception of this subtle logical difference between 'having' and 'seeking'.

III

A. You are considering a horse that has a colour not to be a horse. But it is not the case that there are colourless horses in the world. Then (by your argument) there would be no horses in the world. Is that admissible?

Again, the objection is plain enough. If coloured horses are not horses, then, since there are only coloured horses in the world, there would be no horses in the world. But there plainly are horses in the world. Therefore, by *modus tollens* or *reductio ad absurdum*, it cannot be true that coloured horses – and particularly white horses – are not horses.

---

[1] This point is well made in Fêng Yao-Ming (*1985*).

B. Horses certainly have colour. That is why there are white horses. If horses had no colour, there would (only) be horses plain and simple. How could one then pick out white horses? Therefore[1] 'white' is not the same as 'horse'. 'White horse' is 'horse' combined with 'white'. But 'horse' combined with 'white' is not (the same as) 'horse'.[2] Therefore I maintain: 'a white horse' is not 'a horse'.

Here one might at first sight construe 'horse combined with white' as a combination of parts in a whole. But the sophist says that horses have colour. He does not say that colours have horses. There is no symmetry between 'white' and 'horse'. They are not construed as parts of the same order. Still, one can say that WHITE HORSE is composed of the parts WHITE plus HORSE. One can then go on to say that WHITE plus HORSE does not equal HORSE.[3] That is what the sophist does.

The sophist does not deny the common-sense point of view that horses have colour, or that there are such things as white horses. But in picking out the white ones from among horses, he claims, one acts on the assumption that indeed 'white' is not the same as 'horse', and that 'white horse' is not the same as 'horse' either.

Here the sophist uses the rather subtle argumentative device of showing that the conclusion he wishes to demonstrate is actually implicit in what his opponent is saying. This is again only possible because Kungsun Lung deliberately insists on construing the sentence *pai ma fei ma* 白馬非馬 differently from his opponent.

On hindsight we are tempted to say that the opponent might easily have won the contest by simply calling the bluff and explaining that the sophist is putting a very special interpretation on the sentence *pai ma fei ma* which is not the generally accepted one. However, it is not quite so easy to imagine what an opponent could have said in Classical Chinese to call that particular bluff.[4]

[1] Because one picks out by this criterion.

[2] I follow Wu Fei-Po (*1984*), p. 558, who understands *yeh* 也 as a loan word for the question particle *yeh* 耶. From handbooks on Chinese particles it may appear that *yeh* is used as a loan word for the question particle *yeh* in marked questions like *hsi i chih chhi jan yeh* 奚以知其然也 'how do we know this is so?', and that it is never used in unmarked questions. This impression is mistaken. In *Lü Shih Chhun Chhiu* 10.5 we find *yu i kuo chê i kuo chih chu yeh. Chin wo fei chhi chu yeh* 有一國者一國之主也。今我非其主也 'He who controls a whole state is the ruler of that state. Now am I not its ruler?' Chhen Chhi-Yu (*1984*), p. 563, note 16, remarks that 'the *Lü Shih Chhun Chhiu* often uses *yeh* for *yeh*'. Another entirely convincing example of the use of *yeh* for *yeh* in unmarked questions will be found in *Lü Shih Chhun Chhiu* 9.3, ed. Chhen Chhi-Yu, p. 490, line 12, and in *Chan Kuo Tshe*, ed. Chu Tsu-Keng, p. 326, line 7, both of which are dialogue passages. It seems almost malicious that we should find this usage in a logical context. An error by a scribe who was not following the argument properly seems not unlikely.

[3] White pigment and the biological features that make up a horse could perhaps be imagined to be parts of one whole white horse, although this would involve considerable biological and logical abstractions which I see no reason to attribute to the ancient Chinese. But, as we have noted earlier in this Section, when it comes to a Chhu man, then the state of Chhu can by no stretch of the intellectual imagination be construed as literally a part of one whole Chhu man. Not even 'Chhu-ishness' can be said to be a proper part of a Chhu man. For one can be born in Chhu, and therefore a Chhu man, but actually by upbringing, language, etc., a Chhi man, as the ancient Chinese were well aware.

[4] Of course, on our interpretation Khung Chhuan does bring out the point in the preface to the *Kungsun Lung Tzu*, and the *Khung Tshung Tzu*, ch. 11, quoted above, does elaborate the point. But neither of these sources have convenient simple phrases like 'is not identical with *X*' versus 'is not a case of an *X*' to mark the crucial logical difference.

IV

A. If 'horse' not yet in combination with 'white' counts as 'horse', and if 'white' not yet in combination with 'horse' counts as 'white', then when we put together 'white' and 'horse' to make a combined term 'white horse', we give names that are out of combination to things that are in combination, and that is not quite admissible. Therefore I say that 'a white horse is not a horse' is not quite admissible.

Now the objector abruptly shifts from common-sense objection to accusing the sophist of maintaining an inherently inconsistent philosophical position. Whereas until now the objector maintained the point of view of general common sense against the artificialities of theorizing, he now shows quite a subtle perception of the sophist's point of view. It is as if we are now witnessing one sophist arguing against another.

What is the nature of the objector's objection? He objects to Kungsun Lung's treating 'white' and 'horse' in construction with each other as if they had the same reference as they have separately. But they clearly do not have the same reference inside and outside the construction 'white horse' (*pai ma* 白馬). They change (*pien* 變) when they are used in one syntactic construction, we might say, echoing a chapter of the *Kungsun Lung Tzu* which is not genuine, but which, after all, may in turn be echoing earlier logical reflections.

How, then, does 'white' change when it enters the construction 'white horse'? It refers no longer to the things that are white but only to the white things that are horses.

How does 'horse' change when it enters the construction with 'white'? It refers no longer to the things that are horses but only to the horses that are white.

Thus the objector points out that construing the combination 'white horse' as the sum of things that are white and things that are horses is not quite justified. Such a way of taking the construction would make anything white (also things that are not horses) as well as any horse (also horses that are not white) count as 'white horses'.

As I understand the objector's argument, he claims that the sophist's proposed analyses rest on such an interpretation of the combination 'white horse'. This interpretation is plainly incorrect. Therefore, by *modus tollens* the sophist's analyses cannot be correct.

B. Since you consider 'having a white horse' is 'having a horse', does this mean that 'having a horse'[1] counts as 'having a brown horse'? Is that admissible?
A. It is not quite admissible.

The sophist refuses to tackle the objection directly. He just tries to demonstrate that his account of the relation between 'white horse' and 'horse' does not involve the misunderstanding he is being accused of.

The sophist argues his point by going back to the objector's original submission that 'having a white horse' means 'having a horse' and pointing out that while 'having a white horse' entails 'having a horse', 'having a horse' does not conversely

---

[1] With Than Chieh-Fu, I omit the *pai* 'white' in front of *ma* 'horse'. Without this emendation the link with the reply becomes incomprehensible.

entail 'having a white horse'. Since he is interpreting the sentence *pai ma fei ma* 白馬非馬 symmetrically as '"a white horse" is not (the same as) "a horse"', this asymmetry in the entailment is clearly crucial to his case.

B. When you make a distinction between 'having a horse' and 'having a brown horse', you are making a distinction between 'brown horse' and 'horse'. Since you make a distinction between 'brown horse' and 'horse', then 'brown horse' must count as not being (the same as) 'horse'. To consider 'brown horse' as not (the same as) 'horse', and to consider that in the case of 'white horse' you have[1] (the same as) a 'horse', that is (like the saying) 'flying things move in a pond, inner and outer coffins change places'. These are the most contradictory words and the most confused formulations.

Having established that the objector makes a distinction between 'brown horse' and 'horse', the sophist simply points out that when one makes such a distinction between *X* and *Y*, then *X* cannot be said to be (the same as) *Y*, and if this argument works for 'brown horse' and 'horse', it must, by analogy, also work for 'white horse' and 'horse' if one is to avoid inconsistency.

The sophist gives a triumphantly flowery rhetorical picture of the inconsistency in which he feels the objector would involve himself.

The stylistic flourish is rather important, because it shows that the dialogue as we have it is not entirely thought of as an algebraic disputation but as related to a real situation at court where an objector is trying to show that the sophist cannot maintain his outrageous thesis. The difference with the Mohist logical texts is striking: it seems that the later Mohists operated and wrote in a purely scientific milieu where such rhetorical flourishes would have been out of place.

V

B. 'If one has a white horse, one cannot be said not to have a horse' means that one separates off the 'white'. If the white is not separated off, then having a white horse one cannot be said to have a horse. Therefore, as regards the reason why one counts as having a horse, it is exclusively because of (the word) 'horse' that one counts as having a horse. It is not because of (the expression) 'white horse' that one counts as having a horse. Therefore 'counting as having a horse' cannot refer to any (particular) sort of horse.

Now the sophist launches into a longish final monologue in two parts. He explains why the objection that 'If one has a white horse, one cannot be said not to have a horse' does not prove that (in his sense) '"white horse" is "horse"'. By separating off the 'white' we understand that he has a horse, Kungsun Lung argues, but this does not commit us to the thesis that '"white horse" is "horse"'. Kungsun Lung contends that by disregarding the 'white' and focusing on the 'horse' one is said to 'have a horse' when one 'has a white horse'. Thus, he claims, the inference is not from 'white horse' to 'horse' but from 'horse' to 'horse'. The validity of this inference does not imply that 'white horse' is the same as 'horse'.

---

[1] I regard the *yu* 有 'there is' in the text as an insertion for the sake of superficial parallelism with *fei* 非 'not be'. The opposite of *fei* would, perhaps, be expected to be *wei* 為, but that would in this instance have given an unattractive sequence *wei wei* 為為 in which the two *weis* are taken as synonymous.

Having spoken of the separating off of 'white', Kungsun Lung now goes on to focus on the difference between the two things that are separated off from each other.

B   (continues):[1] 'White' does not fix that which it declares white. That which it declares white may be left out of account. In 'white horse', when one says 'white', one fixes what one declares white. That which fixes what one declares white is not (the same as) the 'white'. 'Horse' does not reject or pick out with respect to colour. That is why one can come up with a brown or black horse (when asked to point out a horse). 'White horse' does reject and pick out with respect to colour. Brown and black horses are all rejected on the grounds of colour. Therefore one can only come up with white horses (when asked to point out a white horse).
    Now that which rejects something is not (the same as) that which does not reject it. That is why I say 'a white horse' is not 'a horse'.

Quite properly, the dialogue ends with the victor concluding with his main thesis. This is indeed a very plausible and natural end to a dialogue concerned with one main thesis.

In his concluding dialogue the sophist discusses the terms 'white' and 'horse' separately and in combination. He distinguishes between a word like *ma* 馬 which fixes something, and a word like *pai* 白 which does not fix anything in that way. Compare the following passage from a possibly +3rd-century source:

Naming the colour first and thereafter the substance (*chih* 質), that is something that applies to all things, and it is a practice which the sages have consistently followed.[2]

That which would fix what is declared white is a word for a substance (*chih*) like 'horse', as in 'white horse'. This is important evidence on early Chinese grammatical sensibilities, but we hasten to add that Kungsun Lung himself never uses the term *chih* 'substance'. What we have here is an early Chinese attempt at an explanation of what is going on in the White Horse paradox.

Kungsun Lung then simply goes on to expound his interpretation of the paradox *pai ma fei ma* 白馬非馬 in the context of what these terms pick out and do not pick out

---

[1]   The *yüeh* 曰 at the beginning of this clause does not introduce a new speaker but a new sequence by the same speaker and is therefore troublesome. Phang Phu (*1974*), p. 16, surmises that *yüeh* is a writing mistake for the graphically similar *i* 以. A. C. Graham takes the *yüeh* as clear evidence for a major dislocation, transposing the final sequence of the dialogue to a position after no. 9 of the original text.
    However, in *Mêng Tzu* 2A1.4, *yüeh* is translatable by 'Mencius continued'. Cf. also *Li Chi*, section *Than Kung* 檀弓, ed. Couvreur (1950), vol. 1, p. 251 for another neat example of sequence of *yüeh* introducing the same speaker. In *Chuang Tzu* 22.61 and 63 we find a similar case of a dialogue which ends with two sequences by the supposed victor: there the same speaker is even named twice, and also in that case he finishes off a dialogue with a longish monologue. The present case differs from the parallels in *Mêng Tzu* and *Chuang Tzu* referred to above in that the speakers in our *Kungsun Lung Tzu* text are not named and in that there are three *yüeh* with the same speaker in a row. It looks as if the editor of the dialogue had added what he had by way of extra comments by the sophist at this point.
    However we take the *yüeh*, and whether we rearrange the text or not, we must take this passage to represent Kung-sun Lung's opinion. This much is uncontroversial.
[2]   *Khung Tshung Tzu*, ch. 12, ed. *SPTK*, 1, 75a. Note that this rare example of a syntactic rule makes it clear that *pai* 'white' and *ma* 'horse' are not perceived as being of the same order. One of them is a substance (*chih* 質), the other is not.

from reality. He thus contrasts the way in which 'white horse' and 'horse' pick out parts of reality, expounding his original position.

The text of Kungsun Lung makes reasonably good sense as it stands and does not need textual surgery to become interpretable as an important document in the history of early Chinese logic. On this view, then, the compiler of the *Kungsun Lung Tzu* has not fundamentally misunderstood the sophist, and neither have his early critics. His was indeed a somewhat flippant interest in argumentation for the purpose of creating entertaining puzzles. It is for the subtlety of argumentative skill which Kungsun Lung deployed that he deserves an important place in Chinese intellectual history.

APPENDIX: The Mass Noun Hypothesis and the Part–whole Analysis of the White Horse Dialogue

The White Horse Dialogue is among the most widely discussed documents in Chinese intellectual history. I shall at this point add a few comments on two of the most recent interpretations. The first of these, C. Hansen (1983) is of a certain interest from a methodological point of view, and his work must be seen against its proper philosophical background.

The logician W. V. O. Quine, in perhaps the most famous part of his most widely read book *Word and Object*, maintains that in some ways we may attribute to others different and mutually contradictory structures of thought without risking ever being refuted by any evidence from their speech. Quine takes as his example the hypothetical 'native' word *gavagai* 'rabbit', and shows that without risk of being refuted one can attribute to a people a notion not of an enduring entity but only of rabbity time-slices. Another alternative is to attribute to the native a notion not of a rabbit as such but only of all and sundry undetached rabbit parts. He then continues:

A further alternative likewise compatible with the same old stimulus meaning (of the word *gavagai*) is to take 'gavagai' as a singular term naming the fusion, in Goodman's sense, of all rabbits: that single though discontinuous portion of the spatiotemporal world that consists of rabbits. Thus even the distinction between general and singular terms is independent of stimulus meaning.[1]

Hansen attributes what Quine reconstructs as the notion of a spatio-temporal fusion to the ancient Chinese – with the difference, admittedly, that the Chinese talk not of a rabbit but a horse:

Then the question, 'Of what is *ma* "horse" the name?' has a natural answer: the mereological set of horses. 'Horse-stuff' is thus an object (substance or thing-kind) scattered in space-time. . . . As a result, Chinese theories of language tend to treat adjectives as terms denoting mass substantives; for example, red is the stuff that covers apples and the sky at sunset.[2]

---

[1] Quine (1969), p. 52.   [2] Hansen (1983), p. 35. Hansen does acknowledge his debt to Quine.

The thought of a mereological set consisting of apples and skies at sunset scattered through space and time is perhaps a formally and physically feasible construct, since the great mathematician Stanislaw Lesniewski developed his very abstract and complex theory of mereology (thus introducing and defining the technical term 'mereological set' which Hansen employs), and since Einstein has been able to establish a relationship between space and time. However, I find nothing in traditional Chinese literature that remotely suggests that the ancient Chinese ever thought of anything like Lesniewski's mereology, or of apples and evening skies as one object scattered through time and space. In any case, Hansen does not provide any of the necessary detailed philological evidence to prove that the Chinese did think in this way.

We note that reactions to Hansen's suggestion vary considerably. A. C. Graham writes:

Chad Hansen in his *Language and logic in ancient China* has opened the first radically new approach to the 'White Horse'. It is an application of his hypothesis that Classical Chinese nouns function like the mass nouns rather than the count nouns of Indo-European languages.[1]
We start from one of Hansen's crucial insights, that thinking with mass nouns is in terms of whole and part of which what for us are class and member are only one variety.[2]

A. C. Graham explicitly takes the mass noun hypothesis as his starting-point for a new interpretation of the White Horse Dialogue. But for all his enthusiasm Graham wisely avoids the attribution of the notion of a mereological set to the ancient Chinese when he writes:

But if he (i.e., Kungsun Lung) is thinking of horse as a mass with discontinuous parts similar in shape, and of white as a mass of discontinuous patches of colour, then for him a white horse is indeed a part of the former mass combined with part of the latter.[3]

Let us call Graham's interpretation the part–whole interpretation of the White Horse Dialogue as opposed to Hansen's mereological interpretation. In view of A. C. Graham's reticence about the concepts of a 'mereological set' and 'object', we can perhaps try to avoid attributing to Kungsun Lung such notions as that of single objects scattered through space and time and still save the essence of the part–whole interpretation for which Graham argues so eloquently. Perhaps we can take the much simpler line that Kungsun Lung treated terms like *pai* 白 'white' and *ma* 馬 'horse' as mass terms *without* thereby anachronistically envisaging an object consisting of all white patches or horsey stuff past, present and future scattered through space and time forming one single whole.

The mass noun hypothesis

We know that 'information' is a count noun in French, German, Danish, Norwegian, Russian and many other languages, but not in English. What most languages

---

[1] Graham (1986a), p. 196. [2] *Ibid.*, p. 199. [2] *Ibid.*, p. 197.

refer to as 'countables' or count nouns, may in some languages be referred to as 'non-countables' or non-count nouns.

The English word 'evidence' is not today a count noun, but in the introduction to Sir Thomas Malory, *Le Morte D'Arthur* we hear of 'many evidences of the contrary'.[1] What in a given language is a non-count noun may well at an earlier stage of that language have been a count noun.

Now one can imagine a language structured in such a way that it treats physical objects as Modern English rather than earlier English 'evidence' treats evidence, and as the English rather than the French 'information' treats information. If Classical Chinese were such a language, then we should read the Classical Chinese *i jên* 一人 ONE MAN as 'one of mankind' or *san ma* 三馬 THREE HORSE as 'three of horse-kind' even when there is no measure word (like *phi* 匹 'horse-like item of') between *san* 三 THREE and *ma* HORSE. One would thus treat *jên* and *ma* as mass nouns of the same order as *shih* 食 'food' and *shui* 水 'water'.[2] Let us call such a hypothesis about Classical Chinese nouns the mass noun hypothesis, and let us note that this mass noun hypothesis does not involve Lesniewski or the advanced mathematical notion of mercology.

A. C. Graham claims that *ma* should be interpreted as a mass term. He does not inquire whether *ma* 'horse' behaves syntactically differently from other common nouns like *shui* 'water' or like *sheng* 牲 'domestic animal', not to speak of proper nouns like *Chung Ni* 仲尼 'Confucius' and pro-nouns like *tzhu* 此 'this item'. Among the common nouns the question of a grammatical distinction between count nouns, generic nouns, and mass nouns in Classical Chinese certainly needs careful attention before we are entitled to decide whether *ma* 馬 is a mass noun or not. Let us try to identify provisionally some of the diagnostic syntactic environments that might bring out into the open any grammatical distinction that might exist between count nouns, generic nouns and mass nouns in Classical Chinese.

(1) *Count nouns*

It turns out that count nouns may not only be quantified by *to* 多 'much/many' and *shao* 少 'little/few' but also by *shu* 數 'a number of', *ko* 各 'each', *chien* 兼 'each of the objects', *mei* 每 'every'. Mass nouns are never quantified by *shu, ko, chien* or *mei*, and when they are counted by ordinary numbers, the semantics of counting is entirely different (e.g., 'three *kinds of* wine', as we shall see below).

If we study the scope of quantifiers like *ko, chien, mei* and *shu* we can make an extensive list of count nouns in Classical Chinese. Let us take *shu* 'a number of' which is frequent with measure phrases. Measure phrases are count nouns. We have *shu jih* 數日 'a number of days'[3] *shu yüeh* 數月 'a number of months',[4] *shu nien*

---

[1] I.e., evidence to the effect that King Arthur was a real historical person. Ed. Janet Cowen, Penguin English Library, Harmondsworth 1969, p. 4.
[2] I disregard the meaning 'river' for *shui* 水.   [3] *Han Fei Tzu* 30.49.2.   [4] *Han Fei Tzu* 32.57.12.

數年 'a number of years'[1] or *shu jên* 數仞 'a number of jên-lengths',[2] *shu chhih* 數尺 'a number of feet'.[3] However, we also have the common *shu hang* 數行 'a number of rows (of tears)',[4] *shu jên* 數人 'a number of people'[4] *pu kuo shu jên* 不過數人 'no more than a few people',[5] *tzhu shu tzû* 此數子 'these several men',[6] *tzhu shu wu* 此數物 'these several things',[7] *chih shu thi* 之數體 'these several limbs',[8] *shu kuei shen* 數鬼神 'the several ghosts and spirits',[9] *shu pai jên* 數百人 'several hundreds of people',[10] *tzhu shu chieh* 此數節 'these several accomplishments',[11] *tzhu shu kuo* 此數國 'these several states',[12] *shih shu chü* 是數具 'these several tools',[13] *shu khou chih chia* 數口之家 'a home of a number of persons',[14] *shu shih* 數世 'a number of generations'.[15]

Huan 患 'disaster' turns out to be a count noun because we have '*tzhu shu huan* 此數患 'these several disasters', but we also have *chhi huan* 七患 'seven kinds of disastrous behaviour'.[16] *Huan* is thus used both generically, and as a count noun.

Some nouns are used both as mass nouns and as count nouns. Thus we have *shu chin* 數金 'a number of units of money'[17] in spite of the fact that *chin* 金 is also often used as a mass noun meaning 'metal'. This, I think, is simply a case of lexical ambiguity.

Chhê 車 'cart' is sometimes used as a classifier indicating a container–measure: 'a cart-load'. As a count noun meaning 'cart' it can in turn be counted with *shu*: *tê chhê shu shêng* 得車數乘 GET CART A-NUMBER-OF ITEMS 'get a number of carts'.[18] We can also count them like ordinary count nouns with an itemising classifier: *ko chhê san pai liang* 革車三百輛 WAR-CHARIOTS THREE HUNDRED VEHICULAR-ITEMS 'three hundred war chariots',[19] not: *san pai liang ko chhê* 三百輛革車. One might at first sight suspect that the complexity of the phrase may be a motive for the choice of this construction. But we regularly have *chhê i shêng* 車一乘 VEHICLE ONE VEHICULAR-ITEM 'one chariot',[20] *i chhê shih shêng chih Chhin* 以車十乘之秦 WITH VEHICLE TEN VEHICULAR-ITEM GO-TO CHHIN 'go to Chhin with ten chariots'.[21]

Some count nouns are naturally counted in pairs, for example *ko lü wu liang* 葛屨五兩 FIVE PAIR KO-SHOE 'five pairs of dolichos shoes'.[22] Others are counted in

[1] *Lun Yü* 7.17.   [2] *Lun Yü* 19.23.   [3] *Mêng Tzu* 7B34.
[4] *Mêng Tzu* 2B2. Cf. *Tso Chuan*, Duke Chao 8, ed. Legge, p. 621, line 5.
[5] *Tso Chuan*, Duke Ai 12, ed. Legge, p. 836, line 10.   [6] *Chuang Tzu* 8.21.   [7] *Chuang Tzu* 23.14.
[8] *Lao Tzu Chia Pên Chüan Hou Ku I Shu Shih Wên* 老子甲本卷後古佚書釋文 p. 19b, bamboo strip no. 318.
[9] *Mo Tzu* 31.58, 31.60, 31.73.   [10] *Han Fei Tzu* 30.39.5.
[11] *Hsün Tzu* 7.9, ed. Liang Ch'i-hsiung, p. 69.
[12] *Hsün Tzu* 15.36.   [13] *Hsün Tzu* 16.67.   [14] *Mêng Tzu* 1A3.
[15] *Tso Chuan*, Duke Hsiang 31, Hsi 33, ed. Legge, p. 222, line 12 *et passim*.
[16] *Chuang Tzu* 20.38 and *Mo Tzu* 5.1.   [17] *Chuang Tzu* 1.39.
[18] *Chuang Tzu* 32.23; cf. Watson (1964), p. 356 who translates: 'four or five carriages'. *Chhê* 車 'carts' are very frequently counted with the itemiser *shêng* 乘 throughout pre-Chhin literature.
[19] *Mêng Tzu* 7B4; cf. also 3B4 and 7B34.
[20] *Tso Chuan*, Duke Chheng 18.2; cf. Couvreur (1951b), vol. 2, p. 167. Cf. also *Shui Hu Ti Chhin Mu Chu Chien* 睡虎地秦墓竹簡 p. 58 *et passim*, for examples of itemized counting with postposed number phrase.
[21] *Lü Shih Chhun Chhiu* 12.5, ed. Chhen Chhi-Yu, p. 641.
[22] *Shih Ching* 101.2. Karlgren (1950), p. 65, misconstrues this as 'the dolichos shoes were five pairs', failing to notice the rules governing the position of individual counting phrases in Classical Chinese with which we are here concerned.

fours: *hsi ma chhien ssu* 繫馬千駟 TIE HORSE 1,000 QUADRUPLETS 'one thousand teams of four horses'.[1]

With count nouns like *ma* 馬 we find a construction like *ma san phi* 馬三匹 HORSE THREE ITEMS 'three horses',[2] but never *san phi ma* 三匹馬 THREE ITEM HORSE. We have *chhi wan phi* 騎萬匹 CAVALRYMAN TEN-THOUSAND ITEM 'ten thousand cavalrymen',[3] but never TEN-THOUSAND ITEM CAVALRYMAN. We have *ko chung êrh ssu* 歌鐘二肆 MUSICAL BELL TWO SET-ITEM 'two sets of musical bells',[4] but never TWO SET-ITEM MUSICAL-BELL.[5] We have common constructions like *yu ma chhien ssu* 有馬千駟 HAVE HORSE 1,000 QUADRUPLETS 'have one thousand teams of four horses each',[6] *yu ma êrh shih shêng* 有馬二十乘 HAVE HORSE TWICE TEN CARRIAGE (-TEAM) 'have twenty carriage-teams of horses'.[7] When count nouns are counted by itemising (or set-identifying) classifiers, the number phrase always comes after the main noun in Classical Chinese. I have found dozens of examples of this, and not a single one where the itemising classifier comes before the main noun.

Container ('mass') classifiers like *pei* 杯 'cup', on the other hand, *can* freely occur in front of the main noun, as in *i pei shui* 一杯水 ONE CUP WATER 'one cup of water', as we shall see below.

The count noun *jên* 人 'man, person, individual' may serve as an itemising classifier, as when 'Shun had five ministers' is expressed by Confucius as *Shun yu chhên wu jên* 舜有臣五人 SHUN HAD MINISTER FIVE INDIVIDUAL.[8] (Compare *Chia hsiao wei ying liu chhih, yu ma i phi* 甲小未盈六尺有馬一匹 'When X was less than six *chhih* tall, he had one horse'.[9]) Cases of this sort where *jên* may be construed as an itemising classifier for humans are very common, and the number phrase always comes after the main noun, e.g., *yung shih i jên* 勇士一人 VALIANT KNIGHT ONE INDIVIDUAL 'one courageous knight',[10] *mei chhieh êrh shih jên* 美妾二十人 BEAUTIFUL CONCUBINE TWENTY INDIVIDUAL 'twenty beautiful concubines',[11] *yu chhieh êrh jên* 有妾二人 HAVE CONCUBINE TWO INDIVIDUAL 'he had two concubines',[12] *yu tzû san jên* 有子三人 HAVE SON THREE INDIVIDUAL 'have three sons'.[13] We also have slightly more complex structures like this: *tshung li wu pai jên i shang* 從吏五百人 已上 ATTENDANT OFFICIAL FIVE HUNDRED ABOVE 'more than five hundred attendant officials'.[14] In the newly discovered law texts of the −3rd century we read *nü tzû i jên tang nan tzû i jên* 女子一人當男子一人 WOMAN ONE PERSON CORRESPOND-TO MAN ONE PERSON 'one woman is reckoned as equivalent to one man'.[15]

---

[1] *Mêng Tzu* 5A7.

[2] Cf. *Tso Chuan*, Duke Chuang 18, for the phrase *ma san phi* 馬三匹. Compare also Duke Hsüan 2, ed. Couvreur (1951b), vol. 1, p. 565, for *wên ma pai ssu* 文馬百駟, again with the postposition of the classifier.

[3] *Chan Kuo Tshê*, no. 393, ed. Chu Tsu-Keng, p. 1364. I find a similar construction in *Wu Tzu* 吳子, ch. 6, ed. Li Shuo-Chih and Wang Shih-Chin, p. 104.

[4] *Tso Chuan*, Duke Hsiang 11.10; cf. Couvreur (1951b), vol. 2, p. 274.

[5] For details see Wang Li (*1958*), pp. 240f., as well as Cikoski (1970), pp. 101f.

[6] *Lun Yü* 16.12.   [7] *Tso Chuan*, Duke Hsiang 22; ed. Legge p. 493, line 18.   [8] *Lun Yü* 8.20.

[9] *Shui Hu Ti Chhin Mu Chu Chien* 睡虎地秦墓竹簡, p. 218.   [10] *Chuang Tzu* 5.11.

[11] *Han Fei Tzu* 34.13.32.   [12] *Han Fei Tzu* 22.31.2.   [13] *Han Fei Tzu* 35.25.5; cf. 38.7.19.

[14] *Wei Liao Tzu* 尉繚子 24, ed. Chung Chao-Hua, p. 77.

[15] *Shui Hu Ti Chhin Mu Chu Chien* 睡虎地秦墓竹簡, p. 75. There are several similar instances on the preceding page.

(Note the 'suffix' *tzû* 子 here, which would seem to be limited to count nouns as it spreads in the language.) The point about *jên* 'man' being used as a classifier is particularly useful in our context, since it is so common and *always* comes *after* the noun counted.

Not all count nouns are countable by classifiers: for example the noun *jên* 'man, individual' itself is very often counted, but never with a classifier, and neither is *shêng* 乘 'carriage', *pu* 步 'pace' or any of the nouns obviously referring to measures, quantities or units of any kind. We have the count noun *yen* 言 'word, sentence'[1] which is counted, but never with a classifier, versus the more general *yü* 語 'talk', which is rarely, if at all, counted.[2] There is ample scope for further subcategorisation of Classical Chinese count nouns. For our present purposes the present rough first orientation must suffice.

(2) *Generic nouns*

Consider the following pairs of Classical Chinese words:

1. *niao* 鳥 (count noun) 'bird', versus *chhu* 畜 (generic noun) 'domestic animal'.
2. *wang* 王 (count noun) 'king' versus *shêng* 牲 (generic noun) 'domestic animal'.
3. *jên* 人 (count noun) 'man' versus *min* 民 (generic noun) 'common people'.[3]

Unlike count nouns, generic nouns are never modified by *shu* 數 'a number of'. Like count nouns, but unlike mass nouns, generic nouns can be modified by *chhün* 群 'the whole flock/crowd/lot of', *chu* 諸 'the various', *chung* 眾 'all the many', *wan* 萬 'the ten thousand, i.e., all the various', *pai* 百 'the one hundred, i.e., all the', etc. We have *chung min* 眾民 'the many people',[4] *shu min* 數民 'the numerous people',[5] and *wan min* 萬民 'the ten thousand people (never: ten thousand common individuals)',[6] *chao min* 兆民 'the innumerable people'.[7] Generic nouns are never counted with classifiers in the manner of mass nouns with container classifiers (*i pei shui* 一杯 水 'one cup of water')[8] or in the manner of count nouns with itemising classifiers (*ma san phi* 馬三匹 'three horses').[9] They thus constitute a proper subcategory of nouns in their own right.

---

[1] Cf. *Chan Kuo Tshe*, ed. Chu Tsu-Keng, p. 475, where *san yen* means 'three words'. Ssuma Chhien speaks of the *Tao Tê Ching* 道德經 consisting of *wu chhien yen* 五千言 'five thousand characters' (*Shih Chi*, ch. 63, ed. Takigawa, p. 6; cf. also ibid., p. 16). *Yen Tzu Chhun Chhiu* 晏子春秋 6.16, ed. Wu Tse-Yü, p. 407, uses *san yen* 三言 to mean 'three sentences', similarly for *Han Shih Wai Chuan* 2.9, ed. Hsü Wei-Yü, p. 41.

[2] The distinction between count nouns and mass nouns in Classical Chinese, like that between ergative and non-ergative verbs, deserves a detailed treatment in its own right. Against the background of such a detailed study the hypothesis that Chinese nouns are mass nouns may appear in a new light.

[3] Note that *min* 民 is not a collective noun for a certain people: *fan min* 凡民 (*Mêng Tzu* 7A10) means 'vulgar people', not 'a common people'.

[4] *Mêng Tzu* 7A2.   [5] *Mêng Tzu* 1A2 *et passim.*

[6] *Kuo Yü*, ed. World Book Co., p. 282; *Han Fei Tzu* 8.4.22.   [7] *Kuo Yü*, ed. World Book Co., p. 405.

[8] This construction will be discussed below under the heading mass nouns.

[9] The apparent counter example, *Tso Chuan* Duke Min 2.7 (ed. Yang Po-Chün, p. 267 line 1), turns out to be spurious. Cf. also Couvreur (1951b), vol. 1, p. 221.

The semantics of counting is different in generic nouns and in count nouns. *Tzhu liu jên chê* 此六人者 means 'these six individuals (or people)',[1] whereas a little further on in the same text *tzhu liu min chê* 此六民者 means 'these six kinds of people'.[2] *Wu min* 五民, again, are 'five kinds of people'[3] and certainly not five individuals of lower rank. *Ssu min* 四民 are explained as 'the four categories of people'.[4] *I min* 一民 is mostly a verb-object phrase meaning 'unify the people'. When *i min* is nominal, it means 'one population' or 'the whole population'.[5] The standard expression for 'one commoner' is not *i min* but *i fu* 一夫.[6] This is what the phrase means in Chia I, *Hsin Shu*.[7] *I jên* 一人, on the other hand, means 'one person'.[8] *Jên* 人 'individuals' are very often counted, and as far as I can see they are never counted generically as so-and-so many 'kinds of persons'. On the other hand I seem to be unable to find *min* 民 'commoner' ever counted as such, except by unspecific numbers such as *wan* 萬 'the ten thousand'.[9]

Liu wang 六王 must mean 'six kings', and *liu ma* 六馬 will normally mean 'six horses', whereas *wu shêng* 五牲 must mean '(the) five kinds of domestic animals' and *liu chhu* 六畜 must mean 'six kinds of domestic animals'.[10]

Already Chêng Hsüan 鄭玄 (+127 to +200) defines *chiu I* 九夷 of *Lun Yü* 9.14 not as nine individual barbarians but as *tung fang chih I, yu chiu chung* 東方之夷有九種 'the I-barbarians of the East of which there are nine kinds'. Similarly *wu Ti* 五狄 has to mean 'the five *kinds of* Ti-barbarians', just as *ssu I* 四夷 in *Mêng Tzu* 1A7 are 'the five *kinds of* I-barbarians', since *Ti* 狄 'Ti-barbarian' and *I* 夷 'I-barbarian' are generic nouns:

The three principal ministers (*san kung* 三公) were in front of the middle stairs . . . As for the states of the various earls (*chu po chih kuo* 諸伯之國), they were positioned to the West of the Western stairs. . . . As for the state of the Nine (kinds of) I-barbarians (*chiu I chih kuo* 九夷之國),[11] they were positioned outside the Eastern Gate. . . . As for the states of the Eight (kinds of) Man-barbarians (*pa Man chih kuo* 八蠻之國), they were positioned outside the Southern

---

[1] *Han Fei Tzu* 44.2.7 and 44.4.2. We have *tzhu êrh jên* 此二人 'these two individuals' (*Han Fei Tzu* 12.5.4), *tzhu san jên* 此三人 'these three individuals' (*Han Fei Tzu* 14.7.72).

[2] *Han Fei Tzu* 46.1.19 and *ibid.*, 46.1.39.   [3] *Shang Chün Shu*, ch. 6, ed. Kao Heng, p. 66.

[4] *Kuo Yü*, ed. World Book Co., p. 161; cf. also *Shu Ching*, ed. Legge, *Chinese Classics*, vol. 3, p. 530.

[5] *Kuan Tzu*, ed. Tai Wang, vol. 2, p. 98, line 3, and *Mêng Tzu* 2A1. Note that in *Mêng Tzu i min* 一民 is quantified by *mo pu* 莫不 and can therefore not be taken to mean 'one humble person'. Legge (1872), p. 518, construes this grammatically correctly: 'There was not one of all the people who was not his subject.'

[6] *Tso Chuan*, Duke Hsi 15.5, ed. Yang Po-chün, p. 355 *et passim*.

[7] Ed. Qi Yuzhang, p. 1044, but in the parallel *Shuo Yüan* 1.6, ed. Chao Shan-I, p. 4, ed. Hsiang Tsung-Lu, p. 5, we have an instance of *i min* meaning 'a single commoner', followed by a strictly parallel *i jên* 一人 'one person'. It would be interesting to investigate whether and when this usage became current.

[8] See, e.g., *Han Fei Tzu* 8.7.30, 15.1.23 and 20.20.3.

[9] The case of *chhên* 臣 (count noun) 'minister' versus *li* 吏 (generic noun) 'official' is especially interesting and puzzling. *Tzhu wu chhên chê* 此五臣者 means 'these five ministers' (*Chan Kuo Tshe*, no. 185, ed. Chu Tsu-Keng, p. 770), whereas *wu li* 五吏 means 'five kinds of officials' (*Tso Chuan*, Duke Hsiang 25). However, I have found one isolated instance where in fact *li* is used as a count noun. In *Lü Shih Chhun Chhiu* 18.8 we find two officials first mentioned as *li êrh jên* 吏二人 OFFICIALS TWO MEN, and then in the end we hear that *êrh li* 二吏 'the two officials' made a report. One needs to study whether *li* is really ambiguous between a generic noun and a count noun reading, or whether this instance is just a stray case motivated by the special context. In any case *wu chhên* 五臣 apparently never has that generic reading.

[10] *Han Fei Tzu* 37.13.45. Cf. also *wu shêng* 五牲 'the five domestic animals' (*Han Fei Tzu* 31.13.39).

[11] Cf. *Han Fei Tzu* 22.25.7.

Gate. . . . As for the states of the Six (kinds of) Jung-barbarians (*liu Jung chih kuo* 六戎之國), they were positioned outside the Western Gate. . . . As for the states of the Five (kinds of) Ti-barbarians (*wu Ti chih kuo* 五狄之國), they were positioned outside the Northern Gate.[1]

Jên 人 'man' and *min* 民 'common people' (as well as *ti* 狄) all refer to humans. But these words refer to humans in radically different ways. *Min* and *ti* are generic nouns which are counted only generically, not by individuals. *Jên* is an ordinary count noun. The distinction between count nouns and generic nouns in Classical Chinese deserves to be studied in more detail.

(3) Mass nouns

Compare the following pairs:

1. *tan* 簞 (count noun) 'basket' versus *jou* 肉 (mass noun) 'meat'.
2. *shu* 樹 (count noun) 'tree' versus *hsin* 薪 (mass noun) 'firewood'.
3. *fu* 斧 (count noun) 'axe' versus *thieh* 鐵 (mass noun) 'iron'.

With mass nouns like *shui* 水 we regularly find container classifiers which are designations of containers: *i pei shui* 一杯水 ONE CUP WATER 'one cup of water',[2] *i phiao yin* 一瓢飲 ONE LADLE DRINK 'one ladleful of drink',[3] *i hu chiu* 一壺酒 ONE POT WINE 'one pot of wine',[4] *i tan shih* 一簞食 ONE BASKET FOOD 'one basketful of food',[5] *i chhieh chin* 一篋錦 ONE BOX BROCADE 'one box of brocade',[6] *i chhê hsin* 一車薪 ONE CART FIREWOOD 'one cart-load of firewood'.[7] We also find other measuring but not itemizing classifiers: *i ku thieh* 一鼓鐵 ONE KU-MEASURE OF IRON 'one *ku* of iron',[8] *shih shu hsin* 十束薪 TEN BUNDLE FIREWOOD 'ten bundles of firewood',[9] *i ping kan* 一秉稈 ONE HANDFUL STRAW 'one handful of straw'.[10]

I suppose one should be able to say *i chhê jên* 一車人 'a cartload of people' or the like, but for some reason I have not been able to find good examples of this kind. If I found such a case, I would be inclined to insist that this syntactic frame converts the count noun into a mass noun, just as a negation *wu* 毋 converts a pronoun *wo* 我 into a verb, as in *wu wo* 毋我 'he avoided being self-centred'.[11] In spite of examples of this sort it remains useful and important to think of *wo* as a pronoun which under certain special circumstances comes to take on a verbal function.

We have passages like this:

He takes one plate of meat (*i tou jou* 一豆肉) and feeds the knights with the rest.[12]
If you taste one piece of meat (*i luan jou* 一臠肉), you know the taste of the whole pot and the flavour of the whole tripod.[13]
Shu Ku-Yang took a beaker of wine (*i shang chiu* 一觴酒) and offered it up.[14]

---

[1] *Li Chi*, ed. S. Couvreur, vol. 1, pp. 725ff.
[2] *Mêng Tzu* 6A18. Note that in *ssu shui* 四水 'the four rivers' we have a different lexical meaning of *shui* 水.
[3] *Lun Yü* 6.11.
[4] *Kuo Yü*, ed. Ku Chi, p. 635. In the same context we also find *êrh hu chiu* 二壺酒 'two pots of wine'.
[5] *Lun Yü* 6.11. [6] *Tso Chuan*, Duke Chao 13; cf. Couvreur (1951b), vol. 3, p. 229.
[7] *Mêng Tzu* 6A18. For further examples see Cikoski (1970), pp. 101ff.
[8] *Tso Chuan*, Duke Chao 29, fu 5; cf. Couvreur (1951b), vol. 3, p. 456. [9] *Chuang Tzu* 4.85.
[10] *Tso Chuan*, Duke Chao 27.4. [11] *Lun Yü* 9.4. [12] *Han Fei Tzu*, 34.7.24; cf. Liao (1939), p. 90.
[13] *Lü Shih Chhun Chhiu* 15.8, ed. Chhen Chhi-Yu, p. 935. [14] *Han Fei Tzu* 10.2.7.

There may be no explicit number in this sort of construction:

If you leave goblets of wine and platters of meat (*chih chiu tou jou* 卮酒豆肉) in the inner court . . .[1]

Some mass nouns frequently occur with fixed numbers, the most well-known being *wu ku* 五穀 'five kinds of grain'. *I ku* 一穀 means 'one kind of (not: corn of) grain'.[2] *San chiu* 三酒 means 'three kinds (not: bottles or measures) of wine',[3] *ssu yin* 四飲 means 'four kinds of drinkables' (not: 'three individual portions of drink'),[4] *wu jou* 五肉 'the five kinds (not: portions) of meat',[5] *wu tu* 五毒 means 'five kinds (not: doses) of poison',[6] *wu chhi* 五氣 'the five kinds (not: portions) of ether',[7] *liu chhi* 六氣 is 'the six kinds (not: portions) of ether'.[8]

It is perfectly true that *chhi* 氣 illustrates the usefulness of the mass noun analysis. But this is because *chhi*, in sharp and clear contrast to the count noun *ma* 馬 is not a count noun. The exact classification of *chhi* remains a problem to be investigated.

We conclude that there is a reasonably clear grammatical distinction in Classical Chinese between count nouns, generic nouns and mass nouns. It remains entirely unclear in what sense *ma* 'horse' can be classified as a mass noun (like *shih* 食 'food'), although in point of fact I have come upon a single isolated and late instance where *ma* is indeed used generically.[9]

The problem of the plural

In English, as in many other languages, the opposition mass noun versus count noun affects the semantics of the plural. 'Horse', unlike 'tea', is a count noun in English. 'Horses' refers to several *individual* horses. 'Teas' is much rarer and would refer to several *kinds* of tea. This difference is essential.

The post-Classical rise of Chinese plural morphemes (or suffixes) roughly coincides with the emergence of the pre-nominal itemizing classifiers like *phi* 匹 as in *san phi ma* 三匹馬 THREE ITEM HORSE 'three horses':[10]

When Wang Tzû-Shen was a few years old, he was once watching his father's pupils (*chu mên shêng* 諸門生) play . . .
The pupils (*mên shêng pei* 門生輩), not showing any respect for him as a small child, said to him: . . .[11]

---

[1] *Han Fei Tzu* 34.29.3; cf. Liao (1939), vol. 2, p. 113. Cf. also *Tso Chuan*, Duke Ai 7.4, *shu chin* 束錦 'bundles of brocade'; *Mêng Tzu* 1A7, *yü hsin* 輿薪 'a cartload's worth of firewood'. We note here an isolated instance in the grammatically very idiosyncratic *Kung Yang* commentary (Duke Hsi 33): *phi ma chih lun wu fan chê* 匹馬隻輪無反者 'not a single horse or a single wheel returned'.

[2] *Kuan Tzu*, ed. *WYWK*, vol. 3, p. 91. Cf. also *Ku Liang Chuan*, Duke Hsiang 24.13, were *ku* 穀 is counted by kinds.

[3] *Chou Li*, ch. 2, ed. Lin Yin, p. 49.   [4] *Ibid*.

[5] *Kuan Tzu*, ed. Tai Wang, vol. 2, p. 75, line 8.   [6] *Chou Li*, ch. 2, ed. Lin Yin, p. 47.

[7] *Chou Li*, ch. 2, ed. Lin Yin, p. 46; cf. also *Ho Kuan Tzu*, ch. Tu Wan, ed. *Tzu Hui*, p. 42.

[8] *Tso Chuan*, Duke Chao 1; *Kuo Yü*, ed. World Book Co., p. 96; *Chuang Tzu* 1.21, 11.46, 11.47; *Kuan Tzu*, ed. Tai Wang, vol. 2, p. 16.

[9] *Chou Li*, ed. Lin Yin, p. 339.

[10] If such classifiers had been obligatory or at least common in Classical Chinese, the interpretation of Chinese count nouns as mass nouns might at least have had some initial plausibility.

[11] *Shih Shuo Hsin Yü* 5.59, ed. Yang Yung, p. 259; cf. Mather (1976), p. 176.

'My little boys (i.e., his nephew, Hsüan, and his younger brother, Shih) have inflicted a crushing defeat on the invader (*hsiao êrh pei ta pho tsei* 小兒輩打破賊).'[1]

Such use of plural suffixes seems uncomfortable for a mass noun analysis of the nouns so modified. For if *mên shêng* 門生 'student' is a mass noun meaning 'student-kind', its plural form (*mên shêng pei*) should presumably come to mean 'student-kinds', 'kinds of students'. Exactly similar observations apply to *hsiao êrh* 小兒 little boy'.

Parts and wholes

A part–whole interpretation of the notion *ma* 馬 'horse' would seem to be doomed to failure because it either has to introduce an anachronistic mathematical construct (the concept of one spatio-temporally discontinuous object, the horsey mass), or because there is no whole and therefore no horsey part. For the notion of a part does not seem to make sense without that of a whole. It is of the essence of a part to be a part of a whole.

A comparison with English will bring out our point. Compare the mass noun *luggage* or the mass noun *tea*. When we use the mass noun *luggage*, we are not thereby explicitly or implicitly committed in any way to an ontology of some giant luggage-like mass-like whole consisting of all manner of luggage past, present and future as proper parts.

If one asks 'Please make me a cup of tea' (in Chinese or in English), one is not committed to an ontology of a discontinuous sea of tea scattered through space and time, of which one wishes to obtain a certain part. A cup of tea may properly only be called a part of that mass of tea which is in the tea-pot, although logicians like Lesniewski would have us think of both the cup of tea and the pot of tea (as well as the tea-cups and the tea-pots) as a proper part of some very abstract object consisting of all manner of tea (tea-cups and tea-pots) past present and future. On the other hand, when one is asking for a cup of tea, one is simply contemplating the *possibility* of some more tea being brought into existence for one's benefit. In no sense is one asking for something conceived as actually existing in an over-arching time-space dimension. One imagines that this cup of tea may never come into existence, but one is presumably hoping that it will.

If, on the other hand, we completely give up the notion of a horsey whole and of horsey parts, and if we simply cling to the central idea that the Chinese always thought of horses as a 'stuff-kind' (and presumably of vegetation rather than of plants, of offspring rather than of children), then such an account is directly refuted by the above preliminary survey of the neat grammatical contrasts between count nouns, generic nouns and mass nouns which shows that the Chinese did make a reasonably clear overt grammatical difference between names of kinds of stuff, names of kinds of objects, and names of individual objects.

---

[1] *Shih Shuo Hsin Yü* 6.35, ed. Yang Yung, p. 286; cf. Mather (1976), p. 192.

The objection that the distinction between mass nouns and other nouns is not always clear in Classical Chinese for all nouns, and that there are many ambiguous cases, carries no weight at all. It applies with equal force to languages like English where the division between count nouns and non-count nouns is recognised as grammatically basic.

Compare the use of 'analysis' in the following dialogue:

'Do you like this analysis?'
'No, I don't like analysis.'[1]

None of this directly affects the possibility of interpreting *pai ma* 白馬 as a whole consisting of one part 'white' and another part 'horse', which I take to be explicitly advocated by Kungsun Lung. The aggregate of 'white' and 'horse' is taken by Kungsun Lung not to be 'horse'. If we detach Graham's interpretation from the untenable mass noun hypothesis which he espouses we can still attribute to Kungsun Lung the view that the aggregate whole (*pai ma*) is not (identical with) its constituent part (*ma* 馬).

The crucial point that then remains is that *pai* 白 and *ma* constitute parts of very different types and that the relation between these is by no means symmetrical.

In any case, however we are to understand the White Horse Dialogue, if Kungsun Lung's aim in writing it was to construct a logically stimulating intellectual teaser, we can all cheerfully agree that he has succeeded.

(3) Hsün Tzu's 荀子 Logic

Hsün Tzu's essay on The Right Use of Names is the most coherent and sustained discursive survey of the problems of logic that has come down to us from ancient Chinese times. In expounding Hsün Tzu's logical thoughts we shall do best to follow his own exposition, adding our own explanatory remarks where appropriate.[2]

Hsün Tzu viewed linguistic conventions as a historical phenomenon and as a social institution. He distinguished between those conventions introduced by decree (the legal terminology of the Shang court, the administrative terminology of the Chou court) and the miscellaneous terms (*san ming* 散名) which he also saw as fixed by kings, but where the kings based themselves on various local customs.

When the later kings established (*chhêng* 成) names, the names for punishment were derived from the Shang; the names for ranks of nobility were derived from the Chou; and the cultural names (*wên ming* 文名) were derived from the (books of) ritual. Now as for the sundry names (*san ming*) applied to (*chia* 加) the various objects, these were haphazardly agreed upon in accordance with the established habits of the Chinese people. Using these names, people from distant regions and of different habits could communicate with each other (*thung* 通).[3]

---

[1] Quirk *et al.* (1985) section 5.4, *Nouns with dual class membership*, p. 247.

[2] There are a number of translations of the chapter on The Right Use of Names (Duyvendak (1924), Dubs (1928), Mei (1951), Köster (1967), and Watson (1963)). None of these serve the needs of our present analysis. We refer the reader to the most conveniently available modern translation by Burton Watson.

[3] *Hsün Tzu* 22.1; cf. Watson (1963), p. 139.

Hsün Tzu already recognises the Chinese language as a *lingua franca*, a kind of Chinese *koinē* 'common language' which was standardised to make communication possible between different regions. Hsün Tzu's interest in language (and logic) was primarily social: he thought that by managing the names (*chih ming* 制名) the ruler could ensure that names were fixed and realities distinguished (correspondingly). The aim of the institution and managing of names, then, was bringing unity and proper organisation to the people. Sorting out reality by means of names was subsidiary to that.

Hsün Tzu thought that the correct managing of names would lead to a *Sprachregelung* 'regulated discourse' where everybody, through speech and action, would articulate the principles inculcated through the managing of names (*chih ming* 制名), through the correct use of names (*chêng ming* 正名).

Hsün Tzu felt he lived in an age of decline from a golden past, but he does not simply look to the golden age for a remedy. He writes:

If a king deserving of that title rises to power, he is bound to follow old names in some cases and to make new ones in other cases.[1]

Hsün Tzu then clearly states the three main directions of his inquiries into names:

Consequently, the purpose (*so wei* 所為) of names, the reasons why we treat things as different or the same, and the crucial points in the management of names (*chih ming chih shu yao* 制名之樞要) must be properly investigated.[2]

Each of these three areas is treated in his opening section. Let us take the purpose of names first. This purpose turns out to be practical rather than theoretical:

If the distinction between noble and base will become unclear, and men will not discriminate properly between things that are the same and those that are different, then their intentions will inevitably not be properly communicated and understood, and undertakings will inevitably be plagued with difficulty and failure. For this reason the wise man is careful to set up the proper distinctions and to institute names so that they point to (the proper) realities. In this way he makes clear the distinction between noble and base and discriminates properly between things that are the same and those that are different. If this is done, there will be no danger that the ruler's intentions will be improperly communicated and understood, and his undertakings will suffer no difficulties or failure. This is the purpose of names.[3]

Hsün Tzu focuses on the need for a common human internal structure of representation if we are to distinguish between things and communicate about them:

On what basis do we treat things as different or as the same? On the basis of the senses. For all people of the same kind or of the same essential make-up (*chhing* 情) the senses will represent (*i* 意) things in the same way. Therefore, by comparing and assessing things men commonly agree on names and thus fix them for the future (*chhi* 期) for each other.[4]

---

[1] *Hsün Tzu* 22.11; cf. Watson (1963), p. 141.   [2] *Hsün Tzu* 22.12; cf. Watson (1963), p. 141.
[3] *Hsün Tzu* 22.13; cf. Watson (1963), p. 141.   [4] *Hsün Tzu* 22.15; cf. Watson (1963), p. 142.

The mind, as opposed to the senses, has discriminative knowledge (*chêng chih* 徵知).

The mind (*hsin* 心) has discriminative knowledge (*chêng chih*). In the case of discriminative knowledge it is acceptable that one knows sounds on the basis of one's ears. It is acceptable that one knows shapes on the basis of one's eyes. But for discriminative knowledge to be acceptable it must necessarily depend upon direct contact of the senses (*thien kuan* 天官) with the kind in question. The Five Senses come in contact with things but do not know. If the mind discriminatively assesses these things (*chêng chih* 徵知) and comes up with no explanation, then everyone will call this ignorance. So much for the question of how we treat things as different or as the same.[1]

We turn to the third question Hsün Tzu has promised to answer: what are the crucial things in the management of names?

Then only (after taking evidence from the senses) do we give names (*ming* 名) to things. If things are the same, we use the same name. If they are different, we use different names.[2]

Hsün Tzu formulates a principle of economy: use complex names only when this is necessary in order to be understood.

If a single (non-composite) name is sufficient to make oneself understood, then we use a single name. If a simple name is not sufficient to make oneself understood, one uses a composite name.

When the simple name and the composite name are compatible with each other (*wu so hsiang pi* 無所相避), then one uses the general (simple) term. Even with the general term (*kung* 公) there will be no contradiction.

To know that different objects have different names, and therefore cause every different object to have a different name, to avoid confusion, is no better than causing every different object to have the same name.[3]

Hsün Tzu is aware that in some cases we fail to distinguish different things but actually refer to all things indiscriminately:

Therefore the myriad things may be many, but sometimes we wish to pick out all of them (*phien chü chih* 遍舉之), and then we call them 'things'. 'Things' is a comprehensive general term (*ta kung ming yeh* 大公名也). We push (*thuei* 推) and generalise, and we go on generalising until we cannot generalise any more: only then do we stop.

Sometimes we wish to pick out only some but not other items (*phien chü chih*). Then we call something a bird or a beast. 'Bird' and 'beast' are comprehensive specific terms (*ta pieh ming yeh* 大別名也). We push on (*thuei*) and specify, and we go on specifying until we cannot specify any more: only then do we stop.[4]

Names have their reference not inherently but by convention:

Names are not inherently appropriate (*ku i* 固宜). We give names by establishing a convention. When the convention is settled and a habit is formed, we call a name 'appropriate' and we call it 'inappropriate' when it is at variance with the convention. Names do not

---

[1] *Hsün Tzu* 22.19; cf. Watson (1963), p. 142. [2] *Hsün Tzu* 22.21; cf. Watson (1963), p. 143.
[3] *Ibid.* [4] *Hsün Tzu* 22.23; cf. Watson (1963), pp. 143f.

inherently have a (corresponding) reality (*ming wu ku shih* 名無固實). It is by convention that we give names to realities. When the convention is settled and a habit is formed, we speak of 'a name of a reality'.[1]

Hsün Tzu then addresses the question whether some names or naming conventions may be better than others.

Some names are inherently good. When they are straight to the point, easy to understand, and not contradictory, we call them 'good names'.[2]

Given naming conventions, Hsün Tzu raises the question what is to count as *one* object or reality to which a name is to be applied or not applied. Hsün Tzu writes:

Of things some are of the same form but in different places; others are of different shape but in the same place. These can be separated. Those things which are of the same shape but in different places, even though they can be combined, are called 'two objects/realities'. When the shape changes and when an object/reality without separating out becomes different, we speak of a transformation. When there is transformation and no separating out we speak of 'one object/reality'. This is how one examines the objects involved in states of affairs and fixes their number. So much about the crucial things in the management of names.[3]

This is where Hsün Tzu's tripartite systematic exposition on naming ends.

Hsün Tzu continues with a brief account of fallacies. These are divided into three groups: 1. those which confound names by confusion in the use of names; 2. those which confound names by confusion in the use of realities; 3. those which confound realities by confusion in the use of names. These fallacies correspond to the 'three areas' mentioned at the beginning of the chapter.

'To be insulted is not disgraceful', 'The sage does not love himself', 'Killing robbers is not killing people', these (claims) confound names by confusion in the use of names. If one tests them by the purpose of having names, and observes which alternative applies generally, one can prevent these confusions.[4]

Hsün Tzu suggests that sophisms of this kind may be solved by applying the principles he has laid out in answer to the first question: what is the purpose of having names? For example, you inquire what the purpose was of instituting the name 'robber' and 'killing people' in order to decide whether 'killing robbers is not killing people' is acceptable.

'Mountains are level with abysses', 'The essential desires are few', 'Fine dishes do not improve the taste', these are cases of confounding names by confusion in the use of realities. If you test them by what one depends on to recognise similarity and difference, and observe which alternative accords, you can prevent these confusions.[5]

Hsün Tzu suggests that sophisms of this second kind may be prevented by reference to his second discipline, his answer to the second question: how does one

---

[1] *Hsün Tzu* 22.25; cf. Watson (1963), p. 144.   [2] *Hsün Tzu* 22.29; cf. Watson (1963), p. 144.
[3] *Hsün Tzu* 22.27; cf. Watson (1963), p. 144.   [4] *Hsün Tzu* 22.29; cf. Watson (1963), p. 145.
[5] *Hsün Tzu* 22.31; cf. Watson (1963), p. 145.

recognise similarity and difference? For example, one decides whether mountains and abysses are on one level by applying one's senses to these objects and deciding the matter on the basis of one's sense impressions.

The third set of sophisms is unfortunately especially hard to understand:

'You introduce yourself by what is not your name', 'The pillar has the ox', 'A horse is not a horse', these are cases of confounding reality by confusion in the use of names. If you test them by the conventions for a name and use what one accepts to show that what is rejected (*so tzhu* 所辭) is incoherent, then one can prevent these confusions.[1]

Hsün Tzu suggests that sophisms of this third kind may be prevented by reference to the conventions governing the use of words as discussed in the third part of his tripartite exposition. For example, if you are Bill, but you introduce yourself as someone else, John, then the conventions governing the application of the names 'Bill' and 'John' are contravened. There is a factual mistake. You are Bill, not John. What you say is factually untrue.[2]

All wicked explanations and devious speeches which go against the correct Way (*chêng tao* 正道) and are initiated without authority may be categorised under these three confusions.[3]

Clearly, Hsün Tzu aspired to give an account of fallacies which he thought of as exhaustive in some sense.

Hsün Tzu turns to the conditions under which we give names to things, define them, give explanations, and construct arguments:

We give names to realities only after they have become unclear; we enter conventions about names only after the names have become unclear; we give explanations of conventions only after the conventions have become unclear; we construct arguments only after the explanations have become unclear.[4]

With astonishing precision Hsün Tzu focuses on the distinction between the semantic content of a name or word and its syntactic capabilities:

The use of names consists in the realities becoming clear when the name is heard. The linking of names consists in names being strung together to make a text (*wên* 文). When the use and the linking of the words are both grasped, that is called 'knowing names'.[5]

In order to understand names one must understand both their semantics and their syntactic links, Hsün Tzu tells us. Unflinchingly, he actually proceeds to give a definition of a name:

The 'name' is that through which by convention (*chhi* 期) we group together (*lei* 類) realities.
The 'sentence' combines the names of different realities to make explicit one communicative intention (*i* 意).[6]

The next definition, of argumentation and explanation, is not as easily understood. Here is an attempt at a literal rendering:

---

[1] *Hsün Tzu* 22.32; cf. Watson (1963), p. 145.   [2] Cf. Graham (1978), pp. 233–5.
[3] *Hsün Tzu* 22.33; cf. Watson (1963), p. 146.   [4] *Hsün Tzu* 22.36; cf. Watson (1963), p. 146.
[5] *Hsün Tzu* 22.38; cf. Watson (1963), p. 147.   [6] *Ibid.*

'Argumentation and explanation' consists in, without varying the name of an object (*pu i shih ming* 不易實名) making plain the way in which it moves or is still (*tung ching chih tao* 動靜之道).

'Conventions and definitions (*chhi ming* 期名)' are what is used in argumentation and explanation.

'Argumentation and exposition' are the mind making an image of the Way.[1]

Hsün Tzu summarises the purpose of the exercise of getting the use of names right as follows:

That right use of one's terminology, the fitting use of sentences (*tzhu* 辭) has as its purpose the explaining of one's intended meaning (*chih i* 志義).

Those 'words and sentences' are messengers of the intended meaning (*chih i* 志意).[2] When they are sufficient to communicate, one dismisses them (the words and sentences).[3]

It has been pointed out that Hsün Tzu's account of names owes a great deal to Later Mohist logical theories.[4] On the other hand we must emphasise that Hsün Tzu's essay on The Right Use of Names is the most disciplined, coherent, and by far the best-organised discussion of naming that has come down to us from ancient China. As we present Hsün Tzu's views on naming, we can simply follow the opening passages of The Right Use of Names as they unfold. We do not have to impose our own extraneous organisation on the text. Hsün Tzu's principles of organisation are explicit enough. He speaks with an almost 'Aristotelian' expository lucidity.

Hsün Tzu was not especially interested in language and logic. But he does give us an inkling of what logical texts by more specialised logicians in ancient China might have looked like if they had not come down to us in such a haphazard and truncated form. What saved Hsün Tzu's account of naming for posterity was the fact that it was embedded and integrated into an important work of Confucian moral philosophy. What prevented the transmission of much of the more professional older logical literature was precisely that it was so specialised and in the end dissociated from the immediate moral and social concerns of the day.

(4) LATER MOHIST LOGIC[5]

(i) *General introduction*

The Dialectical Chapters of the book *Mo Tzu* 墨子,[6] are the most important single document in the history of pre-Chhin science. As Liang Chhi-Chhao 梁啟超 put it with emphatic repetition in 1922:

---

[1] *Hsün Tzu* 22.39; cf. Watson (1963), p. 147.
[2] I take *chih i* 志義 to be the same as *chih i* 志意. Cf. Morohashi (*1955*), p. 4361.
[3] *Hsün Tzu* 22.49; cf. Watson (1963), p. 149.
[4] A. C. Graham (1978), p. 39, goes so far as to say that it 'contains very little not in the Mohist summa'.
[5] In this sub-section all the Mohist texts are referred to according to Graham (1978). Abbreviations like 'A74', 'B14' or 'NO10' refer to A. C. Graham's reconstruction of the text which marks a decisive advance on all earlier work.
[6] Chapters 40 to 45 of the book.

If in ancient Chinese documents you want to look for something that links up with what the modern world calls the spirit of science, then there are only the Dialectical Chapters of *Mo Tzu* and that is all. There are only the Dialectical Chapters of *Mo Tzu* that is all![1]

Sun I-Jang's 孫詒讓 famous commentary on the *Mo Tzu*, first published in 1894, building on Pi Yüan 畢沅's commentary of +1783, represented the first serious Chinese attempt in modern times to grapple with the tremendous problems of the text. A number of distinguished textual studies have appeared in China.[2] Luan Thiao-Fu 欒調甫 stands out as a very systematic early student of Mohist logic. From a philological point of view, Than Chieh-Fu has made outstanding contributions. Wu Fei-Po 伍非百 (*1984*), published nineteen years after the author's death,[3] is a splendid summary and demonstration of what traditional Chinese scholarship can achieve in the interpretation of the Dialectical Chapters. Indeed this awe-inspiring work, representing many decades of dedicated research on early Chinese logic, has been an indispensable tool throughout the writing of this section on explicit logic in early China.

In the West, the Dialectical Chapters were first introduced in a German translation by Alfred Forke (1922).[4] Hu Shih's important study *Development of the Logical Method in Ancient China* which deals with, among other things, some parts of the Dialectical Chapters appeared in Shanghai at about the same time. H. Maspero (1928) made a first attempt at a coherent philological interpretation of Mohist logic in any Western language. Janusz Chmielewski (1969) applied rigorous methods of Western formal logic to Mohist logic. Joseph Needham, in Vol. 2 of *SCC*, made the first serious and sustained attempt at understanding the scientifically most important parts of the Dialectical Chapters.[5] A. C. Graham's book *Later Mohist Logic, Ethics and Science*, published in 1978, is a landmark in the history of the interpretation of Mohist logic, easily superseding anything that had been published on the subject before. Our account of Mohist logic is indebted to his study on more points than it would be practicable to mention in footnotes. We trust that the patient reader with a special interest in the subject will want to compare our conclusions with those presented in Graham's *magnum opus*.

In this Section we shall concentrate only on those parts of the Dialectical Chapters that are of relevance to logic and the philosophy of language.[6] It is thus not our ambition to give a complete account of the scientific efforts of the Later

---

[1] Liang Chhi-Chhao (*1922*), p. 1.

[2] Notably Chang Hui-Yen (*1909*), Chang Hsüan (*1919*), Liang Chhi-Chhao (*1922*), Hu Shih *et al.* (*1923*), Wu Fei-Po (*1923*), Lu Ta-Tung (*1926*), Luan Thiao-Fu (*1926*), Fan Kêng-Yen (*1934*), Than Chieh-Fu (*1935*), Chhen Wu-Chiu (*1935*), Yang Khuan (*1942*), Chan Chien-Fêng (*1979*), Than Chieh-Fu (*1958*), Kao Heng (*1966*), Chang Chhi-Huang (*1960*), Liu Ts'un-yan (*1965*), Li Yü-Shu (*1968*), Than Chieh-Fu (*1981*) and Chhên Mêng-Lin (*1983*), to name but the more important contributions. The history of logic early in the 20th century is studied in Kuo Chan-Pho (*1935*).

[3] See especially pp. 1–446 of this singularly useful work. [4] Pp. 413–526. [5] Cf. Graham (1978), p. xi.

[6] For translations from the scientific sections of the Dialectical Chapters see Vol. 4, Part 1, pp. 17–27 and pp. 81–7. These should be compared with the more recent and more detailed accounts in Graham (1978), particularly pp. 53–9, 201–315, and 369–97. For the optical sections see particularly Graham and Sivin (1973). If Graham and Sivin are right, then the Later Mohist optics represent a much more radical departure from the correlative current scientific thinking prevalent in early China than had been previously recognised.

Mohists. The sort of questions to which we seek answers are these: What were the main analytical achievements of the Later Mohist logicians? What logical questions did they ask? What logical concepts did they apply? How did their methods differ from those of their contemporaries? And why did they ultimately fail to make a significant impact on Chinese intellectual history?

It seems clear that the Later Mohist logicians were on the brink of developing a spirit of rationality and of scientific enquiry that is quite unique in Chinese intellectual history.[1] The Later Mohists deserve our careful attention.

The organisation of the Dialectical Chapters

The Dialectical Chapters of the *Mo Tzu* are divided into two parts: the canons with their explanations covering chapters 40 to 43, and two more discursive chapters, 44 and 45. The all-important division within the canons is that between the definitions (A1 to A75) and the propositions (A76 to B82).[2] The Later Mohists' ability to sustain this rigid distinction throughout the canons is quite remarkable and in itself an achievement of advanced intellectual discipline. It turns out that the order of the items in the canons is far from arbitrary, and that the individual items have to be read within their context.

The definitions and propositions can again be divided into groups, although this division is not always as neat as that between definitions and propositions. We first find six definitions concerning description (A1–A6), then thirty-three definitions concerning action (A7–A39), then twelve definitions concerning knowledge and change (A40–A51), eighteen definitions on geometry (A52–A69), and finally six definitions on disputation (A70–A75).

The first 35 propositions are concerned with procedures of description, including an opening section on ambiguity of terms (A76–B12), followed by knowledge and change (B13–B16), the sciences (B17–B31), and finally a long section concerning problems in disputation (B32–B82).

The organisation of the canons may, at first sight, look arbitrary. Indeed, Than Chieh-Fu (*1981*) has quite recently taken it upon himself to rearrange the material in what he thought was a more coherent manner. However, A. C. Graham has shown that the canons actually have a reasoned structure based on what the Mohists considered as the four objects of knowledge: names, objects, how to relate them, and how to act. Thus, apart from a section on problems of knowledge and change (definitions A40–A51, and propositions B13–B16) we have four sections:

---

[1] One of the significant methodological steps in the direction of rationality was their consistent refusal to attribute opinions to their authors. The Later Mohists insisted on discussing any thesis on its own terms without reference to the person advocating it. Already Mo Tzu himself, in a comment noteworthy for its generosity, quotes Confucius with approval. (*Mo Tzu* 48.58; cf. Mei (1929), p. 238: 'In a discussion with Chheng Tzu, Mo Tzu praised Confucius. Chheng Tzu said: "You are not a Confucian, why do you apply praise to Confucius?" Mo Tzu replied: "This is surely right and must not be changed."') The Later Mohists clearly continued this tradition of regarding propositions each on their merit rather than on the merits of their originators.

[2] I prefer to consider A76 to A87 as a part of the section on description, whereas Graham (1978), p. 32, places these as an appendix to the definitions.

1. explaining how to relate names to objects (definitions A1–A6, and propositions A88–B12);
2. explaining how to act (definitions A7–A39, and Expounding the Canons);[1]
3. explaining objects (definitions A52–A69, and propositions B17–B31);
4. explaining names (definitions A70–A87, and propositions B32–B82).[2]

The first thing we must realise, then, when dealing with the Dialectical Chapters is that they do indeed constitute a structured system and must not be arbitrarily reshuffled or read as stray fragments. When viewed in its systemic context, much of what looks arbitrary and incomprehensible fits into a larger coherent picture. But having said this, we hasten to add that a great deal still remains very obscure indeed, and the Dialectical Chapters still rank among the most difficult texts in the corpus of Classical Chinese literature.

2. A survey of Mohist logic

We shall now turn to the Later Mohist theories in so far as they are relevant to logic or the philosophy of language. We shall first consider the notion of logical analysis (*pien* 辯), i.e., the definition of the Later Mohists' logical endeavour.

The term *pien* is generally translated as 'disputation', and this is indeed the way the term is taken in Graham (1978). Certainly, we are convinced that the word *pien*, when applied to the earlier *pien chê* 辯者 or 'sophists',[3] refers to a battle of words and arguments, a verbal dispute, that in short *pien* refers to disputation.

However, the Later Mohists' logical practice is profoundly different from that of Kungsun Lung and his fellow 'sophists'. As Graham noted, the Mohists were quite systematically uninterested in attributing theses to individuals. They show no special signs of interest in public verbal dispute. For them the essence of *pien* is no longer in a battle of words but in the fitting description of the world and the systematic logical analysis of concepts.

Aristotelian logic is essentially and mainly concerned with the relation between concepts, relations of exclusion, inclusion, etc. Mohist logic, on the other hand, rarely touches on this and focuses almost entirely on the relation between names (*ming* 名) and objects (*shih* 實). The Later Mohists did not show a special interest in meanings of words as such,[4] but primarily in the relation between words and objects.

If one were to regard the Mohists and Kungsun Lung as nominalists on this account, that would be profoundly misleading in so far as it suggests that the

---

[1] Expounding the Canons (*yü ching* 語經) is perhaps the oldest part of the Dialectical Chapters, reconstructed by Graham on the basis of *Mo Tzu*, chs. 44 and 45. (Cf. Graham (1978), pp. 245–59 and 101–8). In addition, Graham has reconstructed from the same chapters what is probably the latest surviving document of the Later Mohist logicians, the section Names and Objects (*ming-shih* 名實) (*Ibid.*, pp. 469–94 and 108–10).

[2] Graham (1978), p. 30.   [3] For the problems surrounding this translation see Reding (1985).

[4] This does not mean that the Later Mohists had no concept of meaning. It just means that – perhaps wisely – they did not show a special philosophical interest in that concept.

Mohists considered an alternative between realism and nominalism and settled for the latter. The Mohists, as well as Kungsun Lung, only look like nominalists to us because they do not seem to consider what – following a medieval European tradition – we should call 'realist' issues involving the objective existence of abstract meanings. In fact, since there was no such thing as a notion of realism in ancient China, it makes no sense to attribute philosophical nominalism to them either.[1]

The general purpose of the Later Mohist logical enterprise is set out coherently in the following way:

The purpose of logical analysis is (1), by clarifying the distinction between right and wrong (*shih fei* 是非),[2] to inquire into the principles of order and misrule; (2), by clarifying points of sameness and difference, to discern the patterns of names and objects; (3), by settling the beneficial and the harmful, to resolve confusion and doubts. Only after that may one by describing summarise what is so of the myriad things, by sorting seek out comparables in the multitude of sayings. (NO6)

A crucial step was to distinguish the traditional notion of 'hearing or having heard (*wên* 聞)' from the notion of articulate discrimination which is the result of the effort of logical analysis. Thus the Later Mohists developed the notion of the discernment or discrimination of the mind (*hsin chih chha* 心之察). Traditionally the Chinese had a notion of 'having heard' which largely took the place of that of knowledge. Having heard something was the predominant way of knowing something. The Later Mohists analyzed the notion of understanding as follows:

Hearing is the sensibility of the ear. To attend to what you hear so that you grasp the intended meaning (*i*) is discernment by the mind (*hsin chih chha*). (NO9)

Strict logical analysis
More specifically, the Mohists define formal logical analysis (*pien* 辯) in a narrower sense as contending on 'contradictory (*fan* 反)' propositions:

Canon: Logical analysis (*pien*) is contending over (claims which are) the contradictories of each other (*chêng fan* 正反). The alternative that prevails (*shêng* 勝)[3] fits the facts (*tang* 當).

Explanation: One calling it 'an ox' and the other calling it 'not an ox', that amounts to contending over (claims which are) the contradictories of each other. Such contradictory (claims) cannot both fit the facts (*tang*). If they do not both fit the facts (*tang*), then one of them necessarily (*pi* 必) does not fit the facts (*tang*).[4] (Not like the case of making *chhüan* 犬 (dog) fit the facts.) (A74)

---

[1] Cf. Graham (1978), p. 33, as well as pp. 29f., 287, 325 and 444.

[2] Perhaps we should also here take *shih fei* 是非 more technically as 'is this' and 'is not'.

[3] The universally accepted interpretation of *shêng* 勝 here is 'to win' as in 'to win a dispute'. But if that were the correct interpretation, we would have to attribute to the Later Mohist an elementary mistake, namely the view that the victor in a verbal dispute is necessarily right. On our view, *shêng* like *pien* 辯 takes on a new meaning in Mohist logic which is somewhat removed from the notion of a verbal dispute and concerns discursive analysis. The subject of *shêng* 'prevail', then, is not a person, as previous translators have had it, but a claim.

[4] Surely this is a purely logical step of reasoning if ever there was one. The sentence sounds as if it were taken straight out of a modern textbook of predicate logic.

The notion of 'contradictory' has been ingeniously defined in the preceding definition:

Canon: Contradictories (*fan*) are what cannot be both regarded as unacceptable (at the same time).
Explanation: In all cases, 'an ox' is marked off[1] from 'not an ox'. These are the two things. There is no criterion by which these two are to be rejected. (A73)

This narrow Mohist notion of logical analysis can only be applied to the purely conceptual questions which dominate the first part of the Dialectical Chapters. Indeed, within the network of definitions, truths must be formally deducible or demonstrable. This is the realm of *pi* 必 or necessity. That the Mohists should have thought up such a realm of conceptual interrelations is a most remarkable intellectual achievement in the history of Chinese science. In the rigour of its methodology it is quite unique in the history of Chinese civilisation.

The Later Mohist scheme clearly met opposition. The objection was simply: what if both of the opponents are wrong? The Taoists would certainly have taken this attitude. The objection was rebutted in the following formal argument:

Canon: To say that neither alternative prevails in logical analysis necessarily (*pi*) does not fit the fact.
Explained by: logical analysis.
Explanation: What something is called is either the same or it is different. In the case where things are (called) the same – one man calling it a whelp and the other man a dog, or where they are different – one man calling it an ox and the other a horse, and neither winning, these are cases of failure to engage in disputation. In (proper) disputation the one calls something 'this' and the other calls it 'not-this', and the one that fits the facts wins. (B35)

Description

The aim of logical analysis was to establish a correct description of the world. Thus a most important part of the Mohist scheme of logic was the sorting out of a proper terminology and method of description. In this process, a number of basic issues in the philosophy of language as well as the philosophy of science had to be faced.

Necessary and sufficient versus necessary reasons. A crucial notion involved in any coherent description of the world is that of a reason or cause.[2] The Later Mohists begin their canons with a definition of this term, and it is a most significant event in the history of Chinese science that they defined the reason in terms that were totally alien to the well-known tendency towards analogical reasoning (*X* is so because something similar *Y* is so) or correlative thinking.

Canon: The 'reason (*ku* 故)' is (such that) if and only if[2] something has got it, it will come about.

---

[1] Cf. *Lun Yü* 19.12 for *chhü* 區 'to mark off'.

[2] It is to be noted that if *ku* 故 is to be taken as a concept in physics, it does not belong in the context of 'explaining how to relate names to objects'.

[3] The notion of 'only if' is expressed in Classical Chinese by the pattern *fei* . . . *tsê pu* 非 . . . 則不 lit: 'if not . . . then not'. *Jan hou* 然後 and *êrh hou* 而後 mean something like 'if . . . then and only then'.

Explanation: Minor reason: having this, it will not necessarily (*pi* 必) be so. Lacking this, it will necessarily (*pi*) not be so . . .

Major reason: having this, it will necessarily be so. Lacking this, necessarily (*pi* 必) it will not be so . . . (A1)

In spite of the mutilated state of this canon, it seems clear that it distinguishes between necessary and sufficient 'major' reasons, and necessary but not sufficient 'minor' reasons. Along with the rejection of correlative and analogical thinking, this must count as a remarkably astute additional logical distinction to make.

However, the Later Mohists do not, unfortunately, go on to use the terms 'major reason' and 'minor reason' coherently in the rest of the Dialectical Chapters: the terms do not recur. More seriously, having defined a reason (*ku* 故) as necessary, they fail to use *ku* exclusively for necessary reasons. They do not seem to realise that somebody's reasons for upholding something (A94) are generally not a necessary (but generally a sufficient) reason for someone's upholding it. They also do not realise that dampness is not a necessary reason for sickness (A77), surely only a sufficient reason. Or are they really thinking about a sickness that can *exclusively* be brought on by dampness? But that surely is not the typical situation. Physical causes should therefore not count as reasons (*ku*) at all according to the Later Mohist definition, since what is physically caused by one thing could in general also have been caused by another thing. Argumentative reasons should not count as *ku* because what is argued for with one reason could in general also have been argued for with another reason.

Sentences, kinds and inferences. Description is by sentences (*tzhu* 辭),[1] and these sentences crucially involve the specification of kinds (*lei* 類). The Later Mohists realised that drawing inferences from a sentence, 'pushing the kind (*thuei lei* 推類)' or seeing whether the sentence 'proceeds (*hsing* 行)' depends on the kinds (*lei*) involved in the sentence:

Sentences (*tzhu*) 'proceed' according to kinds (*lei*). If in putting forward a sentence you are not clear about (the relevant) kinds (*lei*), you are certain to get into trouble. (N010)

This laconic statement gives us one reason why the Mohists had such an extraordinary interest in the precise interdefinition of terms. By these definitions they established the kinds (*lei*) which played a crucial rôle in inference, in the process they called *thuei lei* 'pushing the classes', hence: 'draw inferences'.

The variety of names. Description is by application of names. The Mohists made several fundamental distinctions among names:

Canon: Names (*ming* 名) are unrestricted, classifying or private.
Explanation: 'Thing' is unrestricted. Any object necessarily (*pi* 必) requires this name.

---

[1] When in the appendix to the Expounding the Canons there is a reference to 'outrageous formulations (*chin yin chih tzhu* 浸淫之辭)', I am enclined to follow Wu Fei-Po (*1984*), p. 427, in interpreting the *tzhu* 辭 as 'propositions' (like those advocated by Kungsun Lung). Graham (1978), p. 35, maintains that the discovery of the proposition or sentence by the Later Mohists was a late one only manifest in the Names and Objects. Hansen (1985) maintains that the Mohists never developed a concept of the proposition at all, but he does not address the evidence or arguments provided by Graham.

Naming something 'horse' is classifying. For 'like the object' we necessarily (*pi*) use this name.

Naming something 'Jack' is private. This name stays confined in this object. . . . (A78)

The concept of a name (*ming*) was carefully distinguished from that of its application, the calling (*wei* 謂) a thing by a certain name:

Canon: To call (*wei*) is to transfer, to refer, or to apply.
Explanation: Naming by linking '*kou* 狗 (dog)' and '*chhüan* 犬 (dog)' is calling by transfer. To call something a *kou* (dog) or a *chhüan* (dog) is referring. Hooting at a dog is applying (a name). (A79)

Sense and reference. The Later Mohists showed a practical awareness of the distinction between the semantic content of an expression (the sense) and the object(s) in the world referred to by that expression (the reference). In modern philosophy, Gottlob Frege has made this (problematic) distinction famous. The expressions 'morning star' and 'evening star', he claimed, have different senses but the same reference. The Later Mohists were interested in the paradoxes generated by this situation. For them, the standard pair of words of the same reference but different senses or connotations were *chhüan* 犬 and *kou* 狗, both meaning 'dog'.

Canon: If you know *kou* (dogs), to say of yourself that you do not know *chhüan* (dogs) is a factual mistake (*kuo* 過).
Explained by: identity of objects.
Explanation: If the knowing of *kou* (dogs) is identical with the knowing of *chhüan* (dogs), then there is a mistake. If there is no identity, there is no mistake. (B40)

We can explicate the abstract Mohist point by rewriting the canon, using a more familiar example: If you know Ronald Reagan, to say of yourself that you do not know the first actor to have become president of the United States of America is a factual mistake.

The Later Mohists distinguished carefully between names that involve recognisable properties in things, and names that do not involve such recognisable properties. The formulation of their thought on this topic involved the use of a term that comes close to a 'variable', a notion essential for the development of logical thought. The Later Mohists came closest to the use of variables in their use of *mou* 某 in the following passage:

In cases of naming on the basis of shape and characteristics, we necessarily (*pi* 必) know that this thing is X (*mou*), only then do we know X (*mou*). In cases where naming cannot be on the basis of shape and characteristics, we may know X (*mou*) even if we do not know that this thing is X (*mou*). (NO2)

In order to avoid making a factual mistake, the Later Mohists insist that one must ensure that one understands the reference of the terms involved:

Canon: Understand the communicative intention (*i* 意) before replying.
Explained by: you do not know which item he refers to.

Explanation: The questioner says: 'Do you know *lo*?'[1]
One replies: 'What does "*lo* 騾" refer to?'
When the other says: 'The *lo* (mule) gallops', then it turns out you did know a *lo* (mule).
If without asking what '*lo*' refers to you directly answer that you do not know, then you make a factual mistake. . . . (B41)[2]

Quantification. We have so far only considered names of things. But there are words that are not names of things, and among these there are logically crucial particles like quantifiers, words like 'all' and 'some'. These naturally attracted the Later Mohists' special attention, since the Mohist concept of universal love involved universal object quantification.

At first thought it might seem truly astonishing that the Later Mohists should have seen the importance of defining such seemingly insignificant words like the quantifiers 'all' and 'some', but there was good theoretical reason for their interest. They took the notion of 'none (*mo* 莫)' as their primitive undefined term and proceeded as follows:

Canon: 'All (*chin* 盡)' is none not being so.
Explanation: Something is fixed of all (*chü* 俱)[3] of them. (A43)
As for 'some (*huo* 或)', it is 'not all'. (NO 5)

From a modern point of view one might object that 'not all' is an insufficiently precise definition of 'some', because it does not exclude 'none'. The Later Mohists may have overlooked this point, but their arguments on universal object quantification are anything but unsophisticated:

Canon: Their being countless is not inconsistent with (*hai* 害) doing something to everyone.
Explained by: whether it is filled or not.
Explanation: (Objection:) The south, if limited, is exhaustible; if unlimited is inexhaustible. If whether it is limited or limitless is unknowable '*a priori*', then whether it is exhaustible or not, whether men fill it or not, and whether men are exhaustible or not, these are likewise unknowable '*a priori*'. It is fallacious to treat it as necessary that men can be exhaustively loved.
(Answer to the objection:) If men do not fill the limitless, men are limited, and there is no difficulty about exhausting the limited. If they do fill the limitless, then the limitless has been exhausted, and there is no difficulty about exhausting the limitless. (B73)

The structure of the Mohist reply invites formalisation:

p = Men fill the limitless.
q = There is no difficulty about exhausting the limitless.
Either p or not-p (implicit assumption).

---

[1] *Lo* 騾 is suspected of being a rare character meaning 'mule'.
[2] Our interpretation tentatively follows Wu Fei-Po (*1984*), pp. 152 and 250.
[3] Note that the word *chü* 俱 is not used to define *chin* 盡. Instead, the Mohist very intelligently chooses to give an analysis of the concept of the universal quantifier in terms of negation and the quantifier 'none'.

If *p* then *q*.
If *not–p* then *q*.
Ergo: *q*. (*Quod erat demonstrandum*)

The Later Mohists focused on the thought that in order to know that a universally quantified sentence is true one does not have to know how many items one quantifies over:

> *Canon*: Without knowing their number we know that all are included.
> Explained by: the questioner.
> *Explanation*: Assume that there are two men, then we do know their number. In 'How do we know that the loving of the people applies to all?', one man is presumably left out of his question. If he asks about all men, then one loves all whom he asks about. (B74)

Having demonstrated that an infinite number of men does not create any difficulty for the Mohist doctrine of loving all men, the Later Mohists now go on to consider any finite number of men, for example two. Then, if an objector were to question whether the Mohist was sure that he loves all men, this question would either not refer to all men (in the concrete example: leave out one of the two) and therefore fail to ask the question he wants to ask. Or if indeed it asks the question of all existing men, then the reply to the question is simply that the Mohist loves exactly those that the questioner asked about. Thus the objector can only raise the objection by supplying a ready answer to it.

Having demonstrated that not knowing the precise number of men does not create any difficulty for the Mohist doctrine of loving all men, the Later Mohists go on to consider the case of knowing the number but not the whereabouts:

> *Canon*: Not knowing their whereabouts is not inconsistent with loving them.
> Explained by: having lost a son.
> *Explanation*: Like knowing that love includes all of them when you do know their number; there is no difficulty. (B75)

One can clearly love a son without knowing where he is. And as the explanation remarks, this is similar to the case of loving all objects of a certain kind without necessarily knowing how many they are.

Names, sentences and explanations. Names are seen as only one of different linguistic units relevant to the scientific description of the world:

> One uses names (*ming* 名) to refer to objects (*shih* 實).
> One uses sentences (*tzhu* 辭) to bring out intended meanings (*i* 意).
> One uses explanations (*shuo* 説) to bring out reasons (*ku* 故).
> One accepts according to the kind (*lei* 類), and one proposes according to the kind (*lei*).
> (NO11)

The kinds (*lei*) form the basis of the Mohist topography of the universe: it is when a given thing belongs to the relevant kind (*lei*) that they accept a proposition or that they propose a proposition.

Assertion. Articulating a sentence is not the same as asserting it. The Later Mohist logicians found occasion to develop the notion of assertion or maintaining (*chih* 執):

Saying is the flow of the tongue; to maintain what you say (*chih so yen* 執所言) so that your intended meaning (*i* 意) can be seen is analysis by the mind (*hsin chih pien* 心之辯). (N09)

The opposition here between *yen* 言 'saying' and *chih so yen* 'maintain what one is saying' certainly reminds us of the modern analytical distinction between articulating a sentence and asserting it.[1]

Semantic criteria. In order to ask whether a description applies or does not apply one needs criteria. As we have seen in our Sub-section on definition, Chinese philosophers had for a long time defined meanings of words, but the Later Mohists went on to ask the more abstract general question: how does one determine or fix meanings? In other words, they were concerned with the question by what criteria we apply the words which we do correctly apply. The example they chose was the concept of a circle. The Later Mohists distinguished with unfailing precision between the criterion as applied, which they called the standard (*fa* 法), and the criterion as the property corresponding to the standard applied, which they called criterion (*yin* 因).

> *Canon:* The standard (*fa*) is that in being like which something is so.
> *Explanation:* The mental picture (*i* 意), the compass, a circle, all these may serve as a standard (*fa*). (A70)
> *Canon:* The criterion (*yin*) is that wherein it is so.
> *Explanation:* Being so is the characteristics being like the standard. (A71)

Ambiguity. Even when criteria are fixed and meanings determined, there remains the problem of ambiguity which aroused the Mohists' systematic attention because it has a decisive effect on the procedures of description.[2] It is most significant from the point of view of the history of science that the Later Mohists by no means limited their attention to important philosophical or moral terms. They showed a curious scientific interest in ordinary words as well. Their interest was not only in ambiguity as such, but in the structure of it. In one instance, they show how one word can look as if it had two quite opposite meanings in spite of the fact that these are closely linked.

> *Canon:* To finish (*i* 已) is to bring about or to get rid of.
> *Explanation:* Of making a coat: 'to bring about'.
> Of curing an illness: 'to get rid of'. (A76)

---

[1] Asserting is, of coure, only one of a very large number of speech acts or things one can do with propositions. It happens to be the one most relevant to the practice and philosophy of science. Cf. Searle (1970).
[2] Cf. A76–A87. Graham (1978), p. 32, places this as an appendix to definitions, but it could also be regarded as the opening part of the section on description, as we have preferred to do. Note that A86 and A87 discuss the notions of sameness and difference, and A88 begins with a canon on 'sameness and difference'.

A scientifically important concept like *shih* 使 'make to', the most common verb for causation, is perceptively analysed into an effective and a non-effective version:

Canon: To make (someone or something) do something is to tell or to be the cause.
Explanation: Giving orders is to tell. The thing does not necessarily (*pi* 必) come about.
Dampness is the cause. It is necessarily (*pi*) required that the thing it does comes about. (A77)

Substitution. The Later Mohists evidently recognised a principle of substitution to the effect that if *X* is defined as *Y*, then it must be possible to substitute *Y* for *X* in any sentence *salva veritate*, without that affecting the truth or falsity of that sentence.[1] It is this substitutability which interlocks the various sentences in the canons and makes them into a system. Thus substitutability is a central theoretical issue in the Later Mohist analytical scheme, and as such it attracted their detailed attention.

A white horse is a horse. To ride a white horse is to ride a horse.
A black horse is a horse. To ride a black horse is to ride a horse.
Jack is a person. To love Jack is to love a person.
Jill is a person. To love Jill is to love a person.
In these instances something obtains (or is the case) when a thing is such-and-such (*shih* 是). (NO14)

We take 'a thing' here to refer to 'a white horse', 'Jack', etc., and we take 'such-and-such' to refer to being 'a horse', 'a person', etc. The thing which is said to 'obtain' are sentences like 'to ride a white horse is to ride a horse', 'to love Jack is to love a person'.

Like Aristotle, the Mohists noticed that such apparently impeccable inferences often break down. Noticing this breakdown can be of philosophical importance, as is demonstrated in the following passage which brings out the link between logical analysis and ethical theory.

Jill's parents are *jên* 人 (people). Jill's serving her parents is not serving *jên* (her husband).
Her younger brother is a handsome man. But her loving her younger brother is not 'loving a handsome man (= being in love with someone)'. . . .
Robbers are people. Abounding in robbers is not abounding in people, and being without robbers is not being without people.
How shall we make this clear? Disliking the abundance of robbers is not disliking the abundance of people. Desiring to be without robbers is not desiring to be without people. The whole world agrees that these are right. But if such is the case, there is no longer any difficulty in allowing that although robbers are people, loving robbers is not loving people, not loving robbers is not not loving people, and killing robbers is not killing people. (NO15)

---

[1] Modern philosophers of language are aware of obvious exceptions where *Y* cannot substitute for *X*: in a sentence ' "Blueberry" is a nice word' you cannot substitute the botanical definition of a blueberry. But such finesses as these do not affect our present argument. A principle of substitutability of some sort, though, is as essential to any modern precise science as it is to the Mohist canons.

Epistemology

Scientific description expresses knowledge. And obviously the notion of knowledge is closely linked to that of science. The Later Mohists spent a great deal of time in the logical analysis of concepts within this general area.

Learning. There are those (the Taoists) who maintain that learning (*hsüeh* 學) does not add anything to knowledge. The Later Mohists show this to be an inconsistent, i.e., a self-contradictory, statement.

> *Canon*: By learning we add something (to our knowledge).[1]
> Explained by: the objector himself.
> *Explanation*: He considers that learning does not add anything and accordingly informs the other. This amounts to causing the other to know that learning does not add anything, i.e., it amounts to teaching.
> If one believes that by learning one does not add anything, then it is inconsistent to teach. (B77)

The logical point is that if learning does not add anything, then it is incoherent to try to make someone learn this sentence that learning does not add anything. For if the sentence is true, then the learner cannot add it to his knowledge, and if the learner can add it to his knowledge, it is plainly untrue. This paradox is constructed on lines remarkably similar to Bertrand Russell's, when he constructs the sentence that 'The village barber shaved all those in the village who did not shave themselves'. Here again, we have the paradoxical situation that the barber, if he shaves himself, by virtue of the sentence does not shave himself. But if he shaves himself, then by the same sentence, he does shave himself.

Epistemological concepts. In order to avoid confusion and paradox, the Later Mohists sorted out the concepts around knowledge and learning as follows:

> *Canon*: The intelligence (*chih* 知) is the capability.
> *Explanation*: It being the means by which one knows, one necessarily (*pi* 必) does know. (Like the eyesight.)
> *Canon*: Thinking (*lü* 慮) is the seeking.
> *Explanation*: By means of one's intelligence one seeks something, but one does not necessarily (*pi*) find it. (Like peering.) (A4)
> *Canon*: Knowing (*chih* 知) is the connecting.
> *Explanation*: By means of one's intelligence, having passed the thing, one is able to describe it. (Like seeing.) (A5)
> *Canon*: Wisdom (*chih* 㣴) is the illumination.
> *Explanation*: By means of one's intelligence, in discourse about the thing one's knowledge of it is apparent. (Like clearness of sight.) (A6)

---

[1] We might also translate 'To learn something is advantageous', and in the last sentence: 'If you deem learning to be without advantage, to teach (at all) is inconsistent'. In that case we would have a pragmatic paradox.

Types of knowledge.

Canon: Knowledge is by hearsay, by explanation or by personal experience. Knowledge is of a name, of an object, of a collocation (*ho* 合), or of how to perform an action.

Explanation: Having received it at second hand is knowing by hearsay. Knowing that something square will not rotate is by explanation (*shuo* 説).[1] Having been a witness oneself is knowing by personal experience. What something is called by is its name. What is so called is the object. The mating of name and object is collocation (*ho* 合). To do something with the intention of doing it is an action. (A80)

This important division of types of knowledge is applied as follows:

Canon: Why a thing is so, and how I know that it is so, and how I cause others to know that it is so, are not the same.

Explained by: being injured.

Explanation: That someone wounded him is why it is so. That I saw it is how I know. That I tell them is how I make others know. (B9)

The notion of hearing something is discussed in the ensuing definition:

Canon: Hearing is at second hand or in person.

Explanation: Being told by someone is hearing at second hand. Being a witness oneself is hearing 'in person'. (A81)

The Later Mohists also speak of '*a priori* knowledge (*hsien chih* 先知)' based on definitions as opposed to empirical knowledge.[2]

This subtle differentiation must count as one of their major intellectual achievements.

Knowledge and awareness of knowledge. Another sophisticated distinction in the same area is that between knowledge of a fact and knowledge that one knows:

Canon: It is fallacious that the knowledge of whether one knows something or not is sufficient to act on.

Explained by: lacking what distinguishes knowledge.

Explanation: When we sort out one from the other, the non-knowledge lacks what distinguishes knowledge. (B34)

This is an enigmatic reaction to Confucius's *dictum*:

When you know something, to recognise that you know it, when you do not know something, to recognise that you do not know it, that is knowledge.[3]

Knowledge and perception. The Later Mohists were aware that knowledge is fundamentally different both from having a pictorial idea of a thing and from perception:

Knowing is different from having a pictorial idea. (NO3)

Canon: When one knows, it is not by means of the five senses (*wu lu* 五路, literally: the five paths).

---

[1] Cf.: 'An explanation is the means whereby one makes things transparent (*ming* 明).' (A72)
[2] Cf. Graham (1975), pp. 163–79.   [3] *Lun Yü* 2.17; cf. Lau (1983a), p. 14.

Explained by: the duration (*chiu* 久).[1]

Explanation: The knower uses the eye to see, and the eye uses light[2] to see, but the light (as such) does not see. If the sole source of knowledge were the five senses, then knowledge that has duration (*chiu*) would not be true.

Using eyes to see is like using light to see. (B46)

The point of the final remark is that the eye does not see, just as light does not see. Thus the sophism 'The eye does not see', which at first sight might sound like just another piece of nonsense, turns out to be the result of quite remarkably precise and disciplined philosophical reasoning and is not at all flippant. In this instance, then, the Later Mohists agreed with a sophism.

The objectivity of knowledge. In another case, they certainly disagreed, as in the case of the sophism 'Fire is not hot'. The Later Mohists take up the problem of the objectivity of sensual perception:

Canon: Fire is hot.
Explained by: looking.[3]
Explanation: One calls fire hot. One by no means regards the heat of the fire as belonging to oneself. (Like looking at the sun.) (B47)[4]

Here the Mohist seems to be answering someone who considers the heat of a hot object to reside in the observer who feels it. Thus, fire is not in itself hot. It only *feels* hot to a human observer so that the heat must properly be attributed to the observing subject rather than the observed object.

The Mohist's objection to this is clear enough. He insists that however we perceive heat, we apply the name 'hot' to objects, not to ourselves. Thus when looking at the sun, one calls the sun hot. There is no implication that one is hot oneself. One may indeed feel cold on a freezing winter day and still recognise the fact that the sun is hot.

Knowledge and necessity. The Later Mohists were aware that the reasons why names apply are different from the reasons why things are 'so'. Names apply because of conventions of meaning or definitions, and these are regarded as necessary (*pi* 必). Things are the way they are for natural reasons and these are not necessary:

For 'like the object' one necessarily (*pi*) uses such-and-such a name. (A78)
Such cases as 'elder brother or younger' and 'something so in one respect or not so in one respect' are the necessary and the not-necessary. Being this or not this is not necessary. (A51)

By checking through the present Section the patient reader will be able to confirm the wide use the Later Mohists made of the notion of necessity (*pi*).

Knowledge pertaining to the realm of conventions of meaning or definitions is called *hsien chih* 先知, a term which, following A. C. Graham, we translate as '*a priori*

---

[1] The concept of duration (*chiu* 久) is in turn defined in A40: 'Duration is pervasion of different times.'
[2] Literally: 'fire'.   [3] We follow Wu Fei-Po (*1984*), p. 222.   [4] Cf. Vol. 2, p. 173.

knowledge'.[1] What we cannot *hsien chih* 先知 'know *a priori*' is *wei kho chih* 未可知 'not knowable *a priori*'. Thus we are said to know *a priori* the idea of a pillar, but whether a given pillar is square or round is not knowable *a priori* (*wei kho chih*). This, of course, does not mean that it is not knowable (*pu kho chih* 不可知) whether the pillar is round or square.

What you know *a priori* about an object, you know about that object even though you are separated from it by a wall. Even if we are separated from it by a wall, the circular stays fixed as a concept and remains knowable *a priori*.[2]

When we 'jump the wall' the circular stays fixed. By the things which follow from each other or exclude each other, we may know *a priori* what it is. For the five colours, long and short, before and after, light and heavy, adduce the one to which you are committed. (A98)[3]

How, one may ask, is the circle known *a priori*, by pure definition, to the ancient Chinese whose interests in mathematics were so predominantly algebraic and arithmetic? The Later Mohists construct the notion of a circle in a series of definitions as follows:

'Straight' is in alignment. (A57)
'The same in length' is exhausting each other when laid straight. (A53)
'The centre' is the place from which they are the same in length.[4] (A54)
'Circular' is having the same lengths from a single centre. (A58)

These definitions, then, define a realm of logical necessity.

However, the realm of necessity is not limited to that of mathematical science. Ethics is another area in which the concept was systematically deployed to construct an *a priori* system.

In the case of all things that the sage desires or dislikes *a priori* for the sake of men, men necessarily learn from him by considering their essentials; but in the case of desires and dislikes born from the conditions they encounter, they do not necessarily learn from him by considering their essentials. (EC10)

Consider now the way in which ethical terms are inter-defined by the Later Mohists from the terms 'desire' and 'dislike' and 'benefit' and 'harm':

'Desire (*yü* 欲)' is either directly, or having weighed the benefit. (A84)
'Dislike (*wu* 惡)' is either directly, or having weighed the benefit. (A84)
'Benefit (*li* 利)' is what one is pleased to get. (A26)
'Harm (*hai* 害)' is what one dislikes getting. (A27)

---

[1] This Mohist use of the Chinese term for 'fore-knowledge, knowing beforehand' is highly significant. The Later Mohists were in the habit of introducing a logical dimension into terms belonging to a less abstract realm. If our interpretation is correct, they did the same thing to the notion of *pien* 辯 'disputation'.
[2] For the Mohist use of the concept of the wall see Graham (1978), p. 223.
[3] 'Things which follow one another' are propositions which imply each other, and 'things which exclude one another' are propositions which are inconsistent with each other. In this context of description, the Mohists show an interest in the logical relations between propositions, but this never became their main concern.
[4] In the explanation we read the ingenious further derivation: 'Outward from this they are equal/like each other'. The Mohists take the notion of *jo* 若 'be equal' as the fundamental one from which the others are derived.

'To do for the sake of (*wei* 為)' is to give something most weight in relation to the desires, having taken acount of all that one knows. (A75)

'Love'[1]

'To be benevolent (*jên* 仁)' is to love individually. (A7)

'To be righteous (*i* 義)' is to benefit. (A8)

'To be filial (*hsiao* 孝)' is to benefit parents. (A13)

'Achievement (*kung* 功)' is benefit to the people. (A35)

Paradoxes

One of the important things the system of definitions and sentences in the Dialectical Chapters was designed to achieve was the avoidance or resolution of paradox. The Later Mohists addressed a number of paradoxes. Some of these they showed to be theoretically sound and only apparently paradoxical, others they proved to be based on conceptual confusions.

The paradox of knowledge.

Canon: We know what we do not know.[2]

Explained by: picking out by means of a name.

Explanation: If you mix together what he does know and what he does not know, then he is bound to say: 'This one I know' and 'That one I do not know'. To be able to both pick out and reject amounts to knowing them both. (B48)

The point is a very subtle one: we know what we do not know in the sense that we can identify what we know and what we do not know when challenged to do so. Thus we know what it is we do not know, but we still do not know what we do not know.

Loan words. One place where contradiction (*pei* 誖) or paradox necessarily arises is in the case of loan words, where a given character with a certain pronunciation is used as a phonetic loan for a word with the same or a similar pronunciation.[3] The Later Mohists quite properly link the notion of contradiction (*pei*) between the original meaning and the derived meaning of the character with that of necessity or analytical truth (*pi* 必).

Canon: Using loan words is necessarily (*pi*) contradictory (*pei* 誖).

Explained by: not being so of it.

Explanation: What it is loan-named as it necessarily (*pi*) is not, only then do we have a loan-name. When a dog is loan-named as being a crane, it is as when one gives it the clan-name 'Crane'. (B8)

---

[1] This definition was lost together with all definitions of words in Mo Tzu's ten theses. However, the system of definitions is so tight that one can surmise with reasonable assurance that the definition of love must have run somewhat like this: 'to desire benefit and dislike harm for the sake of the man (not for one's own sake)'.

[2] Graham (1978) fails to construe this canon as a sentence in spite of the fact that he counts it as a proposition. This seems inconsistent, and in any case it does not do justice to the force of this particular paradox.

[3] Cf. Karlgren (1952b).

Token-reflexive expressions. Token-reflexive or deictic expressions are words like 'I', 'you', 'this', 'that'. The logical puzzle here is simply that the person I refer to as 'I' is the person you refer to as 'you', and that the person I refer to as 'you' is the person you refer to as 'I'. What you refer to as 'that' I may refer to as 'this' and what I refer to as 'that', you may refer to as 'this'. We have a similar situation for words like 'here', 'there' etc. The meaning of such words depends on where they are used and by whom. Such is the superficially paradoxical nature of deictic expressions. The Taoists delighted in these as proof of the impossibility of achieving objectivity of description, and the Later Mohists were determined to dissolve the apparent paradoxes:[1]

> *Canon*: You cannot use 'that' for this without using both 'that' for this and 'this' for that.
> Explained by: their being different.
> *Explanation*: It is admissible for the man who uses names rightly to use 'that' for this and 'this' for that. As long as his use of 'that' for that stays confined to that, and his use of 'this' for this stays confined to this, it is admissible to use 'that' for this. When 'this' is about to be used for that, it is likewise admissible to use 'that' for this. If 'that' and 'this' stay confined to that and this, and accepting this condition you use 'that' for this, then 'this' is likewise about to be used for that. (B68)

The *Chuang Tzŭ*, ch. 2, plays with the relativity of deictic terms and surreptitiously seems to transfer the problem of relativity to moral terms. But the Later Mohists insist that the relativity of demonstratives does not invalidate their use as long as one is consistent in their use.

The paradox of non-existence. Paradoxically, one can say that there must be a thing to identify before we can say that this thing does not exist. The Later Mohists approach this puzzling state of affairs by making an important conceptual distinction between the absence of an existing thing on the one hand, and the non-existence in the universe on the other hand.

> *Canon*: Non-existence does not necessarily (*pi*) presuppose existence.
> Explained by: What is referred to.
> *Explanation*: In the case of non-existence of something, the thing has to exist before it is in this way non-existent. In the case of the non-existence of the sky's falling down, it is non-existent without ever having existed. (B49)

The paradox of levels of counting.

> *Canon*: One is less than two, but more than five.
> Explained by: setting up levels.
> *Explanation*: Five contains one in it, but one contains five[2] or twelve[3] in it. (B59)

---

[1] We might of course just as well have quoted this canon in our Section on the Mohist solution of paradoxes.
[2] E.g., one hand contains five fingers.   [3] E.g., one year contains twelve months.

Paradox of tense.

> *Canon*: A thing could have not existed, but when it has occurred it cannot be got rid of. Explained by: having been so.[1] (B61)

Paradoxes of space and time.

> *Canon*: Moving over a spatial extension (*hsiu* 脩) uses temporal extension (*chiu* 久).
> Explained by: before and after.
> *Explanation*: The mover is necessarily (*pi* 必) closer at first and further away only after that. Being close or far (*yüan chin* 遠近) is called spatial extension (*hsiu*). Preceding and succeeding is called temporal extension (*chiu* 久). For someone, in order to move over a long distance, necessarily (*pi* 必) uses duration. (B64)

Here the Mohists meticulously analyze basic concepts with the purpose of clarifying spacial paradoxes. We do not know exactly what paradoxes or paradox the Mohists had in mind. Perhaps it was not Hui Shih's 'I go to Yüeh today but arrived yesterday', but it certainly was some paradox of that order. We can see that in themselves such paradoxes are not necessarily of scientific interest, but their tremendous use is in forcing upon the opponent conceptual clarity in order to sort out the apparent paradox. The sophist who flippantly trades in paradoxes thus becomes the catalyst of scientific development. The dogmatic sceptic gets a hard-headed response, as the self-reflexive paradoxes show.

The self-reflexive paradoxes. If one maintains a sentence, one is thereby committed to denying the negation (or rejecting the denial) of what one says. Consequently, refusing to deny any sentences, as the Taoists are fond of doing, is simply inconsistent:

> *Canon*: To reject denial is inconsistent.
> Explained by: he does not reject it.
> *Explanation*: If he does not reject the denial (of his own thesis that denial is to be rejected) then he does not reject denial. No matter whether the rejection is to be rejected or not, this amounts to not rejecting the denial. (B79)

Suppose someone maintains that one should reject denial. Then, if he means what he says, the thesis that one should not reject denial should be denied. But if one admits that that thesis should indeed be denied, then one is not really maintaining that one should reject denial: In at least one instance (i.e., the thesis in which one rejects denial) one fails to reject denial. One's position is therefore inconsistent.

Let us rephrase the matter more along our usual lines of thinking: Suppose someone (the Relativist) maintains the proposition P ('For all propositions s: s is true'). He is then maintaining that 'P is true.' Now by substituting non-P for s in this formula we get 'non-P is true' as a consequence of P. Thus the Relativist is committed to two propositions:

1. P is true.
2. non-P is true.

---

[1] The explanation is textually too problematic to be of much use.

In other words, this special substitution demonstrates that the Relativist is contradicting himself.

Maintaining *P* implies denying non-*P*. But denying non-*P* violates the principle that no proposition is to be denied. Therefore maintaining *P* is self-contradictory. If he maintains that no propositions are to be denied, the Relativist is committed to denying the proposition that not all propositions are to be denied. He is therefore contradicting himself.

It is extraordinary to have to attribute an advanced argument of this degree of logical subtlety to an ancient Chinese thinker, but in this particular instance the text is quite clear and the interpretation is quite straightforward.

With stunning persistence the Mohist goes on to claim that it makes no difference whether the Relativist is right or wrong in refusing denial. It is not a question of whether the Relativist is right or wrong. The point is that the Relativist position is self-contradictory.

The Mohists make a closely similar proof against considering all saying as contradictory:

> *Canon*: To claim that all saying contradicts itself is self-contradictory.
> Explained by: his saying (this).
> *Explanation*: To be self-contradictory is to be inadmissible. If these words of the man are admissible, then this is not self-contradictory, and consequently in some cases saying is acceptable. If this man's words are not admissible, then to suppose that it fits the facts is necessarily ill-considered. (B71)

The crux in this argument is again the application of a thesis about all propositions to that thesis itself. If the Relativist declares his own thesis to be included among the incoherent theses which are to be rejected, then he both maintains and rejects his thesis and therefore contradicts himself.

If, on the other hand, he does not include his thesis among the theses which are incoherent and to be rejected, then there is at least one thesis which is coherent and not to be rejected so that the Relativist is again contradicting himself.

Whoever constructed arguments of this sort, we may safely conclude, was a thinker of considerable logical sophistication. He represents an early but advanced logical and scientific subculture of which, unfortunately, all too few traces have been thought worth handing down in the Chinese tradition.

(5) CHINESE REACTIONS TO ANCIENT CHINESE DISPUTATION AND LOGIC

In China, as in India, disputation or logic was perceived as an essentially rhetorical discipline. Indeed, the ancient Chinese term for disputation or logic, *pien* 辯, is commonly used to refer to rhetorical skill, as when Confucius's disciple Tzu Kung, who surely knows nothing about disputation, is described as a *pien jên* 辯人 'a man skilful with words'.[1] However, in this Section, we shall be concerned not with rhetoric in

---

[1] *Huai Nan Tzu*, ch. 18, ed. Liu Wên-Tien, p. 22a.

general, but more specifically with ancient Chinese reactions to the practice of Chinese *pien chê* 辯者 'sophists'. Reactions have been predominantly negative. We need to ask why. Moreover we need to know whether the ancient Chinese thinkers dismissed *pien* 'disputation' without paying careful attention to it, or whether they rejected disputation for recognisable reasons.[1]

For the intellectual history of China, our question is important, for what we set out to investigate is whether the failure of disputation in traditional China is due to an intellectual failure to understand it or rather to an intellectual choice to reject it. We intend to present evidence that ancient Chinese thinkers did not ignore logic or disputation but consciously and explicitly rejected it. The evolution of the grounds for this rejection is the subject of the present Section.

Confucius

At the time of Confucius there was, for all we know, not much disputation to react to. But already in the *Lun Yü* we find a negative attitude towards morally indifferent reasoned debate:

If one spends all day together, does not speak of what is right (*i* 義) but shows fondness for petty cleverness, that is a difficult case indeed.[2]

Confucius's prejudice against 'artful words (*chhiao yen* 巧言)' remained proverbial and influential throughout the ages:

Artful/clever words confound/ruin one's charismatic power (*tê* 德).[3] I detest the tunes of Cheng for corrupting classical music. I detest clever talkers who overturn states and noble families.[4]

The music of Cheng, we may surmise, was judged morally depraved and purely entertaining, therefore to be rejected along with pure argumentation.[5] Throughout his dialogues, Confucius displays a consistent disrespect for rhetorical nimbleness (*ning* 佞):

Banish the music of Cheng and keep the nimble-tongued (*ning*) at bay.[6]

This negative attitude towards 'artful words', which reverberates through most of pre-Chhin philosophy, has been a major factor in the rejection of the formalistic

---

[1] In Greece the sophists in general, and Socrates in particular, were widely (though of course not universally) despised and ridiculed. K. J. Dover's excellent edition of Aristophanes' *Clouds* (Oxford 1968) illustrates this point brilliantly. Cf. also Sommerstein (1982). On the other hand the popularity of the sophists is evident e.g., in Plato's *Hippias Major* and *Hippias Minor*.

[2] *Lun Yü* 15.17; cf. Lau (1983a), p. 153.   [3] *Lun Yü* 15.27; cf. Lau (1983a), p. 155.

[4] *Lun Yü* 17.16; cf. Lau (1983a), p. 177.

[5] Archaeology has revealed that in point of fact the music of Cheng indeed had special characteristics which have been reconstructed on the basis of bells, etc., from Cheng. Cf. Chen Cheng-Yih (1987), pp. 155–97, especially pp. 171 and 180. I owe this important reference to Kenneth Robinson.

[6] *Lun Yü* 15.11; cf. Lau (1983a), p. 151.

approach of philosophers like the Later Mohist logicians. Words and language were in the end seen as secondary in importance to the charismatic inner power called *tê*:

He who has charismatic power (*tê*) is bound to find words (to express himself). He who finds words does not necessarily have charismatic power.[1]

The primacy and priority of the moral sphere was undisputed throughout Chinese history. The rejection of logic and disputation is profoundly connected with this.

Mêng Tzu

The Confucian Mêng Tzu (−371 to −289) could no longer afford to summarily dismiss disputation:

Kungtu Tzu said: 'Outsiders all say that you, Master, are fond of disputation. May I ask you why?'
'Surely I am not fond of disputation: I just have no alternative but to practise it.'[2]

Mêng Tzu's practise of disputation and argumentation has been studied in Lau (1983c), pp. 334–56. His theoretical statements on disputation are few, however, although he boasted:

'I have an insight into words. . . .'
'What do you mean by "an insight into words"?'
'From biased words I can see wherein the speaker is blind; from immoderate words, wherein he is ensnared; from heretical words, wherein he has strayed from the right path; from evasive words, wherein he is at his wits' end.'[3]

By 'disputation', Mêng Tzu often means something more like rhetoric than like disputation. He still has not developed a deep theoretical interest in the subject.

Hsün Tzu

The Confucian Hsün Tzu (*c.* −298 to −238) took a much more differentiated view of disputation. Indeed, chapter 22 of his book is, as we have seen, one of the most important contributions to ancient Chinese disputation. He voices his objections to logic-chopping as follows:

The underlying pattern of ritual is truly profound. Argumentative investigations of the type 'are "the hard" and "the white" the same?', if they address themselves to this matter, they drown in it.[4] To split words and conduct argumentations, to discuss material things and practise disputation, these are things that the gentleman holds in low esteem.[5]

---

[1] *Lun Yü* 14.4; cf. Lau (1983a), p. 133.   [2] *Mêng Tzu* 3B9; cf. Lau (1983c), p. 127.
[3] *Mêng Tzu* 2A2; cf. Lau (1983c), pp. 57f.   [4] *Hsün Tzu* 19.30; cf. Köster (1967), p. 247.
[5] *Hsün Tzu* 21.90; cf. Köster (1967), p. 284.

Discussing the dialecticians Têng Hsi and Hui Shih, Hsün Tzu finds their practise not only morally but also politically irrelevant:

They do not take the former kings as their models and do not approve of ritual and duty. On the contrary, they love to sort out strange theories and play around with abstruse expressions. They are very sharp but without generosity, well-argued but without use. They make much fuss but achieve little. They cannot be taken as guidelines for the practice of government.[1]

What particularly irritates Hsün Tzu is the arbitrariness of the subject-matter discussed by the sophists and their predilection for the abstruse and difficult:

The gentleman does not value the arbitrarily difficult action [e.g., swimming with a stone], and he does not value the arbitrarily sharp-witted in explanation [e.g., the claims that mountains and abysses are of the same height] . . . These are difficult explanations to maintain, but Hui Shih and Têng Hsi were able to do it. However, the reason why the gentleman does not set prize by them is because they do not fall within the realm of ritual and duty (*li i* 禮義).[2]
    The difference between hard and white, the distinction between what has thickness and what has no thickness, these are by no means things which should not be investigated discriminatingly, but the gentleman does not argue about these things. He leaves them where they are.[3]

Hsün Tzu rejects disputation as a circus performance of intellectual virtuosity and observes wryly:

Not knowing about these things (the niceties of abstract disputation as practised by the ancient Chinese dialectitians) does not prevent one from being a sage. Knowing about them does not make one less of an insignificant person.[4]

This is not an unfounded judgement of someone who does not know what the disputation of ancient China is about. It is a considered verdict by a very well-informed thinker. The attitude expressed is representative of informed traditional Chinese opinion.

Lü Shih Chhun Chhiu 呂氏春秋

The *Lü Shih Chhun Chhiu*, compiled around −240, offers rather subtle objections to disputation.

When a hole is eight feet deep, then the human arm is bound to be unable to explore it completely. Why is this? Because the arm does not reach (i.e., is not long enough). Knowledge also has things which it will not reach. Theoreticians (*shuo chê* 説者) may be sophisticated in their arguments (*pien* 辯), and they may be subtle in their methods, but they cannot catch sight of these things.[5]

---

[1] *Hsün Tzu* 6.8; cf. Köster (1967), p. 55.   [2] *Hsün Tzu* 3.1; cf. Köster (1967), p. 20.
[3] *Hsün Tzu* 2.30; cf. Köster (1967), p. 16.   [4] *Hsün Tzu* 8.36; cf. Köster (1967), pp. 74f.
[5] *Lü Shih Chhun Chhiu* 16.4, ed. Chhen Chhi-Yu, p. 979; cf. Wilhelm (1928), p. 245.

The limits of disputation are seen in terms of what we today would see as political considerations:

That which confuses names (*ming* 名) is dissolute discourse. When theories (*shuo* 説) are dissolute, the acceptable is made out to be unacceptable, that which is so is made out not to be so, that which is right is made out not to be right, and that which is wrong is made out not to be wrong. Therefore, as far as the gentleman's discourse is concerned, when it is sufficient to bring out the quality of the talented and the lack of quality of the untalented, it stops. When it is sufficient to make clear what obstructs good government, and how chaos arises, it stops. When it is sufficient for an understanding of the true nature of things, and when grasping this understanding allows men to survive, then he stops.[1]

The dissoluteness of disputation is illustrated in an entertaining anecdote:

There was a servant in Chhi. His master got into trouble, but the servant failed to die for him. He then met a friend in the street who said to him: 'So you really did not die for him!'

'You are right. In everything we do we aim for benefit. Death is not a beneficial thing. Therefore I did not die for my master.'

The friend said: 'But can you still face people after this?'

'Well, do you imagine I could face them if I was dead?'

This man, according to several accounts, failed to die for his ruler and superior. He greatly failed in his duty. His speeches were none the less incontrovertible. It is thus clear that language is not a sufficient criterion to decide a matter.[2]

Rigid logical thinking was perceived in the *Lü Shih Chhun Chhiu* to often lead to absurd conclusions:

Kao Yang-Ying was about to build a house. His builder told him: 'It is still too early. The timber is too fresh. If you put clay on it, it will bend. If you build with fresh timber, it will be good for the moment, but it will certainly turn out badly in the end.' Kao Yang-Ying said: 'If one follows your words, a house would never disintegrate. The drier the wood gets, the better it bears. The drier the clay gets, the lighter it becomes. Now if you put something that gets lighter on something that gets firmer no harm can arise.'

The builder did not know what to reply, so he obeyed and built the house. When the house was finished it was very beautiful to look at, but after a while it did indeed collapse.[3]

Chinese intellectuals throughout the ages have felt a profound sympathy for the builder, and a deep suspicion of fluent argumentation like that of his master.

Neither Hsün Tzu nor the *Lü Shih Chhun Chhiu* reject disputation as a whole. The *Hsün Tzu*, as we have seen, contains a most important contribution to Chinese disputation, and also the *Lü Shih Chhun Chhiu* is an important document in the study of the early Chinese sophists. What both these texts criticise is the sophists' tendency to overestimate the range and scope of their skills, their profoundly mistaken leaning towards regarding disputation as a philosophical panacea. What, unfortunately, neither of them realised was that such sophists' mental gymnastics might have indirect

---

[1] *Lü Shih Chhun Chhiu* 16.8, ed. Chhen Chhi-Yu, p. 1019; cf. Wilhelm (1928), p. 258.

[2] *Lü Shih Chhun Chhiu* 18.4, ed. Chhen Chhi-Yu, p. 1178; cf. Wilhelm (1928), p. 302.

[3] *Lü Shih Chhun Chhiu* 13.2, ed. Chhen Chhi-Yu, p. 1642; cf. Wilhelm (1928), p. 436.

scientific uses, as indeed they proved to have in the history of theoretical science in the West. As it turned out, the Chinese common-sense rejection of disputation as morally frivolous, political irrelevant and intellectually sterile won almost universal acceptance, so that after the −3rd century disputation never again became a widely significant force in Chinese intellectual history until the 20th century.

The Legalists

The *Shang Chün Shu* 商君書, dating from the −3rd century, as well as the *Han Fei Tzu* 韓非子, which belongs to the late −3rd century, both frequently deplore what they call *pien* 辯 'disputation or sophistry'. The Lord of Shang complains that one can obtain office and rank at the courts of his time through skilful speech and 'sophistical theories (*pien shuo* 辯說)'.[1] The complaint is quite common:

Nowadays, the rulers of the world are all anxious over the perilous condition of their countries and the weakness of their armies, and they listen at all costs to the theoreticians (*shuo chê* 說者). Though these may form battalions, talk profusely and employ beautiful formulations, it is of no practical use. The rulers are fond of this sort of disputation and they do not seek practical application (*shih* 實). When the theoreticians (*shuo chê*) have their way, expound their crooked disputations in the streets, their various groups become great crowds, and the people, seeing that they succeed in captivating kings, dukes and great men, all imitate them.[2]

The *Han Fei Tzu* is full of diatribes against rhetoric or disputation (*pien*) which the author feels leads to ruin.[3] The crucial insight that dominates Han Fei's treatment of disputation is simply that the truth of a thesis and the quality of the arguments adduced in favour of a thesis are quite independent:

When a stutterer argues a point it is 'doubtful'. When a rhetorician/sophist (*pien chê* 辯者) argues a point it is 'true'.[4]

The Taoists

The Confucians were morally opposed to frivolous disputation. The Legalists were politically opposed to 'purely academic' disputation. The Taoist Chuang Tzu tried to show that disputation was a necessarily fruitless and pointless excercise not for moral or political reasons, but for philosophical reasons. Chuang Tzu was a meta-logician in so far as he philosophised *about* practising logic or disputation.

As far as Chuang Tzu is concerned the sophists or dialecticians are just fools who think that things are what they appear to be from the dialecticians' standpoint. Chuang Tzu argues this out in a rare piece of straight meta-logical reasoning:

---

[1] *Shang Chün Shu*, ed. Kao Heng, p. 33; cf. Duyvendak (1981), p. 187.
[2] *Shang Chün Shu*, ch. 3, ed. Kao Heng, p. 40; cf. Duyvendak (1981), p. 195.
[3] Cf. *Han Fei Tzu* 15.1.37 *et passim*.
[4] *Han Fei Tzu* 48.6.12; cf. Liao (1939), vol. 2, p. 269.

You and I having been made to argue over alternatives, if it is you not I that wins, is it really you who are right, I who am wrong? If it is I not you that wins, is it really I who am right, you who are wrong? Is one of us right and the other of us wrong? Or are both of us right and both of us wrong? If you and I are unable to know where we stand, others will surely partake of our perplexity. Whom shall we call in to decide the matter? If we get someone of your party to decide it, being already of your party how can he decide it? If we get someone of my party to decide it, being already of my party how can he decide it? If we get someone of a party different from either of us to decide it, being already of a party different from either of us how can he decide it? If we get someone of the same party as both of us to decide it, being already of the same party as both of us how can he decide it?
Consequently you and I and he are all unable to know where we stand, and shall we find yet another person to depend on?[1]

One reason why issues one may dispute about are subjective is, according to Chuang Tzu, that the meanings we attach to words are subjective. If I use 'black' to mean 'white', then as far as I am concerned white things are black. But what does it matter? For Chuang Tzu this sort of reflection on the arbitrariness of names, which he is the first to maintain in the history of Chinese philosophy, is symptomatic of the subjectivity of mental acts such as judging issues.

The Way comes about as we walk it; as for a thing, call it something, and the thing is what you call it.[2]

Later attitudes

The Salt and Iron Discourse by Huan Khuan 桓寬 (−1st century) contains an entirely typical statement which represents current feelings in China throughout the ages:

If you do not reach the great Way (*tao* 道) but engage in small disputation/analysis (*hsiao pien* 小辯), then that will only lead to harm to your person.[3]

Yang Hsiung 揚雄 (−53 to +18), in conversation, has a poignant way of putting the logicians in their proper place when asked whether the methods of Kungsun Lung were worth adopting:

Chopping wood to make chessmen and working leather to make footballs all have their proper methods. But since these methods are not in accord with the former kings, the gentleman will not cultivate them as methods (*pu fa yeh* 不法也).[4]

Ideas of this sort became something of a *locus communis* among later Confucians:

Impoverished places will be rich in trifling intellectual endeavours (*chhü hsüeh* 曲學). Small disputation/analysis (*hsiao pien* 小辯) will harm great knowledge. Skilful phrasing (*chhiao yen* 巧言) will take away trustworthiness. Petty generosity will impede great righteousness.[5]

---

[1] *Chuang Tzu* 2.84; cf. Graham (1981), p. 60.   [2] *Chuang Tzu* 2.33; cf. Graham (1981), p. 53.
[3] *Yen Thieh Lun*, ch. 55, ed. Basic Sinological Series, p. 184.
[4] Yang Hsiung, *Fa Yen* 4, ed. Wang Jung-Pao, p. 107.   [5] *Shuo Yüan* 16.122, ed. Chao Shan-I, p. 456.

From a book that took its present shape around +100 we have the following dialogue which was spuriously attributed to Confucius himself:

The Duke (Ai) asked: 'I want to study small disputation/analysis (*hsiao pien*) in order to
understand government.'
  The Master (Confucius) replied: 'No. This is not appropriate. The lord over the altars of
the land and the grain should be careful in his use of time. Days gone by will not come again.
Study must not proceed by small disputation/analysis (*hsiao pien*).
  Therefore in ancient times the former kings studied how to adjust the overall Way in
order to understand government.
  The Son of Heaven studied music, sought clarity about customs and put rituals in order
to practise proper government.
  The feudal lords studied the rituals and distinguished official duties in order to carry out
their administration and to serve the Son of Heaven.
  The great officials studied moral excellence (*tê* 德) and sought clarity on various duties,
and they diligently served their ruler.
  The knights studied service and sought clarity about (the ruler's) words in order to carry
out (the ruler's) intentions.
  The common people obeyed their superiors and sought clarity about legal prohibitions.
  The peasants used their physical strength to go about their business.
  In doing these things those involved still feared that they would not succeed: why should
they be concerned with small disputation/analysis (*hsiao pien* 小辯)?'[1]

In this very important passage we can see how the ordered and regimented hierarchy of Chinese society since Han times, and the division of social functions that
went with it, could get in the way of 'petty disputation (*hsiao pien*)'.

Almost 600 years after the flowering of ancient Chinese disputation, the lone
commentator Lu Shêng 魯勝 (*floruit c.* +300) wrote:

From Têng Hsi down to Chhin times the so-called logicians wrote their chapters and books
for generations on end. But they were all extremely hard to understand and no scholars pass
on the art. Now, over 500 years later, the books have disappeared. Of the Mohist disputation
we have the upper and lower canons, with an explanation to each canon, four chapters
in all. Since they are integrated with the rest of the chapters in the book, they alone have
survived.[2]

Thus the most important classic of disputation survived by coincidence, because
it got mixed up with other texts that were more in the main-stream of Chinese intellectual history.

One of the classics of disputation which did not survive was the *Têng Hsi Tzu*
鄧析子, an apparently faked version of which possibly originated around the time
of Lu Shêng. Its characterisation of proper dialectics is significant.

The so-called 'great disputation (*ta pien* 大辯)' draws up distinctions between ways of acting
in the world and it covers the things of this world. It recommends the good and rejects the
bad. At all times it uses the right words and its success is established, its moral power perfect.

---

[1] *Ta Tai Li Chi* 76, ed. Kao Ming, p. 388; cf. Grynpas (1967), p. 206.
[2] *Chin Shu*, ed. Pai Na, ch. 94, p. 6b.

Small disputation (*hsiao pien* 小辯) is not like that. It quibbles with words, sets up distinctions between ways of action. It uses words like arrows to shoot at the opponent.[1]

As time went on, reactions to the logicians became even less sophisticated and informative, indeed much less well-informed. Here is an example from a Buddhist book dating from +622:

Formerly Kungsun Lung wrote the Discourse on Hard and White, condemning the Three Kings and denying the Five Emperors. Even now people who read it still gnash their teeth.[2]

Even the sound of the gnashing of teeth soon ebbed. Indigenous Chinese disputation receded into the limbo of unperceived cultural tradition. The very advanced Buddhist logicians of the +7th century show no signs that they knew much about any originally Chinese logical tradition.

(6) Logical Thought in the +3rd Century

On the eve of the major intellectual impact of the introduction of Buddhist philosophy into China, the +3rd century saw a certain revival of interest in speculative philosophy and in some aspects of Chinese logic.[3] Thanks to the work of Robert G. Henricks, the practice of constructing arguments in support of theses and in refutation of theses may be conveniently studied in reliable English translations. Juan Khan's 阮侃 essay 'Residence is Devoid of Good and Bad Fortune' and the ensuing controversy with Hsi Khang 嵇康, Chang Liao-Shu's 張遼叔 essay entitled 'People Naturally Delight in Learning' with Hsi Khang's reply, and Hsi Khang's wonderful essay 'On Nourishing Life', with the ensuing controversy with Hsiang Hsiu 向秀,[4] are fine examples of the art of debating in the +3rd century. Given this fashion of public argument, it is not surprising that we also find, during the same period, a certain interest in logical problems as such.

As far as explicit logical reflection is concerned, a crucial figure was the philosophical prodigy Wang Pi 王弼 (+226 to +249),[5] whose recently discovered *Lao Tzu Chih Lüeh* 老子指略[6] turns out to be a rather interesting document in the history of Chinese logical thought. As Dan Daor has confirmed, the *Yin Wên Tzu* 尹文子, spuriously attributed in its title to the pre-Chhin debater Yin Wên 尹文, must probably also be attributed to the +3rd century.[7]

[1] Wilhelm (1947), p. 76.   [2] Cf. Graham (1986a), p. 179.

[3] For the general background see E. Balasz (1964). There are two excellent books on Hsi Khang: Holzman (1957) and Henricks (1981a). Cf. also Shryock (1937). Important Chinese works on the intellectual history of the +3rd century include Fan Shou-Khang (*1936*), Ho Chhi-Min (*1966*). Ho Chhi-Min (*1967*), Liu Ta-Chieh (*1939*), Mou Tsung-San (*1963*) and Thang Yung-Thung (*1957*). Especially useful for our purpose are Daor (1979) and Mou Tsung-San (*1960*).

[4] All these controversies are presented *in extenso* in Henricks (1981a), pp. 21–70, 135–43, and 144–99.

[5] Cf. Petrov (1936) and the review of this, Wright (1947). Lou Yü-Lieh, *Wang Pi chi chiao-shih*, published in 1980, provides a splendid critical edition of Wang Pi's works. The best translation of Wang Pi's commentary is Rump and Chan (1979).

[6] Lou Yü-Lieh (see preceding note), pp. 195–210, translated in Daor (1979), pp. 142–59.

[7] *Tao Tsang*, vol. 46, pp. 279–279. Cf. Masson Oursel (1918), bibl. p. 24, no 97. A convenient recent edition with commentary is Li Shih-Hsi (1977). Daor (1979), pp. 104–41, provides a reliable annotated translation.

It was during the +3rd century that the abstract term *ming li* 名理 became common,[1] a term which remained the vague general word for logic (or perhaps we should rather translate anachronistically: analytical philosophy) until the 20th century.

There are two main logical areas in which the +3rd-century intellectuals showed an interest: the correspondence of names to shapes, and the relation of words to meanings. Already in the *Chuang Tzu* we find a rather cryptic remark that 'names are the "guests" of realities',[2] and in one late Han book we read:

Names are that by means of which we name realities (*shih* 實). The reality is there and the name adapts (*tshung* 從) to it. It is by no means so that the name is there and the shape (*hsing* 形) adapts (*tshung*) to it. Thus when there is a long shape, we call it long. When there is a short shape, we call it short. It is by no means so that the names 'long' and 'short' are there first and that the long and short shapes adapt (*tshung*) to them.[3]

All names have their origin in shapes. It never happens that a shape is born from a name.[4]

Ouyang Chien 歐陽建

Ouyang Chien (+268 to +300) wrote a dialogue between The Conformist Gentleman and The Master Who Defies the Masses entitled *Yen chin i lun* 言盡意論 (On That Words Exhaust Meaning).

The Conformist gentleman said: 'All the debaters of the world consider that words do not exhaust meaning, and this has a long tradition.[5] All men of large talent and penetrating knowledge consider that this is so. . . .'

The Master Who Defies the Masses replied: 'Heaven does not speak, but the Four Seasons follow their perfect path. The Sage does not speak, but his awareness and knowledge follow their natural path. A shape does not depend on a name for its roundness or squareness to become evident. Colours do not depend on an appellation for the black or white to become manifest. Thus names have no active rôle towards things. Speech is inactive (*wu wei* 無為) towards the patterns of things (which it describes).

Why then is it that in ancient and modern times the Sages were unable to get rid of words? It was truly because if you have got hold of a pattern of things in your mind, it is only by words that you can bring it out. When you have pinned down an object, it is only by names that you can mark it out.'

---

[1] Cf. Mou Tsung-San (*1960*), pp. 2 and 7ff. Since Mou Tsung-San wrote this, the discovery of the Ma-wang-tui manuscripts has revealed that the term *ming li* 名理 was not, as Mou Tsung-San claims, coined during the +3rd century, but was in fact known at the very least in early Han times. Cf. Anon. (*1976*), pp. 42f., where a document entitled *ming li* recommends 'investigating the principles of names (*ming li*)'.

[2] *Chuang Tzu* 1.15; cf. Graham (*1981*), p. 45.   [3] *Chung Lun*, ed. *SPTK*, ch. 2, p. 6a; cf. Daor (1979), p. 68.

[4] Wang Pi, *Lao Tzu Chih Lüeh* 老子指略, ed. Lou Yü-Lieh, p. 199.

[5] Compare Lao Tzu's opening line 'The way that can be shown is not the constant way, the name that can be named is not the constant name' and Chuang Tzu's *dictum* 'What can be expounded with words are the crude things'. (*Chuang Tzu* 17.23). For a good account of the debate on meaning in the +3rd century see Mou Tsung-San (*1960*), pp. 12ff. There is also evidence that Hsi Khang had earlier written an essay entitled 'Words do not exhaust meaning', thus defending the famous statement in the Great Appendix to the *I Ching* 易經. One would indeed love to read this essay, but, unfortunately, it is no longer extant. (Cf. Henricks (1981a), p. 22.)

If speech does not bring out what is on the mind, it does not relate to anything. If names do not distinguish things, awareness and knowledge are unelucidated. . . . It is not as though things inherently have names, or that patterns of things have definite appellations. If you want to mark out a (part of) reality, you give it a separate name. If you want to express your thinking (*chih* 志), you set up a way of expressing it. Names change with things, speech is transformed with the patterns of things. This is like a sound being followed by an echo, like a shape being immediately followed by its shadow. These must not be regarded as separate things. When they are unseparated, there is nothing (in meaning) that is not exhaustively articulated.[1]

Yin Wên Tzu 尹文子 *on names*

The *Yin Wên Tzu* is a book representing rather miscellaneous but predominantly political interests. In a passage concerned with logic it sets up a classification of names as follows:

There are three categories of names (*ming* 名) . . .

1. names referring to things, like square, round, white, black.
2. names of (objective) blame and praise, like good, bad, noble, base.
3. non-descriptive names (*khuang ming* 況名),[2] like 'consider as talented', 'consider as stupid', 'love', 'hate'.[3]

The *Yin Wên Tzu* provides one of the rare attempts in ancient China at something approaching grammatical reflections. My translation is tentative:

Now in 'befriending the talented', 'keeping away from the incompetent', 'rewarding the good' and in 'punishing the bad', the appropriateness of the names 'talented', 'incompetent', 'good', 'bad' is in the objects (*pi* 彼), whereas the appropriateness of the designations (*chhêng* 稱) 'keep at a distance from', 'befriend', 'reward', and 'punish' is attached to the subject (*wo* 我). When the subjective (*wo*) and the objective (*pi*) each get their names that is called clear-mindedness about names. When you use the names 'talented', 'incompetent' like 'befriend' and 'keep away from'; when you use the names 'good' and 'bad' like 'reward' and 'punish', i.e., when you scramble the subjective (*wo*) and the objective (*pi*) designations

---

[1] Wang Tien-Chi (*1979*), p. 265, and cf. Mou Tsung-San (*1960*), p. 12. In the +4th-century commentary to the *Lieh Tzu* 列子 we find a short summary of the dialogue: 'Names change with things, speech changes with the pattern of things (*yin li* 因理). These pairs (i.e., names versus things and speech versus patterns of things) must not become separated from each other. As long as they are not separated from each other, words will exhaustively describe things. (Wang Tien-Chi (*1979*), p. 265, who quotes from Chang Chan's (*fl.* +370) commentary to the book *Lieh Tzu*.)

[2] The technical term *khuang* 況 'to compare, how much more so' is hard to understand properly. I rely on the context in this passage for my interpretation. Cf. Wang Pi's statement: 'The Way is a designation for Nothing. There is nothing which does not reach it. There is nothing which does not follow it. As a non-descriptive designation we call it 'the Way' (*khuang chih yüeh tao* 況之日道)' (ed. Lou Yü-Lieh, p. 624). Kuo Hsiang (died +312) writes: 'One must forget the words and look for what is being indirectly expressed (*i wang yen i hsün chhi so khuang* 宜忘言以尋其所況).' Chheng Hsüan-Ying's subcommentary expands: 'One must look for the overall meaning indirectly expressed (*i hsün chhi chih khuang* 以尋其旨況).' (Kuo Chhing-Fan (*1961*), p. 24. Cf. also *ibid.*, pp. 525 and 947. The old commentary to the *Kung-Sun Lung Tzu*, ed. *SPPY*, p. 3b, has a comparable interesting use of *khuang*, where the word is parallel to *yü* 喻 'to illustrate' and means something like referring to something without using its name. Han Khang-Po's commentary on *I Ching*, ed. *SPPY*, 40a10 and 47a3 gives further examples. Cf. Daor (1979), p. 78.

[3] *Yin Wên Tzu*, ed. Li Shih-Hsi, p. 5; *Tao Tsang*, p. 2a; cf. Daor (1979), p. 107.

into one and do not distinguish between them, that is called 'confusion of naming' (*ming chih hun* 名之混).[1]

The dictionary *Shih Ming* 釋名 distinguishes:

Name (*ming* 名): to make plain. One names objects or facts and makes them distinct and clear.
Appellation (*hao* 號): to call, to name according to its good or bad qualities.[2]

Kuo Hsiang makes a distinction between *ming* 'designate, refer to' and *chhêng* 'call something by a name':

Now what one calls (*chhêng*) (the emperors) Yao and Shun only names (*ming*) their dust and dregs.[3]

Another distinction that caught the attention of the anonymous author of the *Yin Wên Tzu* was that between what we would call a transitive verb and its object, and particularly the sharp logical distinction between what we would call the verb/object relation and the subject/predicate relation:

Consider the phrase (*yü* 語) 'loves oxen'. 'Loves' is a designator which reaches out to things (*wu chih thung chhêng yeh* 物之通稱也). 'Oxen' (refers to) a specific shape of things (*wu chih ting hsing* 物之定形). There is no limit to the possibility of putting a designator that reaches out (*thung chhêng* 通稱) and then (a name which) fixes the shape (of the object to which it reaches out). For example if we make the complex phrase 'love horses', we tie (the designator which reaches out) to 'horse'. So there are no bounds to what 'loves' can reach out to. For example, if we make the complex phrase 'loves men', then the object (*pi* 彼) belongs to the class of 'men' (*shu yü jên yeh* 屬於人也), so that 'loves' is not a man, and man is not 'loves', and the (complex) phrases 'loves oxen', 'loves horses', 'loves men' are inherently separate. Therefore it is said that naming (*ming*) and setting up attitudinal distinctions (*fên* 分) must not be confused.[4]

Attitudinal distinctions (*fên*) (as opposed to objective distinctions (*pien* 辯) discussed elsewhere in earlier Chinese logical literature) apparently fascinated the author of the *Yin Wên Tzu* to such a degree that he felt impelled to give us a rich set of examples in an effort to make his meaning clear:

The applicability (*i* 宜) of names attaches to the object (*pi*). The applicability (*i*) of the setting up of attitudinal distinctions (*fên* 分) attaches to me (*wo* 我).
In 'loves white' and 'hates black', 'finds the *shang* 商 note agreeable', 'rejects the *chih* 徵 note', 'likes the smell of goats', 'dislikes the smell of burnt food', 'is fond of sweets', 'avoids bitter taste', the terms 'white', 'black', '*shang* note', '*chih* note', 'goaty smell' 'smell of burnt food', 'sweet taste', and 'bitter taste' are names reaching out to the object (*pi*). 'Loves', 'hates' 'finds agreeable', 'rejects', 'likes', 'dislikes', 'is fond of', 'avoids', are my attitudinal distinctions (*fên*). When the names (*ming* 名) and attitudinal distinctions (*fên*) are fixed, then everything will fall into place.[5]

We have here a rare instance of intensive logical reflection on common words.

[1] *Yin Wên Tzu*, ed. Li Shih-Hsi, p. 9; *Tao Tsang*, p. 3a; cf. Daor (1979), p. 109.
[2] Cheng Tien and Mai Mei-Chhiao (*1965*), p. 286.   [3] Kuo Chhing-Fan (*1961*), p. 33.
[4] *Yin Wên Tzu*, ed. Li Shih-Hsi, p. 9; *Tao Tsang*, p. 3a; cf. Daor (1979), p. 110.
[5] *Yin Wên Tzu*, ed. Li Shih-Hsi, p. 10; *Tao Tsang*, p. 3b; cf. Daor (1979), p. 111.

Wang Pi 王弼 on names

The distinction between the objective and the subjective names is also prominent in the logical part of Wang Pi's (+226 to +249) summary of Lao Tzu's teaching. He tries to work out a distinction as follows:

Names (*ming*) are that which fixes the object.
Designations (*chhêng* 稱) come from the person who uses them (*wei chê* 謂者). Names (*ming*) originate in the object.
Designations (*chhêng*) emerge from oneself. . . . Names (*ming hao* 名號) originate with shapes. Designations (*chhêng wei* 稱謂) emerge through conscious effort.[1]

Wang Pi explains his interest in logical problems in a coherent way:

If someone cannot intelligently discriminate between names (*pien ming* 辯名), then one cannot discuss principles (*li* 理) with him. If someone cannot fix names (*ting ming* 定名), one cannot discuss realities with him. Every name originates in shapes, and shapes do not originate in names. Thus when something has a given name, there must necessarily be a given shape present. When something has a given shape, there must be the corresponding distinction. 'Benevolence' cannot be called 'sageliness', 'wisdom' cannot be called 'benevolence', and in that way every (name) has its corresponding reality (*shih* 實).[2]

Occasionally, Wang Pi carries his urge for precision of definition surprisingly far, as when he produces the following, almost Aristotelian, elaborate discourse on the notion of *chin* 近 'close', which I find in his commentary on the *Lun Yü*:

Confucius says 'by nature they are close'. If they were completely the same, the formulation 'are close' would not arise. If they were completely different, the formulation 'are close' could also not be established. When it now says 'close', this means that there are the same features and different features, and it picks out (*chhü* 取) common features. Their neither being good nor bad is the same. Their having more or less of each (goodness and badness) are different features. Although they are different, they are not far from each other, therefore they are called 'close'.[3]

Wang Pi had the notion of a 'general term':

All these are called gentlemen, and that surely is a general designation for people who have moral charisma (*yu tê chê chih thung*[4] *chhêng yeh* 有德者之通稱也).[5]

Wang Pi was not a logician. He was primarily a commentator. But in the course of his commentatorial work he had occasion for logical reflection and a certain amount of analytical innovation.[6]

---

[1] Wang Pi, *Lao Tzu Chih Lüeh*, ed. Lou Yü-Lieh, pp. 197f.; cf. Daor (1979), p. 152.
[2] Ed. Lou Yü-Lieh, p. 199.
[3] Ed. Lou Yü-Lieh, p. 632. One may still point out that Aristotle's approach to the problem of definition would have been more abstract than this, but the spirit here is clearly the spirit of analytic enquiry.
[4] Kuo Hsiang 郭象 (died +312) defined *thien ti* 天地 as 'a general term for all things (*wan wu chih tsung ming* 萬物之總名)'. (Kuo Chhing-Fan (*1961*), p. 20.)
[5] Ed. Lou Yü-Lieh, p. 624.
[6] For example, we find in his work the notion of a general as opposed to a specific concept: He says that 'gentleman' is a general (or 'pervasive') term (*thung chhêng* 通稱) for a person who has moral charisma. (Ed. Lou Yü-Lieh, p. 624). This concept is also used by Ma Jung in his commentary on the first chapter of *Lun Yü*. However, the explicit distinction between *thung chhêng* 'general term' and *pieh chhêng* 別稱 'specific term' seems to have emerged somewhat later in the commentatorial tradition.

In general, it would be an exaggeration to speak of a school of logicians in the +3rd century. For all we know, interest in logic did remain somewhat incidental in this period. However, in the context of Chinese intellectual history even such a marginal interest in logic is significant because such interest is generally so rare. That is why I have taken the trouble to give some examples of logical reflections of the +3rd century.

(g) CHINESE BUDDHIST LOGIC

(1) THE EVOLUTION OF BUDDHIST LOGIC

The total extent of indigenous Chinese writing on logic which we have discussed so far is quite insignificant when compared with the considerable bulk of Chinese texts on Buddhist logic. Buddhist logic played a marginal rôle in Chinese intellectual history, but we do have a great deal of it, and this has received comparatively little sinological attention in Western languages since Sadajiro Sugiura's and E. A. Singer's first exploratory work *Hindu Logic as Preserved in China and Japan*, published in 1900, and since Giuseppe Tucci's contributions to the subject thirty years later.[1]

How did Buddhist logic get to China? How was it perceived and received by Chinese intellectuals? What were the main logical doctrines expounded by Buddhists using the Chinese language? Was Buddhist logic translatable into a language as different from Sanskrit as Chinese without serious loss of meaning and intellectual perspective? We shall want to discuss these questions separately, and we intend to discuss them in a sinological rather than indological context. We shall look closely at Buddhist logic as it would have appeared to a Chinese of the +7th and +8th centuries in its Chinese linguistic form. We shall use Chinese texts as sources for Chinese thought, not at this stage for the reconstruction of an essentially Indian intellectual history. Finally, and quite separately, we shall pay proper attention to the fascinating questions of cultural transmission and translation from the Sanskrit originals. But first, we must turn to a historical and bibliographic orientation in this difficult terrain.

Historical and bibliographic orientation

Buddhist logic (Chinese: *yin ming* 因明)[2] originated in India, not with the Buddhists but with other philosophical schools, particularly those known as *Vaiśeṣika*, *Mīmāṃsā* and *Nyāya*.[3] In India, the origins of logic were linked to the interpretation of the

---

[1] The one notable exception being Richard S. Y. Chi's highly technical reconstruction of the formalism underlying Buddhist logic as practised in China in his *Buddhist Formal Logic* published in 1969.

[2] The term *yin ming* 因明 is derived and translated from the Sanskrit *hetuvidyā*.

[3] The pioneering account of Indian logic is Vidhyabusana (1971), which remains an outstanding work of reference today. Khuei Chi 闚基, in the opening remarks to his Great Commentary on the *Nyāyapravesa*, creates the misleading impression that *yin ming* originated with the Buddhists.

Vedas and the practice of public philosophical debate, as well as to the science of linguistics as it developed under the influence of Pāṇini. The first autonomous development of logic (known as *nyāya*) can be dated with any certainty at most to the +3rd century. The Yogācāra Buddhist Vasubandhu (+5th century) was an early figure to show a distinct interest in logical problems.

Some texts of a certain logical interest were translated into Chinese by Indian missionaries during the +5th and +6th centuries. Three such works have been introduced in G. Tucci (1929a): The *Tarkaśāstra*,[1] attributed to Vasubandhu and translated as *Ju Shih Lun* 如實論 by the monk Paramārtha between +552 and +557, the *Upāyahṛdaya*,[2] tentatively attributed to Nāgārjuna and translated as *Fang Pien Hsin Lun* 方便新論 between +471 and 476,[3] and the *Vigrahavyāvartanī*,[4] more plausibly attributed to Nāgārjuna, translated in +541 as the *Hui Ching Lun* 廻靜論.

Dignāga (+6th to early +7th century) was the first great practitioner of Buddhist logic as we know it today and was duly reviled for his strong scientific interest in the validity of arguments and for his rejection of argumentation by reference to religious authority.[5]

Two complete works, both representing early stages of Dignāga's logical thinking, were translated into Chinese. Later, more sophisticated, developments of logic in India were not taken over by the Chinese Buddhists. This was not because these later developments were unknown. Hsüan-Tsang 玄奘 must have been familiar with them. These developments were neglected because they were found uninteresting. Names such as that of Dharmakīrti were familiar to practitioners of Chinese Buddhist logic in +8th-century China, but the intense epistemological interest of the later Dignāga as well as his rigid logical formalism did not win the interest of the Chinese Buddhists.

The tradition of Buddhist logic which did win Chinese interest was that of Dharmapāla (Hu Fa 護法, flourished during the early +7th century), who emphasised the use of logic in practical argumentation, and for whom the art of Buddhist logic was not abstractly formalistic but was essentially linked to the social practice of debate and argumentation.

The early history of Indian logic must today be studied largely on the basis of the materials transmitted to Tibet and China and then to Japan (and to Korea).[6] However, while logic struck strong roots in Tibet and flourished within Japanese Buddhism, it mostly remained a marginal phenomenon in Chinese intellectual

---

[1] *Taishō Tripitaka* no. 1633.   [2] *Taishō Tripitaka* no. 1632.

[3] Apparently the work had been translated before or during the Eastern Chin dynasty (+317 to +420).

[4] *Taishō Tripitaka* no. 1631.

[5] For basic orientation on early Buddhist logic in India, see Keith (1921), Randle (1930), Tucci (1929a) and Bochenski (1960), pp. 416–46. Nakamura (1980) provides a singularly useful first bibliographic guide to this vast area of scholarship. Matilal (1971) provides a modern logical perspective on the Indian logical tradition. Potter (1977) is the most authoritative recent account to date of the early development of logic in India. Cf. also Junankar (1978).

[6] There is a good reason why Giuseppe Tucci's classic work *Pre-Dinnāga Buddhist Texts on Logic* has a subtitle *from Chinese Sources*. It turns out that the main sources on Buddhist logic from this early period are indeed Tibetan and Chinese.

life.[1] Only during the second half of the +7th century and during the renaissance of Buddhism in the early +20th century did Buddhist logic make an important impact on Chinese intellectuals.

The crucial figure for the Chinese reception of Buddhist logic was Hsüan-Tsang 玄奘 (+602 to +664)[2] with his disciples, notably Khuei Chi 窺基 (+632 to +682),[3] with whom he founded the school of Consciousness Only (wei shih 唯識) also known as the sect of Consciousness Only (fa hsiang wei shih tsung 法相唯識宗).[4]

Buddhist logic was introduced to China as an integral part, one might even say the methodological organon (hsiao tao 小道), of the Buddhist 'theology' of this school.[5]

Consciousness Only was an intensely intellectual and demanding school of Buddhism. It was not naturally designed for popular consumption. As Consciousness Only was understandably eclipsed by other less obsessively intellectual schools like the Pure Land and Chhan 禪 (Sino-Japanese: Zen), this also spelt the end of the creative phase in Chinese Buddhist logic.

When Consciousness Only was revived during the late 19th and early 20th century, this meant a revival also for Buddhist logic in China.

In +629 Hsüan-Tsang set out for India on an unauthorised and illegal lonely journey in order to enquire about doubtful points of doctrine and in order to find a solution to the conflict between the Southern and the Northern schools of Chinese Buddhism. During many years of study in China with leading Buddhist teachers of his time he had found that they were unable to give clear answers to his doctrinal questions. He decided to go to India which at that time still was the centre of Buddhist learning.

In India, Hsüan-Tsang studied with a number of distinguished masters of Buddhist logic, notably with Śilabhadra, a student of one of the logician Dignāga's students, the colourful learned hermit Jayasena, and Prajñābhadra, an authority on grammar and logic. He returned to Chhang-an sixteen years later in grand style, and with no less than 657 Buddhist treatises in his baggage. This was not a mean load, since the books had to be packed in 520 crates. 37 of these treatises were concerned with the study of yin ming 因明 or Buddhist logic.

Under imperial patronage, Hsüan-Tsang organised a large-scale translation programme, and he found time to translate 74 of the texts he had brought along from India. Among these a small proportion only were on logic.[6] One reason for this

---

[1] Kenneth Chen (1964), the classic Western work on the history of Buddhism in China, does not even mention Buddhist logic or yin ming in its index.

[2] See Julien (1853), Grousset (1929) and Samuel Beal (1911). For a briefer and more readable orientation see Waley (1952), pp. 11–130.

[3] See Weinstein (1959) and Taishō Tripitaka no. 2061 for his biography. Khuei Chi 窺基 was known as Master Chi (Shih Chi 師基) in his own time. The first part of his name, Khuei, is first attested during the Sung dynasty.

[4] For this school of Buddhism see Kenneth Chen (1964), pp. 320–5. Other important translators in the history of Chinese logic include: Shen Thai 神泰 and Wên Pei, 文備 Ching Mai 靖邁, Ming Chüeh 明覺, Wên Kuei 文軌, Hui Chao 慧沼 (+650 to +714) and Chih Chou 智周 (+668 to +723), all of the +7th and +8th centuries.

[5] See Lü Chheng (1982), p. 383, for an illustration for the importance of yin ming 因明 within the scheme of Consciousness Only. This has not always been appreciated properly. Fung Yu-lan (1953) devotes no less than 39 pages (vol. 2, pp. 299–338) to Hsüan-Tsang's great Chhêng Wei Shih Lun 成唯識論. But he does not introduce yin ming at all. Neither does Chan (1963).

[6] Cf. Chi (1969), p. lxxv.

must have been that Buddhist logic in India had its social roots in the wide-spread practice of public philosophical debate, whereas this social practice never quite took root in China. Correspondingly, there was no natural need for logical treatises. Not so surprisingly, Hsüan-Tsang introduced none of Dignāga's advanced works of logic.

In the third year after his return from India Hsüan-Tsang introduced Śaṅkara-svāmin's[1] *Nyāyapraveśa* (translated in +647 as *Yin Ming Ju Chêng Li Lun* 因明入正理論),[2] a brief introductory manual on Dignāga's logic.[3] Whatever its merits in the history of Indian logic, this short work was to become the main source of information and the main focus of interest among Buddhist logicians in China.

In Sung times (+960 to +1279) the Chinese version of the *Nyāyapraveśa* was translated into Tibetan. The Tibetans mistook the book to be Dignāga's famous *Nyāyamukha* of which they had heard, and the same happened to another Sanskrit version which they translated into Tibetan around the +13th century.

In India itself, the Sanskrit original of the *Nyāyapraveśa* was considered as lost even by the great pioneer historian of Indian logic, Vidhyabusana. However, since then a commentary by Haribhadra (+11th century; see Mironov (1927)) and two subcommentaries (B. Dhruva (1968)) have been edited. Haribhadra was a Jain, and the fact that he should have written a commentary on the *Nyāyapraveśa* testifies to the popularity of the work even outside Buddhist circles. Ui Hakuju 宇井伯壽 (1882 to 1963), the great scholar of *yin ming* 因明, has collated the various Chinese, Tibetan and Sanskrit versions at the end of his splendid book *Bukkyō ronrigaku* 佛教論理學, and his general conclusion is that the Chinese translation comes closest to what one must assume to have been the original Sanskrit version.

In +655 Hsüan-Tsang attempted the translation of the *Nyāyamukha* (*Yin Ming Chêng Li Mên Lun* 因明正理門論) by Dignāga.[4] This was to remain the only other important Indian book on Buddhist logic in China. The *Nyāyamukha* in the translation we have of it remains a singularly inaccessible work (in spite of G. Tucci's courageous attempt at a translation) and in any case it never achieved anything like

---

[1] His names translates as 'Skeleton'. His dates are unknown, but he must have been Dignāga's disciple at an early stage, since he makes no references to Dignāga's later logical theories. There are no other works attributed to him. Judging by his name he must have been a brahmin.

[2] Cf. the Sanskrit text in Mironov (1931) ed. and reconstructed. T'oung Pao, 1931, no. 1–2, pp. 1–24), and Mironov (1927) (Dignāga's Nyāyapraveśa and Haribhadra's Commentary on it; originally *Jaina Shasan*, extra number Benares 1911). The most convenient Sanskrit text is Tachikawa (1971), which also includes an annotated English translation from the Sanskrit. Tibetans have attributed the work to Dignāga himself, the Chinese to a disciple of his. On the extensive controversy concerning the authorship see Tubianski (1926), Tucci (1928a), pp. 7ff., Nakamura (1980), p. 300, and Chi (1969), p. lxxiv. By far the best edition of the Chinese translation is Lü Chheng (*1983*). Huo Thao-Hui (*1985*), pp. 43–135, provides more Westernized philosophical perspectives on the Chinese text.

[3] The *Nyāyapraveśa* was also known in China as 'The Little Treatise (*Hsiao Lun* 小論)' to distinguish it from the somewhat longer and more authoritative *Nyāyamukha*, translated into Chinese two years later. The title *Nyāyapraveśa* can perhaps be interpreted as 'An introduction to the *Nyāyamukha*'. Cf. also Chi (1969), p. 124.

[4] Cf. Tucci (1930). The best edition of the Chinese text of this with textual notes is by Lü Chheng and Yin Tshang (*1927*). This work is singularly useful because it provides systematic cross-references to Dignaga's *Pramāṇasamuccaya*. There is another translation of this work by I Ching (+635 to +713, *Taishō Tripitaka* no. 1629). Lü Chheng and Yin Tshang (*1927*), p. 336, suspect that the great traveller I Ching gave up his attempt at translating the book and that disciples filled in Hsüan Tsang's text under I Ching's new translations of the chapter headings. For the date of this work see Lo Chao (*1981*).

the popularity of the *Nyāyapraveśa* for this reason. Giuseppe Tucci, undoubtedly one of the most formidable Western scholars of Buddhist literature, writes on the *Nyāyamukha*: 'The treatise, small as it is, is one of the most difficult books that we have in the Chinese canon.'[1] The *Nyāyamukha* was not the subject of many commentaries, but a fragment of Shen Thai's 神泰 commentary on it, dated +655, survives.[2]

The *Nyāyamukha* and the *Nyāyapraveśa* are often presented as the only treatises of Buddhist logic which were translated into Chinese. But the *Yü Chhie Shih Ti Lun* 瑜伽師地論 (*Yogācārabhūmiśāstra*), chapters 15 and 18, (translated in +645) contains important surveys of Buddhist logic. Moreover, in +649 Hsüan-Tsang also translated a text which exemplifies the practical use of *yin ming* 因明, the *Chang Chên Lun* 常珍論.[3] In +650 he translated the *Kuang Pai Lun Shih* 廣百論釋[4] which is rich in examples of Buddhist logical practice. Finally, *Chhêng Wei Shih Lun* 成唯識論[5] contains material relevant to the study of *yin ming*.

The Chinese commentaries to 'Nyāyapraveśa'

As Hsüan-Tsang 玄奘 translated the *Nyāyamukha* and the *Nyāyapraveśa* he expounded their meaning to his disciples who were very eager to note down the esoteric new teaching they received.

The tremendous difficulty of the subject was no doubt vividly perceived by everyone present. Khuei Chi 窺基, who was to become Hsüan-Tsang's most trusted disciple in the end, writes:

Dignāga made detailed investigations (of the earlier literature on the subject) and produced his *Nyāyamukha* and other works. Although the basic teaching was orderly and lucid, the meaning was obscure and the words recondite. The beginner was unable to fathom its subtleties.[6]

Writing commentaries on *yin ming* became something of a rage.[7] Twelve commentaries to the *Nyāyapraveśa* survive in the *Taishō* 太正 Tripitaka, published in Tokyo, and the Supplementary Tripitaka, published in Kyoto.[8] In practice it is impossible to distinguish in these commentaries between the opinions of Hsüan-Tsang 玄奘 and those of his disciples. These commentaries, in any case, are the main basis for our understanding of the development of Buddhist logic in China.

---

[1] Tucci (1930), p. 2.

[2] *Taishō Tripitaka* no. 1839, *Supplementary Tripitaka* 1, vol. 86, no. 717. Cf. the partial translation in Tucci (1928a).

[3] *Taishō Tripitaka*, vol. 30, pp. 268–78.   [4] *Taishō Tripitaka*, vol. 30, pp. 187–250.

[5] See Wei (1973).   [6] Great Commentary 1.1949.

[7] The earliest commentaries were those by Shen Thai 神泰, Ching Mai 靖邁 and Ming Chüeh 明覺, all of which are lost. The second generation of commentaries was by Wên Pei 文備 and Wên Kuei 文軌 of which by far the more popular was Wên Kuei's commentary which survives today as the *Yin Ming Ju Lun Chuang Yen Shu* 因明入論莊嚴疏.

[8] Cf. Nakamura (1980), p. 300. The commentaries are conveniently listed up in Chi (1969), p. 189. A useful but quite outdated first orientation on Buddhist logic in China and in Japan is Sugiura and Singer (1900). Bibliographies of the relevant Classical Chinese sources may be found in the Japanese *Bussho Kaisetsu daijiten* 佛書解説大詞典, Tokyo 1934, pp. 182–206.

The earliest surviving commentary, and the most disciplined and philosophically inspiring, is that by Wên Kuei 文軌.[1] By far the most comprehensive and encyclopaedic of the commentaries is that by Khuei Chi 窺基, the *Yin Ming Ju Chêng Li Lun Shu* 因明入正理論疏, more conveniently (and very appropriately) known as the *Yin Ming Ta Shu* 因明大疏 'The Great Commentary on Buddhist Logic', or simply as *Ta Shu* 大疏 'The Great Commentary'. We shall follow Chinese practice and refer to the book simply as the Great Commentary.[2]

The textual history of the Great Commentary is complicated. It appears that Khuei Chi only finished five sixths of it, and that Hui Chao (+650 to +714) finished the rest for him. Hui Chao 慧沼 also produced a sub-commentary or an extended version on the Great Commentary, known as the *Hsü Shu* 續疏, which was published in two volumes by the *Chih-na Nei-hsüeh-yüan* 支那內學院 in 1933.[3]

The very bulk and detail of the *Great Commentary*, and the fact that it made extensive use of earlier commentaries by men like Shen Thai 神泰 meant that many of the earlier commentaries were consigned to oblivion. The last of the early commentators was Chih Chou 智周 (+679 to +733), three of whose commentarial works are printed in the Supplementary Tripitaka.[4]

Yin ming 因明 logic spread to Japan, where +7th- and +8th-century material on Buddhist logic was collected in the *Yin Ming Lun Shu Ming Têng Chhao* 因明論疏明燈抄.[5] There also survives a Japanese commentary on the *Nyāyapraveśa*, written in Chinese by Hōtan 鳳潭 (+1653 to +1738),[6] which provides material from now-lost earlier Chinese sources.[7]

The transmission of Indian Buddhist logic to China is linked to Hsüan-Tsang, perhaps the best-known Chinese monk of all time,[8] but this did not ensure a great success for Buddhist logic in that country. Even those doctrines of the School of Consciousness Only (*wei shih* 唯識) which Hsüan-Tsang had felt were suitable for the Chinese context, did not fare well in China itself.[9]

A most interesting question is that of the cultural impact of Buddhist logic in the China beyond the confines of monasteries. In fact, there is one *cause célèbre* in

[1] See Supplementary Tripitaka, series 1, vol. 86, no. 718, and the wonderful expanded re-edition *Yin Ming Ju Lun Chuang Yen Shu* published in 1932, which relates Wên Kuei's comments to parallel passages in the Great Commentary by Khuei Chi. Together with the *Zuigenki* 瑞源記 these are the standard reference editions of the most important texts on Chinese Buddhist logic. On Wên Kuei's commentary see Takemura (1969).

[2] Of all the commentators, Khuei Chi seems to have been the one most interested in the original Sanskrit words. One might be tempted to doubt whether Khuei Chi actually was fully literate in non-Buddhist Chinese. But by the time one reads his introduction to the Great Commentary one is quite certain that Khuei Chi was a learned Chinese scholar well versed in the Confucian classics.

[3] Cf. also *Supplementary Tripitaka*, vol. 87, no. 722.   [4] First Series, nos. 723 to 725.

[5] *Taishō Tripitaka* no. 2270.   [6] Cf. Anon, *Tetsugaku jiten*, Kōdansha, Tokyo 1971, p. 1299.

[7] This work was reprinted in 1928 by the Commercial Press in Shanghai.

[8] The classical novel *Hsi Yu Chi* 西游記 brilliantly translated by Arthur Waley as 'Monkey', deals with the exploits of Hsüan-Tsang (or Tripitaka) on his way to India.

[9] The study of Chinese Buddhist logic has to take account of the many sources written by Japanese Buddhist logicians. Moreover, much of the detailed modern research on Buddhist logic has been conducted by Japanese scholars. Notable Chinese scholars include the philologist Lü Chhêng, the philosopher and logician Chhên Ta-Chhi, and in the younger generation Lo Chao, who continues Lü Chhêng's tradition and works with both Sanskrit and Tibetan sources.

Chinese intellectual history that is directly connected with Buddhist logic. The case involves the celebrated polymath Lü Tshai 呂才 (+600 to +665).[1] Lü Tshai was employed as one of the emperor's two chief physicians. He was also known as an astronomer and as an authority on dance as well as music. By all accounts, he was generally very fond of solving puzzles. Arthur Waley tells the story how the Emperor Thai Tsung 太宗 once came across a certain ancient book called *San Thu Hsiang Ching* 三圖相經 'The Treatise on Three-board Chess' and found he could not understand it. Lü Tshai 呂才, then thirty years old spent a night studying the book and was able to explain it completely and illustrate it with diagrams.[2] Similarly, when challenged to explain the notoriously tricky *Thai Hsüan Ching* 太玄經 by Yang Hsiung 揚雄, he was able to extemporise a reasonable account of the work.

In +655 Lü Tshai came across Dignāga's *Nyāyamukha*. The book was sent to him on account of its notorious obscurity by a Buddhist friend with the following note: 'This treatise is full of profound mysteries and extremely hard to get to the bottom of. Many monks of great intelligence and wide learning have found themselves unable to understand it. If you succeed in doing so, you may consider yourself a complete master of Buddhism no less than of lay studies.'[3]

Clearly, a man of the calibre of Lü Tshai could not dismiss this challenge. He studied the *Nyāyamukha* as well as the commentaries to the *Nyāyapraveśa* by Shen Thai 神泰 and others, and he found the interpretations in the available commentaries by Hsüan-Tsang's 玄奘 disciples contradictory (*so shuo tzu hsiang mao tun* 所説 自相矛盾) though all attributed to Hsüan-Tsang himself.[4] Lü Tshai was not a man to leave things at that. He produced his own critical exposition on the work under the title *Yin Ming Li Pho Chu Chieh I Thu* 因明立破注解意圖 'Explanations and Diagrams on Logical Demonstration and Refutation' in three *chüan* 卷. And as an intellectually fearless man, Lü Tshai offered, in the preface which is the only part of the work that survives,[5] to submit his work to the judgement of the Master Hsüan-Tsang himself.

Lü's preface gives splendid evidence of his independence of mind. He confesses that he had never heard of Buddhist logic before he saw the *Nyāyamukha*. He declares that he does not follow the accumulated wisdom of a master. As for the existing three commentaries,[6] he declares that he will give proper praise to those explanations that are good, but that he will refute those that are questionable. And very reasonably he proposes to make his meaning more intelligible at difficult points by introducing diagrams (*i thu* 意圖 'meaningful pictures').[7]

Buddhist reactions were, of course, fierce. Hsüan-Tsang's biographer Hui Li 慧立 compared Lü Tshai to the mouse which, because it could get to the top of the stove, thought it would have no difficulty in climbing Mt Khun-lang. It was felt to be

---

[1] For a detailed survey of what is known about him see Hou Wai-lu (*1957*), vol. 4, part 1, pp. 108–40. For discussion of Lü Tshai in the context of Chinese logic see Wang Tien-chi (*1979*), pp. 301–4. The episode we are about to relate is splendidly described in Waley (1952), pp. 107–11.

[2] Waley (1952), p. 107. [3] *Ibid.*, p. 109. [4] *Taishō*, vol. 50, p. 263. [5] *Taishō* no. 2053, pp. 262–6.

[6] Of these only a fragment of the commentary of Shen Thai survives, as mentioned above.

[7] Compare the late emergence of logical diagrams in the history of Western logic.

outrageous that Lü Tshai 呂才 as an outsider and non-Buddhist should presume to write on matters of Buddhist logic or indeed on any subject within the special competence of the Buddhist masters. As Hui Li put it, Lü Tshai 'with the qualifications of a commoner misappropriated (*chhieh* 竊) the explanations of many Buddhist masters'.[1] One of Hsüan-Tsang's collaborators, Ming Hsüan, came forward with a letter full of accusations and detailed criticisms, claiming among other things that Lü had misunderstood various Sanskrit terms and even the nature of Sanskrit inflection.[2]

Finally, a confrontation was arranged between Lü Tshai and Hsüan-Tsang. We would love to know what happened on this historical occasion, and particularly what arguments were exchanged. In any case, in the end Lü Tshai was humiliated and made to recant his rash opinions concerning Buddhist logic. Thus ended the single most significant encounter between Buddhist logic and a Chinese layman philosopher in ancient times.

In India, logic flourished until the +11th century and continued as an unbroken tradition (the so-called *navya-nyāya* or 'New Logic' school) right down to the 18th century,[3] while in China the interest in logic remained marginal even under the Thang (+618 to +906) and all but disappeared after the decline of Buddhism during the latter part of that dynasty. Many of the Chinese books on Buddhist logic will have survived for a time, and Yen Shou 延壽 (+904 to +975) did include some materials on logic in his *Tsung Ching Lu* 宗鏡錄.[4] However, during the upheavals of the +14th century, practically all the Chinese material on Buddhist logic was lost. When Ming scholars like Ming Yü 明昱, Chih Hsü 智旭, Chên Chieh 真界, and Wang Khên-Thang 王肯堂 felt a new curiosity for *yin ming* 因明 studies,[5] they had to base themselves on the very difficult texts of Hsüan-Tsang's translations together with second-hand materials such as they found in Yen Shou's *Tsung Ching Lu*.[6] The commentarial tradition was broken, and much of what we have of older Chinese works on Buddhist logic was preserved in Japan only, inaccessible to Chinese scholars during the Ming dynasty.

Wang Khên-Thang (born +1553, graduated +1589), one of the eminent scholars whom Matteo Ricci associated with in Nanking, has a rather moving preface to his commentary on the *Nyāyapraveśa*. With a friend he visits a Buddhist monk and the two are given a little booklet by the monk:

He looked pleased and handed it to us, saying: 'If you want to enter deep into the sea of learning, then this is your rudder and your oars.' We looked at it and it was the *Nyāyapraveśa*. We were all confused and did not know what to say . . . When I visited the Zen master Mi

[1] Yang Pai-Shun, *Xuanzang yu yinming*, p. 310.
[2] For details of the accusations see Waley (1952), pp. 274–5.
[3] The standard work on this is Ingalls (1951). Cf. also Matilal (1966).
[4] The *Tsung Ching Lu* 宗鏡錄 in one hundred *chüan* 卷 by Yen Shou 延壽 (+904 to +975) was not published until between +1078 and +1085. It became a very popular summary of Buddhist doctrines and has been reprinted and revised many times since.
[5] Cf. *ST* 1, vol. 87, nos. 726 to 729. These epigonic *yin ming* 因明 pursuits deserve a study in their own right.
[6] Cf. Shen Chien-ying (*1984*), p. 66.

Tsang, I saw a book on the table in his place. In a desultory way I opened it to see what it was, and it turned out to be a commentary on the *Nyāyapraveśa*.[1]

Without much success, Wang tried to study the *Nyāyapraveśa*, but whenever he asked other people's advice on it, he says, they all answered as follows:

This treatise has been known as the most mysterious and hard-to-read from the times of Hsüan-Tsang onwards. The old manuscripts of the commentaries and manuscripts are lost . . .[2]

Access to Chinese books on Buddhist logic may have been difficult during Ming times, but the situation became even worse during the Thai-phing Rebellion (1851 to 1865), when the Thai-phing leaders aimed at destroying all Buddhist repositories of learning within their control.

The historically crucial figure for the revival of Buddhist logic and Consciousness Only in China was Yang Wên-Hui 楊文會 (1837 to 1911).[3] In 1878, while in London on a diplomatic mission, Yang Wên-Hui met the great indologist Max Müller, and his disciple Nanjō Bunjū, the compiler of the still indispensable bibliographic guide to the Buddhist Tripitaka. In 1890, Yang wrote to Nanjō soliciting his help in collecting Buddhist books for him in Japan. As a result, a relative of Yang's, an official in the Chinese embassy in Japan, brought back hundreds of Buddhist texts that were not in the Chinese Tripitaka as it was then known. Among these happened to be Khuei Chi's 窺基 commentary on Hsüan-Tsang's 玄奘 *magnum opus*, the *Chhêng Wei Shih Lun* 成唯識論, published by Yang Wên-Hui 楊文會 in 1901. The edition we have used of the Great Commentary on *Nyāyapraveśa* was published five years earlier, in 1896, by the legendary Buddhist Press, the *Chin-ling khō-ching chhu* 金陵刻經處. Publications like these sparked off a revival of interest in Buddhism.[4]

Noteworthy among those Buddhist schools of thought that were revived in the early 20th century was the school of Consciousness Only, for which Yang Wên-Hui had shown a special interest.

In 1918, Ouyang Ching-Wu 歐陽竟無 (1871–1943), with Chang Ping-Lin 章炳麟 (1868 to 1936) and others, founded the China Institute of Inner Learning at Nanking to propagate Consciousness Only and *yin ming* 因明. In 1922 the abbot Thai Hsü 太虛 (1890–1947),[5] a student of Yang Wên-Hui and an enthusiast for *yin ming* as well as Consciousness Only, founded the competing but less important 'Wuchang Buddhist Institute'. These associations, and several others like them,[6] were places of considerable learning, and they had a pervasive influence on intellectual life in China.

---

[1] Wang Khen-Thang on *NP*, *ST* vol. 87, p. 53a top 5ff.
[2] *Ibid.*, top 6.   [3] For a biography see Yang Wen-Hui (*1923*).
[4] Cf. Wei Tat (1973), see *Ch'eng Wei Shih Lun*, p. 54. For more general accounts putting the revival of Buddhism into perspective see Brière (1956) and the authoritative Chan (1953). For the general social setting see (1968) (1967).
[5] See Yin-Shun (*1950*) and the less enthusiastic account in Welch (1968).
[6] For the most complete list available to date of such Buddhist seminaries see Welch (1968), pp. 285–7.

The school of Consciousness Only, and with it its central component, Buddhist logic (*yin ming*) attracted special attention from most of the leading intellectuals of the time: the scholar Chang Ping-lin 章炳麟 (1868 to 1936), the philosophers Hsiung Shih-Li 熊十力 (1883 to 1968) and Chhên Ta-Chhi 陳大齊 and the distinguished philologists Chhên Wang-Tao 陳望道, Hsü Ti-Shan 許地山, and Lü Chhêng 呂澂 all wrote treatises on Buddhist logic.[1] Men like Yen Fu 嚴復, Liang Chhi-Chhao 梁啟超, Khang Yu-Wei 康有為, Than Ssu-Thung 譚嗣同, Chhên Yin-Kho 陳寅恪, Thang Yung-Thung 湯用彤, and the grand old man of Confucianism Liang Shu-Ming 梁漱溟 all took a lasting benign interest in Consciousness Only Buddhism and Buddhist logic. The organs of the intellectual revival of Consciousness Only were journals like the *Nei-yüan nien-khan* 內院年刊 and *Nei-yüan Tsa-chih* 內院雜誌, and *Nei-hsüeh* 內學 all published in Nanking, and particularly Thai Hsü's 太虛 *Hai-chhao-yin* 海潮音.[2]

Many wrote about *yin ming* 因明, but those who took a serious philological interest in it, who learned Sanskrit and Tibetan as well as Japanese to be able to consult the most important sources, were few.[3]

The reasons why so many leading Chinese intellectuals turned to the School of Consciousness Only, and particularly *yin ming* logic, were profound: most important was probably the deep-seated desire for a distinctly Eastern logical and methodological identity. *Yin ming* provided a way of being scientific in method and deeply spiritual in purpose, while remaining Chinese – or in any case oriental – in basic outlook.

One participant in this revival of Buddhist logic Yu Yu 虞愚, in Liu Phei-Yu (1982), p. 40, reports: 'But on the one hand the old translations were so abstruse that even those laymen who had the inclination to work with it withdrew. On the other hand, since Buddhist logic lacked precise modern examples and was short of realistic significance, it found it hard to attract the interest of readers.

This is probably a realistic assessment of the reasons why Buddhist logic failed to have a more sustained and lasting intellectual impact in China than it did have.

(2) THE SYSTEM OF BUDDHIST LOGIC

Introductory remarks

In ancient Greece we find at least two competing though in fact complementary traditions of logical enquiry: The Aristotelian peripatetic tradition and the Stoic tradition which had Chrysippus as its main early exponent.[4] In pre-Chhin China,

---

[1] See Chang Ping-Lin (*1917*), Chhen Wang-tao (*1931*), cf. Chou Wen-ying (1979), p. 232. Chou Shu-chia (*n.d.*), Hsiung Shih-li (*1926*), Hui Yüan (*1978*), Lü Chheng (*1935*), Lü Chheng and Shih Tshang-ho (*1927*); cf. Chou Wen-ying, (*1979*), p. 232.

[2] For the most complete bibliography to date of Buddhist journals of the time see Welch (1968), pp. 279–84.

[3] Lü Chhêng 呂澂 is perhaps the most accomplished Chinese scholar in this field since Hsüan-Tsang himself. Lo Chao 羅炤 of the Academy of Social Sciences in Peking continues this great philological tradition.

[4] For standard treatments see Patzig (1969), Mates (1961) as well as Bochenski (1947).

too, there were competing schools of logical thinking: among others Kungsun Lung and the Later Mohists expounded competing logical theories. In Thang China, by contrast, all practitioners of *yin ming* 因明 seem to have seen themselves as practitioners of one and the same art introduced in to China by Hsüan-Tsang 玄奘. Buddhist logic was one system represented by the *Nyāyapraveśa* and the *Nyāyamukha* and expounded in Chinese commentaries on these two works.

In a recent survey of Buddhist studies we read:

Most present-day Japanese scholars consider Hsüan-Tsang's translations as faulty, and Khuei Chi's 窺基 comments miss the original meaning even more. Presumably because Khuei Chi did not quite understand Buddhist logic.[1]

Hajime Nakamura bears out this view:

Indian logic was accepted only in part, and even the part that was accepted was not understood in the sense of the Indian originals. . . .
Tz'u-ên's [i.e., Khuei Chi. Tzhu-ên was the name of the monastery after which his school was called.] work, which was regarded as the highest authority in China and Japan, contains many fallacies in philosophical and logical analysis.[2]

The harshness of Hajime Nakamura's generalisation is remarkable: he goes on to discuss Buddhist Logic in China under the heading: 'Acceptance of Indian Logic in a Distorted Form'.[3]

One way of reading the Chinese logical texts is indeed to look for logical mistakes. One can also read Plato in this way, and Plato's works turn out to be a happy hunting ground indeed for those who delight in logical blunders.

The logical work *De grammatico* by Saint Anselm of Canterbury (+1033 to +1103) suffered abrasive criticism by a scholar of medieval European thought:

. . . Saint Anselm maintains these heights (of theological thought) as long as he remains within the border of Christian metaphysics; but he falls back into the barbarity of his period as soon as he leaves the field of Christianity to take up the philosophy of his period, namely, the Scholastic dialectic. Thus the dialogue *De grammatica* [*sic!!*], which most unfortunately is one of his pieces, turns on a wretched question arising from Aristotle's *De Interpretatione*; it is as useless and insignificant a production as . . .[4]

Victor Cousin's harsh judgement was echoed also in the first large-scale history of logic, C. Prantl (1870). Detailed research since then, however, has shown that Anselm's dialogue *De grammatico* happens to be one of the great masterpieces in the history of Western philosophy of language.[5]

It is worth enquiring in detail whether Hsüan-Tsang and his followers did quite as badly as Hajime Nakamura seems to think. We need to study Hsüan-Tsang's work from the point of view not only of traditional syllogistic theory but of modern philosophical logic.

---

[1] Wu Ju-Chün (*1984*), p. 84.   [2] Nakamura (1964), p. 192.   [3] *Ibid.*, p. 191.
[4] Victor Cousin (1836), p. ciii, as quoted in Henry (1967), p. 32.   [5] Henry (1967).

I do not deny that it will be interesting to know exactly what kinds of mistakes the Chinese logicians were capable of making. (I would be amazed if they weren't making mistakes, of course.) But the more important first questions are the positive ones: Were the Chinese of Thang times incapable of truly proficient logical analysis along the lines of Indian Buddhist logic? Was their language lacking in the precision and logical subtlety necessary for the purpose of rigid formal logical analysis? There is only one effective way of finding out about this: to make a close philological study of the Chinese works on Buddhist logic.

These questions are not easy to answer in a general way. However, the evidence I have seen strongly suggests that the early Chinese Buddhist logicians of the Thang dynasty, particularly for example Wên Kuei 文軌, but also Khuei Chi 窺基, have done remarkably well. It is sometimes treacherously easy to attribute to the Chinese commentators obvious logical blunders in their excruciatingly difficult texts.[1] Moreover, it is perfectly true that the terminology of Chinese logicians is often curiously multivalent. For example, the term *tsung* 宗 can mean 1. the thesis, 2. the subject of the thesis, 3. the predicate of the thesis, and 4. a discussant party. Again, it is facile and misleading to blame this (truly unfortunate) terminological multivalence on the Chinese commentators' incompetence. Similarly, there is a most disturbing ambiguity of *yin* 因 'reason', which can either refer 1. to the property R in the reason, or 2. to the proposition ascribing the property R to the subject S of the thesis. But it turns out that the Chinese commentators in both cases faithfully reproduce (part of) a terminological muddle of equivocation found in the Sanskrit texts themselves![2]

In what follows we shall simply try to reconstruct Buddhist logic on the basis of the earliest Chinese sources we have. We shall see to what extent a coherent picture emerges.

A preliminary simplified survey of the 'yin ming' 因明 *system*

The history of Buddhist logic in China is full of subtle distinctions, subdistinctions and all the formidable appurtenances of a highly developed scholastic systematicity

---

[1] A passage which at first glance looks like a glaring mistake often turns out in the end to be a passage which is insufficiently understood. For example, in Great Commentary 2.10a3ff. Khuei Chi argues, apparently at variance with his Sanskrit sources, that subject and predicate specify or subcategorize each other. Chhen Ta-Chhi (*1970*), pp. 27f., following Lü Chheng, regards this as a major logical blunder. But the thought expressed by Khuei Chi, when properly understood, turns out to maintain, quite sensibly, that the sentence 'The real objects of this world contain no ego' specifies the property in the predicate 'contains no ego' as being exemplified by the real objects of the world. It also specifies 'the real objects of this world' as containing no ego. Wen Kuei argues along somewhat similar lines that the predicate specifies the subject, and that the subject conversely specifies the predicate. His example is the standard 'sound is impermanent'. He claims that the subject 'sound' specifies that what is said to be impermanent is sound and not, for example, other material objects; the predicate 'is impermanent' specifies that what sound is said to be is 'impermanent', and not, for example, 'permanent' or the like. (Wên Kuei on *Nyāyapraveśa*, p. 233a bottom 10ff.)

These may not have been thoughts currently expressed in the Sanskrit literature, but as Hsiung Shih-Li (*1926*), p. 15b, points out, we need not consider them illogical or mistaken. As Wên Kuei is at pains to demonstrate explicitly in the passage mentioned above, it is quite consistent with the standard view in Buddhist logic that in one sense the predicate modifies the subject but that the subject is not taken to modify the predicate.

[2] For the first three meanings of *tsung* see Chi (*1969*), p. 123. For the fourth, rarer meaning see, e.g., Great Commentary 2.2b3.

which are alien, on the whole, to pre-Buddhist China. Introductions to this system of Buddhist logic are often extremely hard to follow. One has to acquire new intellectual habits, and this is not easy.

The first thing one will notice is that just as the Chinese detective story usually exasperates the Western reader by giving away the identity of the culprit at the very beginning and not at the end, so the Buddhist argument surprises us by putting the conclusion before the grounds. Instead of arguing *cogito, ergo sum*, a Buddhist would instead consider an argument like *I exist, because I think*. And the Buddhist logician's enquiry is exactly into the conditions that might make such a way of arguing acceptable. Chinese Buddhist logic is an art of formally justifying a thesis or a tenet.

For an argument to be of an acceptable form, the Chinese Buddhist logician also requires that it should be substantiated by examples. In this instance, what needs to be exemplified is the generalisation which is so crucial to the argument: that wherever there is thought there is an existing subject of thought, and that conversely wherever there is no subject of thought there will be no thought. As long as exemplification of this kind is not forthcoming, the argument is – by the standards of Chinese Buddhist logic – not well-formed. The examples serve to insure that an argument is logically connected with this world as it is, that it does not consist of empty abstraction only.[1]

In this concrete instance, the Chinese Buddhist adherent of the School of Consciousness Only would maintain that the issue is precisely whether the process of thought necessarily must have a neat metaphysical subject in the form of a Self, or an ego to do the thinking. *Cogito, ergo sum*, according to our imagined Chinese Buddhist, begs the question by surreptitiously and fallaciously introducing the *ego* into *cogito*.

Moreover, the Buddhist logician would insist that the argument *cogito, ergo sum* would be unacceptable (because pointless) if used with an audience in which the thesis or conclusion is not in doubt in the first place.

Again, the argument *cogito, ergo sum* would be unacceptable (because ineffective) if the notions of 'person', 'exist' and 'think' were not understood by the arguer and his opponent in the same way.

Finally, a standard argument like 'People exist, because they think' would only be acceptable if in fact

1. every person thinks; (All S are R)
2. some things that think definitely exist; (Some R are P)
3. no things that do not exist think. (All *non-P* are *non-R*)
   These three (meta-logically formulated) requirements in Chinese Buddhist logic are known as the Three Aspects of the Reason. By an elementary rule of logic we may simplify the third Aspect as 3' All R are P.

and deduce from All S are R, and All R are P, that All S are P.

---

[1] Tu Kuo-Hsiang, (*1962a*) and (*1962b*), strongly stresses the epistemological and not purely logical emphasis in *yin ming* 因明.

At this point one may be tempted to conclude with *quod erat demonstrandum* and rejoice at having reduced the Chinese Buddhist form of argument to the most elementary of the Aristotelian syllogisms.[1] Indeed, a great deal of work has been produced using mathematical logic to describe Chinese Buddhist logic.[2] R. S. Y. Chi (1969), p. 15, states the method in his study of the Buddhist logical system in China:

Therefore in my formulation I do not actually translate the Indian authors' words into symbols, but express what they meant to say – i.e., what they would have said if they had used precisely the same language and symbols as we do.

R. S. Y. Chi's method is not entirely without its merits, since it defines many basic logical issues relating to Chinese Buddhist logic, and he does recognise its limitations: 'I shall not list these *nyāya* meta-logical terms and give a comparative study of them, (such a task could be better handled by philologists) . . .' (*Ibid.*, p. xiv)

In the context of our present enterprise it is a philo-logical enquiry which holds our special interest. And for this philological enquiry we need all the formal logical acumen we can muster. Benson Mates, whose textbook *Elementary Logic* I have admired and used for years in teaching logic, provides a shining example of the advantages of formal logical training in philological research. His book *Stoic Logic* is a masterpiece of philology precisely because it shows thorough familiarity with philological detail as well as a proficient command of modern philosophical logic. Sound philology of logical texts requires a strong background in *both* philology and logic. The philological tasks involved require logical decisions at every juncture.

In order to understand Chinese Buddhist logic in its own terms, our first step must be to understand the basic terminology of Buddhist logic in China, not as proto-logical adumbrations of modern logical concepts, but as forming a conceptual scheme in its own right. A conceptual scheme not to be reduced to modern logic, but to be more fully understood by comparison with modern theories.

After this methodological excursion on the formalism of the Three Aspects of the Reason, let us revert to our introductory application of the general scheme of Chinese Buddhist logic to the familiar *cogito, ergo sum*.

It turns out that the Chinese Buddhist logician is far from exclusively concerned with the formal features stated above under nos. 1–3. The Chinese Buddhist logician is also crucially interested in the presuppositions involved in such an argument. For example, he might find the argument *cogito, ergo sum* acceptable as long as it is directed at an audience which accepts the idea that thought must be a process with (or in) a subject. For a Chinese Buddhist of the School of Consciousness Only,

---

[1] In a series of interesting articles on the methodology of the comparative study of logic, the logician Douglas D. Daye has pointed out the dangers of reducing the Indian system to a Western formalism: 'I would suggest (and *only* suggest) that the *descriptive* utility of mathematical logic with *early* Nyāya texts has simply been overrated . . . Therefore, the Procrustean use of mathematical logic . . . should not be viewed, in my eyes, as the *sine qua non rūpa* [*rūpa*: form] from which all expository *bhagas* [*bhagas*: blessings] must flow.' (Daye (1977), p. 231.)

[2] In 1951 there was the pioneering D. H. H. Ingalls. In 1958 came H. Nakamura and J. F. Staal. Other important contributions include Kitagawa (1960), Barlingay (1965), McDermott (1970) and B. K. Matilal's work, especially Matilal (1971).

however, this is not a foregone conclusion. For him it is in fact a mistaken view. The person or Self or ego – for adherents of the School of Consciousness Only – is only a construct, not an existent thing. Thought may be real in some sense, even if only as illusion, but the person, ego, or Self does not exist in any other sense than this: it is an object of imagining. There can be a process of thinking without there necessarily being a thinker as the subject of the thinking, just as there can be blowing without there 'being' a wind that does the blowing.

Thus, the argument *cogito, ergo sum* may be acceptable in unenlightened parts of the world where the doctrine of Consciousness Only is not (yet) understood, but it is quite unacceptable when used to prove anything to an adherent of that school of Buddhism.

For the Chinese Buddhist logician, then, the question of the acceptability of an argument depends on the truth of adduced grounds and on the audience to which it is addressed. He is not concerned with deductive logical validity as such. An argument must be seen in its intellectual context. Its acceptability is not an absolute matter. It is relative to the historical context in which it occurs. Chinese Buddhist logic is not purely formal logic in our Western sense of the word. It is – as we shall see – ultimately concerned not with formal logical validity, but with material truth. Chinese Buddhist logic is therefore always concerned with theoretically important significant propositions and their justification through formalised argument.

The place of Chinese Buddhist logic among the sciences

Yin ming 'the science of grounds' is intended as a comprehensive philosophy of science.

Scientific discourse (*lun* 論) consists of proof (*liang* 量) and disputation (*i* 議). Proof (*liang*) fixes what is true and what is (only) apparent(ly true) (*ting chên ssu* 定真似). Disputation (*i*) goes into detail about establishing (*li* 立) and refuting (*pho* 破).[1]

Quite generally, *yin ming* is concerned with the conditions of acceptability of grounds as proof of a thesis.

Traditionally, *yin ming* is regarded as one of the Five Sciences (*wu ming* 五明), a fact which has guaranteed logic a place in the Indian (but not in the Chinese) general educational curriculum:

Logic (*yin ming*) is one of the Five Sciences (*wu ming*). I have humbly looked for accounts of the Five Sciences and the term goes back a long time.[2]

The five branches of learning were the literary sciences (*shêng ming* 聲明), the technological sciences (*kung chhiao ming* 工巧明) the medical sciences (*fang ming* 方明), the logical sciences (*yin ming*), and finally the only specifically Buddhist branch, the religious sciences (*nei ming* 內明). Apart from the religious sciences (*nei ming*), all the

---

[1] Great Commentary, ed. Sung Yen, 1.6a5.
[2] Wen Kuei on *Nyāyapraveśa*, ed. Lü Chheng, p. 1a, *Supplementary Tripitaka*, vol. 86, p. 331a, top 3.

others were regarded as not linked up with any specific religion. However, the logical sciences were obviously used in the service of Buddhism:

Religious sciences (*nei ming* 內明) are directed towards the believers and their provision. Logical sciences (*yin ming* 因明) are directed towards the heretics (*pang* 謗) and their control.[1] By means of theses (*tsung* 宗), reasons (*yin* 因) and examples (*yü* 諭) one explains dogmas to those who have not yet understood (*wei liao* 未了).[2]

Special importance is attached to Buddhist logic because it helps one to sort out competing views on ultimate truth:

The ultimate truth (*chih li* 至理) is a recondite thing which is not accessible to superficial knowledge (*chhien shih* 淺識). Therefore the competing views arose concerning the various profound dogmas (*ao i* 奧義) . . . If you want to see if your shape is beautiful or ugly, you look at yourself in a clear mirror or in a clear pool.[3] If you are to fix truth or deviation (*chêng hsien* 正邪), you must look at them through logic (*yin ming* 因明), the theories of perception (*hsien* 現) and inference (*pi* 比).[4]

Chinese Buddhist logic is essentially concerned with the justification of orthodox Buddhist claims against unorthodox opponents. It is concerned with promoting reasoned, rational discourse, but within this specific area:

Even if you put forward a thesis (*tsung*), as long as you do not put forward reasons (*yin*) and supporting examples (*yü*), like all those discussants who hold on to the Ego (*chih wo* 執我) and keep thinking along their own lines, that surely is wrong![5]

Question: In order to establish the significance of what you establish (*so chhêng li i* 所成理義), why first establish a thesis (*tsung*)?

Answer: It is in order to first show the significance of the thesis which one likes (*so ai lê tsung i* 所愛樂宗義).

Question: Why does one go on to dispute about reasons (*pien yin* 辯因)?

Answer: In order to make something plain on the basis of observed facts (*i hsien chien shih* 以現見實), and to decide the point (*chüeh thing tao li* 決定道理), to make others accept the import of the thesis propounded.

Question: Why does one go on to pull in examples (*yin yü* 引喻)?

Answer: In order to make plain what that which can establish the point (*tao li* 道理) rests on and fix it on observed facts (*hsien chien shih* 現見實).[6]

The translation of Dignāga's own *Nyāyamukha* opens with this statement:

I write this treatise because I want to grasp simply proof (what can establish) and refutation (what can refute) (*nêng li nêng pho* 能立能破) and the true facts in a dogma (*i chung chên shih* 義中真實).[7]

---

[1] Wên Kuei on *Nyāyapravesa*, ed. Lü Chheng, p. 1a, *Supplementary Tripitaka*, vol. 86, p. 231a, top 9.

[2] *Nyāyamukha*, ed. *Taishō Tripitaka* p. 1a7; *Nyāyapraveśa*, ed. Lü Chheng, p. 8, *Taishō Tripitaka*, p. 11b2.

[3] This echoes, e.g., *Chuang Tzu* 13.3.

[4] Wên Kuei on *Nyāyapraveśa*, ed. Lü Chheng, p. 1af, *Supplementary Tripitaka*, vol. 86, p. 231a top −10.

[5] Khuei Chi, Great Commentary, ch. 1, ed Sung Yen, p. 13a9.

[6] Khuei Chi, Great Commentary, ch. 2, ed Sung Yen, 2.1b1.

[7] *Nyāyamukha*, ed. *Taishō Tripitaka*, p. 1a4; cf. Tucci (1930), p. 5.

Primarily, arguments are adduced to make non-believers understand (*wu tha* 悟他), but they also function to make one understand one's own tenets or theses (*tzu wu* 自悟):

Proof (*nêng li* 能立, lit.: 'what can establish') or refutation (*nêng pho* 能破, lit.: 'what can destroy/refute'), as well as only apparent cases of these (*ssu* 似) are called 'making others understand (*wu tha*)'. Judging a thesis on the basis of perception (*hsien liand* 現量) or judging a thesis on the basis of inference (*pi liang* 比量), as well as only apparent cases of these (*ssu* 似) are called 'making oneself understand (*tzu wu*)'.[1]

Making others understand and making oneself understand were the two benefits (*êrh i* 二益) thought to be accruing to one from the study of Buddhist logic. Establishing, refuting, perceiving, and inferring, together with their only apparent counterparts, make up the eight gates (*pa mên* 八門) of logical knowledge.

Chinese Buddhist logic – like the pre-modern traditional concept of logic – thus includes more than the study of logical entailment or consequence. It also includes what we are today inclined to call epistemology, or the theory of knowledge and perception. The explanation for this is clear: Chinese Buddhist logic is the study of what grounds one may adduce under what circumstances to justify a given thesis. Such grounds may indeed be either logical grounds in our sense (i.e., the thesis follows logically from the publicly accessible adduced ground), or they may be sense data.

Before Dignāga there were four recognised avenues of knowledge:

1. perception (*hsien liang* 現量), 2. inference (*pi liang* 比量), 3. analogical inference (*pi yü liang* 比喻量), and 4. verbal testimony (*shêng liang* 聲量).[2] According to the logician Dignāga, only two of these were legitimate: perception, which was concerned with particular appearances (*tzu hsiang* 自相), and categorization, which was concerned with the application of concepts (*kung hsiang* 共相).[3] It was this rather hard-headed philosophical attitude which was transmitted to China – and which was to some extent applied to the tenets of Buddhism. Knowledge by analogy and by verbal testimony (*shêng* 聲) was rejected.[4] The scheme of Chinese Buddhist logic is designed, then, to explain logical and perceptual grounds which one might successfully adduce in order to justify a thesis.

---

[1] *Nyāyapraveśa*, ed. Lü Chheng, p. 6, *Taishō Tripitaka*, p. 11a–2.

[2] The four sources of knowledge are expounded in the opening passages of the Nyāya Sūtras. These may be usefully compared with the Mohist standards:

> Therefore there must be three standards for speech. What are these? Master Mo Tzu said: 'There must be that which is the basis, that which is the enquiry, and that which is the usefulness. In what way do we provide it with a basis? Looking up, we base it on the ways of the sage kings of antiquity. How do we enquire into it? Looking down, we study it in the testimony of the ordinary people. How do we put it to use? Propagate it as law and governmental policy, observe how it accords with the benefit of the people and the state. This is why we say that there are three standards for speech. (*Mo Tzu* 35.7; cf. Mei (1929), p. 183, and Hansen (1983), pp. 84f.)

[3] Cf. Ting Fu-Pao (*1928*), p. 477. Shen Thai explains: 'Names and words (*ming yen* 名言) only express concepts (*kung hsiang* 共相), they cannot express the particular appearances (*tzu hsiang* 自相) of things. This is because the particular appearances are separate from language and explanation.' (Fa Tsun Fa-shih (*1984*), p. 55.)

[4] The *Nyāyasūtras* define Sanskrit *śabda* (Chinese *shêng* 聲) as follows: 'Verbal traditional authority (Sanskrit *śabda*, Chin. *shêng*) is the instructive assertion of a reliable person.' (Vidhyabhusana (1981, first ed. 1930), p. 5.)

The well-formed argument in Chinese Buddhist logic

In order to understand what constitutes a formally acceptable or well-formed argument in Chinese Buddhist logic, it is best to start out with a standard example. As it happens, the most standard example involves the concept of *shêng* 聲, a word which literally means 'sound', but which in the context of logic, as we have seen, comes to mean something like traditional authority. The standard example of an argument starts with a thesis on which Chinese Buddhists and other traditionalist schools disagree, and which runs as follows:

Thesis: Verbal traditional authority (*S*) is impermanent (*P*) (*shêng shih wu chhang* 聲是無常)[1]

The standard reason adduced in support of the thesis is this:

Reason: By reason of having been produced (*R*) (*so tso ku* 所作故).[2]

Chinese Buddhist logicians were increasingly concerned to make explicit the logical requirements which would guarantee the step from the reason to the conclusion. This – as we shall see – they have in common with logicians in the West. However, their strategies were different. Their first step was to introduce positive examples (*thung yü* 同喻) and negative examples (*i yü* 異喻) to illustrate the systematic link between the reason adduced and the thesis justified. This systematic link was called the substance of the examples (*yü thi* 喻體) and it typically took the form of an implication:

Substance of the positive example: If anything is produced, it is not permanent.

Apart from this, the Chinese Buddhists required an example item (*yü i* 語依). The positive example is typically stated (without the so-called 'substance of the positive example') as follows:

Positive example item: For example bottles and the like (*phi ju phing têng* 譬如瓶等).[3]

The Chinese Buddhist logicians insisted on a negative example intended to show that there was indeed a concrete case which showed a systematic link between *R* and *P* such that if, in some place, *P* (e.g., impermanence) was not found, then *R* (e.g., producedness) would never be found in that place either.

Substance of the negative example: If anything is permanent, it is not produced.

---

[1] For simplicity we shall refer to the subject of the thesis (in this case, for example, 'verbal traditional authority') by the letter *S*, and to the predicate term (in this case, for example, 'impermanent') by the letter *P*.

The standard English translation of this thesis is 'Sound is impermanent'. This is, of course, literally correct, but is does not seem to us to convey the underlying force and intrinsic philosophical interest of the thesis within the context of Nyāya logic.

[2] For simplicity we shall refer to the term in the reason (in this case, for example, 'having been produced') by the letter *R*. Cf. Sallust, *Jugurtha* 2.2: *Postremo corporis et fortunae bonorum ut initium, sic finis est, omniaque orta occidunt* 'Finally there is a beginning and an end to the blessings of the body and of fortune, and everything that has an origin will disintegrate.'

[3] It is amusing to find that the word *phing* 瓶 'bottle' is consistently translated as 'pot' in the literature on Chinese Buddhist logic that I have seen. It is perfectly true that the Indians used a pot as an example, but it is equally true that the Chinese used a bottle!

The negative example is typically stated (also without its 'substance of the negative example') as follows:

Negative example item: For example space and the like (*ju hsü khung têng* 如虛空等).[1]

Both positive and negative examples are – theoretically – required if an argument is to count as well-formed by Chinese Buddhist standards.[2] Here is another typical argument in its full form:

Thesis: Space is eternal (*hsü khung shih chhang* 虛空是常).
Reason: Because of not being produced (*fei so tso ku* 非所作故).
Positive example: The various things that are not produces are each and all of them eternal. Like nirvāṇa (*chu fei so tso chê chieh hsi shih chhang. Ju nieh phan* 諸非所作者皆悉是常。如涅槃).
Negative example: If something is not eternal, then it is produced. Like a bottle or a pot etc. (*jo shih wu chhang chi shih so tso. ju phing phên têng* 若是無常即是所作。如瓶盆等).[3]

In the case of a thesis like 'All things are momentary', for example, the negative leads to evident problems: how is one to find examples of non-momentariness all of which are not things. Since all acceptable examples have to be real things, this turns out to be impossible and has – quite properly – exercised the minds of Chinese Buddhist logicians a great deal.

Presuppositions and the intended audience

Aristotle's syllogistics is an abstract deductive logical theory. Chinese Buddhist logic is a science of argumentation within its social context. That social context becomes crucially relevant to argumentation by way of the logically significant presuppositions it sanctions.

To start with, the Chinese Buddhist logicians find an argument unacceptable if its conclusion is self-evident to all parties in the first place. Thus an argument to the effect that sound is something one hears is held to be unacceptable because it is uncontroversial. The presuppositional requirement on the thesis on the whole, then, is that it must be controversial in such a way that the proponent must hold it to be true while the audience must be less than certain that it is true. This requirement shows that acceptability to the Chinese Buddhist logician is not the same as logical validity to the Western logician. The charge of triviality is not a relevant objection to the logical validity of an argument, but it is a relevant objection to the acceptability of the argument in terms of Chinese Buddhist logic.

Again, to the Chinese Buddhist an argument which justifies a thesis which goes against the teachings of one's own school or against one's own known views is logically unacceptable.

---

[1] Note that the formulations of both examples make nice four-character phrases.
[2] As we shall see in our more detailed treatment of the examples, these are conceived by the Chinese Buddhist logicians to consist of two parts: the illustrative thing, and the conditional illustrated.
[3] Wên Kuei on *Nyāyapraveśa*, ch. 1, ed. Lü Chheng, p. 18a3.

Another requirement is that the relevant meanings of S (the subject of the thesis), P (the predicate of the thesis) and the property R in the reason must be agreed upon (*chi chhêng* 極成) by the proponent and the audience.

Consider an example:

Thesis: You and I are not eternal.
Reason: By reason of being active items.
Example: Like sparks of the lamplight.[1]

Such an argument would only be acceptable in Chinese Buddhist logic if there were agreement between the proponent and his audience on the relevant sense of S ('you and I'). This would not necessarily be a trivial matter, since the Chinese Buddhist does not accept the notion of the self, or of 'I' and 'you' as metaphysically real or enduring entities. On the other hand, contesting parties could agree to use terms like 'I' and 'you' in a non-metaphysical, or – we might say – pre-theoretical ordinary sense.

Another required shared presupposition is that all S must be R, i.e., it must be agreed between the proponent and the audience that you and I are active items.

Further, it must be a shared presupposition that all $+E$ (positive examples, e.g., sparks of lamplight) must be P (e.g., not immortal), and that *not-E* (negative examples, e.g., space) must be *not–P* (e.g., (not not) immortal), i.e., it must be agreed between the proponent and the audience that things like sparks of the lamplight are not immortal, and that things like space are immortal.

An argument is acceptable in terms of Chinese Buddhist logic only if there is agreement on these specific presuppositions between the proponent and his audience.

We have seen that the *Nyāyapraveśa* distinguishes between arguments designed to make oneself understand (*tzu wu* 自悟) and those designed to make others understand (*wu tha* 悟他).

Thus all the arguments (*pi liang* 比量) may generally be divided into three kinds: those directed towards others, those directed towards oneself, and those directed generally at others and at oneself.[2]

In an argument directed towards oneself one can legitimately operate on the assumption that presuppositions of oneself or one's group are justified. In an argument directed towards others one can legitimately operate only on the presuppositions shared by the other person or other group. In an argument directed towards both others and one's own group then one can legitimately operate only on the presuppositions shared by one's own group and the other group.

According to the Chinese Buddhist logician, no argument is in principle detached from this presuppositional context, no matter what audience one is

---

[1] Wên Kuei on *Nyāyapraveśa*; Shen Chien-Ying (*1984*), p. 133.
[2] Khuei Chi, Great Commentary, ch. 6, ed. Sung Yen, p. 12b3; cf. also Khuei Chi, Great Commentary, ch. 5, ed. Sung Yen, p. 4b8.

addressing with one's argument. No argument is valid *per se*. It may only be judged valid in relation to the presuppositions the argument is based on and which participants in the argumentation may or may not share.

These presuppositions can be made explicit in the various parts of an argument, and the purpose of making presuppositions explicit is to obviate objections:

Generally in the method of Buddhist logic (*yin ming* 因明), when there are presuppositional distinguishers (*chien pieh* 簡別) for the thesis and the grounds, one will avoid the logical mistakes (of which one would have been guilty without the presuppositional distinguishers).[1]

If an argument is directed towards one's own group, one marks it as *hsü yen* 許言 to make plain that one's own group accepts it. In that way one avoids the fallacies like 'the truth of one party not being conceded by the other party'.

If an argument is directed towards another group, then one marks it as *ju shü* 汝許. In that way one avoids the fallacies like 'the thesis being inconsistent with one's own claims'.

If an argument is directed towards one's own and another group one marks it as *shêng i* 勝義. In this way one avoids the fallacy of 'contradiction with common knowledge and with what one's own group teaches.[2]

The choice of presuppositional distinguishers can thus make a decisive difference to the acceptability of an argument, as is illustrated by Jayasena's famous proof of the authenticity of the Mahāyāna Sūtras:

Thesis: All the Mahāyāna Sūtras are speeches of the Buddha.
Reason: By reason of their being, as both sides agree, encompassed by what the Buddha said.
Example: Like the *Āgamasūtra* scriptures.[3]

When this argument is used against adherents of Hīnayāna Buddhism, it surely does not carry any weight, since adherents of Hīnayāna will deny that the reason-property *R* applies to the subject *S* of the thesis. Hsüan-Tsang changed the presuppositional distinguisher 'as both sides agree' into 'as our side agrees', converting this argument into one by which adherents of Māhayāna justify their position to each others.

Agreement can apply to propositions or to terms. In the case of a proposition, the Chinese Buddhist logicians take the agreement to be concerned with whether or not the reason-property *R* is agreed to apply to the subject *S* of the thesis. This form of agreement is called *chuan i chi chhêng* 轉依極成 'agreement of application'. The case of terms is more complicated. Here the agreement can be on the extension of the term (*ching chieh chi chhêng* 竟界極成 'agreement on the external world'). For example, there can be agreement on what is and what is not a traditional verbal authority etc. The agreement can also be on the intension of a term (*chha pieh chi chhêng* 差別極成 'agreement on distinguishing characteristics'). For example, there can be agreement on a way of understanding the notion of 'I' as either a perishable constellation without ontological status or as an ultimate ontological item.

---

[1] Khuei Chi, Great Commentary, ch. 5, ed. Sung Yen, p. 2b1.
[2] Khuei Chi, Great Commentary, ch. 5, ed. Sung Yen, p. 2b2.
[3] Khuei Chi, Great Commentary, ch. 6, ed. Sung Yen, p. 3b11; Shen Chien-Ying (*1984*), p. 141.

All agreement, in order to count in Chinese Buddhist logic, must be definite and not uncertain. Agreement on the constituents of the thesis (*tsung* 宗) is agreement on a relevant meaning of these constituents, and on the non-emptiness of the class specified by the constituents as understood in the agreed way.

Presuppositions in Chinese Buddhist logic are not that such-and-such is true, but that such-and-such is agreed among the relevant parties. However, this does not mean that the Chinese Buddhist logicians were unable to distinguish between agreement of opinion and presupposition of fact. For example, they did occasionally distinguish between *chên chi* 真極 'factual presupposition' that a term is non-empty (*yu thi* 有體) on the one hand, and psychological agreement (*kung hsü* 共許) on the other.[1]

Having looked at the general scheme of Chinese Buddhist logic, and at the framework of presuppositions within which it operates, we are now in a position to consider in more detail the constituents of an argument as interpreted by the Chinese Buddhist logicians. We begin with the thesis, turn to the reason, and end up with the examples, just as an argument in Chinese Buddhist logic does.

The thesis

Chinese Buddhist logicians call the thesis as a whole 'what is established (*so li* 所立)' or *tsung thi* 宗體 'body of the thesis', or *tsung tsung* 總宗 'the overall result of the thesis'. The thesis as a whole is systematically and consistently distinguished from its constituents known as the *tsung i* 宗依 'the supports of the thesis' or *pieh tsung* 別宗 'the thesis as separated'.[2] As we have seen, the constituents of the thesis must be agreed upon:

A thesis consists of a commonly agreed (*chi chhêng* 極成) owner of a characteristic (*yu fa* 有法) and a commonly agreed (*chi chhêng*) subcategorizer (*nêng pieh* 能別).[3]

Khuei Chi 窺基 explicates:

The owner of the characteristic (*yu fa*) and that which can subcategorize (*nêng pieh*) are only constituents of the thesis (*tsung i*). They do not constitute the thesis (*yu fa, nêng pieh tan shih tsung i êrh fei shih tsung* 有法，能別但是宗依而非是宗。). The constituents must be completely agreed on by the two parties (*liang tsung* 兩宗). Only when the meanings of the constituents (*i i* 義依) of the thesis are established can the substance of the thesis (*tsung thi* 宗體) take shape. . . . If these constituents are not agreed on in advance (*hsien* 先), then one has to go on to establish these, and in that case (*tzhu* 此) they do not count as constitutents.[4]

The essential import of the thesis as a whole is analysed as the non-separateness (*pu hsiang li hsing* 不相離性) of the two constituents in the thesis, subject and predicate. Khuei Chi writes:

---

[1] For good account of agreement see Chhen Ta-Chhi (*1952*), pp. 29–38.

[2] Occasionally, the thesis as such (*tsung* 宗) is distinguished from the expression used to articulate it, the *tsung yen* 宗言 'the words of the thesis'. Cf. Lü Chheng (*1983*), p. 8.

[3] *Nyāyapraveśa*, ed. Lü Chheng, p. 10; *Taishō Tripiṭaka*, p. 11b3.

[4] Khuei Chi, Great Commentary, ed. Sung Yen 2b2; cf. Hsiung Shih-Li (*1926*), p. 14a.

In modern times Dignāga maintained that the reason and the examples are what can establish a thesis (*nêng li* 能立), and that the thesis is what is established (*so li* 所立). The identifier (*tzu hsing* 自性 lit: 'self-nature') and the predicate (*chha pieh* 差別, lit: 'distinguisher, subcategorizer') are both mutually agreed on. The constituents of the thesis by themselves (*tan shih tsung i* 但是宗依) do not quite make up the issue of the dispute. Only when they are tied together to make up a thesis (*tsung* 宗) and there is non-separateness (*pu hsiang li hsing* 不相離性) do we have the issue of the dispute.[1]

The subtle distinction between the whole of the thesis and the sum of its parts is further explained:

What the other party (*tha* 他) does not agree on is only the thesis as a linkage (*ho tsung* 合宗). This thesis is what is established. The identifier (*tzu hsing* 自性) and the subcategorizer (*chha pieh* 差別) are both constituents of the thesis (*tsung i* 宗依). They are not what is established (*so li* 所立).[2]

Khuei Chi makes a lucid and subtle distinction between the property (*fa* 法) as such in the predicate of the thesis, which must be agreed upon, and the application of that same property to the subject (*yu fa* 有法), which must not be agreed upon:

The property as applied in the thesis (*tsung chung chih fa* 宗中之法) is not agreed on beforehand by the opponent.[3]

The internal structure of the thesis: subject and predicate

Since it is an important issue whether or not Chinese is a subject/predicate language, the discussions concerning subject and predicate within the context of Chinese Buddhist logic holds a special philosophical interest for us. We shall want to look carefully at what the Chinese Buddhist logicians had to say on the matter. Could they make proper sense of the distinction as they found it in their Indian sources?

One common Chinese way of distinguishing between subject and predicate was by simply calling the subject 'that which is put forward first (*hsien chhên* 先陳)' and the predicate 'that which is expressed last (*hou shu* 後述)'.

That which is put forward first (*hsien chhên*) is the 'possessor of a property' or 'that which is distinguished'. That which is expressed last (*hou shu*) is the 'property (*fa* 法) or the subcategorizer (*nêng pieh* 能別).[4]

This way of putting things is unlikely to be of Indian origin, since in fact it is grossly inappropriate to Sanskrit where the predicate quite commonly precedes the subject in an ordinary proposition. However, the Chinese Buddhist logicians were not insensitive to the subtler philosophical aspects of the distinction between subject and predicate:

---

[1] Khuei Chi, Great Commentary, ch. 1, ed. Sung Yen, p. 11b2.
[2] Khuei Chi, Great Commentary, ch. 1, ed. Sung Yen, p. 12b4.
[3] Khuei Chi, Great Commentary, ch. 3, ed. Sung Yen, p. 3a6.
[4] Wên Kuei on *Nyāyapravesa*, ed. Lü Chheng, ch. 1, p. 7a8; *Supplementary Tripitaka*, p. 332b bottom 3.

Expressions like 'eternal' and 'non-eternal' contain a semantic criterion (*kuei* 軌). Therefore we call them properties (*fa*). 'Colour', 'sound' and the like can have (*yu* 有) these properties. Therefore we call them possessors of properties (*yu fa* 有法). Identifiers (*tzu hsing* 自性) like colour are subcategorized by 'non-eternal' and put either inside or outside this category. Therefore we call them subcategorizers (*chha pieh* 差別). Since 'eternal' and 'non-eternal' can subcategorize identifiers like colour we call them 'that which can subcategorize (*nêng pieh* 能別)'. This is like the case of wax receiving a square or round shape from a seal. The shape is called the subcategorizer (*nêng pieh*). This is because it causes this wax to have a square or round shape (*nêng pieh tzhu la chhêng fang yüan* 能別此臘成方圓).[1]

Khuei Chi goes on to reflect subtly:

Among all quiddities/properties (*fa*) there are two kinds: that of substance (*thi* 體) and that of attribute (*i* 義).[2]

Khuei Chi goes on to introduce and explain three technical terms each for subject and predicate. For 'subject' the three terms are *tzu hsing* 自性 'identifier', *yu fa* 有法 'possessor of the property', and *so pieh* 所別 'that which is subcategorized'. For predicate the terms are *chha pieh* 差別 'subcategorizer', *fa* 法 'property', and *nêng pieh* 能別 'that which can subcategorize'. Confusingly, Khuei Chi explains matters by using a fourth pair: 'that which is put forward first' for 'subject' and 'that which is put forward last' for 'predicate'.[3]

(The proponent and the opponent) do not disagree on that which is put forward first (i.e., the subject). They argue about whether in (*shang* 上) that which has been put forward first (*hsien chhên* 先陳) there is that which is said of it afterwards (*hou so shuo* 後所説). By that which is said afterwards (*hou so shuo* 後所説) one subcategorizes that which has been put forward first (*hsien chhên*). One does not by means of that which is put forward first apply subcategorization to what comes last (*pieh yü hou* 別於後). Therefore, the substance (*thi* 體) that comes first is called 'that which is subcategorized (*so pieh* 所別)', and the attribute (*i*) which is put forward last is called 'that which can subcategorize (*nêng pieh*)'.[4]

One might find the choice of Chinese terms less than fortunate. Khuei Chi, in any case, continues:

These three terms all have their faults. What are these faults? The objection (*nan* 難) to the first runs like this: If the substance (*thi*) is called self-nature (*tzu hsing*) and the attribute (*i* 義) is called distinguisher (*chha pieh*), how can one go on further along to say: 'If the various masters of learning establish the thesis that "I think", then "I" is a self-nature (*tzu hsing* 自性), and "think" is the distinguisher (*chha pieh* 差別).' In this text[5] you treat the attribute (think (*ssu* 思)) as the self-nature (*tzu hsing*) and the substance (*thi* 體) as the distinguisher.[6]

---

[1] Wen Kuei on *Nyāyapraveśa*, ed. Lü Chheng ch. 1, p. 7a; *Supplementary Tripitaka*, p. 332b top −6ff.
[2] Khuei Chi, *Great Commentary*, ch. 2, ed. Sung Yen, p. 3a6.
[3] Like medieval European practitioners of philosophical grammar he does not pay attention to such possibilities as a preposed predicate.
[4] Khuei Chi, *Great Commentary*, ch. 2, ed. Sung Yen, pp. 5a3ff.
[5] I.e., in the phrase '"think" is the distinguisher'.
[6] Khuei Chi, *Great Commentary*, ch. 2, ed. Sung Yen, p. 5a.

There is no space to spell out the difficulties Khuei Chi raises with the other terms.

The problem of subject and predicate preoccupies him in many places. Khuei Chi writes:

According to the theory of Buddhist logic all terms (*fa* 法) that refer (*chü* 舉) to some things and only apply to these things without being applicable to anything other than what they are being applied to, are called identifiers (*tzu hsing*). If, as in binding together flowers, you tie a property to something other (the subject), then all such properties subcategorize meanings (*i* 義), and they count as 'subcategorizers (*chha pieh*)'.[1]

Occasionally, Khuei Chi's explanations of the concepts of subject and predicate are written in such a way that they must have been totally incomprehensible to those who were unfamiliar with inflected languages like Sanskrit, as when he refers to the verbal endings (*chhü chhü* 屈曲) in the predicate, a term which must surely have been totally obscure to everyone who did not know how an inflected language works:

There are two attributes/meanings (*i*) of terms (*fa*). The one is able to put forward a self-contained substance (*tzu thi* 自體), the other gives a criterion (*kuei* 軌) to introduce a new predication (*tha chieh* 他解). Therefore the various treatises say: 'The properties (*fa*) are a maintaining of criteria (*kuei chih* 軌執). That which is maintained first is the self-substance (*tzu thi*) which everyone understands. The criterion (*kuei* 軌) that comes afterwards newly produces an explanation, and it needs to have a verbal inflection (*chhü chhü* 屈曲).'

In the case of what is put forward (*chhên* 陳) first, there is no preceding comment (*shuo* 説), it directly puts forward a substance. This term does not have a verbal inflection (*chhü chhü*) and does not introduce a new predication (*tha chieh* 他解). In the case of what is put forward last (*chhên*!) there is a preceding comment and the later addition can subcategorize (*fên pieh* 分別) what has been stated before (in the subject). It is only this second part that has verbal inflection (*chhü chhü*) and introduces a new predication (*tha i chieh* 他異解). The arising of the the predication only depends on what is put forward last.[2]

Khuei Chi is aware that the multiplicity of terminology and of criteria deciding what is the subject and what is the predicate demand an explanation:

First, there is referring (*chü* 舉) versus linking (*thung* 通). That which refers to things is called an identifier (*tzu hsing* 自性) because it is narrower in its application. That which links with other things is said to be the *chha pieh* 差別 'subcategorizer' because it is wider in its application.

Second, there is first versus last. That which is put forward first is called an identifier (*tzu hsing*) because there is nothing before it which it can subcategorize. That which is put forward last is called subcategorizer (*chha pieh*) because there is something before it which it can subcategorize.

---

[1] Khuei Chi, Great Commentary, ch. 2, ed. Sung Yen, p. 3a5; cf. *ibid.*, ch. 2, ed. Sung Yen, p. 4a5. Khuei Chi attributes this explanation to a *Fo Ti Lun* 佛地論, which must be the *Fo Ti Ching Lun* 佛地淨論 translated by Hsüan-Tsang 玄奘. But in the version of that book as we have it today the passage to which Khuei Chi refers does not occur.

[2] Khuei Chi, Great Commentary, ch. 2, ed. Sung Yen, pp. 4b4ff. The obvious question arises to what extent the Chinese practitioners of Buddhist logic wrote in Chinese, but only for an audience that was also familiar with other languages like Sanskrit. The answer to this one is resoundingly clear: very few readers of Chinese indeed could be expected to know Sanskrit. (Cf. on the point van Gulik (1956).)

Third, there is that which is spoken of (*yen* 言) versus that which is declared acceptable (*hsü* 許). That which is carried forward by the words is said to be the identifier (*tzu hsing* 自性), and that which in one's thought is declared acceptable (*hsü*) is said to be the subcategorizer (*chha pieh* 差別) because it is the subcategorizing meaning (*pieh i* 別義) put forward in the words.[1]

The subject *S* of the thesis is described by Khuei Chi as 'the overall lord of the two properties (i.e., of the predicate *P* and of the reason *R*) (*êrh fa tsung chu* 二法總主).[2]

The reason

The Chinese Buddhist logicians distinguished various kinds of what we for want of a better term translate as 'reasons':

There are two kinds (*chung* 種) of reason (*yin* 因). One is the effective reason (*shêng yin* 生因), as in a seed producing sprouts and the like. The other is the epistemic reason (*liao yin* 了因) as in a lamp shedding light on an object.[3]

Each of these kinds of reasons are again subdivided into three categories which all have to do with the proponent:

1. linguistic effective reasons (*yen shêng yin* 言生因), i.e., a proponent's adduced explicit reasons which guarantee the truth of his thesis.
2. intellectual effective reasons (*chih shêng yin* 知生因), i.e., the intellectual skill which allows the proponent to grasp his thesis.
3. semantic effective reasons (*i shêng yin* 義生因), i.e., a fact or general truth to which the proponent refers and which entails his thesis.

We have similar subcategories for the epistemic reasons. They all have to do with the opponent:

1. linguistic epistemic reasons (*yen liao yin* 言了因), i.e., arguments understood as convincing by the opponent.
2. intellectual epistemic reasons (*chih liao yin* 知了因), i.e., intellectual capacity on the part of the opponent which allows him to grasp a thesis.
3. semantic epistemic reasons (*i liao yin* 義了因), i.e., a fact or general truth which convinces the opponent of the truth of a thesis.[4]

All these kinds of reasons have their importance, but Buddhist logic in China as we have it is particularly concerned with only two of them: the linguistic effective reason corresponding roughly to a sufficient logical reason, and the semantic epistemic reason corresponding roughly to an objectively convincing argument.

---

[1] Khuei Chi, Great Commentary, ch. 2, ed. Sung Yen, pp. 4a9ff.
[2] Great Commentary, ch. 3, ed. Sung Yen, p. 1b6.
[3] Wên Kuei on *Nyāyapraveśa*, ed. *Supplementary Tripitaka*, vol. 86, p. 234a top 8; cf. Khuei Chi, Great Commentary, ed. *Supplementary Tripitaka*, vol, 86, p. 358a top −6. When different early commentaries coincide, we may often assume that they all draw on a common oral source, the explanations of Hsüan-Tsang himself.
[4] Great Commentary, ch. 2, ed. Sung Yen, pp. 16a6ff.; cf. Yang Pai-shun (in Liu Phei-Yü (*1984*), pp. 298–312), p. 307.

Although our subdivisions of effective and epistemic reasons yield six (categories of) reasons, we are directly concerned only with the linguistic effective reasons (*yen shêng* 言生) and the semantic epistemic reasons (*chih liao* 知了). On the basis of the linguistic effective reason the other side's (*ti* 敵) reasoned understanding arises. On the basis of the semantic epistemic reasons the meaning (*i* 義) that used to be hidden is now made plain.[1]

Predominantly, Chinese Buddhist logic is concerned with sufficient logical reasons and with fallacious reasons of more vaguely defined kinds.

The meaning of the thesis established follows from the proponent's words that are able to establish (the thesis). The linguistic effective reasons are in a direct sense (*chêng* 正) what can establish (*nêng li* 能立).[2]

The reason (*yin* 因) is described as the main constituent of the grounds (*nêng li*).[3] The reason can be called *chêng nêng li* 正能立 'that which directly is able to establish' as opposed to the examples (*yü* 喻) which can be called 'that which helps to be able to establish'. Or, as Khuei Chi puts it, the reason (*yin*) directly establishes (*chhin chhêng* 親成) the thesis (*tsung* 宗) whereas the examples establish the thesis in a distant way (*chhêng tsung shu yüan* 成宗疏遠).[4]

The reason (*yin*) as a proposition and a constitutive part of the argument is sometimes distinguished from the property *R* mentioned in the reason (*yin thi* 因體).[5]

Reasons (*yin*) in Chinese Buddhist logic are expressed not as sentences or propositions but as terms or properties:

Thesis: Traditional verbal evidence is impermanent.
Reason: Because of its producedness (*chhi so tso hsing ku* 其所作性故).

The reason (*yin*) provides the crucial concept which will establish a firm link between the subject *S* and the predicate *P* of the thesis (*tsung*). It thus loosely corresponds to the middle term in Aristotelian logic. The workings of this main constituent are expounded with unfailing precision by Wên Kuei:

---

[1] Khuei Chi, Great Commentary, ch. 2, ed. Sung Yen, p. 17b2.

[2] Khuei Chi, Great Commentary, ch. 1, ed. Sung Yen, p. 23a1.

[3] Cf. Lü Chheng (*1935*), p. 12a. In the Great Commentary a number of lucid objections to this nomenclature are raised: 1. Why is Buddhist logic called a science of reasons and not of examples and theses, all of which are involved? 2. It is fair enough to call true reasons and true clarifications (*ming* 明) *yin ming* 因明, but only apparent reasons and only apparent clarifications should surely not be called *yin ming*? 3. Reasoned demonstrations and reasoned refutations should be called *yin ming*, but mistaken refutations and only apparent refutations should surely not be called *yin ming*? 4. In establishing or refuting theses there is verbal knowledge involved, and these may be called *yin ming*. But in arguments on the basis of perception no verbal knowledge is involved. They should surely not be called *yin ming*? 5. In the case of intellectual effective grounds and intellectual hermeneutic grounds (see above) one may call arguments involving them *yin ming*, but surely in the case of the other four kinds of grounds where there is no knowledge the word *yin* 因 is appropriate, but the word *ming* 'clarification' is surely inappropriate? 6. The reason and the examples can establish a thesis, and they can be called *yin ming*, but surely the thesis cannot establish a thesis and should not be called *yin ming*. 7. Why is Buddhist logic only called *yin ming* 'clarification of grounds' and not *kuo ming* 果明 'clarification of conclusions'? (Great Commentary 1.7a2) There is no room for Khuei Chi's detailed answers to these questions, but these questions by themselves illustrate the spirit of scientific clarity and probing intellectual perseverance which pervades the Great Commentary.

[4] Cf. Great Commentary, ch. 3, ed. Sung Yen, p. 18a1, and also Shen Chien-Ying (*1984*), p. 72.

[5] For the term *yin thi* 因體 see, e.g., Great Commentary, ch. 3, ed. Sung Yen, p. 1a10.

On the possessor of a property 'traditional verbal evidence' there are two properties: one is the controversial (*pu chhêng* 不成 lit: un-agreed-upon) property 'impermanent'; the other is the agreed upon property 'being produced'.[1] On the basis of the agreed-upon property being on 'traditional verbal evidence' one goes on to infer the un-agreed-upon 'impermanent' and make it also agreed upon.[2]

The property which is established (*so li chih fa* 所立之法) is a non-agreed property on the 'subject' (*yu fa* 有法). If in a place (*chhu* 處)[3] there is the commonly agreed reason (*yin* 因) there is certain (*chüeh ting* 決定) to be this non-agreed property (*pu kung hsü fa* 不共許法) (i.e., the 'predicate of the thesis'). By means of a commonly agreed property one establishes something non-agreed.[4]

The Nine Reasons

Consider now the relation of the property R in the reason to the examples in an argument. The restrictions on this are studied in what Khuei Chi calls the *chiu yin* 九因 'Nine Reasons'.[5] R can either be present, or not be present, or be present in some but not all of the cases (paradoxically expressed as 'both present and absent' in the Chinese logical texts) and this clearly will have an impact on the acceptability of the argument involved.

We are told that there are nine permutations in the relations between R and the examples:

1. R is present in the positive examples
   a. and also present in the negative examples.
   b. but not present in the negative examples.
   c. and both present and absent in the negative examples.
2. R is absent in the positive examples
   a. and present in the negative examples.
   b. and absent in the negative examples.
   c. and both present and absent in the negative examples.
3. R is both present and absent in the positive examples
   a. and present in the negative examples.
   b. and absent in the negative examples.
   c. and both present and absent in the negative examples.

Only cases 1b and 3b are considered acceptable. As an example for 1b, Shen Thai gives the standard argument:

---

[1] The reason (*yin*) is a property which is agreed to be on the subject. (Great Commentary, ch. 4, ed. Sung Yen, p. 2b2.)

[2] Wen Kuei on *Nyāyapraveśa*, ch. 1, ed. Lü Chheng, pp. 13b10ff. Cf. Shen Chien-Ying (*1984*), p. 73.

[3] *Chhu* is a technical term for subjects other than that in the thesis.

[4] Khuei Chi, Great Commentary, ch. 4, ed. Sung Yen, p. 2b2.

[5] Khuei Chi's detailed treatment of these will be found in Great Commentary, ch. 3, ed. Sung Yen, p. 9b1 to 3.11a3. A crucial earlier treatment is in Shen Thai on *Nyāyamukha, Supplementary Tripitaka*, vol. 86, p. 324b bottom 9ff. A paraphrase of Shen Thai's comments will be found in Tucci (1928a), pp. 385f. The Nine Reasons go back to Dignaga and have been extensively studied. See Chi (1969), pp. 1–29, which attempts a coherent formal analysis from the point of view of symbolic logic.

Thesis: Traditional verbal evidence is impermanent.
Reason: Because it is produced.
Examples: like a bottle etc., unlike space etc.
This is the direct reason (*chêng yin yeh* 正因也).[1]

The case of 3b is more interesting. Here the example is:

Thesis: Traditional verbal evidence is impermanent.
Reason: Because it is produced by effort.
Examples: Like a bottle or lightning etc., unlike space. In this context lightning and pot are positive examples of impermanence, and being produced by effort is the reason produced. In pots and the like the reason is present, in lightning and the like it is not. Space and the like are negative examples of impermanence, and the quality of being produced by effort is pervasively absent (*thung wu* 通無).[2] This is the second category (*ti êrh chü* 第二句) and it is also a correct reason (*chêng yin* 正因).[3]

In some cases, conversely, the reason not only fails to establish the thesis, it actually establishes the negation of the thesis. 2a and 2c are cases in point:[4]

Thesis: Traditional verbal evidence is permanent.
Reason: Because it is produced.
Examples: Like space and unlike a bottle.
In this context space is the positive example and a pot, etc., is the negative example of permanence. Producedness is the reason, and in the positive example it is absent, in the negative example it is present.[5]

The example argument for 2c is plain enough:

Thesis: Traditional verbal evidence is permanent.
Reason: Because it is produced by effort.
Examples: Like space and unlike a bottle or lightning.[6]

Khuei Chi declares that all the other five configurations produce inconclusive arguments.[7] Throughout his commentary he frequently alludes to this division of the Nine Reasons (*chiu yin* 九因) into two valid, two incompatible, and five inconclusive patterns.

---

[1] Shen Thai on *Nyāyamukha, Supplementary Tripitaka*, vol. 86, pp. 324b–4 bottom.
[2] Great Commentary, ch. 3, ed. Sung Yen, p. 10b9, writes *i hsiang fei yu.*
[3] Shen Thai on *Nyāyamukha*, ed. *Supplementary Tripitaka*, vol. 86, p. 325a bottom 4.
[4] Great Commentary, ch. 3, ed. Sung Yen, p. 11b1, calls 2a and 2c *hsiang wei yin* 相違因 'contradictory reasons'. Shen Thai on *NM, ST* vol. 86, p. 325, speaks of 2a, 2b, and 2c as the three contradictory reasons (*san hsiang wei yin* 三相違因). Here we seem to catch out Shen Thai on a logical howler.
[5] Shen Thai on *Nyāyamukha*, ed. *Supplementary Tripitaka*, vol. 86, p. 325a top 4.
[6] Shen Thai on *Nyāyamukha*, ed. *Supplementary Tripitaka*, vol. 86, pp. 325a top –7.
[7] Great Commentary, ch. 3, ed. Sung Yen, 11b2. The inconclusive example arguments in which nothing follows are:

  1a: Sound is permanent, because it is knowable. Like space and unlike a pot.
  1c: Sound produced by effort, because it is impermanent. Like a pot and unlike lightning or space.
  2b: Sound is permanent, because it is audible. Like space and unlike a pot.
  3a: Sound is not produced by effort, because it is impermanent. Like lightning or space, and unlike a pot.
  3c: Sound is permanent, because it is incorporeal, like the subtlest substance (*chi wei* 極微) and unlike space.

The Three Aspects of the reason

How does the reason manage to link the 'subject' to the 'predicate' of the thesis? One answer to this is: on the strength of its Three Aspects (*san hsiang* 三相). It is time, now, to look closely at these famous Three Aspects. The literature on these is large.[1] It is easy to lose one's way in it. We shall limit our attention to Chinese accounts of the matter and start out with a literal translation of the celebrated statement of the matter in the *Nyāyapraveśa*:

1. (The reason-property *R*) is pervasively a property of the subject of the thesis (*S*).

E.g., producedness must inhere in every traditional verbal evidence. In other words: if producedness is to count as a reason for any property of traditional verbal evidence, then traditional verbal evidence must have producedness. One can only adduce as reasons what are quite generally properties of the subject of the thesis.

2. The positive examples (*P*) undoubtedly have the property (*R*).

E.g., producedness must undoubtedly be present in (at least some) cases of impermanence. It need not be present in all cases of impermanence because the reason (*yin* 因) is only sufficient, not necessary. There may be reasons other than producedness for impermanence.[2]

3. The negative examples (not–*P*) pervasively lack the property (*R*).

E.g., producedness must be absent in all instances of permanence, or: all things, if permanent, must be unproduced. If producedness is to count as a sufficient reason for impermanence, then clearly no thing can be permanent and yet produced.[3]

The problematic Aspect is the second one: why do we have *ting* 定 'undoubtedly' and not, as in the other two Aspects *pien* 遍 'pervasively'? Part of the answer is that the second Aspect involves positively ascertaining whether there are indubitable instances (other than that mentioned in the subject of the thesis), in which the property in the reason actually entails the property in the predicate of the thesis. The word *ting* emphasises the confirmed existential claim involved.[4]

A brief comparison with the notions of a necessary reason, and that of a sufficient reason, in modern logic will make the point clear. A standard textbook in logic writes:

The notions of necessary and sufficient conditions provide other formulations of conditional statements. For any specified event there are many circumstances necessary for its occurrence. Thus for a car to run it is necessary that there be gas in its tank, its spark plugs

---

[1] See Chi (1969), pp. 30–104.

[2] Cf. Tachikawa (1971), Huo Thao-Hai (*1985*), pp. 58ff., Lü Chheng (*1983*), pp. 3ff., and Chi (1969), pp. 41f.

[3] This formulation of the Three Aspects will be found in *Nyāyapraveśa*, p. 11b1ff.; cf. Tachikawa (1971), p. 121. An alternative formulation of the Three Aspects is in *Nyāyamukha*, pp. 2c–3ff. and 4a3ff.; cf. Tucci (1930), pp. 42 and 44, though going back to Dignaga himself, is less clear. However, it also involves the distinction between *pien* 遍 for the the first and third Aspect, and *ting* 定 for the second.

[4] Note the difference between *ting* 'certainly, assuredly, undoubtedly' and *pi* 必 'necessarily'.

properly gapped, its oil pump working, and so on. So if the event occurs every one of the conditions necessary for its occurrence must be fulfilled . . . For a specified situation there are many alternative cirumstances any one of which is sufficient to produce that situation. Thus for a purse to contain over a dollar it would be sufficient for it to contain one hundred and one cents, twenty-one nickels, eleven dimes, five quarters, and so on. If any one of these circumstances obtains, the specified situation will be realized. Hence to say 'That the purse contains five quarters is a sufficient condition for it to contain over a dollar' is to say the same as 'If the purse contains five quarters then it contains over a dollar.' In general, 'p is a sufficient condition for q' is symbolized as $p \rightarrow q$.[1]

In standard logic '$p \rightarrow q$' is a proposition which is defined as true whenever p is untrue. Thus, according to the definitions in formal logic, a false proposition can be said to imply any other proposition whatever. Translated into our idiom of sufficient conditions: a condition which does not obtain in the universe of discourse counts as a sufficient condition for anything whatever within that universe. The Chinese Buddhist logicians are quite explicitly excluding such very strange sufficient conditions. They are interested only, and precisely, in sufficient conditions that can be ascertained definitely to occur in this concrete world. That is why the existential second Aspect is profoundly significant. It demonstrates that Chinese Buddhist logic is not formally deductive but preoccupied with concrete instances of standardised arguments.[2]

Examples

Wên Kuei, when challenged by a questioner to explain why the examples are necessary, explains:

Producedness (*so tso hsing* 所作性) is the direct reason (*chêng yin* 正因), but it can only bring out into the open the first Aspect (*chhu hsiang* 初相). The second and the third are not yet understood. Therefore one makes them explicit (*ming chih* 明之) by the two examples.[3]

Khuei Chi was aware that the example, while supporting the thesis, was not by itself a justification for its acceptance:

Although it (the example (*yü* 喻)) is in accordance with and in support of (*shun i* 順依) the conclusion, it is not what the thesis directly rests on (*fei shih chêng tsung chih so i* 非是正宗之所依).[4]

The positive and negative examples are stated with various degrees of explicitness. In its most explicit form, the positive example involves a general implication. Khuei Chi reflects on the importance of the generalisation in this scheme:

---

[1] Copi (1953), p. 285.
[2] The issue of whether there are such things as objective facts does not affect this argument.
[3] Wên Kuei on *Nyāyapraveśa*, ch. 1, ed. Lü Chheng, p. 22b; cf. also Khuei Chi, Great Commentary, ch. 4, ed. Sung Yen, p. 4a.
[4] Khuei Chi, Great Commentary, ch. 2, ed. Sung Yen, 18a1.

If you do not say 'All produced things are impermanent, like the bottle, etc. (*chu so tso chê chieh shih wu chhang* 諸所作者皆是無常)', then you have not established that in the place where the produced is, the 'impermanent' is bound to follow.[1]

The general part of this formulation in the form of an implication linking *R* to *P* is called the substance of the example (*yü thi* 喻體). The second, concrete instance is known as the constituent of the example (*yü i* 喻依) or as the example item (*yü phin* 喻品). Sometimes, the Chinese commentators call the specific second part the *so i* 所依, lit.: 'what (the example) relies on'.[2] Very subtly, Wên Kuei speaks of pointing out the substance of the example directly (*chêng chih thi* 正指體) versus 'bringing up the constituent (*chü so i* 舉所依).[3] Both the substance and the constituent of the example are linguistic entities.

There are two kinds of examples (*yü* 喻): one is the positive term (*thung fa* 同法), the other is the negative term (*i fa* 異法). In the case of the positive term (*thung fa*), if in this locus (*chhu* 處) you show a case of a thing similar to the reason (*yin* 因), then there is certainly (*chüeh-ting* 決定) going to be the property ascribed in the thesis (*hsing* 性). I.e., (*wei* 謂 lit.: 'call') if there is producedness, then you will find the impermanence in question,[4] e.g., a bottle, etc.[5]

As for the negative example, if in this locus (*chhu*) you explain there is not the property that is to be established (*so li* 所立),[6] then in all cases the (property in the) reason (*yin*) will not be there. I.e., (*wei*) if a thing is permanent then you will find that it is not produced, like space etc.[7]

Stated in this way, the positive example establishes an explicit link between the reason (*yin* 因) and the predicate of the thesis (*hsing* 性). As the *Nyāyamukha* has it:

The positive example says that the (predicate of the) thesis (*tsung* 宗) follows from the reason. The negative example says that when the (predicate *P* of the) thesis (*tsung*) is absent, so is the reason.[8]

Khuei Chi comments on the positive example as follows:

In all places (*chhu* 處) where there is coming into existence there certainly is going out of existence. Where the mother cow goes the calf is bound to follow. In the place where the

---

[1] Khuei Chi, Great Commentary, ch. 8, ed. Sung Yen, p. 4a.

[2] The thought seems to be that the generalisation relies on the example to be psychologically convincing. Our rather awkward translation 'constituent' is designed to remind us of the use of *i* to refer to the constituents of the thesis, the *tsung i* 宗依.

[3] Cf. Wên Kuei on *Nyāyapraveśa*, p. 337a top −4.

[4] Wên Kuei elaborates this concomitance beautifully and poetically: i.e., 'Wherever the producedness goes, the stupid and the wise in this world alike know, impermanence is bound to follow it; like the case of a mother cow: wherever she goes, the little calf will follow her.' (Wen Kuei on *Nyāyapraveśa*, p. 337a top −1.)

[5] Wên Kuei expounds: 'This exemplification (*yü* 喻) signifies: When the "produced" goes to pots, then the quality of impermanence (*wu chhang hsing* 無常性) is in the bottles. Therefore we know that when the "produced" goes to sound, the quality of impermanence will also be in sound.' (Wên Kuei on *Nyāyapraveśa*, p. 337a bottom +2.)

[6] The concept *so li* 所立 'what is established' often refers generally to the thesis, but sometimes – as here – to the controversial part of the thesis – its 'predicate' – only. The term *tsung* 宗 'thesis' has a similar ambiguity.

[7] *Nyāyapraveśa*, ed. Lü Chheng, p. 15. Wen Kuei expounds: 'The meaning of this exemplification (*yü*) is as follows: on space, etc., there is not the "impermanent" which is the property attributed in the thesis, and consequently there is not the quality of producedness from the reason. Since on sound there is the quality of producedness, then we know that there certainly must be the "impermanent" of the thesis in it.' (Wen Kuei on *Nyāyapraveśa*, p. 338a top −7.)

[8] *Nyāyamukha*, ed. *Taishō Tripitaka*, p. 2c3; cf. Tucci (1930), p. 36.

reason (*yin*) is, the (predicate *P* of the) thesis (*tsung*) is bound to follow. This is the connection. If there is producedness, then the opponent as well as witnesses will see the impermanence.[1]

From a Western logical point of view, the generalisation in the positive example is the essential element (and it will be remembered that the Chinese call it 'the substance of the exemplification (*yü thi* 喻體)'). Indeed, Khuei Chi quotes Dignāga as saying that the positive example item as such does not show anything about the subject *S* of the thesis.

You may say that the substance of the bottle (*phing thi* 瓶體) and the substance of space (*khung thi* 空體) constitute the example (*yü* 喻), but you must assimilate (*lei* 類) the bottle to the meaning 'impermanent' in what is being established. Since you do not say that 'The various produced things are all of them impermanent', then bringing up (*chü* 舉) the bottle to show something about traditional authority (*shêng* 聲) is ineffective.[2]

Curiously, the spelling out of this generalisation continues to be regarded as optional in Chinese Buddhist logic. Even the negative exemplification as a whole is often omitted. On the other hand an argument without the concrete positive example item would seem to be simply not acceptable. Khuei Chi writes:

When a member of the argument (*chih* 支) defaults, we say it does not have the quality of being able to establish (*fei nêng li hsing* 非能立性).[3]

It is customary to refer to the rôle of the positive instance as 'inductive'. This is profoundly misleading in so far as the question of how far the single example guarantees the generalisation (or 'substance of the example') is never at issue among Chinese Buddhist logicians.

Finally, let us turn to a rather curious formulation of an argument in which Khuei Chi seems to come close to using the *modus ponens*.

Suppose we establish a thesis (*tsung* 宗) that traditional verbal evidence is impermanent on the grounds that it is produced. 'Impermanent' means 'liable to go out of existence'. 'Produced' means 'liable to come into existence'. (The thesis then runs:) 'Traditional verbal evidence has (the property) "liable to go out of existence" on the grounds that it has (the property) "liable to come into existence", for all things which are liable to come into existence are liable to go out of existence.' Since traditional verbal evidence by virtue of the reason comes into existence, then it is clear that it has the resulting (property of) 'going out of existence'.

If the 'produced' of the reason does not pervade the subject 'traditional verbal evidence' of the thesis, then how could one conclude (*chhi tê* 豈得) that it (the traditional verbal evidence) was pervasively characterised by 'impermanent'?[4]

In this remarkable passage, Khuei Chi disregards the rôle of the examples. Moreover, he makes the logical structure of the argument very clear:

[1] Khuei Chi, Great Commentary, ch. 4, ed. Sung Yen, p. 3b7.
[2] Khuei Chi, Great Commentary, ch. 4, ed. Sung Yen, p. 5b2.
[3] Khuei Chi, Great Commentary, ch. 1, ed. Sung Yen, p. 25a3.
[4] Khuei Chi, Great Commentary, ch. 3, ed. Sung Yen, p. 3b1.

1. All traditional verbal evidence is liable to come into existence.
2. Everything that comes into existence goes out of existence.
3. (Traditional verbal evidence comes into existence.)
4. It is clear traditional verbal evidence must go out of existence.

Affirmation and negation

All words have something they block out/negate and something they display/affirm (*i chhieh ming yen yu chê yu piao* 一切名言有遮有表). If you say 'fire' you block out 'is not fire'. It is not that you have got hold of the concept of fire itself, you have got hold of a fire occurring.[1]

Chinese Buddhist logicians distinguished between affirmative statements (*piao chhüan* 表詮) and negative statements (*chê chhüan* 遮詮):

Negation (*chê* 遮 lit. 'hide, block, keep out') refers to the exclusion of what it negates (*fei* 非).[2]

Suppose you speak of salt. If you say it is not tasteless, that is negative (*chê*), and if you say it is salty, that is positive (*piao* 表). Or suppose you speak of water. If you say it is not dry, that is negative, (*chê*), and if you say it is wet, that is positive (*piao*).[3]

The Chinese Buddhist notion of negation was a subtle one. You cannot look at a sentence, see whether it has a negation in it and on that basis decide whether it is positive or negative. In a very puzzling but significant passage, Khuei Chi writes:

There are two modes of establishing the predicate of a thesis: one is to purely negate and not affirm, as in 'The I does not exist'. This merely wants to negate 'the I' (*tan yü chê wo* 但欲遮我) and does not subcategorize the I as 'non-existent'. . . . The second modus is both negative and positive, as in 'The I is eternal'. This not only negates the non-eternality; it also asserts that there is the substance (*thi* 體)[4] of eternity.[5]

It appears that the notion of negation (*chê* 遮) is applied to predicate terms and not to sentences or propositions. 'Permanent' can be described, and often is described, as the 'negation' of 'impermanent'.

Contradiction

The notion of contradiction is discussed in Chinese Buddhist logic. The Chinese Buddhist technical term is *wei* 違 'contradictories':

Eternal and non-eternal are direct contradictories (*chêng hsiang wei* 正相違).[6]

For example, if one sets up 'good' as the (predicate *P* of the) thesis (*tsung* 宗), 'not good' is in contradiction with it (*wei hai* 違害) and these are therefore called contradictory (*hsiang wei*

---

[1] Khuei Chi, Great Commentary, ch. 8, ed. Sung Yen, p. 14a2.
[2] See Shen Chien-Ying (*1984*), p. 173.   [3] See Shen Chien-Ying (*1984*), p. 173.
[4] The use of the term *thi* 體 'substance' here is disconcerting. I have no explanation why Khuei Chi should want to use the term here.
[5] Khuei Chi, Great Commentary, ch. 8, ed. Sung Yen, p. 3a.
[6] Wên Kuei on *Nyāyapraveśa, Supplementary Tripitaka*, vol. 86, p. 325b, bottom 13.

相違). 'Bitter' and 'joyful', 'bright' and 'dark', 'cold' and 'hot', 'large' and 'small', 'permanent' and 'impermanent', etc., all are like this (*i chhieh chieh êrh* 一切皆爾).[1]

The distinction between contradictories and contraries does not seem to have
struck Khuei Chi in the context. Or perhaps we should say: Khuei Chi disregarded
the difference, just as the Greeks often disregarded the difference when they used
the word *enantion*.

A little further along on the same page the notion of contradiction is expressed in
a slightly different way:

The property of non-eternality in the bottle has a directly incompatible relation to (*chêng
nêng wei hai* 正能違害) the property of eternity in space. Therefore it is called a negative
example.[2]

(3) THE ARGUMENT FOR CONSCIOUSNESS ONLY

Having surveyed the abstract system of Buddhist logic as taught in China, we shall
now briefly turn to an outline of the one most famous historical example of the art
of Chinese Buddhist argumentation.[3] Inevitably, we enter deep and treacherous
waters of Buddhist reasoning at this point. As outside observers without the necessary life-time's exposure to the doctrinal intricacies of the School of Consciousness
Only, we can only provide an outside observer's account from a logical perspective.

The background story is well told in Arthur Waley's splendid biography of
Hsüan-Tsang.

Harsha, accompanied by Tripitaka (=Hsüan-Tsang) reached Kanauj at the end of the year
(AD 642). They were joined by the kings of eighteen vassal countries, by some three thousand
monks and some two thousand Hindus and Jains . . . Tripitaka was then called upon to take
the chair and state the case of the Mahayana as against the doctrines of the Lesser Vehicle.

Arthur Waley translates the formal argument Hsüan-Tsang used as follows:

Philosophically speaking, such forms of matter as are accepted both by us and by you are
inseparable from the eye-consciousness (Proposition). Because according to the eighteen-
fold classification that we too accept, these forms of matter are not included under the heading 'Organ of Sight' (Proof). Like eye-consciousness itself (Analogy).

Arthur Waley concludes his assessment of the argument thus:

It is certain at any rate that the famous syllogism comes perilously near to being nonsense;[4]
otherwise Tripitaka's disciples would not have needed to devote page after page to proving

---

[1] Khuei Chi, Great Commentary, ch. 3, ed. Sung Yen, p. 13a7.
[2] Wen Kuei on *Nyāyapraveśa*, ed. *Supplementary Tripitaka*, vol. 86, p. 325b, bottom 19.
[3] For detailed analyses of this argument cf. particularly Lü Chheng (*1983*), pp. 62–79, Lo Chao (*1988*) and
Shen Chien-Ying (*1984*). The argument was originally called The Argument for Consciousness Only in Khuei
Chi's Great Commentary, ch. 5, where it is first summarised. A more detailed account of it will be found in Hui
Li's biography of Khuei Chi, chs 4 and 5.
[4] (Footnote by Arthur Waley:) I mean, of course, nonsense from the standpoint of his own contemporaries;
whether it does or does not conform to our conventions of thought is irrelevant.

that it makes sense. At the risk of suffering the same fate as Shun-ch'iung who, as a punishment for his views on Tripitaka's syllogism, dropped headlong to Hell, through a hole that opened in the ground at his feet, I must confess that I don't find the apologias of his disciples K'uei-chi and Shen-t'ai very convincing.[1]

Hsüan-Tsang's Argument was designed to show that the external world (*ching* 竟) was inseparable from awareness of it (*shih* 識). As an example of external things he chooses visible things (*sê* 色 lit. 'colour'). This much is uncontroversial. But once that is said, one has to admit that Hsüan-Tsang's senior disciples differed widely in their interpretations of the Argument. The literature is vast, and it deserves a specialised monograph in its own right.

Hsüan-Tsang's followers, of course, regarded the Argument as 'the ten thousand generation guideline by which to produce arguments'. It deserves our detailed attention as an historically significant test case of logic in action. It is a valiant attempt at *metaphysica in ordine logico demonstrata* preceding Spinoza's *ethica in ordine geometrico demonstrata* by many centuries.

I shall not and cannot at this point enter deeply into the history of the various interpretations attached to the Argument, or indeed into the detailed interpretation of the Argument in its Buddhist context.[2] Instead, I shall make a tentative attempt to use the concepts of *yin ming* 因明 introduced in the last Sub-section to see whether what looked like indubitable near-nonsense to Arthur Waley will yield to a somewhat more charitable interpretation in the light of Buddhist logical theory.

After all, when an argument is as celebrated as The Argument for Consciousness Only it seems perfectly legitimate to declare it seriously wrong and mistaken. But to declare such a *locus classicus* 'clearly near-nonsense' is no more helpful than saying about Anselm's proof of the existence of God that it is clearly near-logical rubbish. One must first find out what gave Anselm's argument its apparent plausibility and force, and only then proceed to criticise it. Exactly the same applies to the Argument of Consciousness Only. We must first understand what appeared to make it tick. Only after that can we sensibly declare that it doesn't work. Since I have been very much preoccupied with this argument, let me present a brief and summary account of it; at the risk of appearing indeed to be retailing nonsense.

Hsüan-Tsang proposes something that may remind one of Bishop Berkeley's *esse est percipi* 'being is being perceived', but it should rather be expressed as *esse est aestimari esse* 'to be is to be deemed to be', and a detailed comparison with Arthur Schopenhauer's position on this issue is most instructive.[3] Objective existence is doomed to be exactly what it is deemed to be. We have no other access to it than our deeming. Hsüan-Tsang's position emphatically does not assume the existence of a perceiver or a deemer. As we have seen before: an argument of the sort *cogito, ergo sum* would, for a Buddhist of the School of Consciousness Only, carry no weight because in the one word *cogito* it begs the crucial question at issue: could there be

---

[1] Waley (1952), pp. 62f. For a more detailed account of the historical circumstances of this famous debate see Lü Chheng (*1983*), pp. 62ff.

[2] By far the best treatment to date is Lo Chao (*1988*).　　[3] Cf. e.g., Schopenhauer (1960), vol. 2, pp. 11ff.

thought without that thought being the thought of a thinking subject? Is there a metaphysically real *ego* in any conceivable *cogitatio*?

There is a problem about the visible objects we think about: what exactly do we mean by claiming that they exist? And moreover: do they in fact exist? But there is also a question about the subject which does the thinking: what exactly do we mean by saying that it does exist? And does it in fact exist? The Argument of Consciousness Only addresses a question regarding the nature of the existence of objective things.

Here is how I have come to understand the essence of Hsüan-Tsang's argument as paraphrased in my own terms.

Sense impressions are not directly of objective things. Sense impressions are *construed* by 'eye-consciousness' (the thinking as applied to the visible world) to be of objective things. Things are constructs of eye-consciousness. They are therefore inseparable from eye-consciousness in so far as they belong to the visual sphere and eye-consciousness defines their ontological status.

Hsüan-Tsang argues that visible things are by common agreement in the visual sphere (consisting of the eye, eye-consciousness and visual things). But how, one may ask, do visible things get attached to the visual sphere? Not, Hsüan-Tsang would insist, because the eye directly perceives them. (Because what meets the eye are sense impressions, not visible things themselves. When the eye sees a stone, what hits the eye is not the stone but sense impressions construed to be of the stone.)

But how, one may persist, are visible things attached to the visual sphere, if not because they are directly perceived by the eye? There is only one link left: it must be through eye-consciousness which construes and judges the sense impressions to be *of* visible things. Visible things, therefore, though not identical with a judgement by eye-consciousness that they are visible objects, are inseparable from eye-consciousness, essentially linked with it.

A preliminary translation of the celebrated Argument might run like this:

Thesis: Speaking philosophically, and assuming a meaning of the terms ('visible things' and 'be not separable from eye-consciousness') accepted by the disputing parties: VISIBLE THINGS ARE DEFINITELY NOT SEPARABLE FROM EYE-CONSCIOUSNESS.

Reason: Because, using the term ('visible things') in the proponent's sense of the term: THOUGH BEING IN THE VISUAL SPHERE THEY ARE NOT PERCEIVED BY THE VISUAL ORGAN.

Example: (Everything which, though in the visual sphere, is not perceived by the visual organ, is not separable from eye-consciousness.)[1] AS FOR EXAMPLE EYE-CONSCIOUSNESS.

Let us keep in mind that this is an argument belonging not to the 20th but to the +7th century. It belongs not to modern analytical philosophy but to an *early* phase in the history of Buddhist logic.

It is beyond the scope of this survey to attempt to fully expound the buddhological significance of the argument, or to give a detailed justification of our interpretation

---

[1] This addition in brackets makes explicit what, according to the Chinese logicians, the example served to exemplify.

– if indeed it can be properly justified. But, however we must understand the Argument, it does represent a valiant attempt at a *metaphysica in ordine logico demonstrata* deserving a close logical scrutiny in the light of modern logic, which to my knowledge it has yet to receive, and which in any case it has not received here and now.

A number of formal features of the Argument must be noted here nonetheless. First, we have already seen that the optional explicit part (*thi* 體) of the positive example – which is added in brackets – is omitted in the original formulation. Moreover, we note with surprise that the example item used occurs in the thesis itself. But there was apparently no rule to prevent this and it is not clear what other example Hsüan-Tsang should or could have chosen.

Second: The negative example is completely omitted. This is a fine example of how the crucial link in an argument can be left implicit in Buddhist logic.

Third: the Argument adds limiting clauses (*chien pieh* 簡別) at the beginnings of both thesis and reason.

Why did Hsüan-Tsang introduce the qualification 'philosophically speaking (*chên ku* 真故)'? Hsüan-Tsang deploys a distinction between levels of discourse here. He intends his thesis understood on the level of *shêng i* 勝義 'highest meaning'. Such hierarchies of meaning are not unknown in China (especially to the Taoists), but they play a much more central part in formal Indian philosophy.

Why, then, did Hsüan-Tsang introduce this qualification? The answer is that he thereby avoided the recognised fallacy of *shih chien hsiang wei* 世間相違 'being in conflict with plain fact'. The common-sense objection that plainly a table, though visible, is not perception but a physical object, is ruled out. As indeed it would be for someone discussing bishop Berkeley's *esse est percipi*. At a philosophical level, the question is not as simple as that.

Why did Hsüan-Tsang introduce the qualification 'on the common acceptance of the term (*chi chhêng* 極成)'? The answer is again that he is avoiding a fallacy which in this case would consist in arguing a thesis the concepts of which are understood in incompatible ways by proponent and opponent. As we have seen, one convention in Buddhist logic is that a sense of the concepts in the thesis must be agreed upon by proponent and opponent, and that the thesis as a whole must not be commonly agreed beforehand. What Hsüan-Tsang is claiming in his qualification is that he is using the terms *sê* 色 'visual object' and *li yü yen shih* 離於眼識 'separate from optical perception' in a sense acceptable to the proponent himself and his opponent.

Why did Hsüan-Tsang introduce the qualification 'in the proponent's sense of these expressions'? This is a tricky and highly controversial question on which I have changed my mind many times. Let me submit my current interpretation. In the thesis Hsüan-Tsang has assumed that the term 'visible things' is to be understood in a commonly agreed 'pre-theoretical' sense. In the reason he now stipulates that 'visible things' as understood by adherents of the School of Consciousness Only, though in the visible sphere, are not perceived by the visual organ. The commonly agreed 'visible things' are included in, or part of, the things taken by the

School of Consciousness Only to be the 'visible things', so that this is seen as a technically legitimate qualification in terms of the rules of *yin ming* logic.

Why did Hsüan-Tsang introduce the qualification 'though in the visual sphere'? Clearly, if he had not added this qualification, the thesis would not follow at all. The point of the argument is that the visible object (as everyone admits) *is* in the visible sphere but *is not* (directly) perceived by the visual organ. The visible object, then, since it is not the eyeball itself, must belong to eye-consciousness, i.e., it follows that the visible object is not separable from eye-consciousness.

Our biased sources indicate that nobody dared raise any objections against the formidable Hsüan-Tsang on that occasion in India in the year +642. Naturally, though, his Argument on Consciousness Only did not go unchallenged. One opponent constructed a most ingenious competing argument which might almost look like a parody of the original one. Undaunted by Hsüan-Tsang's argument, Yüan Hsiao 元肖 (Korean: Wŏnhyo, +617 to +686), a formidable Buddhist from Korea (writing in Chinese),[1] declared that like the eyeball itself, the visual object was separate from visual perception. Here is the Korean's challenge to Hsüan-Tsang:[2]

Thesis: Philosophically speaking, and taking the term visible thing in the commonly accepted meaning: Visible things are separate from eye-consciousness.
Reason: Because, using the term 'visible thing' in the proponent's sense, though in the visual sphere they are not perceived by the eye.
Example: (Everything which, though in the visual sphere is not perceived by the eye, is separate from eye-consciousness.) Like the eyeball.[3]

At the time when Wŏnhyo put forward this argument, Hsüan-Tsang was already dead, but his foremost disciple Khuei Chi, as one can imagine, did not let the matter rest at that. In his Great Commentary (ch. 5) he pointed out six serious fallacies committed by the Korean. The debate continues even today.[4] I cannot and will not pretend to be the ultimate arbiter of its rights and wrongs. But I wish to insist that it is far from being philosophically pointless or theoretically inane.

By way of a conclusion of this discussion of Buddhist logic, I shall highlight some fundamental contrasts between Chinese Buddhist logic and Aristotelian logic in the following Sub-section. But first I turn to problems of translation.

(4) The Translation of Logic from Sanskrit to Chinese[5]

To what extent, one may ask oneself, is the nature of Greek logic linked to the nature of the Greek language? Or for that matter: to what extent is Western logical

---

[1] Wŏnhyo was quite a colourful personality. After he turned Buddhist he had a child with a princess and preferred to be known thereafter by the name Lay Practitioner of Small Nature (*Sosŏng kŏsa*). 25 volumes of his works are extant. For an account of his life and thought see Hong Jung-shik (1982). The attribution of this argument to Wŏnhyo is, however controversial. The evidence on this is collected in Lo Chao (*1988*), p. 28.

[2] It must be noted that Wŏnhyo did not present the argument in China himself, and some sources attribute the syllogism to a messenger and disciple of his.

[3] Yüan Hsiao, *Phan pi liang lun, ST*, vol. 94, ch. 4.

[4] Shen Chien-Ying (*1984*) argues vigorously but quite superficially against Hsüan-Tsang as an 'idealist'. Lü Chheng (*1983*), pp. 62–79, carefully reconstructs the plausibility of the Argument for Consciousness Only.

[5] Cf. Fuchs (1930), van Gulik (1956), pp. 3–36, Chen (1960) and Hurvitz (1963).

thinking conditioned by the fact that it was articulated and developed in the medium of Indo-European languages?

Similarly, one may ask: to what extent was Indian logical thinking, and particularly Buddhist logic, conditioned by the Sanskrit language in which it was articulated? How much of the Sanskrit Buddhist logic can be adequately and naturally conveyed in a language as totally different from Sanskrit as Classical Chinese?[1]

In general: what is the relation between logical theories and the natural languages in the medium of which they are developed and ultimately defined?

All these are crucial questions for the history of science. They are questions any philosopher of language worth his salt should take seriously, indeed should have taken seriously a long time ago. They are of more than purely historical interest. Unfortunately, they cannot find any definitive answer here. What we can investigate empirically is not the question whether the Chinese might have invented Buddhist logic of the kind the Indians invented. That is a matter for pure speculation. What, however, we can investigate is the question to what extent Indian logic can be translated into Chinese and transported into China.

A brief survey of the technical aspects of Buddhist translations into Chinese will help to put things into perspective. For when we speak of Hsüan-Tsang's translations, we really mean a translation made by a large imperially sanctioned office where translation was done by no less than eight individuals:

1. The head of the translation team (*i chu* 譯主), in our case Hsüan-Tsang, under whose name the translation would appear.
2. The transmitter of the words (*chhuan yü* 傳語 or *i yü* 譯語), who produced an oral first interpretation of the Sanskrit text.
3. The verifier of the meaning of the Sanskrit text (*chêng fan i* 證梵義), who ensured that the oral translation was in accordance with the Sanskrit.
4. The scribe (*pi shou* 筆受).
5. The verifier of the meaning of the Chinese translation (*chêng i* 證義), who checked that the Chinese made sense as it stood.
6. The embellisher of the style (*jun wên* 潤文).
7. The proof-reader (*chiao khan* 校勘 or *tsung khan* 總勘), who ensured that the Chinese text was complete.
8. The proof-reader (*chêng tzu* 證字), who ensured that all the Chinese characters were written correctly.[2]

---

[1] For Kumārajīva's methods and achievements of translation of Buddhist texts see Robinson (1967), pp. 77–87. Robinson begins his summary as follows: 'To summarize the virtues and faults of this translation: The Chinese is often more explicit than the Sanskrit. It relies less heavily on anaphora and so is clearer. It sometimes supplies explanatory phrases such as one finds in the prose paraphrases of Sanskrit commentaries. . . . The Chinese copes successfully with syntactic features such as the locative absolute and statements of reasons by means of ablative noun compounds.' Robinson notes important defects, and concludes: 'This confusion of the existential, the modal, the logical, and the epistemological prevents everyone who does not know the Sanskrit from grasping the subtler points of the text. The substitution of rhetorical questions for negative statements and vice versa is neither a virtue nor a defect. . . . The defects of such a translation are bothersome to the scholar who wishes to reconstruct a Sanskrit original but, with the exception of the mishandling of logical terms, I do not think that the mistranslations prevent the reader from understanding the Mādhyamaka system as an aggregate. . . . (*Ibid.*, pp. 87f.)

[2] For the organisation of Buddhist translation work see Fuchs (1930) and van Gulik (1956).

Our present enquiry will be based on a close comparison of the Sanskrit, Tibetan and Chinese versions of the systematic expository part of the *Nyāyapraveśa* only.[1] So far, we have chosen to consider the Chinese texts on Buddhist logic in their own right and to reconstruct Chinese Buddhist logic on this basis. We must now turn to the crucial question: how far were the Chinese able to represent in their language the nuances, subtleties, and technicalities of the Sanskrit original of *Nyāyapraveśa*? How, for example, does their achievement in this particular instance compare with that of the Tibetans?

One might suspect that the Sanskrit version of *Nyāyapraveśa* was much more precise and explicit than its Chinese translation, and that in general the Chinese – given the very different nature of their language – were only able to convey such parts of the messages in the Sanskrit original as could be naturally conveyed in the Chinese medium – a kind of watered-down version of *nyāya* logic. Again, one might imagine that the Tibetans were much more proficient in their adaptation to the Sanskrit logical discourse. However, this is not an occasion for abstract speculation or ratiocination on the ultimate relation between language and logic in Sanskrit, Chinese and Tibetan.[2] It is an occasion for simply addressing the philological facts of the historical case at hand.

Technical terms

Consider *nyāya*, the technical term most commonly used for Indian logic. One way of translating this into Chinese is simple phonetic transcription. Since the literal meaning of *nyāya* in Sanskrit is less than immediately obvious in the first place, this would not seem to put the Chinese at a significant disadvantage. However, Hsüan-Tsang preferred a more analytical rendering: he called *nyāya* 'the science of reasons (*yin ming* 因明)', modelling himself on the more explicit Sanskrit term *hetuvidyā* which is one of the many less widely used Sanskrit terms for logic.[3]

In order to avoid arbitrary eclecticism in our choice of topics in translation to discuss, let us now simply turn to the technical terms in the opening lines of the *Nyāyapraveśa*.

Consider, then, the main subjects treated in *nyāya* logic and listed at the outset of our little primer of Buddhist logic: *sādhanam* '(means of) establishing: proof', *dūṣaṇam* '(means of) destroying: refutation', *pratyakṣa* 'what lies before one's eyes: perception' and *anumāna* 'what is measured according to something else: inference'.

Hsüan-Tsang treats these as two structural pairs: he calls proof *nêng li* 能立 '(that which) can establish', refutation *nêng pho* 能破 '(that which) can destroy', perception

---

[1] For a systematic treatment of the techniques of translation of Buddhist texts into Chinese see Fuchs (1930).

[2] There has been an enormous amount of fascinating philosophical speculation on this area within the framework of analytical philosophy, inspired by men like Gottlob Frege and Ludwig Wittgenstein.

[3] In order to mark off clearly all works on logic, Hsüan-Tsang as well as his successors throughout Chinese history have tended to insure that any work specifically concerned with Buddhist logic would contain *yin ming* 因明 in its title, preferably at the beginning of the title. The result is that Chinese (and, incidentally, Japanese) works on Buddhist logic can be conveniently identified because their titles begin with the characters *yin ming*.

hsien liang 現量 'measuring (on the basis of) what is at hand'[1] and inference *pi liang* 比量 'measuring (on the basis of) comparison'. The Chinese and Sanskrit terms have, of course, this in common that they are technical terms used in specific senses and according to definitions. But the difference which strikes the eye is that Hsüan-Tsang's terminology here is more structurally transparent than the Sanskrit. In Chinese we have, palpably, two pairs of concepts. In Sanskrit one may remember, if one recalls the technical meanings, that the four terms fall into two groups, but the pairing is not made morphologically transparent. The logical terminology which Hsüan-Tsang constructed was often more transparent and coherent than the Indian original terms he aimed to translate, because he had the need – *and the free-dom* – to create a terminology through stipulative definition much of the time.[2]

To each pair the Sanskrit text adds *sabhāsa* 'and fallacies', which Hsüan-Tsang most aptly translates as *chi ssu* 及似 'and (only) apparent cases (of these)'. *Ssu* 似 is basically a transitive verb 'be like, be only like (cf. *ssu êrh fei* 似而非 'resemble something but not really be it'). The transitivity of *ssu* makes it plain that the text refers to only apparent cases of proof and refutation, of arguments from perception and inference.

The fact that *ssu* is basically a verb, and that *nêng li* 能立 as well as *nêng pho* 能破 are verb phrases, but that both are here used as a subject-nominalization ('the subject of the verb (phrase)'), is no disadvantage in the context because the particles *chi* 及 and *yü* 與 can stand only between noun phrases and thus act as effective and quite unambiguous nominalizers in the context.

Proof and refutation are said in the Sanskrit to be *parasaṃvide* 'designed to make others understand', perception and inference are said to be *ātmasaṃvide* 'designed to make oneself understand'.[3] Hsüan-Tsang adds here the restrictive copula *wei* 為 to indicate that these and only these things constitute the relevant realm of making others understand and making oneself understand respectively in the context of *nyāya* logic.[4]

One might think that the plural forms in the Sanskrit could present a problem when the Chinese plural morpheme *mên* 們 cannot be used to convey the plural in the form of Classical Chinese used by the Chinese Buddhist logicians. However, when appropriate and necessary the Chinese Buddhists simply broke the rules of Classical Chinese syntax and introduced the word *to* 多 in front of the Chinese word standing for a Sanskrit word in the plural, and this, though not making exactly elegant Chinese, adequately encoded the Sanskrit nuance and made it accessible to the Chinese reader. This introduction of *to* in front of words in the plural, however,

---

[1] The crucial point is that the Chinese makes it *explicit* that what is being spoken about is not the reality before one's eyes (*hsien* 現 expresses that idea), but a conclusion or judgement of some sort (called *liang* 量 by Hsüan-Tsang), based on what is being perceived.

[2] The development of terminology can be compared to that of technology. When, after the Second World War the Japanese had to construct an industrial base from scratch, they were able to develop more streamline solutions to technological problems than the countries that were their industrialized models.

[3] The Tibetan text, fascinatingly, uses two different words for 'understand' here.

[4] For all we know Hsüan-Tsang *may*, of course, have added the restrictive copula only for the sake of metre, but it seems wrong to assume that at the outset.

was not used indiscriminately but only in cases where it was felt to be helpful or important by the Chinese translators.

Hitherto, we have considered only some terminological and grammatical details. But there are larger and philosophically much more central issues raised in the Chinese translation of *Nyāyapraveśa*. There are crucial passages in which Hsüan-Tsang made substantial improvements on the Sanskrit original.

First, the very definition of the thesis or proposition. In Sanskrit *śabdo 'nityah* can either mean 'impermanent sound' or 'sound is impermanent'. Correspondingly, there is a devastating uncertainty whether *paksa* is 'thesis' or 'the subject of the thesis as qualified by the predicate'. Hsüan-Tsang's translations, and the early Chinese commentaries, make it perfectly plain that the thesis is a proposition, not a complex noun phrase. They translate quite unambiguously and perfectly unanimously *shêng shih wu chhang* 聲是無常 'Sound is impermanent.' (Yes: They did add that quite surprising 'copula' *shih* 是.) In making it clear that *tsung* 宗 is not a qualified noun but a proposition, the Chinese translation marks a distinct advance over the Sanskrit. In Chinese one does not say anything that can either mean 'impermanent sound' or 'sound is impermanent'. The Chinese therefore said nothing of the kind. At this crucial point, Chinese is less ambiguous than Sanskrit. The Chinese were at an advantage because of their language. Where the Sanskrit leaves you with a lurking suspicion that the Indian thinkers had not sorted out for themselves the significant distinction between a modified noun and a subject predicate structure, the Chinese forces us to make a clear and unambiguous choice.

There is another most extraordinary detail regarding the definition of the thesis. All Sanskrit versions of this definition involve a clause that a thesis must not go against *pratyaksa* 'evidence from direct perception'. This clause is logically problematic because it presupposes that a thesis, in order to be a thesis, must be acceptable. But a thesis does not cease to be a thesis because it can be refuted by reference to direct perception.[1]

Quantification in the Three Aspects

Consider a second example. That of the famous Three Aspects (*san hsiang* 三相) which are generally treated and recognised as the theoretical core of *yin ming* 因明 logic.

Suppose that Aristotle had written:

Greeks are men,
Men are mortal,
Therefore Greeks are mortal.

---

[1] The case is different for the requirement that the meanings or references of the terms in the thesis must be agreed upon: for if there is no agreement on this matter, there is no thesis to discuss or disagree about.

Suppose then that a translator had come along and translated:

All Greeks are human,
All humans are mortal,
Therefore *all* Greeks are mortal.

The added precision of such a translation, we would surely say, makes *all* the difference to the logical point at hand.

Hsüan-Tsang's translation made just this sort of contribution to the formulation of the crucial (meta-)logical rule of *yin ming* 因明 logic. He added that crucial quantifier ('all') to get the rule into logical shape.

The standard stock-in-trade example of an argument in *yin ming* runs – as will be remembered – like this:

Sound is impermanent
Because it is produced.
Like a bottle, but unlike space or time.

The Indian early formulations of the three conditions which the reason must fulfil in order to guarantee the truth of the thesis:

1. The reason must be predicable of the subject of the thesis. (Sound must be produced.)
2. The reason must be present in positive examples. (Impermanent things must be produced.)
3. The reason must be absent in negative examples. (Permanent things must not be produced.) (Cf. Tachikawa (1971), p. 121.)

Presumably the Sanskrit version of the original text available to Hsüan-Tsang was something like the above.

Hsüan-Tsang's contribution consists in the addition of the quantifiers which are needed to make the principle logically transparent and important:

1. *All* sound must be produced.
2. (*At least some*) impermanent things must *indubitably* be produced.
3. *No* permanent things *at all* must be produced.[1]

---

[1] A literal translation would run like this:

1. (The reason-property R) is pervasively a property of the subject of the thesis (S).

E.g., producedness must inhere in every sound. In other words: if producedness is to count as a reason for any property of sound, then sound must have producedness. One can only adduce as reasons what are quite generally properties of the subject of the thesis.

2. The positive examples (P) undoubtedly have the property (R).

E.g., producedness must undoubtedly be present in (at least some) cases of impermanence. It need not be present in all cases of impermanence because the reason (*yin* 因) is only sufficient, not necessary. There may be reasons other than producedness for impermanence.

3. The negative examples (not–P) pervasively lack the property (R).

E.g., producedness must be absent in all instances of permanence, or: all things, if permanent, must be unproduced. If producedness is to count as a sufficient reason for impermanence, then clearly no thing can be permanent and yet produced.

For all we know, the addition of quantifiers in the formulation of the Three Aspects in Hsüan-Tsang's translation seems to be his original contribution which he added as he was translating. In any case, the consistent addition of the quantifiers is unattested in Indian logical texts earlier than Hsüan-Tsang's translation.[1] What is more, Hsüan-Tsang adds these quantifiers throughout his translation wherever the Three Aspects overtly or indirectly become relevant again.

Hsüan-Tsang could only add his quantifiers where he did because he was far from mechanically translating a text the intellectual import of which he understood only imperfectly. He was an advanced practitioner of the art of logic an introduction to which he was translating into his mother tongue.

If the systematic expository first part of *Nyāyapraveśa* which I have examined closely is anything to go by, Hsüan-Tsang's Chinese translation is not only often an improvement on the Sanskrit original, it has turned out – to my great surprise – to be generally easier to read as well. I have been reading the *Nyāyapraveśa* together with three Sanskritists who between them have spent over sixty years reading Buddhist Sanskrit texts. The Sanskrit *Nyāyapraveśa* presented quite considerable concrete problems of grammatical interpretation for all of us. By comparison my task of expounding the literal meaning of the Chinese translation was considerably easier. Of course Hsüan-Tsang's translation raised a host of problems of philosophical interpretation. But unlike the Sanskrit it was not so full of problems of literal grammatical comprehension. It is as if the analytical nature of the Chinese language enforced a plainness of construction which made my task so much easier.

The problem of parallelism

Consider the following Sanskrit formulation:

Prasiddho dharmī prasiddhaviśeṣeṇa viśiṣṭatayā
'the by the agreed-upon property subcategorized agreed-upon subject'

The slightly complex construction is rendered by Hsüan-Tsang into what looks like a series of three four-character phrases: *chi chhêng yu fa, chi chhêng nêng pieh, chha pieh hsing ku* 極成有法極成能別差別性故 'the agreed-upon subject and the agreed-upon predicate, in so far as (the latter) is a subcategorizing characteristic (of the former)'. Hsüan-Tsang expounds the *tsung* 宗 'thesis' as a proposition, not as a modified noun. This is a crucial point.

Hsüan-Tsang imposes a four-character rhythm which is entirely consonant with the stylistic standards of his time, but which does make comprehension more difficult than it need be. Other instances of the obtrusive use of parallelism are

---

[1] When it was first systematically and clearly focused on within the Indian tradition is of no concern to us here. The Indian logicians may obviously have developed this refinement themselves without inspiration from Hsüan-Tsang. Indeed, this seems most likely, although one cannot exclude the possibility that Hsüan-Tsang did, after all, play a certain rôle in the development of indigenous Indian logic. He was, undoubtedly, recognised by the Indians as a leading practitioner of the art.

common in the commentarial literature where at times one is uncertain whether a character has crucial meaning and weight or whether it is used to fill out the four-character pattern.

Lü Chheng (*1983*), p. 10, finds the Chinese opaque at this point, and he claims a plainer translation into Classical Chinese would have been simply to write: *yu chi chhêng nêng pieh chih so chha pieh* 有極成能別之所差別 'that which is subcategorized by the agreed-upon predicate'. This would indeed have been a more literal rendering of the Sanskrit, but the problem is that the Sanskrit formulation expounds the 'thesis' as a modified noun, which is not at all the same thing. At this point I am inclined to insist that Hsüan-Tsang's more elaborate version adds a significant nuance: we are not just talking of a 'subcategorized subject' but of a subject 'in so far as it is subcategorized'. Hsüan-Tsang surely could have rendered this literally, but he tries to subtly and unobtrusively improve on the original.

Wên Kuei, perhaps the most brilliant of Hsüan-Tsang's Chinese disciples, read the last phrase differently: *chha pieh wei hsing* 差別為性 'the subcategorization counts as the issue', (and, incidentally, the non-Buddhist polymath Lü Tshai emphatically agrees with him). I find that this makes smoother reading, and indeed it puts the issue into much better focus. We need not enter this particular philological brawl. Suffice it to say that both the alternatives considered by the Chinese logicians in this particular instance seem to be clear improvements on the Sanskrit text we have today.[1]

Wên Kuei's commentary, as usual, is strictly to the point. He comments on the phrase *chha pieh wei hsing* separately and introduces his comments with the pithy remark: 'This explains the substance (*thi* 體) (of the thesis).'[2] The substance of the thesis is the main predication itself. Wên Kuei knows that in taking this stance he differs from what he calls 'the old masters of Buddhist logic'. But he also declares and demonstrates with a quotation that he has the support of Dignāga himself.

Wên Kuei writes with confidence. Khuei Chi, on the other hand, indulges in a venomous attack on Wen Kuei's reading: 'Some do not understand where this comes from, and they change the text to read: "*chha pieh wei hsing* 差別為性", but this not only goes against the rules of Buddhist logic, it also shows ignorance of the local dialects (*fang yen* 方言) of Thang 唐 and Fan 梵 (China and India). Constantly changing the original texts deserves profound criticism. . . . When latter day laymen change the superb subtle words of a thousand generations, that is like the silly habits persevering under the Hundred Kings, it surely is painful. And how much more so when the culprit is not a translator of sutras!'[3] Khuei Chi is emotional at this point, but unfortunately he offers no reasoning of any logical substance to support his case.

---

[1] We cannot, of course, exclude the possibility that Hsüan-Tsang was working with a better Sanskrit original than those we work with today.

[2] *Chuang Yen Shu* 1.9b. In this the Great Commentary, 2.10a3, agrees with him.

[3] Great Commentary 2.11a5. Note that Khuei Chi's translations are full of quotations from the works of Dignāga, and that they also mention such things as cases in Sanskrit.

When one turns from Thang times to the later Ming dynasty accounts of *yin ming* 因明 logic one does get a sense of muddled incomprehension on the part of the Chinese commentators, which could have been pervasive throughout the Chinese tradition but isn't. The weakness of Ming dynasty *yin ming* logic, we must conclude because of the earlier evidence, is because the Ming commentators did not understand *yin ming* logic well. It is not because of inherent deficiencies in the Classical Chinese language.

(5) Contrasts Between '*Yin Ming* 因明' and Aristotelian Logic

By way of a conclusion to our discussion of Chinese Buddhist logic,[1] we shall now turn to a brief summary of the main contrasts between *yin ming* and Aristotelian logic.

'Yin ming' is material, not formal logic

The standard thesis for Chinese Buddhist logicians is standardly translated as 'Sound is impermanent'. This may sound as arbitrary as 'All Greeks are mortal', but in fact it is not at all comparable. For what we have translated as 'traditional verbal authority is impermanent' is the thesis that the Vedic truths (*śabda*), rather than being eternal and incidentally also temporarily manifested, are as ephemeral as anything else. The thesis is of concrete philosophical interest.

Aristotelian syllogistic is concerned only with the formal validity or invalidity of a very limited set of formal arguments. In all of the *Prior Analytics* of Aristotle there are plenty of examples of terms, but the syllogism is discussed as a form with the use of variables. Western logical theory is in principle indifferent to the question what concretely the premises of an argument are, or whether the premises of an argument as a matter of fact are true or not. Aristotelian and later more general formal theories of logic in the West are interested in premises only in a conditional way: *if* the premises were true, would this logically imply that the conclusion also was true? The true Aristotelian syllogism is not

> All men are mortal.
> All Greeks are men.
> Therefore all Greeks are mortal.

Rather, it tends to go something like this:

> *If* all men are mortal
> and all Greeks are men,
> then all Greeks are mortal.

---

[1] Chinese Buddhist logic must, of course, not be confused with Buddhist logic in general. We shall not be concerned with developments in India, and particularly not with later developments from Dharmakīrti onwards. The comparison of these later developments with Aristotelian traditional logic must obviously be left to Indologists. Our focus of comparison is only Buddhist logic as it emerges in the Chinese sources here under discussion.

A Chinese Buddhist logician is concerned with acceptable proof and refutation of controversial propositions in a given historical context. The mortality of Greeks (or Indians) is of no 'logical' interest to him. A valid argument with a blatantly untrue conclusion counts as unacceptable. So does an argument with a blatantly or obviously true conclusion.

For Aristotle, as for medieval European logicians, a blatantly untrue conclusion would make no difference whatsoever to the logical problems they are concerned with.

Medieval European logicians happily amused themselves arguing from blasphemous premises like 'all popes are asses' to equally outrageous conclusions. For example, we find:

> *Quicumque sunt episcopi sunt sacerdotes*
> *isti asini sunt episcopi*
> *Ergo isti asini sunt sacerdotes*[1]

Quoting this piece of reasoning in their wonderful survey of the history of Western logic, William and Martha Kneale add: 'Asses, and Brunellus [the quintessential ass] in particular, were always a favourite subject with medieval logicians.'

To Chinese Buddhists such arguments are not just frivolous or flippant, but literally unacceptable. They are logically unacceptable, according to Chinese Buddhist logicians, because they are based on blatantly untrue grounds.

There is no room in Chinese Buddhist logic for the sort of dignified flippancy in the choice of example arguments which characterises much of the Western (and, as we have seen, the early Chinese) logical traditions.

Chinese Buddhist logicians use concrete examples to exemplify the truth of certain premises. If they were unconcerned with factual truth about this world, and of their premises, their examples would be quite out of place.

'Yin ming' does not systematically distinguish between factual truth and logical validity

'If all *A* are *B*, and all *B* are *C*, then all *A* are *C*' is a valid schema of inference. It is *logically* valid. The factual question whether in point of fact all *A* are *B* does not arise. Formal logic is concerned with validity, not with truth. The Chinese Buddhist logicians fail to focus on formal validity as such. They discuss the truth of the premises together with the acceptability of theses without noticing the crucial distinction between questions of factual truth and questions of logical consequence or formal validity. In Aristotle, the art of plausible reasoning and the legitimate justification of philosophical claims is discussed in the *Rhetorics* and *Topics*, quite separately from the *Analytics* which are exclusively concerned with validity.

---

[1] 'Whoever is a bishop is a priest. These asses are bishops. Therefore these asses are priests. (W. and M. Kneale (1962), p. 233.)

'Yin ming' has no variables

Examples like 'If all men are mortal, and all Greeks are men, then all Greeks are mortal' do not, of course, belong to the realm of formal logic at all. For formal logic is neither about mortality, nor about Greeks, nor even about men. The way Aristotle tends to express himself throughout is as follows:

> If all Beta is Alpha
> and all Gamma is Beta,
> then all Gamma is Alpha.

Aristotle's formulations absolutely crucially and quite invariably involve *variables*. Chinese Buddhist formulations, as far as I have been able to ascertain, never involve any variables.[1]

The Chinese Buddhists were eminently capable of creating new concepts by stipulative definition, but unlike Aristotle they did not have occasion to fashion such very specialised linguistic tools as variables.[2] This they had in common with the early Indian logicians.[3]

Suppose someone was discussing arithmetic in terms of theorems like 'two apples and two apples makes four apples' and 'two pears plus two pears makes four pears' without ever generalising his insights into some form like 'two and two makes four'. Such a person can, of course, record a great many insights, and these insights will be closely related to those arrived at in the science of arithmetic. But they remain essentially on a less formalised and less theoretically transparent and lucid level. Buddhist logic in China suffers from this sort of limitation. It cannot systematically and freely abstract from the concrete terms in its proposition and formulate the formal laws governing valid inference.[4] It invariably discusses concrete examples, and moreover always examples that are of Buddhist doctrinal significance. It is as if one were to write a handbook of Western logic and discuss exclusively arguments of the sort of controversial nature of a *cogito, ergo sum*.

---

[1] For the history of variables in early China see Section (*f*,7).

[2] The Later Mohists came closest to using variables when they used terms like 'horse' and 'ox' to stand just for any property whatever. In principle, this would be a perfectly equivalent notational variant of variables. One could write modern textbooks of mathematical logic by using 'ox', 'horse' and the like for *f*, *g* and the like without this making the slightest philosophical difference. The crucial achievement of the Mohists here was the ability to abstract from the concrete content of 'horse', the ability to discuss 'horse' without showing any interest whatever in horses or the particular meaning of the word 'horse'.

[3] As the great Polish mathematical logician and historian of logic Jan Lukasiewicz has pointed out in his wonderful *Aristotle's Syllogistic* (first ed. 1951), and as my teacher Günther Patzig has pointed out in his important work *Aristotle's Theory of the Syllogism* (first ed. 1958), the introduction of variables *X*, *Y*, etc., is an absolutely crucial step in the history of logic. The Buddhists did not take this step. This limited their logical range. They were unable to abstract completely from the concrete example arguments they discussed. Every new argument had to be considered in its own terms.

[4] The nearest the Chinese Buddhist logicians come to variables is the abstract description of the terms in their propositions. They thus refer to the subject and the predicate of the thesis and to the property in the reason, and they have formulated – in the Three Aspects (*san hsiang* 三相) – general laws which do indeed guarantee the validity of the arguments conforming to them. There was also a tendency for the Chinese Buddhist logicians to use stock examples like 'sound', etc., almost like variables standing for any property. But the Buddhist logicians did not carry this abstraction from the concrete subjects, predicates and reasons far enough for us to want to say that they used variables.

'Yin ming' is sensitive to pragmatic context

A *yin ming* argument can be acceptable when used against one opponent, but invalid when used against another because of intricate rules concerning what disputing parties have to agree and to disagree on. They must, for example, disagree on the thesis as a whole, but agree on the meaning of the constituents of the thesis.

Chinese Buddhist logicians chose to concentrate not so much on formal logical presuppositions but on actual agreement among the contestants on premises. While Aristotle in his *Rhetoric* and *Topics* is concerned with plausible proof, the Chinese Buddhists are concerned with plausible victory in debate.

A *yin ming* argument can be acceptable when used by one proponent, and unacceptable when used by another, because by the rules of *yin ming* the proponent can only acceptably argue for theses that are in accordance with his own general position. He can only acceptably use terms which according to his own general position have the meaning and reference which he attributes to them in the argument.

In Aristotelian logic it does not matter at all how subject term *S*, predicate term *P*, and middle term *M* are understood, as long as they are understood consistently throughout an argument. That is why Aristotle used letters of the Greek alphabet to represent these terms. You can understand them as you like, as long as you understand them consistently. There is absolutely no reason to agree on how to understand the terms. A Chinese Buddhist logician considers an argument not as a formal structure but as a concrete semantic structure which communicates a certain content. (At the same time, it will be remembered that the Chinese Buddhist logicians formulated strict formal rules for the construction of arguments, as in the theory of the Three Forms of the Reason.)

'Yin ming' arguments are not systematically explicit

Another contrast between Aristotelian logic and Chinese Buddhist *yin ming* concerns the notion of explicitness. The whole point of Aristotelian logic was that it insisted on being absolutely and perfectly explicit on every logically relevant aspect of an argument. It may not always have achieved its aim, but the ambition is clear enough.

Chinese Buddhist logic was not explicit in the sense that you are not supposed to be able to look at the surface structure of an argument and assume that it directly represents the underlying logical structure. For a Chinese Buddhist logician, an argument may be perfectly acceptable although it is highly implicit and leaves essential parts to be understood by the reader. Having defined a well-formed argument as containing both a positive and a negative example, for example, the Chinese practitioners of Buddhist logic – like their Indian colleagues – omitted the negative example much of the time. Again, having explained that the import of the negative examples contains a general conditional ('If anything is eternal, it is unproduced. Like space.'), they proceed to almost invariably leave out the conditional clause as obviously understood by the audience. At the same time they continue to theoretically insist that such omission constitutes a formal fallacy. On account of the tendency to leave out things that are obviously understood one clearly cannot regard Chinese Buddhist logic as formalistic in any Aristotelian or modern sense.

Overall difference between 'yin ming' and syllogistic

The crucial overall difference is that Chinese Buddhist logic always remains a set of rules for concrete debate, whereas in Aristotelian logic rules of debate and rules of logic are kept quite separate. (Indeed, one should perhaps compare much more of Aristotle's rhetorical theory with Buddhist logic than I have chosen to do in this limited space.) The Chinese Buddhist logicians insisted on looking upon logic as a profoundly *social* science, the science of rules and conventions governing argumentation as a social institution. Chinese Buddhist logic is about how to refute an opponent's thesis, and about how to establish one's own thesis. It is essentially 'rhetorical'.

Syllogistic and its successor, formal logic, are quite abstractly about formal logical relations between terms, and formal relations (e.g., of consequence) between ordinary propositions. Non-empirical philosophical theses predominate among *yin ming* theses. Non-philosophical arbitrary propositions like 'there is fire' play a subsidiary part only in Chinese Buddhist logic. They are not what this logic is all about. Buddhist logic is a tool for religious insight. If Buddhists had the notion of God, one might say '*logica ancilla theologiae*'.

(h) CONCLUDING REFLECTIONS

However we define highly problematic words like 'science' and 'scientific' for the traditional Chinese context, the medium of 'scientific' discourse will turn out to be predominantly (but not exclusively) Classical Chinese enriched by a technical vocabulary. The technical language of scientific texts as well as the artificial symbolism of certain disciplines are defined and built up in terms of the current language. To the extent there is science in traditional China, Classical Chinese is the linguistic medium of that science. To the extent that the medium of this language has shaped the messages expressed in it, the study of the relevant aspects of Classical Chinese is a fundamental part of Chinese history of science.

Logic is concerned with the organisation of discourse about the world into patterns of argument, of conceptual coherence, of scientific discipline. The study of the logic inherent in Classical Chinese is the study of the structural elements which give Chinese scientific thought its argumentative architecture.

Language and science

Classical Chinese is not only an important medium of scientific communication. It is also a medium of communication about itself: The study of writing systems was nowhere pursued with more enthusiasm and with more scientific rigour than in traditional China. The graphological analysis of Chinese characters presented in the *Shuo Wên Chieh Tzu* in the +1st century does not represent natural science, but it certainly provided an outstanding example of rigid scientific methodology and systematic classification for natural scientists of later ages. Dictionaries like the *Erh Ya* explore and systematise the conceptual schemes the ancient Chinese used to explain the world.

The sounds of a language provide a less permanent and much more elusive subject for study. But again, the systematic and comprehensive phonological analysis of the sounds of the Chinese language achieved in China around +600 is a significant example of sustained and systematic scientific observation of the physical facts of the language on the one hand, and of a highly disciplined ability to assemble these observations into a logically coherent and highly structured whole.

It makes sense to speak of a technology of information retrieval systems. Alphabetisation, for example, was a major breakthrough in the organisation of dictionaries in the West. The ability of Chinese lexicographers since the +1st century to organise large dictionaries in such a way that most items in the dictionaries became increasingly easy to retrieve by those who had learnt the principle of the compilation was in my view a major contribution to the technology of information management in the ancient world.

Chinese civilisation paid persistent attention to the problem of the systematic definition of the meanings of words, and to the semantic evolution of concepts. In its monumental mono-lingual Chinese-Chinese dictionaries the culture has a varied tradition of systematic reflection upon its own conceptual world. It is a measure of the intellectual maturity of traditional Chinese civilisation that it problematised its own conceptual world, discussed, systematically analysed and defined its own words in such a long series of Chinese-Chinese dictionaries. It is a measure of the scientific spirit of the traditional Chinese lexicographers that they habitually insisted on explicit evidence for all the meanings they provided, both in terms of earliest examples of usage and in terms of the earliest semantic gloss attributing a certain meaning to a given word.

One aspect of Chinese linguistic civilisation that was potentially relevant to the development of natural sciences was the tradition of defining problematic concepts in terms of simpler or less problematic concepts. This tradition was one of the factors encouraging the development of intricate systematic taxonomies of inter-defined terminology in various branches of Chinese natural sciences.

The Chinese reflections on their own language – particularly phonology, graphology, and lexicography – provide us with the only example that is known to us of a non-Indo-European tradition of scientific inquiry into language. Though natural science and technology formed a highly specialised and articulated field containing a large number of scientific subcultures and sub-subcultures, these were all held together by a common cultural classical heritage of which the graphological and the semantic dictionaries were a formative part.

Grammar in Chinese culture

Given the Chinese fascination with the systematic analysis of their own conceptual world it is interesting that the field of grammatical analysis did not arouse their special attention, even after they had been exposed to the world of Buddhism which came from India, where systematic grammar and the philosophy of grammar was nothing less than an intellectual rage. Traditionally, the Chinese rarely showed much interest in the inner morphology of their words. Neither did they ever begin to write systematic accounts of Chinese syntax comparable with their accounts of Chinese phonology. It is true enough that the first grammar of Chinese was published by a Latin-trained Chinese in 1898, but what preceded this grammar, the traditional systematic studies of grammatical particles and the classifications of characters into such general categories as lexical items on the one hand (full characters) and grammatical words on the other (empty characters) deserves our close attention. After all, here again, the Chinese civilisation is the only one to have paid systematic attention to the problem of the grammatical categorisation of its own lexical items, and to the detailed study of grammatical words. Chinese is only just beginning to take its place in the history of the science of linguistics.

China is the only non-Indo-European civilisation from which we have a wide range of general reflections on the rôle of language in society, on the nature of language change and so forth. Chinese perceptions on these matters, however, as in the case of grammar, were not systematised. No extensive canon of scientific argumentation or justification of views on language was developed in traditional times. But precisely because of their non-systemic character, Chinese perceptions on these matters are of special interest to the serious student of general linguistics.

There is much to be learnt from the Chinese about human language, but one thing that I have not found in classical Chinese is any kind of special Chinese logic. The logical space within which that language articulates itself is the space determined in terms of negations, sentence connectives, quantifiers and the like. It is no more good form, no more bad form, to contradict oneself in Chinese than in English or Greek. The syllogism – rarely used in argumentative practice in Greece and Rome – does occasionally occur in Chinese, and arguments of that sort are by no means alien to the Chinese way of thinking. None the less, there are striking contrasts between, say, Latin and Classical Chinese with respect to logical structure. One thing is the absence of disjunction 'or' down to Han times, an absence which in cases of multiple disjunction 'either . . . or . . . or . . . or . . .' is not effectively compensated by the use of 'if not . . . then . . .' in Classical Chinese. On the other hand, the very currency of the form 'if not . . . then' where we should expect 'or' in Classical Chinese indicates clearly how 'Western' logic is inherent in the linguistic structure of the Chinese language.

This point about disjunction and multiple disjunction in Chinese brings out the connection of logic with the practice of natural science. Multiple disjunctions are current in Western scientific discourse, and the invention of an equivalent tool in China, a standard expression meaning 'or', added a significant form to traditional Chinese scientific discourse.

Rhetoric and science

One might feel that the rhetorical conventions of Classical Chinese, the literary traditionalism of Chinese prose, might stand in the way of the development of science and technology in China. The point may be conveniently illustrated through the +15th-century carpenter's manual *Lu Pan Ching*. This important text, which is full of detailed and very explicit quantified descriptions of traditional Chinese wooden architecture, starts out with a three-page biography of the legendary carpenter Lu Pan (born −507), about whom we are told:

Thereupon, when his age had reached forty, he took up the life of a recluse again, this time near Mount Li. In the end he met an immortal, who taught him secret formulas to travel through the world riding on a cloud. In broad daylight he ascended, leaving behind only his axe and his saw in the Cave of the Immortal of the White Deer. These remnants can still be seen clearly up to the present day. Therefore in the Warring States period, since he was greatly loyal, he was given the title 'Eternally perfect loyal gentleman expecting to serve'. Three years later the Marquis of Chên added the title 'Wise and benign master of magic'. During the Han, Thang and Sung dynasties he could still help the country many times by manifesting himself, and was repeatedly enfeoffed with an honorary title. When in the Yunglo period of our Ming dynasty the imperial palace was erected in Peking, the ten thousand corvée workers relied without exception, and full of awe, on the divine guidance of the master, and only then did they succeed in completing the work. Thereupon a temple was built to worship him. On the tablet over the entrance was written 'Gate of Lu Ban'. He was given the title of 'Marquis of Beichêng, Grand Master, support of the state, expecting to serve'. At the sacrifices of spring and autumn the *suovetaurilia* (swine, sheep and bull sacrifice) was used. Every time when the workers of our present age pray to him, he responds to their prayers without delay. Truly, as his image radiates from high above, it will be worshipped for evermore.

Thus ends the opening chapter of *Lu Pan Ching*. From this folkloric biography, or hagiography, the book immediately proceeds not only to a precise and technically circumstantial description of how to set about building a house; religious rules for the choice of auspicious days and the like are presented alongside technical instructions on how to construct the house: a house may be ruined either because the timber was procured on an inauspicious day or because it was wrongly constructed. Both explanations belong to one world view. The technical sphere is not isolated from the religious sphere. In the course of history it gains more and more relative independence, but throughout traditional Chinese history it would be profoundly uncongenial to separate out the technological from the religious-folkloric aspects of traditional Chinese architecture. In the tradition these things are interwoven. And this means that the study of science and technology cannot be dissociated from the study of the culture and the civilisation.

The study of the conceptual foundations of Chinese science must not be dissociated from the study of ancient Chinese grammar and rhetoric, because traditional Chinese scientific discourse is interwoven into the fabric of the less specialised and more general literary tradition. This is particularly evident in the case of medical discourse, but to varying extents it is true of all the sciences. Even in the mathematical commentaries we find quotations from Confucius as a natural part of the scientific rhetoric.

When we reconstruct some rhetorical and grammatical constraints that operate in classical Chinese, for example down to Han times, we have at the same time identified constraints upon scientific discourse in so far as this is articulated in Classical Chinese.

When we note that expressions like 'that is to say' are not available in Classical Chinese at a certain stage, we have identified an intellectually significant area where Chinese scientific texts differ from Latin ones, where *scilicet* and *id est* are intellectually important and perfectly current forms.

When we discover that the mode of distancing oneself as a writer from one's own choice of words through phrases like 'so to speak' was not available in classical Chinese at a certain stage, this implies that Chinese scientific texts at that stage of their cultural development differ significantly from Latin ones where *ut ita dicam* 'so to speak' is as common as is the phrase *hōs epos eipein* is in Classical Greek.

Turning now from what the ancient Chinese were logically, grammatically and rhetorically equipped to do to the question of logical practice, I found that there is what one might call an architectural difference. In principle, basic forms of logical argumentation are pervasively present throughout classical Chinese literature. The strict and complex internal logic of the *I Ching* may be controversial in its details, but there is no doubt whatsoever that we are faced in this case with a strictly organised, and logically structured whole. On the other hand there is a fundamental difference in the ways in which logic enters into traditional Chinese discourse. The inner structural complexity of arguments explicitly presented as such in the texts that have come down to us from pre-Buddhist times at least, is strikingly less great than that of contemporary Greece and Rome. Traditional Chinese arguments do not tend to have the complex explicit and sustained logical architecture of a Platonic dialogue. It is not that the Chinese arguments have a different logic. It is that their arguments tend to be shorter and simpler. Reasoning was not widely used in ancient China as a tool for main truths. It tended always to remain in a subservient rôle.

Reasoning in science

Let me take up the case of mathematics. Here the secondary rôle of reasoning and proof has significant consequences. The validity of procedures and theorems in Chinese mathematics is of vital importance to the traditional Chinese mathematician. But the systemic and logical connections between theorems and procedures, the distinction between axiomatic statements and logically derived theorems, the procedures of proof rather than calculation or problem-solution, were not in the main focus in the history of traditional Chinese mathematics. Jean-Claude Martzloff (1988), p. 66 writes:

For those who undertake to write the history of a science in which reasoning is the very essence – how, after all, can we imagine mathematics without demonstrations, or, at the very least, without argumentation? – the question of logic is the most fundamental question that there is. And yet, in spite of this obvious fact the general histories of Chinese mathematics rarely take up this question.

Martzloff raises a problem that is crucial for our present discussion. The five-and-a-half pages he devotes to the matter of the importance of logical method and argumentation in the history of Chinese mathematics, however, are intensely disappointing. He does not present important methods of proof or logically significant presentation. The strength of the Chinese mathematical tradition was in the discovery of effective and useful mathematical procedures providing reliable solutions for a wide range of problems of calculation and analysis. The establishment of a complex system of geometrical or arithmetical theorems, and their derivation by logical procedures of proof from axioms, was evidently not part of the Chinese tradition, and perhaps it is quite wrong to take the specificities of Greek mathematics and simply look for their counterparts in China. There was, as Karine Chemla has shown in a number of articles, a considerable amount of methodological discipline implicit in the very form of the presentation of mathematical results in Chinese. But what matters to us here is that mathematics was not explicitly cultivated as a logically structured hierarchical system of logically interrelated and hierarchically organised theorems. Mathematics was cultivated in China as a range of effective mathematical procedures to solve given problems that were found accessible to mathematical treatment.

The history and anthropology of logic

It turns out that the traditional Chinese had two periods of intellectual innovation and of remarkable progress. One is well known, very sparsely documented, but exceedingly well researched. It occurred during the Warring States Period, especially the −4th and −3rd centuries. The significance of this period for the general history of science lies in the fact that this period provides clear evidence of an indigenously Chinese and entirely autochthonous intellectual movement in the direction of a critical rationalism and of systematic logical reflection. The problem with this first period is that its documents are so scantily and poorly preserved, although its intellectual effects have been shown to be far-reaching in the literature of that period.

However, explicitly logical discourse remained marginal during the literature of the Warring States Period, and its direct influence on the development of scientific discourse would appear to have been limited.

The second golden age of logic in traditional China (the +7th to +8th centuries) is extensively documented, and the textual quality of the material we have is generally good. On the other hand this logical corpus which is of great importance for the study of the the history of scientific thought in China, has still not received anything like the intellectual attention by historians of science that it deserves. The extensive treatises of the *yin ming* school of Chinese Buddhist logic constitute an extensive corpus of highly disciplined and technical scientific literature which contributes considerably to our picture of Chinese intellectual history. This scientific corpus shows that in the very abstract realm of systematic discourse logic the ancient Chinese were capable of a rigid intellectual discipline and a high degree of theoretical explicitness which would seem to have been fully abreast with contemporary logical theories in India, and which in its systematicity of approach is quite as sophisticated as anything that was being done in that area during the same period in Europe.

One may speculate why this remarkable logical flourish in China remained as marginal as it did to the Chinese intellectual tradition as a whole. Obvious perennial questions re-emerge from these summary considerations: Why did Buddhist logic not catch on even among Chinese Buddhists, not to speak of Chinese thinkers within other traditions? Why, for that matter, do we not find a sustained presence of a significant intellectual subculture cultivating the traditions of *yin ming* and of Mohist logic for that matter? Why did no one want to read the *yin ming* literature? Why did those who did read it in later times tend to misunderstand it? Why did the practice of *yin ming* decline whereas Aristotelian logic was revived and developed into a central discipline within the European educational curriculum? These are questions that belong properly to the anthropology of logic. They concern the societal and cultural conditions that may or may not favour the cultural and sociological success of the intellectual practice of the science of logic.

No simple distinction between the different rôles of logic in despotic versus civil societies will help; occidental despotism (including the special variety of spiritual, psychological despotism directed against thought and inner conscience) was rife in many places where logic flourished. Republican Rome with its freedoms was not noted for its advances in theoretical logic. The story here, when it is fully told, will be extremely complex. It will not be a simple matter of occidental civil society in Greece or Rome versus oriental despotism in China.

Again, the structure and nature of the Chinese language – for all the constraints these impose on Chinese rhetoric – cannot be the decisive factor, since the extant logical *yin ming* texts are about as clear or as unclear as their contemporary and comparable European counterparts within the field of Aristotelian logic.

In the end I agree on this point with my much-missed teacher and friend Angus Graham when he used to complain that these questions were misdirected. Instead of asking why things did not develop in China we should, at this stage, allow ourselves the luxury of investigating as closely as we can what exactly did develop, and under what internal and external intellectual conditions developments did take place in China.

From the point of view of the anthropology of logic the case of ancient Greece and Rome is unique. What were the conditions that made a Stoic like Chrysippus want to write so many books about logic that their titles can begin to match in length all the strictly logical texts that have come down to us from ancient China? How could Stoic logic be so largely neglected by comparison with the much inferior Aristotelian system in the West? How could logic come to occupy such a central place in the Western intellectual curriculum?

Scribal reticence and the epitomising scribal mode

Viewing the European case through the Chinese looking glass, one major aspect strikes me as particularly relevant to the history of science. It has to do with the evolution of the psychology of writing in China and in the West. In Ancient Greece and Rome certain writers came to affect the expression of inner processes of thought and inner arguments directly in their writings. Writing came to transcribe – or at least affected to transcribe – thinking processes and inner argument simultaneous with the writing process, the dialogue of a person with himself. Aristotle's lectures are full of argumentative parentheses and deliberative asides. He comments on his thoughts as he formulates them. These lectures may not really be transcripts of on-going thought. But they are written as if they were. And *that* is the point that matters here.

Literary products are in a sense always summaries after the event, after the occurrence of a thought. They are deliberate literary acts and never in any strict and direct, unmediated sense transcripts of inner thought-processes. The difference is this: some cultures cultivate texts that pretend to transcribe current on-going thought, other cultures do not pretend to do this. To what extent any culture *in fact* ever does cultivate the transcribing of current thought is not something one needs to go into at this point.

The anthropology of writing is markedly different in traditional China from what we have described in Greece and what we could have described for Rome. The difference is marked, but one must remember that it still is a difference only in degree. It is in the nature of things that one would not find absolute differences on matters of this sort.

There are obvious constraints on what would be written down on the oracle bones. The scribe who recorded the speeches recorded in the old parts of the Book of Documents, *Shu Ching*, summarised and condensed these speeches, I would argue, boiled them down to their essentials. He never cares to aim at creating an illusion that he is recording an actual flow of words, the actual historical flow of words. He was not in the business of creating a historical illusion, like Plato in the *Apology*. Hesitation, fumbling for words, correcting oneself, off-the-cuff incidental, thrown-in remarks are never recorded in the Documents. The cultural act of writing down these documents was an act of condensation. In a rhetorically deep sense writing down is making a digest in pre-Buddhist China. The hesitations, grammatical inconsistencies, asides, etc., that characterise Plato's *Apology* are not just rare in pre-Buddhist China. They are culturally ungrammatical. A digest would never tend to be anacoluthic. It never could be. Except when it deliberately breaks the very rules of the medium.

At the same time one must always remember that the appearance of spontaneity of writing that meets us in Cicero's or in Seneca's letters, the apparent transcription of on-going thought in writing, is a very conscious literary device. Many Greek and Roman writers learnt to write as if they thought spontaneously as they wrote, and very few pre-Buddhist writers in China cultivated such a style.

The Analects are an interesting case in point. There is no hint of any parenthetic remark in the Analects. There is no case of Confucius trying to find a word, correcting his own use of language. There are no incidental irrelevant asides. These are boiled-down precious snippets of conversation, summaries of their essence through what stylistically clearly purports to be *verbatim* quotation of the words, often without any indication of context. For this latter feature of credible purported *verbatim* authenticity, the Analects are indeed a unique document from pre-Buddhist China. But the Analects are still very clearly condensations of conversational episodes, not transcriptions of conversations.

Epitaphs constitute an excellent test case for our reflections on the anthropology of writing. Tombstone inscriptions or epitaphs are one kind of writing where cultures are comparable. Epitaphs constitute a privileged form because in many cases we have such inscriptions as archaeological finds, and not just as late reports on what they say. We have inscribed tombstones in China and in Greece. And yet, how total is the difference between the tombstone inscriptions that we have from Greece on the one hand and those from China! How totally different the style of communication represented in writing on these tombstones. Here is an interesting example dating from the −3rd century:

ἔνθαδ᾽ἐγὼ κεῖμαι Ῥόδιος. τὰ γελοῖα σιωπῶ
καὶ σπαλάκων ὄλεθρον λείπω κατὰ γαῖαν ἅπασαν.
αἰ δέ τις ἀντιλέγει, καταβὰς δεῦρ᾽ἀντιλογείτω.

'Here I lie, Rhodios. About those ridiculous things [which you expect to read on tombstones] I keep silent;
    And what the moles destroy all over the earth I leave aside [will not speak about how beautiful I am even in death]
    If anyone speaks up against this, let him speak up against it after he has come down here [to the Yellow Springs known in Greece as Hades].' (Peek no. 219.)

Insult from the tomb is not unheard of, as in a tombstone from the +2nd to +3rd century:

ζῶν ἔτι Ἀπολλῶνις τὸ μνῆμ᾽ ἐποησαθ᾽ ἑαυτῷ,
κληρονόμων εἰδὼς τὴν ὑποληομοσύνην.

'Still living Apollonis made this monument for himself,
  Of the inheritors he knew the forgetfulness.' (Peek no. 242.)

It will be evident to any student of tombstone literature that different cultures differ widely with respect to how much of their speaking and thinking they encourage their members to write down. The Chinese did make jokes about death, but they certainly did not inscribe these on any tombstones.

One might say that this contrast shows no more than just this: the Chinese were more restrictive with respect to what they wrote on tombstones than the Greeks. But the argument is that tombstone literature may in a sense turn out to be symptomatic of literature on bamboo, silk and later paper. Similarly different cultural constraints may turn out to operate in all media of writing. And I do think that the act of writing literature in ancient China is a very different linguistic and anthropological act from that of writing literature in ancient Greece.

I am prepared to argue in detail that what the pre-Buddhist Chinese writer purports to write down is very rarely the thought as it occurs just after it has occurred. It is the result of a thought that has occurred and that is summarised in a balanced way through text. One tends to record a thought-process that is over. Being typically epitomes or summaries of thought, pre-Buddhist Chinese texts very rarely indeed include representations of the tangled and contradictory process of deliberative thought. Many of Plato's dialogues purport to be essays in disentanglement of such inner tangles.

In Classical Chinese texts we get the upshot of thought and speech. Things are put in linguistic nutshells.

Since Classical Chinese writing summarises thought and speech, there is no need for parentheses, that is why there are hardly any in all of traditional Chinese literature. Since writing is something that is presumed to come after thought, and rarely pretends to be concurrent with thought, there is a natural threshold between current thought and current written expression. Direct speech in ancient China right down to the 19th century tends to be presumed to be summary of direct speech. Very often it is written in a Classical medium which simply does not constitute sayable Chinese of the time at all. Thus a 19th-century person may be quoted in direct speech as having used a language which had been dead for more than a thousand years when he spoke. The reader learns to translate such Classical Chinese direct speech into the colloquial language of the time, and he reads the direct speech as summary. Since direct speech is treated in this way, there is no need for indirect speech. And indeed, the rôle of indirect speech in Classical Chinese is negligable. The summarising function of indirect speech is taken over by the very conventions of writing things down.

There is an endemic – but far from universal – scribal reticence with respect to what may be written down, even in personal notes like the *pi-chi* from Thang times onwards. Corresponding to the scribal reticence and the epitomising scribal mode, there is a reading culture which has perfected the art of teasing out what underlies the reticent text. The reading of a Classical Chinese text is a very active cultural process.

And yet, it would be profoundly wrong to take the general culture of literacy in China as determining the style of scientific prose. As I have shown in the preceding chapters, the various subcultures of scientific prose style can be as straightforward and explicit in China as anywhere else. What carries over into the scientific prose, on the other hand, is the general summarising effect of writing, the absence of syntactic parentheses, the scarcity of the metalinguistic perspectives, etc. Just as narrative literature in pre-Buddhist times is characterised by the virtual absence of the extended inner dialogue, so in the science of logic there is a corresponding and related absence of the scientific inner dialogue made explicit through text. Such inner dialogues will no doubt have occurred, but they have not been recorded.

Logic and the anthropology of writing

In the context of logic the consequences of this scribal reticence are incisive. If inner deliberation and argumentation is not customarily externalised and made manifest in writing, then it becomes harder to make it the object of rigorous scientific analysis. If the writing culture encourages the summarising and digesting of scientific results rather than the delineation of the intellectual path by which they were achieved, then there is nothing objectively textual for the logician to pick up and analyse in detail. The usefulness of logic as an intellectual tool for the inner life of deliberation and reasoning cannot be made manifest on texts.

However, scribal reticence is not a scribal incapacity. Two strategies have commonly presented themselves to Chinese writers. The first was to create a transparent summarising reticent surface through which the reader can see some of the world of continuing inner reasoning and deliberation. This tremendously important strategy reveals itself only to a very close study of Classical Chinese texts. It is cultivated to consummate perfection in poetry. This explains the truly extraordinary position of poetry in Chinese culture. And it turns out that a large number of Chinese prose texts avail themselves of this important strategy of the transparent reticent surface. At times one is sure that it even carries over into scientific texts. In any case, a historical text like the *Shih Chi* is full of such transparent reticence.

The second strategy is simply to revolt and break out of the cultural pattern, to defy the rules. This would appear to have been Chuang Tzu's way. He has had many admirers and quite a few epigones.

What matters to us in the context of the history of science and of intellectual history is that the scribal reticence of the culture at large had a considerable impact on the logical form of scientific writing. This writing continued to be summary-orientated, result-orientated, not reasoning-orientated. Even in such cases as that of mathematics. This is a profound difference in logical and scientific style between China and Europe.

The pragmatism inherent in this Chinese result-orientation had a certain well-defined positive influence on the progress of technology. For the user of any mathematical theorems or procedures it is indeed of no consequence how these results were achieved or whether they can be proved to be correct. From the user's point of view it is quite enough that the theorems are correct and that the procedures are effective. Invention of important procedures and discovery of useful theorems are technologically decisive. Here the Chinese proved to be extraordinarily effective.

Up to a certain point the systematic theoretical justification of theorems and procedures has little direct technological effect. But when that point was reached where further progress in technology depended no longer on discoveries about numbers, physical objects and the like, but on discoveries about how one discovers new truths, then the Chinese mode became markedly less successful. It stands to reason that in a broad sense the logic of Chinese written discourse played a crucial rôle in these developments.

Two major obstacles to comparative intellectual history

When comparing scientific achievements between widely different cultures we are under especially severe constraints regarding the demands of a presupposed, unquestioned, and mostly unanalysed feeling for what is politically correct and for what is desirable in cultural and scientific international relations inside and outside the academic world. The unperceived and therefore all the more powerful obligatory paradigms of scientific discourse on intellectual history often seem to crowd out straightforward unflinching contrastive analytical observation. This intellectually pernicious internalised censorship exercised by our eager sense for what is politically correct and culturally acceptable easily produces the sort of argumentation and rhetoric which belongs more to the realm of cultural politics than to that of objectively reasoned intellectual analysis. Political correctness, which is, of course, a factor in all sciences, constitutes a particularly serious problem for comparative studies because comparative studies must count as a highly 'politicised' and sensitive area.

The case of comparative intellectual history has certain structural similarities with that of environmental studies. The environmental sciences naturally aim to promote balanced respect for the environment, if necessary at the expense of man's own natural but morally indefensible ontological arrogance. Western comparative intellectual history aims to promote respect for other civilisations, if necessary at the expense of the Westerners' natural but morally indefensible intellectual arrogance. The writing of the history of Chinese science and thought has been a central part of a campaign to gain respect for the Chinese intellectual tradition which is still ill-understood and grossly underestimated in the West. It still remains important to put China on the intellectual map of the world, but thanks to Joseph Needham more than anyone else, China is now emphatically on the world map of the history of science and of intellectual history. Thanks to his efforts, the campaign for the vindication of China as an intellectual power is entering a new phase where sinologists can no longer remain advocates for the case of China but must at long last begin to become fair judges of the relative strengths and weaknesses of the Chinese intellectual tradition. The weaknesses have to emerge from the footnotes, the subordinate clauses, perfunctory asides and from the sinological table talk into the focus of critical analysis. Moreover, having shown convincingly that the judgement of the case of traditional China crucially depends on familiarity with the Classical Chinese language and culture, sinologists must begin to appreciate and accept that their comparative judgements on the Western tradition will crucially depend on familiarity with the classical languages of traditional Europe, Latin and Greek, in addition to the main European vernacular languages.

The comparative study of logic makes extraordinary intellectual demands on those who embark on it. Modern formal and philosophical logic are in themselves intellectually difficult enough. To the unwary they constitute a treacherous terrain full of intellectual land-mines. To the philosophically wary there are even more logical land-mines. Moreover, it turns out that our own classical European tradition is not much less exotic to modern Europeans than that of China. Balanced comparison demands balanced familiarity not only with the Classical Chinese but also with the classical European languages and intellectual heritage. Co-operation on this point between scholars of different specialisation is necessary and desirable, but in the end it is not enough. Here, as in life generally, primary exposure is crucial. One needs to be familiar with the ancient traditions and texts one wishes to compare. It is as simple as that. And it is as difficult as that.

BIBLIOGRAPHIES

A   CHINESE AND JAPANESE BOOKS BEFORE +1800, AND ALL GAZETTEERS

B   CHINESE AND JAPANESE BOOKS AND JOURNAL ARTICLES SINCE +1800,
    EXCLUDING GAZETTEERS

C   BOOKS AND JOURNAL ARTICLES IN WESTERN LANGUAGES

In Bibliographies A and B there are two modifications of the Roman alphabetical sequence: transliterated *Chh-* comes after all other entries under *Ch-*, and transliterated *Hs-* comes after all other entries under *H-*. Thus *Chhên* comes after *Chung* and *Hsi* comes after *Huai*. This system applies only to the first words of the titles. Moreover, where *Chh-* and *Hs-* occur in words used in Bibliography C, i.e., in a Western language context, the normal sequence of the Roman alphabet is observed.

When obsolete or unusual romanisations of Chinese words occur in entries in Bibliography C, they are followed, wherever possible, by the romanisations adopted as standard in the present work. If inserted in the title, these are enclosed in square brackets; if they follow it, in round brackets. When Chinese words or phrases occur romanised according to the Wade–Giles system or related systems, they are assimilated to the system here adopted (cf. Vol. 1, p. 26) without indication of any change. Additional notes are added in round brackets.

Korean and Vietnamese books and papers are included in Bibliographies A and B. Dates in italics imply that the work is in one or other of the East Asian languages.

Please note that the old system of reference numbers has been discontinued in favour of giving year of publication. This has become necessary in view of the difficulty of eliminating inconsistencies in the application of the previous system.

ABBREVIATIONS

See also p. xiii

AM	Asia Major	JCP	Journal of Chinese Philosophy
AN	Anthropos (Hungary)	JOSHK	Journal of Oriental Studies (Hong Kong University)
ARO	Archiv Orientalní (Prague)		
ASEA	Asiatische Studien: Etudes Asiatiques	JRAS/NCB	Journal of the Royal Asiatic Society, North China Branch
BABEL	Babel; Revue Internationale de la Traduction	LAN	Language
BEFEO	Bulletin de l'Ecole Française d'Extrême-Orient (Hanoi)	LA	Logique et Analyse
		MIT	Massachusetts Institute of Technology
BLSOAS(BSOAS)	Bulletin of the London School of Oriental & African Studies	MN	Monumenta Nipponica
		MS	Monumenta Serica
BMFEA	Bulletin of the Museum of Far Eastern Antiquities (Stockholm)	MSOS	Mitteilungen d. Seminars f. orientalische Sprachen (Berlin)
BSN	Behaviour Science Notes	NAJ	New Asia Journal
BUA	Bulletin de l'Université de l'Aurore (Shanghai)	NS	New Scientist
		OAZ	Ostasiatische Zeitschrift
CAAL	Computer Analyses of Asian and African Languages (Tokyo)	ORE	Oriens Extremus
		PEW	Philosophy East and West (University of Hawaii)
CCUL	Chinese Culture		
CS	Chinese Science	RO	Rocznik Orientalistyczny (Warsaw)
EC	Early China	SA	Sinica (originally Chinesische Blätter f. Wissenschaft u. Kunst)
HJAS	Harvard Journal of Asiatic Studies		
HYSIS	Harvard Yenching Sinological Index Series	SY	Syntese
JA	Journal Asiatique	TP	T'oung Pao (Archives concernant l'Histoire, les Langues, la Geographie, l'Ethnographie et les Arts de l'Asie Orientale, Leiden)
JAS	Journal of Asian Studies		
JAOS	Journal of the American Oriental Society		
JCL	Journal of Chinese Linguistics	WARC	World Archaeology

A. CHINESE AND JAPANESE BOOKS BEFORE +1800

Each entry gives particulars in the following order:

(a) title, alphabetically arranged, with characters;
(b) alternative title, if any;
(c) translation of title;
(d) cross-reference to closely related book, if any;
(e) dynasty;
(f) date as accurate as possible;
(g) name of author or editor, with characters;
(h) title of other book, if the text of the work now exists only incorporated therein; or, in special cases, references to sinological studies of it;
(i) references to translations, if any, given by the name of the translator in Bibliography C;
(j) notice of any index or concordance to the book if such a work exists;
(k) reference to the number of the book in the *Tao Tsang* catalogue of Wieger (6), if applicable;
(l) reference to the number of the book in the *San Tsang* (Tripitaka) catalogues of Nanjio (1) and Takakusu & Watanabe, if applicable.

Words which assist in the translation of titles are added in round brackets.

Alternative titles or explanatory additions to the titles are added in square brackets.

It will be remembered (p. 421 above) that in Chinese indexes words beginning *Chh-* are all listed together after *Ch-*, and *Hs-* after *H-*, but that this applies to initial words of titles only.

Chan Kuo Tshe 戰國策.
    Records of the Warring States.
    Han compilation based on earlier material concerning the period –454 to –221.
    Writer unknown. Compiled by Liu Hsiang 劉向 (–79 to –8).
    Commented ed. by Kao Yu 高誘 (–170 *c.* 220).
    Ed. Chu Tsu-Keng 諸祖耿, Chiang-su Ku Chi, Yang-chou, 1985. Tr. J. Crump (1970). See Tsuen-Hsuin Tsien's bibliographic survey in Loewe (1993) pp. 1–11.

Chêng Tzu Thung 正字通.
    Pervasive Description of Correct Characters. (Dictionary arranged according to the 214 radicals).
    Preface +1670.
    Chang Tzu-Lieh 張自烈.
    Ed. *SKCS* 四庫全書.

Chi Chiu Phien 急就篇.
    The Chapter for Quick Success. (An abecedarium, made perhaps for language learning).
    Shih Yu 史游 (flourished –40 to –33).
    There is a commentary by Yen Shih-Ku 顏師古 (+581 to +645). Ed. Yü Yüeh-Hêng 喻岳衡, Yüeh Lu Shu She, Chhang-sha, 1989.

Chi Yün 集韻.
    Collected Rhymes.
    Sung, +1067.
    Ed. Ssuma Kuang 司馬光 (+1019 to +1096).
    SPPY. Cf. Chhiu Chhi-Yang 邱棨鐊, *Chi Yün Yen Chiu* 集韻研究, Cho Shao Lan, Taipei, 1974.

Chin Shu 晉書.
    History of the Chin Dynasty (+265 to +419).
    Thang, +635.

Where there are any differences between the entries in these bibliographies and those in Vols. 1–4, the information here given is to be taken as more correct.

An interim list of references to the editions used in the present work, and to the *tshung-shu* collections in which books are available, has been given in Vol. 4, pt 3, pp. 913ff., and is available as a separate brochure.

ABBREVIATIONS

C/Han	Former Han.
E/Wei	Eastern Wei.
H/Han	Later Han.
H/Shu	Later Shu (Wu Tai).
H/Thang	Later Thang (Wu Tai).
H/Chin	Later Chin (Wu Tai).
S/Han	Southern Han (Wu Tai).
S/Phing	Southern Phing (Wu Tai).
J/Chin	Jurchen Chin.
L/Sung	Liu Sung.
N/Chou	Northern Chou.
N/Chhi	Northern Chhi.
N/Sung	Northern Sung (before the removal of the capital to Hangchow).
N/Wei	Northern Wei.
S/Chhi	Southern Chhi.
S/Sung	Southern Sung (after the removal of the capital to Hangchow).
W/Wei	Western Wei.

    Fang Hsüan-Ling 房玄齡 *et al.*
    Ed. Chung Hua, Peking, 1962.

Ching Chi Tsuan Ku 經集纂詁.
    Glosses on the Classics.
    +1798.
    Juan Yüan (+1764 to +1849).
    Ed. Chung Hua, Beijing, 1982.

Ching Chuan Shih Tzhu 經傳釋詞.
    Explanations of the Grammatical Particles in the Classics and Commentaries.
    Preface +1798.
    Wang Yin-Chih 王引之 (+1766 to +1834).
    Ed. World Book Co., Taipei, 1965.

Ching Fa 經法.
    Regular Models. (Recently discovered silk manuscript).
    –3rd.
    Anonymous.
    Ed. *Ma Wang Tui Han Mu Po Shu Ching Fa* 馬王堆漢墓帛書經法 The Ma-Wang-Tui Han tomb silk book Laws of government, Wen-wu, Peking, 1976.

Ching Tien Shih Wen 經典釋文.
    Explanations of the Texts of Classics.
    +7th cent.
    Lu Te-ming 陸德明 (+556 to +627).
    Ed. *SPTK.*

Chiu Chang Suan Shu 九章算術.
    The Nine Chapter Calculating Technique.
    c. –2nd century.
    Anonymous, commentary by Hsü Yüeh (*fl.* +220).
    Ed. Kuo Shu-Chhun 郭書春, Chiu Chang Suan Shu 九章算術, Liaoning Chiao Yü, Shen-yang, 1990. Cf. Loewe (1993), pp. 16–23.

Chiu Thang Shu 舊唐書.
  Old History of the Thang.
  Wu Tai, +945.
  Liu Hsü 劉昫, Ed. Chung Hua, Beijing, 1963.
Chou Li 周禮.
  Records of the Rites of the Chou dynasty.
  Han compilation, first mentioned in *Shih Chi as Chou Kuan* 周官 'Offices of the Chou'.
  Anonymous.
  Ed. *SPTK* and *SSCCS*.
  Tr. E. Biot, 1851. See William G. Boltz's bibliographic survey in Loewe (1993), pp. 24–32.
Chu Tzu Pien Lüeh 助字辨略.
  A Discriminating Survey of Subsidiary Characters.
  +1711.
  Liu Chhi 劉淇.
  Ed. Chung Hua, Shanghai, 1955.
Chu Tzu Yü Lei 朱子語類.
  Classified Conversations of Chu Hsi.
  +1270.
  Chu Hsi 朱熹.
  Ed. Chung Hua, Beijing, 1986.
Chuang Tzu 莊子.
  The Book of Master Chuang.
  Chou, −4th to −3rd century.
  Chuang Chou 莊周 (*c.* −350 to *c.* −280).
  Ed. Harvard Yenching Sinological Index Series. Cf. also ed. Kuo Chhing-Fan 郭慶藩 Chung Hua, Beijing, 1961. See H. D. Roth's bibliographic survey in Loewe (1993), pp. 56–66.
Chung Lun 中論.
  Discourse of the Middle Way. (Tr. of the *Mūlamadhyamakaśāstra* by Nāgārjuna into Chinese by Kumārajīva.
  +409.
  Attrib. Nāgārjuna.
  Ed. Taishō Tripitaka 30, no. 1564.
  For a survey of translations see David Seyfort Ruegg, *The literature of the Madhyāmaka school of philosophy in India in A history of Indian literature*, Harrassowitz, Wiesbaden, 1981.
Chung Yung 中庸.
  Doctrine of the Mean.
  −4th to −3rd centuries.
  Attrib. to Confucius's disciple Tzu Ssu 子思 (*c.* −493 to −406).
  Ed. Legge, *Chinese Classics* and *SSCCS*.
Chhang Chen Lun 常珍論.
  Constant Jewel Account.
  +649.
  Original by Bhavya (*c.* +490 to +570), tr. by Hsüan-Tsang 玄奘 (+602 to +664).
  Taishō Tripitaka no. 1578, tr. L. de la Vallée Poussin, *MCB*, vol. 2 (1933), pp. 60–146. In Mélanges Chinoises et Bouddhiques.
Chhéng Wei Shih Lun 誠唯識論.
  The True Account of Consciousness Only.
  +7th century.
  Hsüan Tsang 玄奘 (+602 to +664).
  Ed. Taishō Tripitaka Tr. Louis de la Vallée Poussin, *Vijñaptimātratāsiddhi. La Siddhi de Hiuan-Tsang*, 2 vols., Paris 1928/9. English tr. by Wei Tat, *Ch'eng Wei-Shih Lun, Doctrine of Mere-Consciousness by*

Tripitaka-Master Hsüan Tsang, Ch'eng Wei Shih Lun Publication Committee, Hong Kong, 1987.
Chhieh Yün 切韻.
  Carved Rhymes.
  +601.
  Originally compiled by Lu Fa-Yen 陸法言 (fl. *c.* +581 to +601), current editions contain much later material.
  Ed. Chou Tsu-Mo 周祖謨 Kuang-yün chiao-pen 廣韻校本 2 vols., Peking 1958.
Chhien Han Shu 前漢書.
  History of the Former Han Dynasty [−206 to +9] and its Hsin Dynasty [+9 to +23].
  c. +100.
  Pan Ku 班固 (+32 to +92), incorporating work by Pan Piao 班彪 (died +54) and after his death continued by his sister Pan Chao 班昭 (+48 to +116).
  Ed. Chung-hua, Peking, 1962. Partial tr. H. Dubs (1938–55). Cf. Loewe (1993), pp. 129–37.
Chhien Tzu Wên 千字文.
  Thousand Character Text.
  Before +521.
  Chou Hsing-Ssu 周興嗣.
  Ed. *Ch'ien tzu wên: the thousand character classic. A Chinese primer*. Unger, New York, 1963.
Chhu Tzhu 楚辭.
  Elegies of Chhu.
  Late Chou, with many Han additions, −4th to −3rd centuries.
  By Chhü Yüan 屈原 (*c.* −343 to *c.* −277) *et al*.
  Tr. D. Hawkes (1959), revised and reannotated edition 1985.
  Ed. *SPPY*. See David Hawkes's bibliographic survey in Loewe (1993), pp. 48–55.
Chhun Chhiu 春秋.
  Spring and Autumn Annals.
  Chou: chronicle of the State of Lu for the years −722 to −481.
  Attrib. Confucius.
  Tr. S. Couvreur (1951) and J. Legge, *Chinese Classics*.
  Ed. *HYSIS* Index and *SSCCS*. Cf. Loewe (1993), pp. 67–76.
Chhun Chhiu Fan Lu 春秋繁露.
  String of Pearls on the Spring and Autumn Annals.
  Early Han, incorporating a fair amount of later material added.
  Attrib. Tung Chung-Shu 董仲舒 (+176 to +104).
  Ed. *SPTK* and World Book Co., Taipei, 1970.
  Tr. Robert H. Gassmann, *Tung Chung-shu, Ch'un-ch'iu fan-lu Üppiger Tau des Frühling-und Herbst-Klassikers. Übersetzung und Annotation der Kapitel eins bis sechs = Etudes Asiatiques Suisses*, vol. 8, Bern, etc., 1988. See Steve Davidson's and Michael Loewe's bibliographic survey in Loewe (1993), pp. 77–87.

Dainihon zokuzōkyō 大日本續藏經, Zōkyō shoin, Kyoto 1905–12.
  Supplementary Tripitaka.

Erh Ya 爾雅.
  Getting Close to What is Correct.
  Chou material, stabilised in Chhin and Early Han, *c.* −3rd century.

Enlarged and commented on *c.* +300 by Kuo Phu
    郭璞 (+276 to +324).
Ed. *SPTK* and *HYSIS* Index. See South Coblin's
    bibliographic survey in Loewe (1993), pp. 94–9.

Erh Ya I 爾雅翼.
    Wings to the Literary Expositor.
    +1174, preface by Wang Ying-Lin 王應麟, dated
        +1270.
    Luo Yüan 羅願.
    Ed. *SPTK*.

Fa Yen 法言.
    Model Words.
    –1st century.
    Yang Hsiung 揚雄 (–53 to +18).
    Ed. Wang Jung-Pao 王榮寶, World Book Co., Taipei
        1958.
    Tr. Erwin von Zach, *Yang Hsiungs Fa-yen* (*Worte
        strenger Ermahnung*): Batavia 1939; reprint San
        Francisco: Chinese Materials Center, 1976. See
        D. R. Knechtges' bibliographic survey in Loewe
        (1993), pp. 100–4.

Fan Chiu Phien 凡將篇.
    General Guide.
    Han, –2nd century.
    Sima Hsiang-Ju 司馬相如 (–179 to –117).
    Not extant.

Fang Pien Hsin Lun 方便新論.
    New Treatise on Upaya.
    Between +471 and +476.
    Original attributed to Nāgārjuna.
    Ed. Taishō Tripitaka no. 1632.

Fang Yen 方言.
    Local Words.
    Han, –1st century.
    Yang Hsiung 揚雄 (–53 to –18).
    Ed. *SPTK*.

Fen Shu, Hsü Fen Shu 焚書, 續焚書.
    Book for the Burning, Continuation of the Book
        for the Burning.
    Ming, +16th century.
    Li Chih 李贄 (+1527 to +1602).
    Ed. Chung-hua, Peking, 1975.

Han Fei Tzu 韓非子.
    The Book of Master Han Fei.
    Early –3rd century, with some possible later addi-
        tions.
    Han Fei 韓非 (died –233).
    Ed. Chou Chung-Ling 周鍾靈, *Han Fei Tzu Suo Yin*
        韓非子索引, Chung Hua, Peking, 1982. See Jean
        Levi's bibliographic survey in Loewe (1993), pp.
        115–24.

Han Shih Wai Chuan 韓詩外傳.
    Supplementary Commentaries on the Han Songs.
    –2nd century.
    Han Ying 韓嬰.
    Ed. Hsü Wei-Yü 許維遹, Chung Hua,
        Peking, 1980. Tr. Hightower (1951). See J. R.
        Hightower's bibliographic survey in Loewe
        (1993), pp. 125–8.

Han Shu 漢書.
    See Chhien Han Shu.

Hsi Khang Chi 嵇康集.
    Collected Works of Hsi Khang.
    Hsi Khang 嵇康 (+223 to +262).
    Ed. Tai Ming-Yang 戴明揚, Chung Hua, Peking,
        1962.
    Tr. Robert G. Henricks, *Philosophy and Argumen-
        tation in Third-Century China, The Essays of Hsi
        K'ang*, Princeton University Press 1983.

Hsiao Ching 孝經.
    Classic of Filial Piety.
    Allegedly Chou, probably –1st century.
    Attrib. Tseng Shen 曾參, disciple of Confucius.
    Ed. *SPTK* and *HYSIS*. See William Boltz's
        bibliographic survey in Loewe (1993), pp.
        141–53.

Hsiao Erh Ya 小爾雅.
    Little Erh Ya.
    San-kuo, +3rd century.
    Preserved as ch. 11 of the *Khung Tshung Tzu* 孔叢子
        by Wang Su 王肅 (+195 to +256).
    Ed. *SPTK*.

Hsin Shu 新書.
    New Book.
    Large parts are –2nd century, but there are many
        later additions.
    Chia I 賈誼 (–200 to –168).
    Ed. Chhi Yü-Chang 祁玉章, *Hsin Shu Chi Shih*
        新書集釋, Chung Kuo Wen Hua Tsa Chih,
        Taipei, 1974. See Michael Nylan's bibliographic
        survey in Loewe (1993), pp. 161–70.

Hsin Thang Shu 新唐書.
    New History of the Thang Dynasty.
    Sung, +1061.
    Ouyang Hsiu 歐陽修 (+1007 to +1072) *et al.*
    Ed. Chung-hua, Peking 1962.

Hsü Tzu Shuo 虛字説.
    Explanation of Empty Characters.
    Preface dated +1747.
    Yüan Jen-Lin 袁仁林.
    Ed. Chieh Hui-Chhüan 解惠全, Chung Hua,
        Peking, 1989.

Hsün Tzu 荀子.
    The Book of Master Hsün.
    Chou, *c.* –240, with later additions.
    Hsün Chhing 荀卿 (*c.* –298 to *c.* –238).
    Ed. *HYSIS*. Cf. *Hsün-Tzu Hsin Chu* 荀子新注,
        Chung Hua, Peking, 1979; Liang Chhi-Hsiung
        梁啟雄 Shang Wu, Taipei, 1975.
    Tr. H. Köster, *Hsün Tzu*, Kaldenkirchen, 1967, and
        J. H. Knoblock. A Translation and Study of the
        Complete Works, Stanford University Press,
        1988, 1990. See Michael Loewe's bibliographic
        survey in Loewe (1993), pp. 178–88.

Hua I I Yü 華夷譯語.
    Translations between Chinese and the Yi-
        Barbarians.
    +1389.
    Huo Yuan-Chieh 火源潔.
    Tr. Marian Lewicki, *La langue mongole des trans-
        criptions chinoises du 14ème siècle*. Wroclawskie
        Towarzystwo Naukowe Prace, Series A, no. 29;
        A. Mostaert ed., Igor de Rachewiltz, Bruxelles
        1977.

Huai Nan Tzu 淮南子.
:   Master of Huai -Nan.
:   Han, *c.* −139.
:   Compiled under the auspices of Liu An 劉安 (−179 to −122).
:   Ed. Liu Wên-Tien 劉文典, Shang Wu, Shanghai, 1923. See Charles LeBlanc's bibliographic survey in Loewe (1993), pp. 189–95.

Huang Ti Nei Ching 黃帝內經.
:   The Yellow Emperor Inner Classic.
:   Anonymous.
:   Han original text now lost. Current version certainly post-Han, probably as late as +7th century.
:   Ed. *Huang Ti Nei Ching Su Wên. Fu Ling Shu Ching* 黃帝內經素問，附靈樞經, Shang Wu, Shanghai, 1955. See Nathan Sivin's bibliographic survey in Loewe (1993), pp. 196–216.

Hui Ching Lun 迴諍論.
:   *Vigrahavyārtanī.*
:   Translation +541.
:   Original Nāgārjuna.
:   Taishō Tripitaka no. 1631.

I Ching 易經.
:   Book of Changes.
:   Chou, with later additions of the 'Ten Wings' from up to Han times.
:   Compiler unknown.
:   Ed. *HYSIS*. Cf. *SSCCS*, and Shchutskii, Iulian K. *Researches on the I Ching*, tr. by William L. MacDonald and Tsuyoshi Hasegawa with Hellmut Wilhelm (Princeton Univ. Press, Princeton, 1979). See E. L. Shaughnessy's bibliographic survey in Loewe (1993), pp. 216–28.

I Li 儀禮.
:   Forms and Ritual.
:   −3rd century.
:   Anonymous.
:   Ed. and tr. S. Couvreur, *I Li, Cérémonial*, Cathasia, Paris, 1951 and *SSCCS*. See William G. Boltz's bibliographic survey in Loewe (1993), pp. 234–43.

I Chhieh Ching Yin I 一切經音義.
:   Pronounciation and Meaning of all Classics.
:   *c.* +645.
:   Hsüan Ying 玄應.
:   Ed. Sun Hsing-Yen 孫行衍, Shang Wu, Shanghai, 1936.

Ju Shih Lun 如實論.
:   Tarkasastra.
:   Translation from +552 to +557.
:   Original attrib. Vasubandhu, translated by Paramartha.
:   Ed. Taishō Tripitaka no. 1633.

Khang Hsi Tzu Tien 康熙字典.
:   Khang Hsi Dictionary.
:   +1716.
:   Chang Yü-Shu 張玉書.
:   Ed. reprint Chung Yang Shu Tien, Shanghai, 1942.

Khung Tshung Tzu 孔叢子.
:   The Khung Family Master's anthology.
:   +3rd century.
:   Probably by Wangsu 王肅 (+195 to +256).
:   Ed. *SPTK*. Cf. Yen Chhin-Nan 閻琴南, *Khung Tshung Tzu Chiao Chêng* 孔叢子斠證, Privately published, Taipei, 1975.
:   Tr. Yoav Ariel, *K'ung-Ts'ung-Tzu, The K'ung Family Anthology. A Study and Translation of Chapters 1–10, 12–14*, Princeton University Press, Princeton, 1989.

Khung Tzu Chia Yü 孔子家語.
:   Family Sayings of Confucius.
:   +3rd century.
:   Wangsu 王肅 (+195 to +256).
:   Ed. Wan-yu-wen-khu 萬有文庫, Shang Wu, Shanghai, 1936.
:   Partial trs. R. P. Kramers, *K'ung Tzu Chia Yü, The School Sayings of Confucius*, Brill, Leiden 1950, and Richard Wilhelm, *Kung Futse, Schulgespräche*, Eugen Diederichs Verlag, Düsseldorf/Köln, 1961. See R. P. Kramers' bibliographic survey in Loewe (1993), pp. 258–62.

Ku Liang Chuan 穀梁傳.
:   Kuliang's Commentary [on the *Spring and Autumn Annals*].
:   −3rd to −2nd centuries.
:   Attrib. Kuliang Chhih 穀梁赤.
:   Ed. *HYSIS*, cf. *SSCCS*. See Anne Cheng's bibliographic survey in Loewe (1993), pp. 67–76.

Kuan Tzu 管子.
:   Book of Master Kuan.
:   Miscellaneous material dating from −5th to −2nd centuries.
:   Attrib. Kuan Chung 管仲 (died −645).
:   Ed. Tai Wang 戴望, Shang Wu, Shanghai, 1936. Cf. Kuo Mo-Jo 郭沫若 *et al.*, *Kuan Tzu Chi Chiao* 管子集校, Kho Hsüeh, Peking, 1957.
:   Tr. W. Allyn Rickett, *Kuan-tzu, A Repository of Early Chinese Thought*, Hong Kong University Press, Hong Kong, 1965; and *idem, Guanzi: Political, Economic and Philosophical Essays from Early China; a Study and Translation.* Vol. 1 tr. by W. Allyn Rickett, Princeton Univ Press, Princeton, 1985. See W. Allyn Rickett's bibliographic survey in Loewe (1993), pp. 244–57.

Kuang Pai Lun Shih 廣百論釋.
:   Expanded Explanation of the Hundred Theories.
:   +650.
:   Hsüan-Tsang 玄奘 (+602 to +664).
:   Taishō Tripitaka vol. 30, no. 1571.

Kuang Ya 廣雅.
:   Expanded Erh Ya.
:   +3rd century.
:   Chang I 張揖 (*c.* 220 to +265).
:   Ed. Wang Nien-Sun 王念孫, Ting-wen, Taipei, 1972.

Kuang Yun 廣韻.
:   Expanded Rhymes.
:   +1107.
:   Chêng Phêng-Nien 陳彭年.
:   Ed. Chou Tsu-Mo 周祖謨, *Kuang Yün Chiao Pên* 廣韻校本, Chung Hua, Peking, 1960.

Kung Yang Chuan 公羊傳.
> Master Kungyang's Commentary [on the Spring and Autumn Annals].
> −2nd century, transmitted orally earlier.
> Attrib. Kungyang Kao 公羊高.
> Ed. *HYSIS*, cf. *SSCCS*. See Anne Cheng's bibliographic survey in Loewe (1993), pp. 67–76.

Kungsun Lung Tzu 公孫龍子.
> Book of Master Kungsun Lung.
> Pai-ma-lun and Chih Wu Lun are genuine −4th century, the other chapters contain much later material. The earliest commentary probably dating to the +7th century.
> Kungsun Lung 公孫龍 (late −4th century).
> Ed. Luan Hsing 欒星, Chung-chou Shu Hua She, Ho-nan, 1982.
> Tr. Y. P. Mei, *HJAS* 16 (1953), 404–37; Kou Pao-koh, *Deux sophistes chinois: Houei Che et Kong-Souen Long*, Paris: Presses Universitaires de France, 1953. See A. C. Graham's bibliographic survey in Loewe (1993), pp. 252–7.

Kuo Yü 國語.
> Discourses on the States.
> −4th/−3rd centuries, treating the period −990 to −453.
> Attrib. Tso Chhiu-Ming 左丘明 (−5th century).
> Ed. *SPPY* and World Book Co., Taipei, 1975. Cf. Ku Chi, Shanghai, 1978.
> Partial tr. Rémi Mathieu, *Guo Yü*, Collège de France, 1985. See the bibliographic survey in Loewe (1993), pp. 263–8.

Lao Tzu 老子.
> The Book of Master Lao.
> −3rd century.
> Attrib. Lao Tan 老聃.
> Ed. and tr. D. C. Lau, Hong Kong (1984). For editions see Yen Ling-Feng 嚴靈峰, *Lao Tzu Chi Chhêng* 老子集成 I Wên, Taipei, undated. See Shima Kunio 島邦男, *Rōsi kōsei* 老子校正, Kyūko shoin, Tokyo, 1973, for an outstanding collation of traditional editions, and Robert G. Henricks, *Lao Tzu, Tao Te Ching*, Bodley Head, London, 1990, for the recently discovered Ma Wang Tui manuscript version with a translation. See also Lao Tzu Chia Pên Chüan Hou Ku I Shu Shih Wên 老子甲本卷後古佚書釋文.
> I have used Anon., *Konkordanz zum Lao-tzu*, Publikationen der Fachschaft Sinologie München no. 19, 1968. See William G. Boltz's bibliographic survey in Loewe (1993), pp. 269–92.

Lao Tzu Chih Lüeh 老子指略.
> Summary of Lao Tzu.
> +3rd century.
> Wang Pi 王弼 (+226 to +249).
> Ed. Lou Yü-Lieh 樓宇烈, Peking: Chung-hua, 1980; tr. R. Wagner in *T'oung-Pao* 72 (1986), 92–129.

Lei Phien 類篇.
> Classifications.
> +1066.

Compiled and finally edited by Ssuma Kuang 司馬光 and others between +1039 and +1066.
> Ed. Shanghai Kuchi, Shanghai 1988.

Li Chi 禮記.
> Record of Rites.
> Compiled *c.* −50, containing earlier material.
> Ed. S. Couvreur, Cathasia, Paris, 1951. Cf. *SSCCS*. See Jeffrey K. Riegel's bibliographic survey in Loewe (1993), pp. 293–7.

Liang Chu-Ko, Pai Chü-I Chi 兩朱閣, 白居易集.
> The collection of Liang Chu-Ko and Pai Chü-I.
> Ed. Ku Hsüeh-Chieh.

Liao Chai Chih I 聊齋志異.
> Strange Stories from the Studio of Phu Sung-Ling.
> Phu Sung-Ling 蒲松齡 (+1640 to +1714).
> Ed. Chang Yu-Ho 張友鶴. Ku-chi, Shanghai, 1962.

Lieh Tzu 列子.
> Book of Master Lieh.
> +3rd/+4th centuries.
> Attrib. Lieh Yü-khou 列禦寇.
> Ed. Yang Po-Chün 楊伯峻, *Lieh Tzu Chi Shih* 列子集釋, reprint Hongkong: Taiping, 1965. See T. H. Barrett's bibliographic survey in Loewe (1993, pp. 298–309.

Liu Shu Ku 六書故.
> Exposition of the Six Modes of Writing.
> Late Sung, published +1320.
> Tai Thung 戴侗.
> Ed. Li Ting-Yüan 李鼎元, Shih-chu-chai 師竹齊本, 1784.
> Tr. L. C. Hopkins, *The Six Scripts – or the Principles of Chinese Writing*. Cambridge: Cambridge University Press, 1954.

Liu Tzu 劉子.
> +6th century.
> Probably by Liu Hsieh 劉勰 (+465 to +522).
> Lin Chhi-Than 林其錟 and Chhen Feng-Chin 陳鳳金, *Liu Tzu Chi Chiao* 劉子集校, Shanghai, Ku-chi, 1986, and Lin Chhi-Than 林其錟 and Chhen Feng-Chin 陳鳳金, *Tun Huang I Shu Liu Tzu Tshan Chüan Chi Lu* 敦煌遺書劉子殘卷集錄, Shanghai: Shanghai shu-tien, 1988.

Lü Shih Chhun Chhiu 呂氏春秋.
> Mr Lü's Spring and Autumn Annals.
> −239.
> Compiled under the auspices of Lü Pu-Wei 呂不韋.
> Ed. Chhen Chhi-Yu 陳奇猷, Shanghai: Hsüeh-lin, 1984. See Michael Carson's and Michael Loewe's bibliographic survey in Loewe (1993), pp. 324–30.

Lun Heng 論衡.
> Discourses Weighed in the Balance.
> *c.* +83.
> Wang Chhung 王充 (+27 to *c.* +100).
> Ed. *Lun Hêng Chu Shih* 論衡注釋, Chung Hua, Peking, 1979, and Huang Hui 黃暉, *Lun Hêng Chiao Shih* 論衡校釋, Chung Hua, Peking, 1990. See Timoteus Pokora's and Michael Loewe's bibliographic survey in Loewe (1993), pp. 309–12.

Lun Yü 論語.

Conversations of Confucius; Analects.

Chou (Lu), *c.* –450, with some slightly later additions.

Compiled by the disciples of Confucius.

Ed. *HYSIS* and Liu Pao-Nan 劉寶楠 Shang Wu, Shanghai, 1936.

Tr. Lau (1983a). See Anne Cheng's bibliographic survey in Loewe (1993), pp. 313–23.

Mêng Tzu 孟子.

The Book of Master Mêng (Mencius).

Chou, –4th cent. *c.* –290?.

Meng Kho 孟軻 (*c.* –372 to –289).

Ed. and tr. Lau (1983c). Cf. ed. Chiao Hsün 焦循, Chung Hua, Beijing, 1983. See D. C. Lau's bibliographic survey in Loewe (1993), pp. 331–5.

Mo Tzu 墨子.

The Book of Master Mo.

Chou, –5th to –3rd centuries.

Oldest parts by Mo Tï 墨翟 (late –5th century), many later additions.

Ed. *HYSIS.* See the excellent new edition Wu Yü-Chiang 吳毓江, *Mo Tzu Chiao Chu* 墨子校注, Hsi Nan Shih Fan Ta Hsüeh, Chhung Chhing, no date. See A. C. Graham's bibliographic survey in Loewe (1993), pp. 336–41.

Pai Chü I Chi 白居易集.

Collected Works of Pai Chü-I.

Pai Chü-I 白居易 (+772 to +824).

Ed. Ku Hsüeh-Chieh 顧學頡, Chung-hua, Peking, 1982.

Pai Hu Thung 白虎通.

Comprehensive Discussions at the Tiger Lodge.

c. +80.

Pan Ku 班固 (+32 to +92).

Ed. *SPTK,* and 白虎通索引, 伊東倫厚等編, 東豐書店, Tokyo 昭和 54 (1980).

Tr. Tjan Tjoe Som, 1949ff. See Michael Loewe's bibliographic survey in Loewe (1993), pp. 347–56.

Pao Phu Tzu 抱樸子.

Preservation-of-simplicity Master.

+4th century.

Ko Hung 葛洪.

Ed. Shang Wu, Shanghai, 1936 (*WYWK*). Cf. Wang Ming 王明 *Pao Phu Tzu Nei Phien Chiao Shih* 抱樸子內篇校釋 rev. ed. Peking: Chung-hua, 1985. Cf. Jay Sailey, *The Master who Embraces Simplicity: A Study of the Philosopher Ko Hung, AD 283–343,* San Francisco: Chinese Materials Center, 1978, and J. Ware, A*lchemy, Medicine and Religion in the China of AD 320,* tr. and ed. by J. R. Ware, Cambridge Mass.: MIT Press, 1966.

Index K. Schipper, *Concordance du Pao-p'ou-tseu nei-p'ien* (Paris, 1965) *and* K. Schipper, *Concordance du Pao-p'ou-tseu wai-p'ien* (Paris, 1970).

Pei Hsi Tzu I 北溪字義.

Character Meanings by Pei-hsi.

Before +1226.

Chhen Chhun 陳淳 (+1159 to + 1223).

Ed. Chung Hua, Beijing, 1983. Tr. Wing-tsit Chan, *Neo-Confucian Terms Explained (The Pei-hsi*

tzu-i) by Ch'en Ch'un, 1159–1223, Columbia University Press, New York, 1986.

Phan Pi Liang Lun 判比量論.

Criticising the Argument (for Consciousness Only).

+7th century.

Yüan Hsiao 元曉 (+617 to +686).

Ed. *Dainihon zokuzōkyō* 大日本續藏經, 1.94/4.

Phei Wen Yün Fu 佩文韻府.

Rhyme Treasury for Matching Texts.

+1711.

Chang Yü-Shu 張玉書.

Ed. Ku Chi, Shanghai, 1983.

Phi Ya 埤雅.

Additional Standard.

Postface +1125.

Lu Tien 陸佃.

Ed. *SPTK.*

Phien Ya 駢雅.

Additional Standard.

Preface dated +1587.

Chu Mou-Wei 朱謀瑋.

Ed. *SPTK.*

Shang Chün Shu 商君書.

Book of the Lord of Shang.

–4th to –3rd centuries.

Attrib. Shang Yang 商鞅 or Kungsun Yang 公孫鞅.

Ed. Kao Hêng 高亨, S*hang Chün Shu Chu I* 商君書注譯, Chung Hua, Peking, 1974.

Tr. Duyvendak, Jean Lévy, Shang Yang, *Le livre du prince Shang,* Paris, 1981; L. S. Perelomov, *Kniga pravitelja oblasti Shan,* Moscow: Nauka, 1968. See Jean Levi's bibliographic survey in Loewe (1993), pp. 368–75.

Shên Chien 申鑒.

Precepts Presented.

c. +190.

Hsün Yüeh 荀悦.

Ed. *SPTK.* See Ch'en Ch'i-Yün's bibliographic survey in Loewe (1993), pp. 390–3.

Shên Pu Hai 申不害.

Fragments of Shen Pu-Hai.

Shen Pu-Hai (died –337).

Ed. Herrlee G. Creel, *Shen Pu-hai,* University of Chicago Press, Chicago, 1974.

Shên Tzu 慎子.

Fragments from the *Shen Tzu.*

Shen Tao (*c.* –350 to *c.* –275).

–4th century.

Ed. P. M. Thompson, *The Shen Tzu Fragments,* Oxford University Press, London, 1979. See P. Thompson's bibliographic survey in Loewe (1993), pp. 399–404.

Shih Chi 史記.

Records of the Historian.

Han, –1st century.

Ssuma Chhien 司馬遷 (*c.* –145 to *c.* –85), but with certain later additions.

Ed. Takigawa Kametarō 瀧川龜太郎, *Shiki kaichū kōshō* 史記會注考證, 10 vols., Tokyo 1932–4, reprint Peking 1957. Tr. E. Chavannes, *Les mémoires historiques de Se-ma Ts'ien,* Paris

1895–1905. See A. F. P. Hulsewé's bibliographic survey in Loewe (1993), pp. 405–15.

Shih Ching 詩經.
  Book of Songs.
  Chou, –9th to –5th centuries.
  Anonymous.
  Ed. and tr. B. Karlgren (1950).
  HYSIS Index. See Michael Loewe's bibliographic survey in Loewe (1993), pp. 415–23.

Shih Ming 釋名.
  Explaining names.
  +2nd/3rd centuries.
  Liu Hsi 劉熙.
  Ed. *SPTK*. See Roy Andrew Miller's bibliographic survey in Loewe (1993), pp. 424–8.

Shih Shuo Hsin Yü 世說新語.
  A New Account of Tales of the World.
  +5th century.
  Liu I-Chhing 劉義慶 (+403 to +444).
  Ed. Yang Yung 楊勇, Ta Chung, Hong Kong, 1969.
  Trans R. Mather, *Shih shuo hsin yü: A New Account of Tales of the World*, Univ. of Minnesota Press, Minneapolis, 1976; 世說新語索引(附原文) 高橋清編藍星書鋪, Tokyo, Preface 昭和 34 (1960).

Shu Ching 書經.
  Book of Documents.
  –10th to +3rd centuries.
  Various authors.
  Ed. and tr. Bernhard Karlgren, *Book of Documents*, *BMFEA* **22** (1950), 1–81. Cf. ed. Chhü Wan-Li 屈萬里, Shang Wu, Taipei, 1977. See E. L. Shaughnessy's bibliographic survey in Loewe (1993), pp. 376–89.

Shui Hu Ti Chhin Mu Chu Chien 睡虎地秦墓竹簡.
  Bamboo Slips from the Chhin Tomb at Shui Hu Ti. Wen Wu, Peking, 1978.

Shuo Wên Chieh Tzu 說文解字.
  Explaining Graphs and Characters.
  +100/ +121, with later additions.
  Hsü Shen 許慎 (died *c.* +149).
  Ed. Tuan Yü-Tshai 段玉裁, Ku Chi, Shanghai, 1981. See William G. Boltz's bibliographic survey in Loewe (1993), pp. 429–42.

Shuo Wên Chieh Tzu Chuan Yün Phu 說文解字傳韻譜.
  Rhyme List to the Explaining Graphs and Characters.
  +986.
  Hsü Hsüan 徐鉉.
  Ed. Ting Fu-Pao 丁福保, *Shuo Wên Chieh Tzu Ku Lin* 說文解字詁林 I Hsüeh, Shanghai, 1930.

Shuo Yuan 說苑.
  A Garden of Talk.
  –20.
  Liu Hsiang 劉向 (–79 to –8).
  Ed. Chao Shan-I 趙善怡, *Shuo Yüan Shu Chêng* 說苑疏證, Hua Tung Shih Fan Ta Hsüeh, Shanghai, 1985 and Hsiang Tsung-Lu 向宗魯, *Shuo Yüan Chiao Chêng* 說苑校證 Chung-hua, Peking 1987. See David Knechtges' bibliographic survey in Loewe (1993), pp. 443–5.

Ssu I Kuan Khao 四夷官考.
  Investigations into the Four Barbarian Offices.
  Preface +1540.
  Wang Tsung-Tsai 王宗載.
  Ed. Tung Fang Hsüeh Hui, no place (Shanghai?), 1924.
  Supplementary Tripitaka: see *Dainihon zokuzōkyō* 大日本續藏經.

Ta Hsüeh 大學.
  The Great Learning.
  c. –200, part of Li Chi 禮記.
  Authorship uncertain.
  Ed. and tr. J. Legge, *Chinese Classics*. Cf. *SSCCS*.

Ta Tai Li Chi 大戴禮記.
  Record of Rites by the Elder Tai.
  Stabilised around +80 and +100.
  Ed. *SSCCS*. Cf. Kao Ming 高明, *Ta Tai Li Chi Chin Chu Chin I* 大戴禮記今注今譯, Shang Wu, Taipei, 1975. See Jeffrey K. Riegel's bibliographic survey in Loewe (1993), pp. 456–9.

Taishō Tripitaka: *Taishō Shinshū Daizōkyō*, 大正新脩大藏經, ed. Junjirō Takakusu and Kaikyoku Watanabe, Daizō shuppan kabushiki kaisha, Tokyo, 1924–34.

Tao Tsang 道藏.
  The Taoist Patrology.
  First collected in Sung, first available printed edition *c.* +1445.
  Ed. *Chêng Thung Tao Tsang* 正統道藏, reprinted Taipei, Yiwen, 1977.
  Cf. C. Schipper, *Cheng Thung Tao Tsang Mu Lu Suo Yin* 正統道藏目錄索引, Yiwen, Taipei, 1977.

Têng Hsi Tzu 鄧析子.
  Book of Master Têng Hsi.
  Han?
  Attrib. to Teng Hsi 鄧析 (died –501).
  Ed. and tr. H. Wilhelm, *Monumenta Serica* **12** (1947), 41ff.

Thai Hsüan Ching 太玄經.
  Classic of Great Mystery.
  Han, –1st century.
  Yang Hsiung 揚雄 (–53 to +18).
  Ed. *SPTK*.

Thung Ya 通雅.
  Comprehensive Standard.
  Finished *c.* +1579.
  Fang Yi-Chih 方以智 (died +1671).
  Ed. *SPTK*.

Tshang Chieh Phien 倉頡篇.
  Book of Tshang-Chieh.
  Pre-Han.
  Anonymous.
  Ed. Sun Hsing-Yen 孫星衍, I Wen, Taipei, 1967.

Tso Chuan 左傳.
  Tso Commentary.
  Between –400 and –250, but with certain later additions.
  Attrib. Tso Chhiu-Ming 左丘明.
  Ed. Yang Po-Chün 楊伯峻, Chung Hua, Peking, 1982; tr. J. Legge, *Chinese Classics*, and S. Couvreur, *La chronique de la principauté de Lòu*,

Cathasia, Paris 1951. See Anne Cheng's bibliographic survey in in Loewe (1993), pp. 67–77.

Tsung Ching Lu 宗鏡錄.
     Records of the Mirror of the School.
     Published between +1078 and +1085.
     Yen Shou 延壽 (+904 to +975).
     Taishō Tripitaka no. 2016.

Tzu Chih Thung Chien 資治通鑑.
     Comprehensive Mirror in Aid of Government
     (covering period –403 to +959).
     +1084.
     Ssuma Kuang 司馬光 (+1019 to +1086) *et al.*
     Ed. Chung-hua, reprint Hong Kong 1971.

Tzu Hui 字匯.
     Collection of Characters.
     +1615.
     Mei Ying-Tso 梅膺祚 (+1570 to +1615).
     Ed. *SPTK.*

Wei Liao Tzu 尉繚子.
     Master Wei Liao.
     Original version –3rd century, later additions,
     partial early Han manuscript found.
     Anonymous.
     Ed. Chung Chao-Hua 鍾兆華, *Wei Liao Tzu* 尉繚
     子校注, Chung-chou Shu Hua She, Ho-nan,
     1982.

Wên Hsin Tiao Lung 文心調龍.
     The Carving of the Literary Dragon.
     Liu Hsieh 劉勰 (+465 to +522).
     Ed. Lu Khan-Ju 陸侃如, Qi Lu, Chi-nan, 1981.
     Tr. Vincent Shih, *The Literary Mind and the Carving of
     Dragons*, bilingual ed. Hong Kong Univ. Press
     1983.

Wên Tsê 文則.
     Patterns of Literature.
     +1170.
     Chhen Khuei 陳騤.
     Ed. Liu Yen-Chheng 劉彥成, Shu Mu Wên Hsien,
     Peking, 1988.

Wu Chhê Yün Jui 五車韻瑞.
     A Jade Vessel for Five Cart-Loads of Rhymes.
     Lost work of the Ming dynasty.

Wu Neng Tzu 無能子.
     The Incompetent Classic.
     +887.
     Anonymous.
     Ed. Wang Ming 王明, Chung Hua, Peking, 1981.

Yen Shih Chia Hsün 顏氏家訓.
     Family Instructions by Mr Yen.
     +6th century.
     Yen Chih-Thui 顏之推 (+531 to after +590).
     Ed. *SPPY*, also ed. Chou Fa-Kao 周法高 Thai
     Lien Kuo Feng, Taipei, 1975.

Yen Thieh Lun 鹽鐵論.
     Salt and Iron Discourse.
     –80.
     Huan Khuan 桓寬 (flourished –73 to –49).
     Ed. Basic Sinological Series, and ed. Wang
     Li-Chhi 王利器, 2nd. ed. Chung Hua, Peking,
     1992.

Cf. partial translation in Esson Gale, *Discourses on
Salt and Iron*, Brill, Leiden, 1931, and the forthcoming profusely annotated complete translation by J. Kroll. See Micheal Loewe's bibliographic
survey in Loewe (1993), pp. 477–82.

Yen Tzu Chhun Chhiu 晏子春秋.
     Annals of Yen-Tzu.
     –3rd century.
     Attrib. Yen Ying 晏嬰.
     Ed. Wu Tsê-Yü 吳則虞, Chung Hua, Peking,
     1982. See Stephen W. Durrant's bibliographic
     survey in Loewe (1993), pp. 483–9.

Yin Ming Chêng Li Men Lun 因明正理門論.
     Nyāyamukha.
     +7th century.
     Hsüan Tsang 玄奘 (+602 to +664).
     Ed. Taishō Tripitaka nos. 1628/9, tr. Giuseppe
     Tucci, *The Nyāyamukha of Dignāga*, Heidelberg
     1930.

Yin Ming Ju Chêng Li Lun 因明入正理論.
     Nyāyapraveśa.
     +7th century.
     Tr. Hsüan-Tsang 玄奘.
     Ed. Taishō Tripitaka no. 1630. Cf. Lü Chheng,
     呂徵, 因明入正理論講解, Chung Hua, Peking
     1983.

Yin Ming Ju Chêng Li Lun Shu 因明入正理論疏.
     Commentary to *Nyāyapraveśa.*
     +7th century.
     Khuei Chi 窺基.
     ST vol. 87, no. 722 ed. Lü Chheng 呂徵 支那內學
     院, Nanking 1933.

Yin Ming Ju Lun Chuang Yen Shu 因明入論莊嚴疏.
     Commentary on the *Nyāyapraveśa.*
     +7th century.
     Wengui 文軌.
     Ed. Lü Chheng 呂徵, Chin-ling kho-ching-chhu,
     Nanking, 1932.

Yin Ming Li Pho Chu Chieh I Thu 因明立破注解意圖.
     Nyāya Proof and Refutation with Comments and
     Illustrations.
     +7th century.
     Lü Tshai 呂才.
     Not extant.

Yin Ming Lun Shu Ming Têng Chao 因明論述明燈抄.
     Collected Commentaries on *Nyāyapraveśa.*
     善珠 Zenshū.
     +8th century.
     Ed. Taishō Tripitaka no. 2270.

Yin Wên Tzu 尹文子.
     Book of Master Yin Wen.
     +3rd century.
     Attrib. Yin Wên 尹文.
     Ed. Li Shih-Hsi 厲施熙 *Yin Wên Tzu Chien Chu* 尹
     文子簡注 Jen-min wen-hsüeh, Shanghai, 1977.
     Tr. Dan Daor (1979).

Yü Chu 語助.
     Aid towards Speech.
     Preface +1324.
     Lu I-Wei 盧以緯.
     Ed. Wang Kho-Chung 王克仲, Chung Hua,
     Peking, 1988.

Yü Chhieh Shih Ti Lun 瑜伽師地論.
    Yogācarabhumiśāstra.
    Translated +645.
    Tr. by Hsüan-Tsang 玄奘 (+602 to +664.)
    Ed. *Taishō Tripitaka* no. 1839. Cf. A. Wayman, in
    JAOS **78** (1958), pp. 29–40.
Yü Phien 玉篇.
    Jade Book.
    First compiled +548, later reworkings.

Ku Ye-Wang 顧野王 (+519 to +581), Chêng
    Phêng-Nien 陳彭年 (+961 to +1017).
    Ed. Ku Chi, Shanghai, 1989.
Yüan Shih 元史.
    History of the Yüan Dynasty.
    Ming, *c.* +1370.
    Sung Lien 宋濂.
    Ed. Chung Hua, Peking, 1963.

B. CHINESE AND JAPANESE BOOKS AND JOURNAL ARTICLES SINCE +1800

Anon. (*1976*)
Ma Wang Tui Han Mu Po Shu, Ching Fa 馬王堆漢墓帛書經法.
The Ching Fa silk manuscript from the Han dynasty Ma Wang Tui tomb.
Wên Wu, Peking.

Anon (*1979a*)
Chhüan Kuo Lo Chi Thao Lun Hui Lun Wên Hsüan Chi 全國邏輯討論會論文選.
A selection of papers from the all China meeting for the discussion of logic.
Chung Kuo Shê Hui Kho Hsüeh, Peking.

Anon (*1979b*)
Hsün-Tzu Hsin Chu 荀子新注.
New Annotation of *Hsün-Tzu*.
Chung Hua, Peking, 1979.

Anon (*1981*)
Chung Kuo Lo Chi Ssu Hsiang Lun Wên Hsüan 1949–1979 中國邏輯思想論文選 1949–1979.
San Lian, Peking.

Ao Ching-Hao (*1984*)
'Pai Ma Lun Chêng I yü Chin I' 白馬論正義與今義.
Commentary and modern translation of the White Horse Dialogue.
In Ku Han Yü Yen Chiu Lun Wên Chi 古漢語研究論文集, vol. 2, Pei-ching chhu-pan-shê, Beijing.

Chan Chien-Fêng (*1979*) 詹劍峰
Mo Chia ti Hsing Shih Lo Chi 墨家的形式邏輯.
Mohist formal logic.
Hu-pei Jen Min, Hupeh.

Chang Chhi-Huang (*1960*) 張其鍠
Mo Ching Thung Chieh 墨經通解.
Complete commentary on the Dialectical Chapters of the *Mo Tzu*.
Shang Wu, Taipei.

Chang Hsin-Chhêng (*1957*) 張心澂
Wei Shu Thung Khao 偽書通考.
Comprehensive investigation into forged books.
2 vols., Shang Wu, Shanghai.

Chang Hsüan (*1919*) 張煊
Mo Tzu Ching Shuo Hsin Chieh 墨子經説新解.
New commentary on the Dialectical Chapters of the *Mo Tzu*.
Kuo Ku Yüeh Khan 國故月刊, Shanghai, vols. 2 and 3 (*Mo Tzu Chi Chhêng* 墨子集成 vol. 21).

Chang Hui-Yen (*1909*) 張惠言
Mo Tzu Ching Shuo Chieh 墨子經説解.
Commentary on the Dialectical Chapters of the *Mo Tzu*.
Reprint Mo Tzu Chi Chhêng 墨子集成 vol. 9 (first ed. Shanghai, 1909).

Chang I-Jen (*1968*) 張以仁
Kuo Yü Yin Tê 國語引得.
Index to the *Kuo Yü*.
Academia Sinica, Taipei.

Chang Man-Thao (*1978*) 張曼濤
Fo Chiao Lo Chi Chih Fa Chan 佛教邏輯之發展.
The development of Buddhist logic.
Mi Le, Taipei (=*Hsien Tai Fo Chiao Hsüeh Shu Tshung Khan* 現代佛教學術叢刊 vol. 42).

Chang Man-Thao (*1978a*) 張曼濤
Hsüan-Tsang Ta Shih Yen Chiu 玄奘大師研究.
Studies in Hsüan-Tsang.
2 vols., Mi Le, Taipei (=*Hsien Tai Fo Chiao Hsüeh Shu Tshung Khan* 現代佛教學術叢刊 vols. 8 and 16).

Chang Phei (*1981*) 章沛
'Hsün Khuang ti Lo Chi Ssu Hsiang Yen Chiu' 荀況的邏輯思想研究.
Studies in the logic of Hsün Tzu.
In Anon (*1981*) Chung Kuo Lo Chi Ssu Hsiang Lun Wên Hsüan 1949–1979 中國邏輯思想論文選 1949–1979, pp. 514–526 and 527–536.

Chang Ping-Lin (*1917*) 章炳麟
Yüan Ming 原名.
The basics of names.
In Chang Shih Tshung Shu 章氏叢書, Chekiang Tu Shu Kuan.

Chang Shih-Chao (*1907*) 章士釗
Chung Têng Kuo Wên Tien 中等國文典.
An intermediate grammar of Chinese.
Shang Wu, Shanghai.

Chang Shih-Chu (tr.) (*1933*) 張師竹
Po La Thu Tui Hua Chi Liu Chung 柏拉圖對話集六種.
Six Dialogues by Plato.
Shang Wu, Shanghai.

Chang Tai-Nien (*1934*) 張岱年
'Chung Kuo Chih Lun Ta Yao' 中國知論大要.
A general survey of Chinese epistemology.
Chhing Hua Hsüeh Pao 清華學報 9, 385 0 409.

Chang Ti-Hua et al. (*1988*)
Han Yü Yü Fa Hsiu Tzhu Tzhu Tien 漢語語法修辭詞典.
A dictionary of Chinese grammar and rhetoric.
An-Hui Chiao Yü, Hofei.

Chao Chheng (*1979*) 趙誠
Chung Kuo Ku Tai Yün Shu 中國古代韻書.
The ancient Chinese rhyme books.
Chung Hua, Peking.

Chao Shan-I (*1985*) 趙善詒
Shuo Yüan Shu Chêng 説苑疏證.
Commentary on the *Shuo Yüan*.
Hua Tung Shih Fan Ta Hsüeh, Shanghai.

Cheng Liang-Shu (*1984*) 鄭良樹
    Hsü Wei Shu Thung Khao 續偽疏通考.
    A continuation of the Comprehensive study of the non-genuine books.
    3 vols., Student Book Company, Taipei.

Chêng Tien and Mai Mei-Chhiao (*1965*) 鄭奠, 麥梅翹
    Ku Han Yü Yü Fa Hsüeh Tzu Liao Hui Pien 古漢語語法學資料彙編.
    Materials on ancient Chinese views on grammar.
    Chung Hua, Peking.

Cheng Tien and Than Chhüan-Chi (*1988*) 鄭奠, 譚全基
    Ku Han-yü hsiu-tzhu-hsüeh tzu-liao hui-pien 古漢語修辭學資料匯編
    Shang Wu, Peking.

Chiang Tsu-I (*1965*) 蔣祖怡
    Wang Chhung ti Wên Hsüeh Li Lun 王充的文學理論.
    Wang Chhung's theory of literature.
    Chung Hua, Peking.

Chiao Hsün 焦循 (*1983*) (ed.)
    Mêng Tzu 孟子.
    The Book of Master Mêng (Mencius).
    Chung Hua, Peking.

Chou Chhao-Hsien (*1966*) 周超賢
    Wei Chin Chhing Than Shu Lun 魏晉清談術論.
    Discussion on the 'pure talk' of the Wei and Chin dynasties.
    Shang Wu, Taipei.

Chou Shu-Chia 周叔迦
    Yin Ming Hsin Li 因明新例.
    New examples of yin ming logic.
    Shanghai, no publisher, no date.

Chou Tsu-Mo (*1951*) 周祖謨
    Fang Yen Chiao Chien Fu Thung Chien 方言校箋附通檢.
    Critical edition of Fang Yen with Index.
    Centre franco-chinois détudes sinologiques, Peking.

Chou Tsu-Mo (*1960*) 周祖謨
    Kuang Yün Chiao Pên 廣韻校本.
    Critical edition of the *Kuang Yün*.
    Chung Hua, Peking.

Chou Tsu-Mo (*1966*) 周祖謨
    '"Erh Ya" Chih Tso Chê Chi Chhi Chhêng Shu Chih Nien Tai' 爾雅之作者及其成書之年代.
    The author and the date of the *Erh Ya*.
    in *Wen Hsüeh Chi* 文學集, Peking, pp. 670–5.

Chou Tsu-Mo and Lo Chhang-Phei (*1958*) 周祖謨, 羅常培
    Han Wei Chin Nan Pei Chhao Yun Pu Yen Pien Yen Chiu 漢魏晉南北朝韻部演變研究.
    Studies in the evolution of the rhyme categories during the Han, Wei, Chin and Southern and Northern dynasty periods.
    Chung Hua, Peking.

Chou Wen-Ying (*1979*) 周文英
    Chung Kuo Lo Chi Ssu Hsiang Shih Kao 中國邏輯思想史稿.
    A sketch of Chinese logical thought.
    Jen Min, Peking.

Chou Wên-Ying (*1982*) 周文英
    '*Sui Thang Shih Chhi Yin Ming ti Shu Ju*' 隋唐時期因明的輸入.
    in Liu Phei-Yü (1982), pp. 240–7.

Chou Yin-Thung (*1983*) 周蘿同
    Wên Yen Hsü Tzu Shih I 文言虛字實義.
    The lexical meanings of classical Chinese particles.
    Shan Hsi Jen Min, Hsi-an.

Chou Yu-Kuang (*1980*) 周有光
    Han Tzu Shêng Phang Tu Yin Pien Chha 漢字聲旁讀音便查.
    Investigations into the pronunciation of the phonetic parts of characters.
    Chi Lin, Chang-chhun.

Chou Yün-Chih (*1981*) 周云之
    'Kungsun Lung Kuan Yü Ming (Kai Nien) ti Lo Chi Ssu Hsiang: "Pai Ma Fei Ma" Chhun Shu Kuei Pien Ma?' 公孫龍關於名(概念)的邏輯思想: 白馬非馬純屬詭辯嗎?.
    Kungsun Lung's logical reflections on the concept *ming* (concept): is 'a white horse is not a horse' simply a sophism?
    in Anon. (1981) *Chung Kuo Lo Chi Ssu Hsiang Lun Wên Hsüan*, pp. 200–5.

Chu Chhien-Chih (*1962*) 朱謙之
    Lao Tzu Chiao Shi 老子校釋.
    Commentary on the *Lao Tzu*.
    Thai Phing, Hong Kong.

Chu Chhien-Chih (*1983*) 朱謙之
    Chung Kuo Chê Hsüeh Tui Yü Ou Chou ti Ying Hsiang 中國哲學對於歐洲的影響.
    The influence of Chinese philosophy on Europe.
    Fu-chien Jen Min, Fuchow.

Chu Chün-Shêng (*1984*) 朱駿聲
    Shuo Wên Thung Hsün Ting Shêng 説文通訓定聲.
    Comprehensive explanation and phonetic interpretation of the *Shuo Wên*.
    Chung Hua, Peking.

Chu Hsing (*1980*) 朱星
    '*Ma Shih Wên Thung ti Tso Chê Chiu Ching shih shei*' 馬氏文通的作者究竟是誰.
    She Hui Kho Hsüeh Chan Hsien 社會科學戰線, no. 3. p. 80.

Chu Kuang-Chhien (*1963*) 朱光潛
    Po La Thu Wên I Tui Hua Chi 柏拉圖文藝對話集.
    Platonic dialogues on artistic themes.
    Chung Hua, Peking.

Chhen Chhêng-Tsê (*1957*) 陳承澤
    Kuo Wên Fa Tshao Chhuang 國文法草創.
    A draft on Chinese grammar.
    Shang Wu, Shanghai.

Chhen Chhi-Yu (*1984*) 陳奇猷
    Lü Shih Chhun Chhiu Chiao Shih 呂氏春秋校釋.
    Commentary on the *Lü Shih Chhun Chhiu*.
    2 vols., Hsüeh Lin, Shanghai.

Chhen Ching-Ho (*1953*) 陳荊和
    '*An Nan Yi Yü Kao Shih*' 安南譯語考釋.
    Investigative commentary on *An Nan Yi Yü* in.
    Wên Shih Chê Hsüeh Bao 文史哲學報 Taipei, no. 5 (1953), 149–240; 6 (1954), 161–227.

Chhên Chu (*1947*) 陳柱
Kungsun Lung Tzu Chi Chieh 公孫龍子集解.
Collected explanations of the *Kungsun Lung Tzu*.
Shang Wu, No. 140, Shanghai.

Chhên Kao-Chhun *et al.* (*1986*) 陳高春
Chung Kuo Yü Wên Hsüeh Chia Tzhu Tien 中國語
文學家辭典.
Dictionary of Chinese linguists.
Honan Jen Min, Honan.

Chhên Khang (*1944*) 陳康
Pa Man Ni Tê Ssu Phien 巴曼尼得斯篇.
Plato's Parmenides.
Chungking, reprint Shang Wu, Peking, 1982.

Chhên Mêng-Lin (*1983*) 陳夢麟
Mo Pien Lo Chi Hsüeh 墨辯邏輯學.
Mohist logic.
Revised edition, Chhi Lu, Chinan.

Chhên Ta-Chhi (*1951*) 陳大齊
'Hsün Tzu ti Ming Hsüeh Fa Wei Chhu Kao' 荀子
的名學發微初稿.
A preliminary study of Hsün Tzu's logic.
Wên Shih Chê Hsüeh Pao 文史哲學報 no. 2, 1–66.

Chhên Ta-Chhi (*1952*) 陳大齊
Yin Tu Li Tse Hsüeh 印度理則學.
Indian logic.
Chung Hua Wên Hua, Taipei.

Chhên Ta-Chhi (*1969*) 陳大齊
*Mêng Tzu ti Ming Li Ssu Hsiang Chi Chhi Pien Shuo
Shih Khuang* 孟子的名理思想及其辯說實況.
Mencius's logical thought and his argumentative
practice.
Jen Jen Wên Khu, Shang Wu, Taipei.

Chhên Ta-Chhi (*1970*) 陳大齊
Yin Ming Ju Chêng Li Lun Wu tha Mên Chhien Shih
因明入正理論悟他門淺釋.
A simple explanation of the section on arguments
to convince opponents in *Nyāyapraveśa*.
Chung Hua, Taipei.

Chhên Wang-Tao (*1931*) 陳望道
Yin Ming Hsüeh 因明學.
Buddhist logic in China.
Shang Wu, Shanghai.

Chhên Wang-Tao (*1980*) 陳望道
Chhên Wang-Tao Yü Wên Lun Chi 陳望道語文論集.
Collected pieces on language by Chhên Wang-
Tao, No. 143.
Shanghai Chiao Yü, Shanghai.

Chhên Wu-Chiu (*1935*) 陳無咎
Mo Ching Hsüan Chieh 墨經懸解.
A commentary on the Dialectical Chapters of the
Mo Tzu.
Shanghai (=*Mo Tzu Chi Chhêng* vol. 38).

Chhien Chien-Fu (*1986*) 錢劍夫
Chung Kuo Ku Tai Tzu Tien Tzhu Tien Kai Lun 中國古
代字典詞典概論.
A survey of ancient Chinese dictionaries of char-
acters and words.
Shang Wu, Peking.

Chhien Ta-Hsin (*1989*) 錢大昕
Chhien Yen Thang Wên Chi 潛研堂文集.
Collected works of Chhien Ta-Hsin.
Ku Chi, Shanghai.

Chhü Chih-Chhing (*1981*) 屈志清
Kungsun Lung Tzu Hsin Chu 公孫龍子新注.
A new commentary on *Kungsun Lung Tzu*.
Hu-pei Jen Min, Wuhan.

Fa Tsun Fa-Shih (*1984*) 法尊法師
Shih Liang Lun Lüe Chieh 釋量論略解.
A brief commentary on *Shih Liang Lun*.
Fo Chiao, Taipei.

Fan Kêng-Yen (*1934*) 范耕研
Mo Pien Shu Chêng 墨辯疏證.
Commentary on the Dialectical Chapters of the
Mo Tzu.
Shanghai (=*Mo Tzu Chi Chhêng* vol. 28)

Fan Ming-Shêng (*1984*) 范明生
Po La Thu Chê Hsüeh Shu Phing 柏拉圖哲學述評.
A critical account of Plato's philosophy.
Jen Min, Shanghai.

Fan Shou-Khang (*1936*) 范壽康
Wei Chin chih Chhing Than 魏晉之清談.
The 'pure talk' of the Wei and Chin dynasties.
Shang Wu, Shanghai.

Fang Hou-Shu (*1979*) 方厚樞
'*Chung Kuo Tzhu Shu Shih Hua*' 中國詞書史話.
A survey of China's dictionaries.
Tzhu Shu Yen Chiu 辭書研究 1, 217–28 and 2,
219–29.

Fêng Tso-Min (*1983*) 馮作民
Pai Hua Chan Kuo Tshê 白話戰國策.
Modern Chinese translation of *Chan Kuo Tshê*.
3 vols., Hsin Kuang, Taipei.

Fêng Yao-Ming (*1985*) 馮耀明
'Pai Ma Lun ti Lo Chi Chieh Kou Chi Chhi Chê
Hsüeh I Han' 白馬論的邏輯結構及其哲學意
含.
The logical structure of the White Horse Dialogue
and its philosophical content.
Legein Monthly (*O Hu Yüeh Khan*), 117, 1985, 34–
43.

Fu Phei-Han (*1986*) 傅佩韓
Chung Kuo Ku Tien Wên Hsüeh ti Tuei Ou I Shu 中國古
典文學的對偶藝術.
The art of parallelism in traditional Chinese liter-
ature.
Kuang Ming Jih Pao Chu Pan Shê, Peking.

Fukuda Jōnosuke (*1979*) 福田之介
Chūgoku jisho shi no kenkyū, No. 138.
Tokyo.

Ho Chhi-Min (*1966*) 何啟民
Chu Lin Chhi Hsien Yen Chiu 竹林七賢研究.
Shang Wu, Taipei.

Ho Chhi-Min (*1967*) 何啟民
Kungsun Lung Yü Kungsun Lung Tzu 公孫龍與公孫龍
子.
Shang Wu, Taipei.

Ho Chhi-Min (*1967*) 何啟民
Wei Chin Ssu Hsiang Yü Than Fêng 魏晉思想與
談風.
Intellectual life during the Wei and Chin dynasties
and the conventions of conversation.
Shang Wu, Taipei.

Ho Wên-Hui (*1985*) 何文匯
    Tsa Thi Shih Shih Li 雜體詩釋例.
    Examples of mixed style poetry explained.
    . . . . , Hong Kong.
Hou Pao-Lin *et al.* (*1985*) 侯寶林
    Chi Chih Yü Yu Mo 機智與幽默.
    Wit and humour.
    Chi-lin Chiao Yu, Chhang-chhun.
Hou Wai-Lu *et al.* (*1957*) 侯外盧
    Chung Kuo Ssu Hsiang Thung Shih 中國思想通史.
    Comprehensive history of Chinese philosophy.
    6 vols., Shang Wu, Peking.
Hsiang Hsia (*1974*) 向夏
    Shuo Wên Chieh Tzu Hsü Chiang Shu 説文解字序
    講疏.
    Commentary on the preface to *Shuo Wên Chieh Tzu*.
    Chung Hua, Hong Kong.
Hsiao Têng-Fu (*1984*) 蕭登福
    Kungsun Lung Tzu Yü Ming Chia 公孫龍子與名家.
    The *Kungsun Lung Tzu* and the School of Names.
    Wên Chin, Taipei.
Hsiung Shih-Li (*1926*) 熊十力
    Yin Ming Da Shu Shan Chu 因明大疏刪注.
    Abbreviation and commentary on *Nyāyapraveśa*.
    Shanghai.
Hsü Fu-Kuan (*1966*) 徐復觀
    Kungsun Lung Tzu Chiang Shu 公孫龍子講疏.
    Commentary on *Kungsun Lung Tzu*.
    Tung Hai Ta Hsüeh, Taipei.
Hsü Fu-Kuan (*1975*) 徐復觀
    Liang Han Ssu Hsiang Shih 兩漢思想史
    History of Han Thought, Hong Kong.
Hsü Wei-Yü (*1980*) 許維遹
    Han Shih Wai Chuan Chi Shih 韓氏外傳集釋.
    Chung Hua, Peking.
Hu Chhêng-Kung (*1936*) 胡承珙
    Hsiao Erh Ya I Cheng 小爾雅義證.
    Commentary on *Hsiao Erh Ya*.
    Shang Wu, ed. SPPP, Shanghai.
Hu Chhi-Kuang (*1987*) 胡奇光
    Chung Kuo Hsiao Hsüeh Shih 中國小學史.
    History of *hsiao hsüeh* philology.
    Shang-hai Jen Min, Shanghai.
Hu Chhu-Shêng (*1964*) 胡楚生
    'Shih Ming Khao' 釋名考.
    A study of *Shih Ming*.
    *Thai-wan Shêng Shih Fan Ta Hsüeh Kuo Wên Yen Chiu
    So Chi Khan* 台灣省師范大學國文研究所集刊,
    Taipei, 8, 139–361.
Hu Shih (*1923*) 胡適
    Ming Hsüeh Chi Ku 名學稽古.
    An inquiry into logic.
    Shang Wu, Shanghai.
Hu Shih (*1929*) 胡適
    '"Erh Ju Phien", "Wu Wo Phien"' 爾汝篇, 吾我
    篇.
    Hu Shih Wên Tshun 胡適文存, 12th ed. Shang Wu,
    Shanghai, vol. 2, pp. 7–14.
Hu Tao-Ching (*1934*) 胡道靜
    Kungsun Lung Tzu Khao 公孫龍子考.
    A study of the *Kungsun Lung Tzu*.
    Shang Wu, Shanghai.

Hui Yüan (*1978*) 慧圓
    Yin Ming Ru Chêng Li Lun Chiang I 因明入正理論講
    義.
    Commentary on *Nyāyapraveśa*.
    Shang Wu, Shanghai.
Hung Chhêng (*1982*) 洪誠
    Chung Kuo Li Tai Yü Yen Wên Tzu Hsüeh 中國歷代語
    言文字學.
    The history of the study of Chinese language and
    writing.
    Chiang-su Jen Min, Nanking.
Huo Thao-Hui (*1985*) 霍韜晦
    Fo Chia Lo Chi Yen Chiu 佛家邏輯研究.
    Studies in Buddhist logic.
    Fo Kuang, Hong Kong (first ed. 1979).

Jên Chi-Yü (*1963*) 任繼愈
    Han Thang Fo Chiao Ssu Hsiang Lun Chi 漢唐佛教思
    想論集.
    Studies in Buddhist thought from Han to Thang.
    San Lien, Peking.
Jên Hung-Chün (*1915*)
    The Reason for China's Lack of Science Kho-
    hsüeh, 1915: 8–13.

Kan-su sheng po wu kuan.
    Wu-wei Han chien 甘肅博物館 (1962).
    武威漢簡. Wen wu ch'u pan she, Peking.
Kao Hêng (*1966*) 高亨
    Mo Ching Chiao Chhüan 墨經校詮.
    Commentary to the Dialectical Chapters of the
    Mo Tzu.
    Reprint Thai Ping, Hong Kong.
Kao Heng (*1979*) 高亨, *Chou I Ta Chuan Chin Chu*, 周易大
    傳今注 Chinan.
Kao Heng (*1984*) 高亨
    Chou I Ku Ching Chin I 周易古經今譯.
    A new translation of the old Book of Changes.
    Chung Hua, Peking.
Kao Ming-Khai *et al.* (*1984*) 高明凱
    Han Yü Wai Lai Tzhu Tzhu Tien 漢語外來詞詞典.
    Dictionary of foreign words in Chinese.
    Shang Wu, Shanghai.
Khuei Chi (*1896*) 窺基
    Yin Ming Ju Chêng Li Lun Shu 因明入正理論疏.
    Commentary on *Nyāyapraveśa*.
    Chin Ling Kho Ching Chhu, Nanking.
Ku Chieh Kang (*1983*) 顧頡剛
    Tshui Tung Pi I Shu 崔東壁遺書.
    The Works of Tshui Tung Pi.
    Ku Chi, Shanghai.
Ku Fang (*1981*) 谷方
    'Ho Shang Kung "Lao Tzu Chang Chü" Khao
    Chêng' 河上公老子章句考證.
    A Philological Inquiry into Hoshang Kung's
    Commentary on *Lao Tzu*.
    in *Chung Kuo Chê Hsüeh* 中國哲學 7 (1981), 41–57.
    See also vol. 9 (1983), 137–68 for an important
    response to this paper.
Kuan Han-Hêng (*1981*) 關漢亨
    'Kuan Yü Tung Chung-Shu ti Hsien Thien Kai
    Nien Shuo' 關於董仲舒的先天概念説.

On Tung Chung-Shu's notion of a priori concepts.
In Anon. (1981) *Chung Kuo Lo Chi Ssu Hsiang Lun Wên Hsüan*, pp. 587–92.

Kung Chhien-Yen (*1987*) 襲千炎
Chung Kuo Yü Fa Hsüeh Shih Kao 中國語法學史稿.
A draft on the history of Chinese grammar.
Yü Wen, Peking.

Kuo Chan-Po (*1935*) 郭湛波
Chin Wu Shih Nian Chung Kuo Ssu Hsiang Shih 近五十年中國思想史.
The history of thought over the last 50 years.
Jen Wen Shu Tien, Peiping.

Kuo Chhing-Fan (*1961*) 郭慶藩
Chuang Tzu Chi Shih 莊子集釋.
Collected commentaries on *Chuang Tzu*.
Chung Hua, Peking.

Kuo Mo-Jo 郭沫若, Wen I-Tuo 聞一多, and Hsü Wei-Yu 許維遹 (*1956*).
Kuan-Tzu Chi Chiao 管子集校.
Kho-hsüeh-chhu-pan-she, Peking.

Kuo Têng-Fêng (*1965*) 郭登峰
Li Tai Tzu Hsü Chhuan Wen Chhao 歷代字序傳文鈔.
Collection of prefaces and autobiographic notes.
Shang Wu, Taipei.

Li Fang-Kuei (*1971*) 李方桂
'Shang Ku Yin Yen Chiu' 上古音研究.
Studies in the pronunciation in high antiquity.
Tsinghua Journal of Chinese Studies, 9, 1–61.

Li Jung (*1985*) 李榮
'Chung Kuo Ku tai ti Yü Yen Hsüeh' 中國古代的語言學.
In Li Jung (1985), *Yü Wên Lun Hêng* 語文論衡, Chung Hua, Peking, 1985, pp. 1–4.

Li Yü-Shu (*1968*) 李魚叔
Mo Pien Hsin Chu 墨辯新註.
New commentary on the Dialectical Chapters of the *Mo Tzu*.
Shang Wu, Taipei.

Liang Chêng-Thing (*1973*) 梁正延
Mêng Tzu Chin Ku 孟子今詁.
New commentary on the *Mêng Tzu*.
Shang-hai Yin Shu Kuan, Hong Kong.

Liang Chhi-Chhao (*1917*) 梁啟超
Mo Hsüeh Wei 墨學微.
The subtleties of Mohism.
Shang Wu, Shanghai.

Liang Chhi-Chhao (*1922*) 梁啟超
Mo Ching Chiao Shih 墨經校釋.
Shang Wu, Shanghai.

Liang Chhi-Hsiung (*1975*) 梁啟雄
Shang Wu, Taipei.

Liang Chhi-Hsiung (*1982*) 梁啟雄
Han Fei Tzu Chhien Chieh 韓非子淺解.
Easy commentary on *Han Fei Tzu*.
Chung Hua, Peking.

Liang Shih-Chhiu (*1934*) 梁實秋
Hsi Sai Lo Wên Lu 西賽羅文錄.
A collection of Cicero's works.
Wan Yu Wên Khu, Shang Wu, Shanghai.

Lin Yü-Shan (*1983*) 林玉山
Han Yü Yü Fa Hsüeh Shih 漢語語法學史.
History of the study of Chinese grammar.
Hunan Chiao Yü, Chhang-sha.

Liu Chhi (*1966*) 劉奇
Lun Li Ku Li 論理古例.
Examples of argumentation in early China.
Shang Wu, Taipei, 1966 (first edition Shanghai 1943).

Liu Phei-Yü *et al.* (*1984*) 劉培育
Yin Ming Lun Wên Chi 因明論文集.
Collection of essays on yin ming.
Jen min, Lan-chou.

Liu Ta-Chieh (*1939*) 劉大杰
Wei Chin Ssu Hsiang Lun 魏晉思想論.
A discourse on intellectual life in Wei and Chin times.
Chung Hua, Shanghai.

Liu Tshun-Jen (=Liu Ts'un-yan) (*1964*) 劉存仁
Mo Ching Chien I 墨經箋疑.
'Comments on difficult points of the Dialectical Chapters of the *Mo Tzu*' *HYHP*, 6/1 (1964), 45–140; 7/1 (1965), 1–134.

Liu Yeh-Chhiu (*1983*) 劉葉秋
Chung Kuo Tzu Tien Shih Lüeh 中國字典史略.
Chung Hua, Peking.

Lo Chao (*1981*) 羅炤
'*Hsüan Tsang "Yin Ming Chêng Li Men Lun Pen" Nien Tai Kao* 玄奘 因明正理門論本 年化考.
On some questions concerning the date of Hsüan-Tsang's *Nyāyamukha*.
Shih Chieh Tsung Chiao Yen Chiu 世界宗教研究, no. 2, 29–36.

Lo Chao (*1988*) 羅炤
'*Yu Kuan "Chên Wei Shih Liang" ti Chi Ko Wên Thi*' 有關真唯識量的幾個問題.
Some questions concerning Consciousness Only.
Shih Chieh Tsung Chiao Yen Chiu 世界宗教研究, no. 3, 28–50.

Lü Chheng (*1935*) 呂澂
Yin Ming Hsüeh Kang Yao 因明學綱要.
A summary of yin ming.
Shang Wu, Shanghai.

Lü Chheng (*1982*) 呂澂
Chung Kuo Fo Hsüeh Ssu Hsiang Kai Lun 中國佛學思想概論.
An outline of Chinese Buddhist thought.
Thian Hua, Taipei, (Reprint).

Lü Chheng (*1983*) 呂澂
Yin Ming Ju chêng Li Lun Chiang Chieh 因明入正理論講解.
An explanation of *Nyāyapraveśa*.
Chung Hua, Peking.

Lü Chheng (*1984*) 呂澂
'Yin Ming Hsüeh Shuo Tsai Chung Kuo ti Tsui Chhu Fa Chan' 因明學説在中國的最初發展.
The earliest evolution of Buddhist logic in China.
In Liu Phei-Yü (1984) pp. 232–9.

Lü Chheng *et al.* (*1927*) 呂澂
Yin Ming Chêng Li Mên Lun Pen Chêng Wên 因明正理門論本正文.
The text of the *Nyāyamukha*.
Chin Ling Kho Ching Chhu, Nanking.

Lü Shu-Hsiang (*1958*) 呂叔湘
Han Yü Yü Fa Lun Wên Chi 漢語語法論文集.
Collected papers on Chinese grammar.
Kho Hsüeh, Peking.

Lü Shu-Hsiang (*1980*) 呂叔湘
Yü Wên Chhang Than 語文常談.
Informal remarks on language.
Chung Hua, Peking.

Lü Shu-Hsiang (*1984*) 呂叔湘
Yü Wên Tsa Chi 語文雜記.
Mixed remarks on language.
Shang Wu, Shanghai.

Lü Shu-Hsiang *et al.* (*1986*) 呂叔湘
Ma Shih Wên Thung Tu Pên 馬氏文通讀本.
A reader of *Ma Shih Wên Thung*.
Shanghai Chiao Yu, Shanghai.

Lü Ssu-Mien (*1930*) 呂思勉
Chang Chü Lun 章句論.
A treatise on sentences and paragraphs.
Shanghai, reprint Shang Wu, Taipei, 1977.

Lu Ta-Tung (*1926*) 魯大東
Mo Pien Hsin Chu 墨辯新註.
A new commentary on the Dialectical Chapters of
the *Mo Tzu*.
Chung Hua, Peking.

Luan Hsing (*1982*) 欒星
Kungsun Lung Tzu Chhang Chien 公孫龍子長箋.
A detailed commentary on *Kungsun Lung Tzu*.
Hsin Hua, Cheng-chou.

Luan Thiao-Fu (*1926*) 欒調甫
Mo Pien Thao Lun 墨辯討論.
A discussion of the Dialectical Chapters of the *Mo
Tzu*.
Shang Wu, Shanghai.

Ma Chien-Chung (*1904*) 馬建忠 (and Ma Hsiang-Po
馬相伯?)
Ma Shih Wên Thung 馬氏文通.
Mr Ma's comprehensive treatment of language.
Preface dated 1898, complete work published in
1904. Shao-hsing-fu-hsüeh-thang, Shao-hsing
1904. See Lü Shu-Hsiang *et al.* (1986) for a most
convenient indexed and commented edition.

Ma Hsü-Lun (*1957*) 馬敍倫
Shuo Wên Chieh Tzu Liu Shu Shu Chêng 說文解字六
書疏證.
Commentary on the *Shuo Wên* and the six cate-
gories of characters, Cho Hsüeh Chhu Pan She
Peking, reprint Ting Wen, Taipei, 1975.

Mei Kuang-Hsi (*1935*) 梅光羲
Yin Ming Ju Chêng Li Lun Chieh Lu Chi Chu 因明入正
理論節錄集注.
Selected passages and collected commentaries on
Nyāyapraveśa (no place, no date,) Shanghai (*c.* 1935).

Mi Lin (*1935*) 密林
Yin Ming Ju Chêng Li Lun I Chieh 因明入正理論易解.
Easy commentary on *Nyāyapraveśa*.
(no publisher) Shanghai.

Morohashi Tetsuji (*1955*) 諸橋轍次
Dai-Kanwajiten 大漢和辭典.
Dictionary of Chinese Characters in Japanese.
13 vols., Taishūkan, Tokyo, 1955–60.

Mou Tsung-San (*1960*) 牟宗三
'Wei Chin Ming Li Chêng Ming' 魏晉名理正名.
Logical debate during the Wei and Chin periods.
Hsin Ya Shu Yüan Hsüeh Shu Nien Khan 新亞書院
學術年刊 2, 1–43.

Mou Tsung-San (*1963*) 牟宗三
Wei Chin Hsüan Hsüeh 魏晉玄學.
'Dark Learning' in Wei and Chin times.
Jen Sheng, Taipei.

Nishida Tatsuo (*1970*) 西田龍雄
Seibankan yakugo no kenkyū 西番館譯語の研究.
Studies in the translations by the Office of the
Western Barbarians.
Shokado, Kyōtō.

Niu Hung-En *et al.* (*1984*) 牛鴻恩
Chan Kuo Tshê Hsüan Chu 戰國策選注.
Selection from and commentary on *Chan Kuo
Tshê*.
Ku Chi, Tien-chin.

Ono Gemmyō (*1936*) 小野玄妙
Bussho kaisetsudaijiten 佛書解説大辭典.
Explanatory dictionary of Buddhist books.
2nd ed., 12 vols. reprint Tokyo 1964–78 in 13 vols.
plus supplement.

Ōta Tatsuō (*1958*) 太田辰夫
Chūgokugo rekishi bumpō 中國語歷史文法.
Tokyo, Chinese translation: *Chuang Kuo Yü Li
Shih Yü Fa* 中國語歷史語法, Chung Hua,
Peking, 1987.

Phang Phu (*1974*) 龐樸
Kungsun Lung Tzu I Chu 公孫龍子譯注.
Translation and commentary of *Kungsun Lung
Tzu*.
Chung Hua, Peking.

Phang Phu (*1979*) 龐樸
Kungsun Lung Tzu Yen Chiu 公孫龍子研究.
Studies in *Kungsun Lung Tzu*.
Chung Hua, Peking.

Phang Phu (*1982*) 龐樸
'Kungsun Lung Phing Chuan'
A critical biography of Kungsun Lung.
In *Chhên Ssu Lu* 沉思錄, Jen Min, Shanghai, pp.
226–152.

Pi Hua-Chên (??)
Lun Wên Chhien Shuo 論文淺説.
A simple explanation of text.
No edition located, but the work was reviewed by
A. V. Rosthorn.

Pi Hua-Chên (??)
Lun Wên Hsü Shuo 論文續説.
Continued explanation of text.
No edition located, but the work was reviewed by
A. V. Rosthorn.

Shen Chien-Ying (*1984*) 沈劍英
'"Chên Wei Shih Liang" Lüeh Lun' 真唯識量
略論.
In Liu Phei-Yü (1984), pp. 312–8; see also *ibid.* pp.
63–83, 131–48, and 165–79.

Shen Wen-Cho (*1987*) 沈文倬 ed.
　Chiao Hsün 焦循
　Mêng-Tzu cheng-i 孟子正義.
　Chung Hua, Peking.
Shih Tshun (*1981*) 石村
　Yin Ming Shu Yao 因明述要.
　A summary of yin ming logic.
　Chung Hua, Peking.
Shima Kuniō (*1973*) 島邦男
　Rōshi kōsei 老子校正.
　Kyūkō shōin, Tokyo.

Tai Ming-Yang (*1962*) 戴明楊
　Hsi Khang Chi Chiao Chu 嵇康集校注.
　Commentary on Hsi Khang's collected works.
　Chung Hua, Peking.
Than Chieh-Fu (*1957*) 譚戒甫
　Kungsun Lung Tzu Hsing Ming Fa Wei 公孫龍子形名
　發微
　The subtleties in the logic of the *Kungsun Lung Tzu*.
　1st ed., new ed., Chung Hua, Peking, 1963.
Than Chieh-Fu (*1958*) 譚戒甫
　Mo Pien Fa Wei 墨辯發微.
　The subtleties in Mohist logic.
　Chung Hua, Peking.
Than Chieh-Fu (*1981*) 譚戒甫
　Mo Ching Fên Lei I Chu 墨經分類譯注.
　Classification, commentaries and translation of
　the Dialectical Chapters of the *Mo Tzu*.
　Chung Hua, Peking.
Thang Chün-I (*1963*) 唐君毅
　'Hsün Tzu Chêng Ming yü Hsien Chhin Ming
　Hsüeh San Tsung' 荀子正名與先秦名學三宗.
　Hsün Tzu's logic and the three schools of pre-
　Chhin logic.
　HYHP, 5, no. 2, 1–22.
Thang Yung-Thung (*1957*) 湯用彤
　Wei Chin Hsüan Hsüeh Lun Kao 魏晉玄學論稿.
　A draft on Wei and Chin 'Dark Learning'.
　Jen Min, Peking.
Ting Fu-Pao (*1928*) 丁福保
　Fo Hsüeh Ta Tzhu Tien 佛學大辭典.
　Large dictionary of Buddhism.
　Reprint, Wen Wu, Peking, 1984.
Ting Fu-Pao (*1930*) See Bibliography A *sub Shuo Wên
　Chieh Tzu Chuan Yün Phu.*
Ting Sheng-Shu (*1935*) 丁聲樹
　'Shih fou-ting tzhu "fu" "pu"' 'Explaining the
　negations pu and fu' 釋否定詞弗不, in: *Chhing
　Chu Tshai Yüan-Phei Hsien Sheng Liu Shih Wu Sui
　Lun Wen Chi* 慶祝蔡元培先生六十五歲論文集,
　vol. 2, pp. 967–97.
Ting Yen-Po (*1961*) 丁彥博
　'Lüeh Lun Yin Ming Chêng Li ti Hsien Tai I I'
　略論因明正理的現代意義 A brief account of
　the contemporary significance of the *Nyāyapraveśa*.
　Wên Hui Pao 文匯報, Shanghai, March 28.
Tso Ching-Chhüan (*1987*) 左景權
　'Man Than Hsi La Ku Tien Ming Chu ti Fan I'
　慢談希臘古典名著的翻譯.
　Informal discussion on the translation of Greek
　classics.

Shê Hui Kho Hsüeh Chan Hsien 社會科學戰線 1987.
　1, pp. 224–32.
Tu Kuo-Hsiang (*1962a*) 杜國庠
　'Kuan Yü Yin Ming' 關於因明.
　Regarding yin ming logic.
　In *Tu Kuo-Hsiang Wên Chi* 杜國庠文集, Jen Min,
　Peking.
Tu Kuo-Hsiang (*1962b*) 杜國庠
　'Yin Ming Mei You Tho Li Jen Shih Lun ti Li
　Chhang' 因明沒有脱離認識論的立場.
　In *Tu Kuo-Hsiang Wên Chi* 杜國庠文集, Jen Min
　Peking.
Tuan Yü-Tshai (*1981*) 段玉裁
　Shuo Wên Chieh Tzu Chu 説文解字注.
　Reprint, Ku Chi, Shanghai.

Ushijima Tokuji (*1971*) 牛導德次
　Kango bumpō ron: chūko hen 漢語文法論: 中古篇.
　A study of Chinese grammar. The middle ages.
　Taishokan, Tokyo.

Wang Jung-Pao (*1985*) 王榮寶
　Fa Yen Shu I 法言疏義.
　Reprint World Book Company, Taipei, 1985 (pre-
　face 1873).
Wang Khai-Luan (*1935a*) 王愷鑾
　Têng Hsi Tzu Chiao Chêng 鄧析子校證.
　Commentary on *Têng Hsi Tzu*.
　Shang Wu, Shanghai.
Wang Khai-Luan (*1935b*) 王愷鑾
　Yin Wên Tzu Chiao Chêng 尹文子校證.
　Commentary on *Yin Wên Tzu*.
　Shang Wu, Shanghai.
Wang Li (*1958*) 王力
　Han Yü Shih Kao 漢語史稿.
　A sketch of the history of the Chinese language.
　3 vols., Chung Hua, Peking.
Wang Li (*1959*) 王力
　Chung Kuo Hsien Tai Yü Fa 中國現代語法.
　Modern Chinese Grammar.
　2 vols., reprint, Chung Hua, Hong Kong.
Wang Li (*1981*) 王力
　Chung Kuo Yü Yen Hsüeh Shih 中國語言學史.
　History of linguistics in China.
　Shan-hsi Jen Min, Tai-yüan.
Wang Li (*1979*) 王力
　Ku Han Yü Chhang Yung Tzu Tzu Tien 古漢語常用字
　字典.
　Dictionary of current ancient Chinese charac-
　ters.
　Shang Wu, Peking.
Wang Li (*1984*) 王力
　Wang Li Wên Chi, 王力文集.
　Shanghai: Commercial Press.
Wang Li (*1987*) 王力
　Thung Yüan Tzu Tien 同源字典.
　Dictionary of cognate characters.
　Shang Wu, Peking.
Wang Li-Chhi (*1981*) 王利器
　Li Tai Hsiao Hua Chi 歷代笑話集.
　A collection of jokes through the ages.
　Ku Chi, Shanghai.

Wang Sung-Mao ed. (*1983*) 王松茂
    Han Yü Yü Fa Yen Chiu Tshan Khao Tzu Liao 漢語語法研究參考資料.
    Materials for the study of Chinese grammar.
    Shê Hui Kho Hsüeh, Peking.

Wang Tien-Chi (*1979*) 王奠基
    Chung Kuo Lo Chi Ssu Hsiang Shih 中國邏輯思想史.
    History of Chinese logical thought.
    Shanghai Jên Min, Shanghai.

Wang Tien-Chi (*1981*) 王奠基
    'Lüeh Than Chung Kuo Ku Tai "Thui Lei" yü "Lien Chu Shih"' 略談中國古代推類與聯珠式.
    A general discussion on the Chinese method of generalising by analogy and on the "form of interlinked pearls'.
    In Anon. (1981) *Chung Kuo Lo Chi Ssu Hsiang Lun Wên Hsüan*, pp. 471–507.

Wang Yin-Chih (*1831*) 王引之
    Ching Chuan Shih Tzhu 經傳釋詞.
    Explanation of grammatical particles in the classics.
    First ed. 1819, many reprints. Very convenient is *Chu Tzu Pian Lüeh Têng Liu Chung* 助字辨略等六種, Shih Chieh, Taipei, 1974.

Wang Yin-Chih (*1831*) 王引之
    Tzu Tien Kao Chêng 字典考證.
    An investigation into the (Khang Hsi) dictionary.
    12 *chüan*, reprint I Wên, Taipei, 1959.

Wu Chien-Kuo (*1963*) 吳建國
    Chung Kuo Lo Chi Shih Shang Lei Kai Nien ti Fa Sheng ho Fa Chan 中國邏輯史上類概念德發生和發展.
    The origin and development of the concept of a category in Chinese logic.
    In *Chung Kuo Che Hsüeh Shih Lun Wen Chi, Erh Chi* 中國哲學史論文集，二集, Chung Hua, Peking, pp. 100–25.

Wu Ching-Hsiung (*1948*) 吳經熊
    Hsin Ching Chhüan Chi 新經全集.
    Complete New Testament.
    Kung Chiao Chên Li Hsüeh Hui, Hong Kong.

Wu Fei-Po (*1923*) 伍非百
    Mo Pien Chieh Ku 墨辯解故.
    Explanation of Mohist dialectic.
    Chung Hua, Peking.

Wu Fei-Po (*1984*) 伍非百
    Chung Kuo Ku Ming Chia Yen 中國古名家言.
    Ancient logical texts from China.
    Chung Kuo Kho Hsüeh, Peking.

Wu Hsien-Shu, tr. (*1929*) 吳獻書
    Li Hsiang Kuo 理想國.
    (Plato's) Republic.
    Shang Wu, Shanghai, many reprints.

Wu Ju-Chün (*1984*) 吳汝鈞
    Fo Hsüeh Yen Chiu Fang Fa Lun 佛學研究方法論.
    The methodology of Buddhist studies.
    Hsüeh Sheng Shu Chü, Taipei.

Wu Kuo-I (*1982*) 鄔國義
    '"Ma Shih Wên Thung" ti Shih Chi tso Chê Shih Ma Hsiang-Po Ma?' 馬氏文通的實際作者是馬相伯嗎?.

Was the real author of the *Ma Shih Wên Thung* Ma Hsiang-Po?
    In *Hsüeh Lin Man Lu* Chung Hua, Beijing, 5, pp. 186–9.

Wu Tsung (*1988*) 伍宗
    Ku Tai Han Yü Thi Chieh Tzhu Tien 古代漢語題解辭典.
    Topical dictionary of ancient Chinese.
    Ssu-chhuan tzhu-shu, Chheng-tu.

Wu Yü-Chiang (*undated*) 吳毓江
    Mo Tzu Chiao Chu 墨子校注.
    Comparative Notes on *Mo Tzu*.
    Hsi Nan Shih Fan Ta Hsüeh, Chhung Chhing.

Yang Khuan (*1942*) 楊寬
    Mo Ching Chê Hsüeh 墨經哲學.
    The philosophy of the Dialectical Chapters of the *Mo Tzu*.
    Shanghai, reprint Cheng Chung, Shanghai, 1959.

Yang Po-Chun (*1961*) 楊伯峻, *Mêng Tzu I Chu*, 孟子譯注.
    Translation and commentary on Mencius.
    Chung Hua, Peking.

Yang Po-Chün (*1984*) 楊伯峻
    Chhun Chhiu Tso Chuan Chu 春秋左傳注.
    Commentary to the Chhun Chhiu and the Tso Chuan.
    Chung Hua, Peking.

Yang Shu-Ta (*1954*) 楊樹達
    Ku Shu Chü Tou Shih Li 古書句讀釋例.
    Chung Hua, Peking.

Yang Wên-Hui (*1923*) 楊文會
    Yang Jen-Shan Chü Shih I Chu 楊仁山居士遺著.
    Posthumous works by Yang Jen-Shan.
    Privately published, no place.

Yen Chhung-Hsin (*1977*) 閆崇信
    Mo Tzu Ta Chhü Phien Chiao Shih 墨子大取篇校釋.
    Commentary on the Great Pick in the *Mo Tzu*.
    Wên Shih Chê, Taipei.

Yen Ling-Fêng (*undated*) 嚴靈峰
    Lao Tzu Chi Chhêng 老子集成.
    The Collected Works of Master Lao.
    I Wen, Taipei.

Yen Phei (*1978*) 演培
    Chheng Wei Shih Lun Chiang Chi 誠唯識論講記.
    Commentary on the treatise on Consciousness Only.
    5 vols., Hui Jih Chiang Thang, Taipei.

Yin Shun (*1950*) 印順
    Thai Hsü Ta Shih Nien Phu 太虛大師年譜.
    Chronological biography of Thai Hsü.
    Hong Kong, reprint Hai Chhao, Taipei.

Yü Hsing-Wu (*1972*) 于省吾
    'Kuan Yü Ku Tai Wên Tzu Yen Chiu ti Jo Kan Wên Thi' 關於古代文字研究的若干問題.
    Some questions regarding the study of ancient characters.
    Wên Wu, no. 2, 32–5.

Yü Min (*1987*) 俞敏
    Ching Chuan Shih Tzhu Cha Chi 經傳釋詞札記.
    Miscellaneous notes on the *Ching Chuan Shih Tzhu*.
    Hu-nan Chiao Yu, Chhang-sha.

C. BOOKS AND JOURNAL ARTICLES IN WESTERN LANGUAGES

ABBOT, E. A. (1913). *A Shakespearean Grammar. An Attempt to Illustrate some of the Differences Between Elizabethan and Modern English*. Macmillan, London.

ABEGG, LILY (1952). *The Mind of East Asia*. Thames and Hudson, London.

ABEL-RÉMUSAT, J. P. (1822). *Elémens de la grammaire chinoise*. Maisonneuve, Paris.

ABEL-RÉMUSAT, J. P. (1826). 'De l'étude des langues étrangères chez les chinois', in *Mélanges asiatigues*, 2 vols. Dondey–Dupré, Paris, 1825/6.

ALBRECHT, MICHAEL VON (1964). 'Die Parenthese in Ovids Metamorphosen und ihre dichterische Funktion.' *Spudasmata* vol. vii, Hildesheim, Olms.

ALEKSEEV, V. M. (1978). *Kitajskaja literatura*. Nauka, Moskow.

AMES, ROGER T. (1983). *The Art of Rulership: A Study in Ancient Chinese Political Thought*. University of Hawaii Press, Honolulu.

AMES, ROGER T. & HALL, D. L. (1987). *Thinking Through Confucius*. Suny Press, New York.

ANON. (1616). *Commentarii Collegii Conimbricensis e Societate Jesu in universam Dialecticam Aristotelis Stagiritae*. Cardon, Lugduni.

ANON. (1717). *Lettres édifiantes et Curieuses écrites des missions étrangères par quelques missionaires de la Compagnie de Jésus*. 31 vols. N. Le Clerk, Paris, 1717–74.

ANSELM OF CANTERBURY (1964). *De Grammatico*. Ed. D. P. Henry, Univ. of Notre Dame Press, Notre Dame.

AQUINAS, THOMAS (1961). *Summa theologiae*. Ed. L. R. Carcedo, O. P., La Editorial Catolica, Madrid.

AQUINAS, THOMAS (1967). *Summa contra gentiles*. Ed. L. R. Carcedo, O. P. La Editorial Catolica, Madrid.

AREND, CARL (1989). 'Has Chinese a Grammar, and, if so, is it Worth being Studied?'. In *Proceedings of the 12th International Orientalist Congress*. Vol. 2, pp. 41–7. Einaudi, Rome.

ARENS, H. (1955). *Sprachwissenschaft: der Gang ihrer Entwicklung von der Antike bis zur Gegenwart*. K. Alber, Freiburg/ Munich.

ARISTOPHANES (1968). *Clouds*. Ed. K. J. Dover. Oxford University Press, Oxford.

ARISTOTLE (1894). *Nichomachean Ethics*. Ed. I. Bywater, Oxford University Press, London.

ARISTOTLE (1918). *Prior Analytics*. Ed. H. Tredennick. Loeb Classical Library, Heinemann, London, 1918.

ARISTOTLE (1922). *Eudemian Ethics*. Ed. R. R. Walzer, Oxford University Press, London.

ARISTOTLE (1957). *Metaphysics*. Ed. W. D. Ross, Oxford University Press, London.

ARISTOTLE (1958). *Sophistici Elenchi*. Ed. W. D. Ross, Oxford University Press, London.

ARISTOTLE (1959). *Rhetoric*. Ed. W. D. Ross. Oxford University Press, Oxford.

ARISTOTLE (1963). *Categories and De Interpretatione*, translated with notes and glossary by J. L. Ackrill. Clarendon Press, Oxford.

ARISTOTLE (1970). *Problemata*. Ed. W. S. Hett, *et al.*, Loeb Classical Library. Heinemann, London.

AX, W. (1965) (ed.). *M. Tullius Cicero, De Divinatione, De Fato, Timaeus*. Teubner, Stuttgart.

AYLMER, CHARLES (1981). *Origins of the Chinese Script*. East Asia Books, London.

BACON, FRANCIS (1905). *The Advancement of Learning*. Ed. G. W. Kitchin. Everyman's Library XI. 1, J. M. Dent and Sons, London.

BAGCHI, PRABODH CHANDRA (1937). *Deux Lexiques Sanskrit-Chinois*. Paul Genthner, Paris, 1929 & 1937.

BALASZ, E. (1964). *Chinese Civilization and Bureaucracy*. Yale University Press, New Haven & London.

BALDWIN, J. M. (1902). *Dictionary of Philosophy and Psychology*. Macmillan, London & New York.

BALLARD, W. L. (1979). *Chinese, a Bastard of the Sino-Tibetan Family Reunion*. Maisonneuve, Paris.

BALLMER, T. T. & PINKAL, M. (1983) (eds.). *Approaching Vagueness*. North Holland Publishing Company, Amsterdam, New York & Oxford.

BARLINGAY, S. S. (1965). *A Modern Introduction to Indian Logic*. National Publishing House, Delhi.

BARTOLD, V. V. (1925). *Istorija izuchenija Vostoka v Evrope i Rossii*. Nauka, Leningrad.

BAXTER, WILLIAM III. (1986). 'Some Proposals on Old Chinese Phonology'. In McCoy & Light (1986), pp. 1–34.

BAYER, T. S. (1730). *Museum Sinicum, in quo Sinicae linguae et literaturae ratio explicatur*. Typographia Imperatoria, St. Petersburg.

BAZIN, ANTOINE PIERRE LOUIS (1856). *Grammaire mandarine*. Imprimerie Impériale, Paris.

BEAL, SAMUEL (1911). *The Life of Hsüan-tsang*. Kegan Paul, Trench and Trubner, London.

BEASLEY, W. G. & PULLEYBLANK, E. G. (1961). *Historians of China and Japan*. Oxford University Press, London.

BECKER, CARL B. (1986). 'Reasons for the Lack of Argumentation and Debate in the Far East'. In *International Journal of Intercultural Relations*, **10**, 75–92.

BENEDICT, PAUL K. (1972). *Sino-Tibetan, A Conspectus*. Cambridge University Press, Cambridge.

BENEDICT, PAUL K. (1975). *Austro-Thai*. Yale University Press, New Haven.

BENVENISTE, EMILE (1971). *Problems in General Linguistics*. University of Miami Press, Coral Gables.

BHIMACCARYA, JHALIKAR (1928). *Nyāyakoṣa, or Dictionary of the Technical Terms of the Nyāya Philosophy*. Bandarkar Oriental Research Institute, Poona.

BIALLAS, F. X. (1934). 'Recent Studies in Chinese Lexicography'. *Bulletin of the Catholic University of Peking*, **9**, 183–6.

BICHURIN, NIKITA JAKOJALEVICH (1832). *Kitajskaja grammatika, sochinennaja monakhom Iakinfom*. Mission Press, Peking.

BIELING, ALEXANDER (1880). *Das Prinzip der deutschen Interpunktion, nebst einer übersichtlichen Darstellung ihrer Geschichte*. Wiedmann, Berlin.

BILFINGER, G. B. (G. B. BÜLFFINGERIUS) (1724). *Specimen doctrinae veterum Sinarum moralis et politicae*. Theodor Mezler, Frankfurt.

BISHOP, J. L. (1965) (ed.). *Studies in Chinese Literature*. Harvard University Press, Cambridge, Mass.

BLASS, F., DEBRUNNER, A. & REHKOPF, F. (1979). *Grammatik des neutestamentlichen Griechisch*. 15th ed. Göttingen. 1st ed. by F. Blass, 1896.

BLOOM, ALFRED (1981). *The Linguistic Shaping of Thought. A Study in the Impact of Language on Thinking in China and the West*. Hillsdale, New Jersey.

BOCHENSKI, I. M. (1947). *La logique de Théophraste*. Librairie de L'Université, Fribourg.

BOCHENSKI, I. M. (1951). *Ancient Formal Logic*. North Holland Publishing Company, Amsterdam.

BOCHENSKI, I. M. (1956). *Formale Logik*. K. Alber, Freiburg, 1956. Eng. tr. University of Notre Dame Press, Notre Dame, Ill.

BOCHENSKI, I. M. (1960). *A History of Formal Logic*. University of Notre Dame Press, Notre Dame, Ill.

BODDE, DERK (1936). 'The Attitude toward Science and Scientific Method in Ancient China.' *TH*, **2**, 139, 160.

BODDE, DERK (1938). *Li Ssu, China's First Unifier, a Study of the Ch'in Dynasty as seen in the life of Li Ssu (–280–208)*. (Sinica Leidensia No. 3). Brill, Leiden.

BODDE, DERK (1939). 'Types of Chinese Categorical Thinking'. *JAOS*, **59**, 200–19.

BODMAN, NICHOLAS C. (1954). *A Linguistic Study of the Shih Ming, Initials and Consonant Clusters*. Harvard-Yenching Institute Studies XI, Harvard University Press, Cambridge, Mass.

BODMAN, NICHOLAS C. (1967). 'Historical linguistics'. In T. A. Sebeok (ed.) (1967), pp. 3–58.

BODMAN, NICHOLAS C. (1968). 'Chinese and Sino-Tibetan: evidence towards establishing the nature of their relationship'. In T. Light ed., (1968), pp. 34–199.

BOEHNER, PHILOTHEUS (1966). *Medieval Logic, An outline of its development from 1250–c. 1400*. Manchester University Press, Manchester.

BOETHIUS, DACIUS (1969). *De modis significandi*. Ed. J. Pinborg, Gad, Copenhagen.

BOLTZ, W. G. (1986). 'Early Chinese Writing'. In *World Archaeology*, **17**, 3, 420–35.

BOODBERG, PETER A. (1940). ' "Ideography" or iconolatry?', *TP*, **35**, 266–288.

BOODBERG, PETER A. (1957). 'The Chinese script: An essay on nomenclature (The first hecaton)' *BIHP*, **29**, 113–20.

BORST, ARNO (1963). *Der Turmbau zu Babel. Geschichte der Meinungen über Ursprung und Vielfalt der Sprachen und Völker*. 4 vols. Klett Verlag, Stuttgart, 1957–63.

BRACE, A. G. (1925). *Five Hundred Proverbs Commonly Used in West China*. Commercial Press, Shanghai.

BRAINE, M. D. S. (1978). 'On the relation between the natural logic of reasoning and standard logic', *Psychological Review*, **85**, 1–21.

BRIERE, O. (1956). *Fifty years of Chinese philosophy (1898–1950)*. Praeger, New York.

BROUGH, J. (1951). *Selections from Classical Sanskrit Literature. With English Translation and Notes*. Luzac, London.

BROWN, T. JULIAN (1984). 'Punctuation'. In *Encyclopaedia Britannica*, 15th ed., vol. 15, pp. 274–7.

BURY, J. B. (1955). *The Idea of Progress: An Inquiry into its Origin and Growth*. Dover, New York.

CALEPINO, AMBROGIO (1567). *Ambrosii Calepini Dictionarium*. Aldus, Venice.

CAMMANN, SHUYLER (1953). 'Old Chinese Magic Squares', *Sinologica*, **7**, 14–53.

CARDONA, GEORGE (1976). *Pāṇini, A Survey of Research*. Mouton, The Hague, 1976.

CHAN, WING-TSIT (1953). *Religious Trends in Modern China*. Columbia University Press, New York.

CHAN, WING-TSIT (1963). *A Source Book in Chinese Philosophy*. Tr. and compiled by Wing-tsit Chan. Princeton University Press, Princeton.

CHAN, WING-TSIT (1986). *Neo-Confucian Terms Explained: the Pei Hsi Tzu-I by Ch'en Ch'un, 1159–1223*. Columbia University Press, New York.

CHANG, KHUN (1967). 'National Languages'. In T. A. Sebeok (ed.), (1967), pp. 151–76.

CHANG, K. C. (1983). *Art, Myth and Ritual. The Path to Political Authority in Ancient China*. Harvard University Press, Cambridge, Mass.

CHANG, TSUNG-TUNG (1970). *Der Kult der Shang-Dynastie im Spiegel der Orakelinschriften. Eine paläographische Studie zur Religion im archaischen China*. Harrassowitz, Wiesbaden.

CHANG, TSUNG-TUNG (1972). *Die Bildungsregeln und Strukturen der altchinesischen Schriftzeichen*. Kitzinger, München.

CHANG, TSUNG-TUNG (1982). *Metaphysik, Erkenntnis und praktische Philosophie im Chuang Tzû, Zur Neu-Interpretation und systematischen Darstellung der klassischen chinesischen Philosophie*. Klostermann, Frankfurt am Main.

CHANG, TSUNG-TUNG (1988). *Indo-European Vocabulary in Old Chinese. A New Thesis on the Emergence of the Chinese Langauge and Civilisation in the Late Neolithic Age*. Sino-Platonic Papers no. 7, Pennsylvania.

CHANG, TUNG-SUN (1931). 'A Chinese philosopher's theory of knowledge', *Teaching Journal of Social Studies*, **1**, 155ff.

CHAO, Y. R. (1940). 'A Note on an Early Logographic Theory of Chinese Writing', *HJAS*, **5**, 189–91.

CHAO, Y. R. (1946). 'The Logical Structure of Chinese Words', *Language*, **22**, 4–13.

CHAO, Y. R. (1955). 'Notes on Chinese Logic and Grammar', *Philosophy of Science*, **2**, 31–41.

CHAO, Y. R. (1968). *Language and Symbolic Systems*. Cambridge University Press, Cambridge.

CHAO, Y. R. (1976). *Aspects of Chinese Socio-Linguistics*. Stanford University Press, Stanford.

CHATTERJI, S. K. (1939). 'Two Sanskrit Chinese Lexicons of the 7th to 8th Centuries, and some Aspects of Indo-Aryan Linguistics', *New Indian Antiquary*, 1939–40, **2**, 740–7.

CHAVANNES, E. (1969). *Les Mémoires Historiques de Se-ma Ts'ien*, vols. 1–6. Leroux et Maisonneuve, Paris, 1895–1905 and 1969.

CHEN, CHENG-YIH (1987). *Science and Technology in China*. World Scientific Publishers, Singapore.

CHEN, KENNETH (1960). 'Some Problems in the Translations of the Buddhist Canon', *Tsing Hua Journal of Chinese Studies*, New Series, **1**, May.

CHEN, KENNETH (1964). *Buddhism in China*. Princeton University Press, Princeton.

CH'EN, SHOU-YI (1935). 'John Webb: 17th-century English Sinologist. A Forgotten Page in the Early History of Sinology in Europe'. In *Chinese Social and Political Science Review*, **19**, 295–430.

CHENG, CHUNG-YING (1965). 'Inquiries into Classical Chinese Logic', *Philosophy East and West*, **15**, 195–216.

CHENG, CHUNG-YING (1970). 'Logic and ontology in the Chih Wu Lun of Kung-Sun Lung-Tzu' in *Philosophy East and West*, **20**, 2, 137–54.

CHENG, CHUNG-YING (1971). 'Toward a Theory of Subject Structure in Language with Application to Late Archaic Chinese', *JOAS*, **91**, 1, 1–13.

CHENG, CHUNG-YING (1972). 'A Generative Unity: Chinese Language and Chinese Philosophy'. Paper presented at the 1972 annual meeting of the Association for Asian Studies in New York City, March 29.

CHENG, CHUNG-YING (1975). 'On Implication (tse) and Inference (ku) in Chinese Grammar and Chinese Logic', *JCP*, **2**, 3, June, pp. 225–44.

CHENG, YAT-SHING (1976). 'Word Order Change in Chinese: some Contributing Factors and Implications', PhD thesis, University of California, San Diego.

CHEUNG, KWONG-YUE (1983). 'Recent Archeological Evidence Relating to the Origin of Chinese Characters'. In *The Origins of Chinese Civilization*, D. N. Keightley (ed.), Berkeley, pp. 323–91.

CHI, RICHARD S. Y. (1969). *Buddhist Formal Logic*. Luzac, London.

CHMIELEWSKI, JANUSZ (1949). 'The Typological Evolution of the Chinese Language', *RO*, **15**, 371–429.

CHMIELEWSKI, JANUSZ (1957). 'The Problem of Syntax and Morphology in Chinese', *RO*, **21**, 71–84.

CHMIELEWSKI, JANUSZ (1957). 'Remarques sur le problème des mots dissyllabiques en chinois archaïque'. In *Mélanges publiés par l'Institut des Hautes Etudes Chinoises*, Paris, vol. 1, pp. 423–45.

CHMIELEWSKI, JANUSZ (1958). 'The Problem of Early Loan Words in Chinese as Illustrated by the Word *p'u t'ao*' *RO*, **22**, 7–45.

CHMIELEWSKI, JANUSZ (1961). 'Two Early Loan-words in Chinese', *RO*, **24**, 65–86.

CHMIELEWSKI, JANUSZ (1964a). 'Syntax and Word-formation in Chinese', *RO*, **28**, 107–25.

CHMIELEWSKI, JANUSZ (1964b). 'Jesyzk starochinski jako narzedzie rozumowania'. In *Sprawozdania s prac naukowych, Wydzialu 1, Polskiej Akademii Nauk*, **8**, 2, 108–33.

CHMIELEWSKI, JANUSZ (1969). 'Notes on Early Chinese Logic', *RO*, Part I: 26, no. 1 (1962), 7–22; Part II: 26, no. 2 (1963), 91–105; Part III: 27, no. 1 (1963), 103–21; Part IV: 28, no. 2 (1965), 87–111; Part V: 29, no. 2 (1965), 117–38; Part VI: 30, no. 1 (1966), 31–52; Part VII: 31, no. 1 (1968), 117–36; Part VIII: 32, no. 2 (1969), 83–103.

CHMIELEWSKI, JANUSZ (1979). 'Concerning the Problem of Analogic Reasoning in Ancient China', *RO*, **40**, 2, 64–78.

CHOMSKY, NOAM (1965). *Aspects of the Theory of Syntax*. M.I.T. Press, Cambridge, Mass.

CHOU, FA-KAO (1964). 'Word Classes in Classical Chinese'. In *Proceedings of the 9th International Congress of Linguists 1962*, The Hague, pp. 594–8.

CHOU, FA-KAO (1973). *Pronouncing Dictionary of Chinese*. Chinese University of Hong Kong Press, Hong Kong.

CHOW, TSE-TSUNG (1989). *Wen-Lin*, vol. 2, Chinese University Press, Hong Kong.

CIBOT, PIERRE MARTIAL (1773). *Lettre de Pékin, sur le génie de la langue chinoise, et la nature de leur écriture symbolique, comparée avec celle des anciens égyptiens*. J. L. de Boubers, Brussels.

CICERO. (1914). *De finibus bonorum et malorum*. Ed. H. Rackham, Loeb Classical Library, Heinemann, London, etc.

CICERO. (1965). *Tusculanae disputationes*. Ed. M. Pohlwu, Teubner, Leipzig.

CICERO. (1973). *Ad Marcum Brutum Orator*. Ed. J. E. Sandys, Olms, New York.

CIKOSKI, JOHN S. (1970). 'Classical Chinese Word Classes'. PhD thesis, Yale University.

CIKOSKI, JOHN S. (1975). 'On Standards of Analytic Reasoning in the Late Chou', *JCP*, **2**, 3, 325–57.

CIKOSKI, JOHN S. (1978). 'Three Essays on Classical Chinese Grammar'. In *Computational Analyses of Asian and African Languages*, Tokyo, **8**, 17–51 & **9**, 77–208.

CLARK, E. V. & CLARK, H. (1979). 'When Nouns Surface as Verbs', *Language*, **55**, 4, 767–811.

CLEMOES, PETER (1952). *Liturgical influence on punctuation in Late Old English and Early English manuscripts*. Cambridge University, Department of Anglo-Saxon, Occasional Papers no. 1.

COBLIN, W. S., JR. (1972). 'An Introductory Study of Textual and Linguistic Problems in Erh-ya'. PhD thesis, Seattle.

COHEN, BENJAMIN CHARLES (1982). 'Understanding Natural Kinds'. PhD thesis, Stanford.

COMRIE, BERNHARD (1981). *Language Universals and Language Typology*. Blackwell, Oxford.

CONRADY, A. (1933). 'Über einige altchinesische Hilfswörter', *AM*, 1926, **3**, 491–525, & 1933, **8**, 510–18.

COOPER, ARTHUR (1978). *The Creation of the Chinese Script*. China Society, London.

COPI, IRVING M. (1953). *Introduction to Logic*. 5th ed. Macmillan, New York.

CORDIER, H. (1895a). *Bibliotheca Sinica; Dictionnaire bibliographique des Ouvrages relatifs à l'Empire Chinois*. 3 vols. Ed. des Laugues Orientales Vivantes, Paris, 1878–95. In ed. 5 vols. Pr. Vienna, 1904–24.

CORDIER, HENRI (1895b). 'Fragments d'une histoire des études chinoises au XVIIIe siècle'. In *Centenaire de l'Ecole des Langues Orientales Vivantes 1775–1875*. Maisonneuve, Paris.

CORNELIUS, PAUL (1965). *Languages in Seventeenth- and Early Eighteenth-Century Imaginary Voyages*. Librairie Drot, Geneva.

COUSIN, VICTOR (1836). *Oeuvres inédits d'Abélard*. Imprimerie royale, Paris.

COUVREUR, F. S. (1947). *Dictionnaire classique de la langue chinoise*. Henry Vetch, Peiping, 1947.

COUVREUR, F. S. (1951). '*Liki ou Mémoires sur les Bienséances et les Cévémonies (Li Chi.)* 2 vols. Repr. Belles Lettres, Paris, 1951.

COUVREUR, F. S. (1951a). *Les quatre livres*. Cathasia, Paris.

COUVREUR, F. S. (1951b). *Tch'ouen Ts'iou et Tso Tchouan. La chronique de la principauté de Lou*. 3 vols. Cathasia, Paris.

CREEL, H. G. (1936). 'On the Nature of Chinese Ideography', *TP*, **32**, 85–161.

CREEL, H. G. (1938). 'On the Ideographic Element in Ancient Chinese', *TP*, **2.34**, 269–94.

CREEL, H. G. (1970). *What is Taoism? and other Studies in Chinese Cultural History*. University of Chicago Press, Chicago.

CREEL, H. G. (1974). *Shen Pu-hai. A Chinese Political Philosopher of the Fourth Century BC*. University of Chicago Press, Chicago.

CRUMP, J. I. (1950). 'Some Problems in the Language of Shin Bian Wu Day Shyy Pyng Huah'. PhD thesis, Yale University.

CRUMP, J. I. (1970) (Tr.) *Chan-Kuo Ts'e*. Oxford University Press, Oxford.

CSONGOR, B. (1960). 'Remarks on a Problem of Negation in Middle Chinese', *AO(Hungary)*, **11**, 69–74.

CUA, A. S. (1985). *Ethical Argumentation. A Study in Hsün Tsu's Moral Epistemology*. University of Hawaii Press, Honolulu.

CUSA, NICOLAS (1964). *De docta ignorantia*. Ed. Paul Wilpert. Meiner, Hamburg.

DAOR, DAN (1979). 'The Yin Wenzi and the Renaissance of Philosophy in Wei-Jin China'. PhD thesis, London.

DAVID, JOHN FRANCIS (1823). *Hien Wun Shoo. Chinese Moral Maxims. With a Free and Verbal Translation; Offering examples of the Grammatical Structure of the Language*. John Murray, London.

DAVID, MADELEINE V. (1965). *Le débat sur les écritures et l'hiéroglyphe aux XVIIe et XVIIIe siècles*. S.E.F.P.E.N., Paris.

DAYE, DOUGLAS D. (1975). 'Buddhist Logic'. In C. S. Prebish, *Buddhism: a Modern Perspective*. Pennsylvania University Press, pp. 127–32.

DAYE, DOUGLAS D. (1977). 'Metalogical incompatibilities in the formal description of Buddhist logic (Nyāya)', *Notre Dame Journal of Formal Logic*, **18**, 2, 221–31.

DAYE, DOUGLAS D. (1979). 'Empirical Falsifiability and the Frequence of Darsana Relevance in the Sixth-century Buddhist Logic of Sankarasvamin', *Logique et Analyse*, **85**, 6, 223–37.

DAYE, DOUGLAS D. (1981). 'Aspects of the Indian and Western Traditions of Formal Logic and their Comparisons'. In N. Katz (1981), pp. 54–79.

DAYE, DOUGLAS D. (1985). 'Some Epistemologically Misleading Expressions: "Inference" and "Anumana", "Perception" and "Pratyaksa" '. In B. K. Matilal & J. L. Shaw (eds.), (1985), pp. 231–52.

DE FRANCIS, JOHN (1984). *The Chinese Language, Fact and Fantasy*. University of Hawaii Press, Honolulu.

DEHERGNE, JOSEPH (1973). *Répertoire des Jésuites de Chine de 1552 à 1800*. Biblioteca Institute Historici S. J., vol. XXXVII, Letouzey & Ane, Rome & Paris.

DEMIÉVILLE, P. (1943). 'La langue chinoise'. In *Conférence de l'Institut de Linguistique de l'Université de Paris, 1933*. Paris. pp. 33–70.

DEMIÉVILLE, P. (1948). 'Le chinois à l'Ecole Nationale des Langues Orientales Vivantes'. In *Cent-cinquantenaire de l'Ecole des Langues Orientales*. Maisonneuve, Paris, pp. 129–61.

DEMIÉVILLE, P. (1950). 'Archaismes de prononciation en chinois vulgaire', *TP*, **40**, 1–59.

DEMOSTHENES (1953). *De corona*. Ed. C. A. Since and J. H. Vince, Loeb Classical Library, Heinemann, London.

DENNISTON, J. D. (1934). *The Greek Particles*. 1st ed. Oxford University Press, Oxford.

DEVERGE, MICHEL (1983). 'Le classique des trois caractères. Un compendium du rudiment'. Dissertation, Paris VII.

DEVÉRIA, JEAN, G. (1896). 'Histoire du Collège des interprètes de Pékin'. In *Mélanges Charles de Harlez*. Brill, Leiden.

DEW, J. E. (1965). 'The Verb Phrase Construction in the Dialogue of Yüan Tzarjiuh. A Description of the Arrangement of Verbal Elements in an Early Modern Form of Colloquial Chinese', PhD thesis, University of Michigan.

DHRUVA, B. (1968). *The Nyāyapraveśa*. Oriental Institute, Baroda.

DICKSON, M. G. (1992). *The Andaman Islanders*. Hauteville Press, London.

DIELS, HERMANN (1954) (ed.). *Die Fragmente der Vorsokratiker*. 7th ed. Berlin.

DIOGENES LAERTIUS (1925). *Vitae philosophorum*. Ed. R. D. Hicks, Loeb Classical Library, Heinemann, London.

DOBSON, W. A. C. H. (1959). *Late Archaic Chinese. A Grammatical Study*. University of Toronto Press, Toronto.

DOBSON, W. A. C. H. (1962). *Early Archaic Chinese. A Descriptive Grammar*. University of Toronto Press, Toronto.

DOBSON, W. A. C. H. (1963). *Mencius: A New Translation*. Oxford University Press, London.

DOBSON, W. A. C. H. (1964). *Late Han Chinese; A Study of the Archaic-Han Shift*. University of Toronto Press, Toronto.

DOBSON, W. A. C. H. (1966). 'Classes de mots ou classes distributionelles en chinois archaïque'. In *Mélanges de sinologie offerts à M. Paul Demiéville*. vol. 1, Presses Universitaires, Paris, pp. 26–33.

DOBSON, W. A. C. H. (1968). *The Language of the Book of Songs*. University of Toronto Press, Toronto.

DOBSON, W. A. C. H. (1974). *A Dictionary of the Chinese Particles. With a Prolegomenon in which the Problems of the Particles are Considered and they are Classified by their Grammatical Functions*. University of Toronto Press, Toronto.

DODDS, E. R. (1951). *The Greeks and the Irrational*. Oxford University Press, Oxford.

DORNSEIFF, F. (1970). *Der deutsche Wortschatz nach Sachgruppen*. 7th ed. Gruyter, Berlin.

DOWNER, GORDON (1959). 'Derivation by Tone Change in Classical Chinese', *BLSOAS*, **22**, 258–90.

DRAGUNOV, A. A. & DRAGUNOV, K. (1936). 'Über die dunganische Sprache', *ARO*, **8**, 34–8.

DUBS, HOMER H. (1928). *The Works of Hsün Tzu*, tr. from the Chinese with notes, Probsthain, London.

DUBS, HOMER H. (1929). 'The Failure of the Chinese to Produce Philosophical Systems'. *TP*, **26**, 96.

DUBS, HOMER H. (1956). 'Y. R. Chao on Chinese Grammar and Logic', *PEW*, **5**, 167–68.

DUYVENDAK, J. J. L. (1924). 'Hsün Tzu on the Rectification of Names.' *TP*, **23**, 221–54.

DUYVENDAK, J. J. L. (1928). *The Book of Lord Shang*. Probsthain, London.

DUYVENDAK, J. J. L. (1936). 'Early Chinese Studies in Holland', *TP*, **32**, 293–344.

DUYVENDAK, J. J. L. (1950). *Holland's Contribution to Chinese Studies*. The China Society, London.

EDKINS, JOSEPH (1871). *China's Place in Philology. An Attempt to Show that the Languages of Europe and Asia have a Common Origin*. Trench, Trubner and Company, London.

EDWARDS, E. D. & BLAGDEN, C. O. (1931). 'A Vocabulary of Malacca – Malay Words and Phrases Collected between AD 1403 and 1511', *BLSOAS*, **6**, 715–49.

EDWARDS, E. D. & BLAGDEN, C. O. (1939). 'Chinese Vocabulary of Cham Words and Phrases', *BLSOAS*, 1939–42, **20**, 53–91.

EGEROD, S. (1967). 'Dialectology'. In T. A. Sebeok (1967), 91–129.

EGEROD, S. (1971). 'The Typology of Archaic Chinese'. In *A Symposium on Chinese Grammar*, Scandinavian Institute of Asian Studies Monograph Series no. 6, Lund, pp. 157–74.

EGEROD, S. (1972). 'Les particularités de la grammaire chinoise'. In J. M. C. Thomas & L, Bernot eds., *Langues et techniques (Festschrift Haudricourt)*. Maisonneuve, Paris, 101–9.

EGEROD, S. (1983). 'China: Melting pot of Languages or Cauldron of Tongues?'. In *Hong Kong-Denmark Lectures on Science and Humanities*. Hong Kong University Press, Hong Kong.

EIFRING, H. (1988). 'The Chinese Counterfactual', *Journal of Chinese Linguistics*, **16**, 2, 193–218.

D'ELIA, PASQUALE M. (1949). *Fonti Ricciane*, 3 Vols. Libreria dello Stato, Rome, 1942–49.

ELMAN, B. A. (1984). *From Philosophy to Philology. Intellectual and Social Aspects of Change in Late Imperial China*. Harvard University Press, Cambridge, Mass.

ENGEMANN, W. (1932). 'Voltaire und China. Beiträge zur Geschichte der Völkerkunde und zur Geschichte der Geschichtsschreibung sowie zu ihren gegenseitigen Beziehungen'. Dissertation, Leipzig.

ERASMUS, DIDIER (1967). *Colloquia familiaria*. Wissenschaftliche Buchgesellschaft, Darmstadt.

ERASMUS, DIDIER (1975). *Morias encomion sive stultitiae laus*. Wissenschaftliche Buchgesellschaft, Darmstadt.

ERKES, E. (1918). 'Zur Geschichte der eurpäischen Singologie'. *Ostasiastische Zeitschrift*, 1916–18, **6**, 105–15.

ERKES, E. (1941). 'The Use of Writing in Ancient China', *JAOS*, **61**, 127–30.

ERKES, E. (1950). *Ho-shang-kung's Commentary on Lao Tse*. Artibus Asiae, Ascona.

FANG, ACHILLES (1953). 'Some Reflections on the Difficulty of Translation'. In A. F. Wright, (1953), pp. 263–85.

FANG, ACHILLES (1957). 'Fenollosa and Pound', *HJAS*, **20**, 213–38.

FANG, THOMÉ (1981). *Chinese Philosophy: Its Spirit and Its Development*. Linking Publishing Co., Taipei.

FANG, WAN-CHUAN (1984). 'Chinese Language and Theoretical Thinking.' *JOS*, **22**, 1, 25–32.

FÊNG, YU-LAN (1922). *Why China has no Science – An Interpretation of the History and Consequences of Chinese Philosophy. International Journal of Ethics*, 1919, **30**, 32 and 1922, **33**, 237–63.

FÊNG, YU-LAN (1953). *A History of Chinese Philosophy*. Tr. Derk Bodde, Princeton University Press, Princeton, N. J., 1952–3.

FÊNG, YU-LAN (1958). *A Short History of Chinese Philosophy*. Tr. Derk Bodde, Macmillan, New York.

FÊNG, YU-LAN (1962). *The Spirit of Chinese Philosophy*. Tr. E. R. Hughes, Kegan Paul, London.

FENOLLOSA, E. F. (1936). *The Chinese Written Character as a Medium for Poetry. An Ars Poetica with a Foreword and Notes by Ezra Pound*. New York: Arrow Editions.

FLEW, ANTHONY G. N. (1979). 'The Cultural Roots of Analytical Philosophy'. *JCP*, **6**, 1–4.

FORKE, ALFRED (1902). 'The Chinese Sophists'. (Includes complete tr. of *Têng Hsi Tsz, Hui Tzu* and other paradoxes, *Kungsun Lung Tzu*). *JRAS/NCB*, **34**, 1.

FORKE, ALFRED (1911). *Lun Hêng, Philosophical Essays of Wang Chhung*. Vol. I, 1907, Kelley & Walsh, Shanghai; Luzac, London; Harrassowitz, Leipzig. Vol. II, 1911 (with the addition of Reimer, Berlin). (*MSOS*, Beiband **10** and **14**.) (Orig. pub. 1906, **9**, 181; 1907, **10**, 1; 1908, **11**, 1; 1911, **14**, 1.)

FORKE, ALFRED (1922). *Mo-Ti des Sozialethikers und seiner Schüler philosophische Werke*. Mitteilungeu des Seminars für Ostasiatische Sprachen, Beiband zum Jahrgang 23–5, Berlin.

FORKE, ALFRED (1939). 'Die chinesischen Skeptiker'. *SA*, **14**, 98.

FORKE, ALFRED (1964). *Geschichte der chinesischen Philosophie*, 3 vols. De Gruyter, Hamburg.

FORREST, R. A. D. (1973). *The Chinese Language*. Faber and Faber, London. (1st ed. 1948).

FOURMONT, S. (1737). *Meditationes Sinicae*. Musier, Paris.

FOURMONT, S. (1742). *Linguae Sinarum mandarinicae hieroglyphicae grammatica duplex*. H. L. Guérin, Paris.

FRANKE, HERBERT (1955). 'Bemerkungen zum Problem der Struktur der chinesischen Schriftsprache', *ORE*, **2**, 135–41.

FRANKE, HERBERT (1987). 'Chinese Patterned Texts'. In Dick Higgins (ed.), *Pattern Poetry*. Albany, University of Albany Press, New York, pp. 210–19.

FRÄNKEL, H. (1969). 'The Formulaic Language of the Chinese Ballad "Southeast Fly the Peacocks" ', *BIHP*, **69**, 219–44.

FRAUWALLNER, E. (1961). 'Landmarks in the History of Indian Logic', *Wiener Zeitschrift*, **5**, 125–48.

FU, SHEN C. Y. (1977). *Traces of the Brush, Studies in Chinese Calligraphy*. Yale University Art Gallery, New Haven & London.

FUCHS, W. (1930). 'Zur technischen Organisation der Übersetzungen buddhistischer Schriften ins Chinesische', *AM*, **6**, 84–103.

FUCHS, W. (1931). 'Remarks on a New Hua-i I-yü', *Bulletin of the Catholic University of Peking*, **8**, 91–7.

FUCHS, W. (1937). 'Das erste Deutsch-Chinesische Vokabular vom P. Florian Bahr', *Sinica*, **1**, 68–72.

FUNG YU-LAN. See Fêng Yu-lan.

GABELENTZ, GEORG VON DER (1881). *Chinesische Grammatik*. T. O. Weigel, Leipzig.

GABELENTZ, GEORG VON DER (1883). *Anfangsgründe der chinesischen Grammatik*. T. O. Weigel, Leipzig.

GARRETT, MARY M. (1983). 'The Mo-Tzu and the Lü-shih Ch'un-ch'iu: A Case Study of Classical Chinese Theory and Practice of Argument'. PhD thesis, University of California, Berkeley.

GARRETT, MARY M. (1985). 'Theoretical buffalo, conceptual kangaroos, and counterfactual fish: A review of Alfred Bloom's *The Linguistic Shaping of Thought*', *EC*, 1983–85, **9–10**, 220–36.

GASPARDONNE, EMILE (1953). 'Le lexique annamite des Ming', *JA*, **241**, 355–97.

GASSMANN, ROBERT H. (1980). *Das grammatische Morphem ye*. Lang, Bern.

GASSMANN, ROBERT H. (1982). *Das Problem der Einbettungsstrukturen*. Lang, Bern.

GASSMANN, ROBERT H. (1984). 'Eine textorientierte Interpretation der Pronomina *wu* und *wo* im *Meng-tzu*', *Asiatische Studien*, **38**, no. 2, 129–53.

GAUTAMA (1974). *The Nyāya Sutras of Gotama*. Ed. and tr. S. C. Vidyabhushana. Reprint, AMS Press, New York.

GERNET, JACQUES (1985). *China and the Christian Impact*. Cambridge University Press, Cambridge Eng. tr. of: *La Chine et le Christianisme*. Gallimard, Paris, 1983.

GIBBON, EDWARD (1787). *Decline and Fall of the Roman Empire*. Random House, New York, 1932.

GIGLIOLI, PIER PAOLO (1972). *Language and Social Context*. Penguin, Harmondsworth.

GILES, HERBERT A. (1892). *A Chinese-English Dictionary*. Quaritch, London, and Kelly and Walsh, Shanghai, Hong Kong, Yokohama and Singapore. (2nd ed. 1912).

GILES, H. A. (1873). *The Three Character Classic*. Chinese Materials Center, A. H. de Carvalho, Shanghai.

GILES, LIONEL (1945). Review of E. R. Hughes, 'Chinese Philosophy in Classical Times'. *BLSOAS* 1943–46, **11**, 235–39.

GILLON, B. S. & LOVE, M. L. (1980). 'Indian Logic Revisited: Nyāyapraveśa Reviewed', *Journal of Indian Philosophy*, **8**, 349–84.

GODWIN, F. (1638). *The Man in the Moone: or a Discourse of a Voyage Thither by Domingo Gonzales, the Speedy Messenger*. J. Norton, London.

GODWIN, JOCELYN (1979). *Athanasius Kircher – A Renaissance Man in Quest for Lost Knowledge*. Thames and Hudson, London.

GOODRICH, L. C. (1976). *Dictionary of Ming Biography*. Columbia University Press, New York & London.

GRAHAM, A. C. (1952). 'A Probable Fusion Word', *BLSOAS*, **14**, 139–48.

GRAHAM, A. C. (1955). 'Kung-sun Lung's Essay on Meanings and Things', *JOS*, Hong Kong, **2**, 2, 282–301.

GRAHAM, A. C. (1958). *Two Chinese Philosophers. Ch'eng Ming-tao and Ch'en Yi-ch'uan*. Lund Humphries, London.

GRAHAM, A. C. (1959a). 'Observations on a New Classical Chinese Grammar', *BSOAS*, **22**, 556–71.

GRAHAM, A. C. (1959b). ' "Being" in Western Philosophy Compared with *shih/fei* and *yu/wu* in Chinese Philosophy', *AM, NS*, **7**, 1–2, 79–122.

GRAHAM, A. C. (1960). *The Book of Lieh Tzu*. Murray, London.

GRAHAM, A. C. (1962). 'The "Hard" and "White" Disputations of the Chinese Sophists', *BLSOAS*, **30**, 2, 282–301.

GRAHAM, A. C. (1964). 'The Place of Reason in the Chinese Philosophical Tradition'. In R. Dawson (ed.), *The Legacy of China*. Oxford University Press, Oxford.

GRAHAM, A. C. (1965). 'Two Dialogues in the *Kung-sun Lung-Tzu*', *AM*, NS, **11**, 2, 128–52.

GRAHAM, A. C. (1967). 'The Background of the Mencian Theory of Human Nature', *Tsing Hua Journal of Chinese Studies*, 1967, **6**, 214–71; and A. C. Graham (1986), pp. 216–82.

GRAHAM, A. C. (1969). 'China, Europe and the Origins of Modern Science'. In Needham (1969).

GRAHAM, A. C. (1970). 'On Seeing Things as Equal', *History of Religions*, **9**, 2/3, 137–60.

GRAHAM, A. C. (1973), 'China, Europe and the Origins of Modern Science: The Grand Titration'. In Nakayama & Sivin (1973), pp. 45–69.

GRAHAM, A. C. (1975). 'The Concepts of Necessity and the "a priori" in Later Mohist Disputation' *AM, NS*, **19**, 2, 163–90.

GRAHAM, A. C. (1978). *Later Mohist Logic, Ethics and Science*. The Chinese University Press and the School of Oriental & African Studies, London & Hong Kong.

GRAHAM, A. C. (1981). *Chuang-tzu. The Seven Inner Chapters and Other Writings from the Book Chuang-tzu*. George Allen and Unwin London.

GRAHAM, A. C. (1985). *Divisions in Early Mohism Reflected in the Core Chapters of Mo-Tzu*. Institute of East Asian Philosophies Occasional Paper and Monograph Series no. 1, Singapore.

GRAHAM, A. C. (1986a). *Studies in Chinese Philosophy and Philosophical Literature*. Institute of East Asian Philosophies Occasional Paper and Monograph Series, no. 3, Singapore.

GRAHAM, A. C. (1986b). 'The Disputation of Kung-Sun Lung as Argument about Whole or Part'. In Graham (1986a), pp. 193–215.

GRAHAM, A. C. (1986c). 'The Date and Composition of *Lieh-tzu*'. In Graham (1986a), pp. 216–82.

GRAHAM, A. C. (1986d). *Yin-Yang and the Nature of Correlative Thinking*. Institute of East Asian Philosophies Occasional Paper and Monograph Series, no. 6, Singapore.

GRAHAM, A. C. (1989). *Disputers of the Tao: Philosophical Argument in Ancient China*. Open Court Pub. Co., La Salle, Illinois.

GRAHAM, A. C. & NATHAN SIVIN (1973). 'A Systematic Approach to the Mohist Optics'. In *Chinese Science*, Nakayama & Sivin (eds.), (1973). MIT East Asian Science Series, vol. 2, pp. 105–90.

GRAHAM, WILLIAM T. (1980). *The Lament for the South*. Cambridge University Press, Cambridge.

GRANEI, MARCEL (1920). 'Quelques particularités de la langue et de la pensée chinoise', in his *Etudes sociologiques sur la Chine*, PUF, Paris.

GRANET, MARCEL (1934). *La pensée chinoise*. Albin Michel, Paris.

GREBENSHCHIKOV, A. V. (1913). *Ocherk izuchenija man'chzhurskogo jazyka v Kitae*. Izvestija Vostochnogo Instituta, vol. 32, Moskva, 1913.

GROVSSET, R. (1929). *Sur les traces du Bouddha*. Plon, Paris.

GRUBE, WILHELM (1896). *Die Sprache und Schrift der Jučen*. Harrassowitz, Leipzig, 1896.

DE GUIGNES, JOSEPH (1760). *Mémoire dans lequel on prouve, que les chinois sont une colonie égyptienne*. Desaint et Saillant, Paris.

VAN GULIK, R. (1956). *Siddham. The Study of Sanskrit in China and Japan*. Monograph, Delhi.

GUREVICH, I. S. (1974). *Strednekitajskij jazyk*. Ocherk grammatiki kitajskogo yazyka, vols. III–V (Outline of the grammar of the Chinese language of the +3rd to +5th centuries). Nauka, Moscow.

GRYNPAS, B. (1967). *Un Legs Confucéen: Fragments du Ta Tai Li Kī*. Brussel.

HAGER, JOSEPH (1801). *An Explanation of the Elementary Characters of the Chinese; with an Analysis of their Ancient Symbols and Hieroglyphics*. R. Phillips, London.

HALLIDAY, MICHAEL ALEXANDER KIRKWOOD (1959). *The Language of 'The Secret History of the Mongols'*. Oxford University Press, Oxford.

HAMILTON, CLARENCE H. (1938). *Wei Shih Erh Shih Lun: The Treatise in Twenty Stanzas*. American Oriental Society Monograph, Connecticut.

HANSEN, CHAD (1983). *Language and Logic in Ancient China*. University of Michigan Press, Ann Arbor.

HANSEN, CHAD (1985). 'Chinese Language, Chinese Philosophy and "Truth" ', *JAS*, **44**, 491–517.

HARBSMEIER, CHRISTOPH (1979). *Wilhelm von Humboldts Brief an Abel Rémusat und die philosophische Grammatik des Altchinesischen*. Frederick Frommann, Stuttgart.

HARBSMEIER, CHRISTOPH (1981). *Aspects of Classical Chinese Syntax*. Curzon Press, London.

HARBSMEIER, CHRISTOPH (1985). 'Where do Classical Chinese Nouns come from?', *EC*, 1983–85, **9–10**, 77–163.

HARBSMEIER, CHRISTOPH (1986). 'The Grammatical Particle *i* in Classical Chinese'. Papers presented at the Second International Conference of Sinology, December 28–30.

HARBSMEIER, CHRISTOPH (1989). 'Marginalia Sino-logical'. In R. Allinson, *Understanding the Chinese Mind*. Oxford University Press, Hong Kong.

HARBSMEIER, C. (1993). 'The grammatical particle *fú* in Classical Chinese'. In *From classical fú to 'three inches high'. Studies on Chinese in Honor of Erik Zürcher* Louvain and Apeldoorn: Garant Publishers, pp. 1–60.

HARPER, E. H. (1881). 'Characterless Chinese Words', *China Review*, **9**, 85–8.

HARRIS, ROY (1980). *The Language Makers*. Cornell University Press, Ithaca.

HÄRTEL, HERBERT (1956). *Karmavācanā. Formulare für den Gebrauch im buddhistischen Gemeindeleben aus ostturkistanischen Sanskrit-Handschriften*. Berlin.

HASHIMOTO, ANNE YUE (1976). 'Southern Chinese Dialects: the Thai Connection'. In *Computational Analyses of Asian Languages*, **6**, 1–9.

HASHIMOTO, MANTARO (1976a). 'Language Diffusion on the Asian Continent – Problems of Typological Diversity in Sino-Tibetan', *CAAAL*, **3**, 49–97.

HASHIMOTO, MANTARO (1976b). The Agrarian and Pastoral Diffusion of Language'. In *Genetic Relationship, Diffusion and Typological Similarities of East and Southeast Asian Languages*, Japan Society for the Promotion of Science, Tokyo, pp. 1–14.

HASHIMOTO, MANTARO (1978). *The Phonology of Ancient Chinese*. Institute for the Study of Languages and Cultures of Asia and Africa, Tokyo.

HASHIMOTO, MANTARO (1986). 'The Altaicization of Northern Chinese'. In *Contributions to Sino-Tibetan Studies* (Festschrift Bodman). Brill, Leiden, pp. 76–97.

HAUDRICOURT, A. G. (1954). 'De l'Origine des Tons en Vietnamien', *JA*, **242**, 68–82.

HEARNE, JAMES WILLIAM (1980). 'Classical Chinese as an Instrument of Deduction'. Dissertation, University of California, Riverside.

HEARNE, JAMES WILLIAM (1985). 'Formal Treatments of Chih Wu Lun', *JCP*, **12**, 4, 419–29.

HENRICKS, ROBERT G. (1981a). *Philosophy and Argumentation in Third-Century China. The Essays of Hsi K'ang*. Princeton University Press, Princeton.

HENRICKS, ROBERT G. (1981b). 'Hsi K'ang and Argumentation in the Wei', *JCP*, **8**, 2, 169–221.

HENRICKS, ROBERT G. (1990). *Lao Tzu, Te Tao Ching*. Bodley Head, London.

HENRY, DESMOND PAUL (1964). *The De Grammatico of St Anselm: the Theory of Paronymy*. University of Notre Dame Press, Notre Dame.

HENRY, DESMOND PAUL (1967). *The Logic of Saint Anselm*. Oxford University Press, Oxford.

HERDAN, GUSTAV (1967). 'Chinese – a Conceptual or a Notational Language?', *MS*, **24**, 47 75.

HESS, R. G. (1972). 'Semantic Concepts in Ancient China', *A Review of General Semantics*, **29**, 243–9.

HIGHTOWER, J. R. (1951). *Han Shih Wai Chuan*. Harvard University Press, Cambridge, Mass.

HIGHTOWER, J. R. (1965). 'Some Characteristics of Parallel Prose'. In John L. Bishop (ed.), *Studies in Chinese Literature*. Harvard University Press, Cambridge, Mass., pp. 108–39.

HIRTH, F. (1909). *Notes on the Chinese Documentary Style*. Kelly and Walsh, Shanghai.

HOCKETT, C. F. (1954). 'Chinese versus English: an Exploration of Whorfian Theses,' *Language and Culture. American Anthropological Association Memoir*, **79**, 106–23 & 247–62.

HOLZMAN, DONALD (1957). *La Vie et la pensée de Hi K'ang* (223–262 ap. J.-C.). Harvard-Yenching Institute, Brill, Leiden.

HONG JUNG-SHIK (1982). 'The Thought and Life of Wonhyo'. In Chun Shin-yong (ed.), *Buddhist Culture in Korea*. Korean Culture Series, Vol. 3. The Si-Sa-Yong-O-Sa Publishers Inc., Seoul.

HOPKINS, L. C. (1954). *The Six Scripts or, The Principles of Chinese Writing by Tai T'ung, 13th century AD*. Cambridge University Press Cambridge. (1st ed. Amoy, 1881).

HOVDHAUGEN, E. (1982). *Foundations of Western Linguistics*. Universitetsforlaget, Oslo.

HRDLICKOVA, V. (1985). 'The First Translations of Buddhist Sutras in Chinese Literature and their Place in the Development of Story Telling', *ARO*, **26**, 114–44.

HSIA, A. (1985). *Deutsche Denker über China*. Suhrkamp, Frankfurt.

HSU CHO-YUN (1965). *Ancient China in Transition*. Stanford University Press, Stanford.

HU SHIH (1922). *The Development of Logical Method in Ancient China*. Shanghai: Oriental Book Company, Shanghai.

HUBRECHT, ALPHONSE (1932). *Etymologie des caractères chinois*. H. Vetch, Peking.

HULSEWÉ, A. F. P. (1985). *Remnants of Ch'in Law*. Brill, Leiden.

HUMBOLDT, WILHELM VON (1827). *Lettre à M. Abel-Rémusat sur la nature des formes grammaticales en général, et sur le génie de la langue chinoise en particulier*. Dondey-Dupré, Paris.

HUMBOLDT, WILHELM VON (1969). *De l'origine des formes grammaticales, suivi de Lettre à M. Abel Rémusat*. Ducros, Bordeaux.

HUMMEL, A. W. (1975). *Eminent Chinese of the Ch'ing Period*. Cheng Wen, Taipei.

HURVITZ, L. (1963). 'The Problem of Translating Buddhist Canonical Texts into Chinese', *Babel*, **9**, 48–52.

HUSSEY, E. (1972). *The Pre-Socratics*. Duckworth, London.

IMAZOV, M. H. (1982). *Ocherki po morfologii dunganskogo jazyka* (A Survey of the morphology of the Dungan language). Frunze, Izdatelstvo Ilim.

INGALLS, D. H. H. (1951). *Materials for the study of Navya-Nyâya Logic.* Harvard Oriental Series, vol. 40, Cambridge, Mass.

ISAENKO, B. (1957). *Opyt kitajsko-russkogo foneticheskogo slovarja.* Gosudarstvennoe izdtel'stvo natsional'nyhh i inostrannykh slovarej, Moskow.

ISAIJA, POLIKIN (1867). *Russko-Kitajskij Slovar' razgovornogo jazyka, Pekinskogo narechija.* (Russian–Chinese Dictionary.) Missionary Press, Peking.

IVERSEN, ERIK (1963). *The Myth of Egypt.* Munksgaard, Copenhagen.

JANERT, KLAUS LUDWIG (1972). *Abstände und Schlussvokalverzeichnungen in Asoka-Inschriften.* Verzeichnis der orientalischen Handschriften in Deutschland, Supplementband 10, Wiesbaden.

JENSEN, H. (1936). 'Indogermanisch und Chinesisch'. In H. Arntz, ed. *Germanen und Indogermanen.* C. Winter, Heidelberg.

JESPERSEN, O. (1929). *The Philosophy of Grammar.* Holt and Co., New York.

JONSON, BEN (1640). *English Grammar.* London; repr. Scholar Press, Menston, 1972.

JULIEN, STANISLAS (1824). *Men Tseu, vel Mencium inter Sineses philosophos ingenio, doctrina, nominisque claritate Confucio proximum, edidit, Latina interpretatione, ad interpretationem utramque recensita, instruxit, et perpetuo commentario, e Sinicis deprompto, illustravit Stanislaus Julien, Societatis Asiaticae et comites de Lasteyrie impensis.* Maisonneuve, Paris.

JULIEN, STANISLAS (1853). *Histoire de la vie de Hiouen-tsang.* Impr. Impériale, Paris.

JULIEN, STANISLAS (1870). *Syntaxe nouvelle de la langue chinoise fondée sur la position des mots.* Maisonneuve, Paris, 1869/70.

JUNANKAR, N. S. (1978). *The Nyāya Philosophy.* Motilal, Delhi.

KADEN, KLAUS (1964). *Der Ausdruck von Mehrzahlverhältnissen in der modernen chinesischen Sprache.* Akademieverlag, Berlin.

KALFF, L. (1954). *Einführung in die chinesischen Schriftzeichen.* Stegler, Kaldenkirchen.

KALIMOR, A. (1968). 'Dunganskij jazyk'. In *Jazyki narodov SSSR,* Leningrad, vol. 5, pp. 475–88.

KALLGREN, GERTY (1958). *Studies in Sung Time Colloquial Chinese as Revealed in Chu Hi's Ts'üanshu. BMFEA.*

KAM, TAK HIM (1977). 'Derivation by Tone Change in Cantonese', *Journal of Chinese Linguistics,* **5**, 186–210.

KANDEL, J. (1974). 'Die Lehren des Kung-sun Lung und deren Aufnahme in der Tradition. Ein Beitrag zur Interpretationsgeschichte des abstrakten Denkens in China'. Dissertation, Würzburg.

KAO, KUNG-YI & OBENCHAIN, D. B. (1975). 'Kung-sun Lung's Ch'ih wu lun and Semantics of Reference and Predications', *JCP,* **2/3** (1975), 285–324.

KARLGREN, BERNHARD (1919). *Études sur la Phonologie Chinoise.* 3 fasc. Brill, Leiden, 1915–19.

KARLGREN, BERNHARD (1923). *Analytic Dictionary of Chinese & Sino-Japanese.* P. Genthner, Paris.

KARLGREN, BERNHARD (1929). *The Authority of Chinese Texts. BMFEA.*

KARLGREN, BERNHARD (1932). 'The Poetical Parts of the Lao-tsi'. In *Göteborgs Högskolas Aarsskrift,* 38.

KARLGREN, BERNHARD (1950). *The Book of Odes.* Museum of Far Eastern Antiquities, Stockholm.

KARLGREN, BERNHARD (1952a). 'New Excursions in Chinese Grammar.' *BMFEA,* **24**, 51–80.

KARLGREN, BERNHARD (1952b). *Grammata Serica Recensa.* (Repr. in *BMFEA,* **29**, 1957.)

KARLGREN, BERNHARD (1961). 'The Parts of Speech and the Chinese Language', in *Language and Society. Essays Presented to Arthur M. Jensen on his 70th Birthday.* Munksgaard, Copenhagen, pp. 73–8.

KARLGREN, BERNHARD (1963). 'Loan Characters in Pre-Han Texts,' *BMFEA,* **35**, 1963, 1–120 and **36**, 1964, 1–105.

KATZ, N. (1981). *Buddhist and Western Philosophy.* Sterling Publishers, New Delhi.

KEIGHTLEY, D. N. (1983). *The Origins of Chinese Civilisation,* University of California Press, Berkeley.

KEIL, H. (1961). *Grammatici Latini.* 8 vols., Teubner, Leipzig, 1855–80. Repr., Olms, Hilesheim.

KEITH, A. B. (1921). *Indian Logic and Atomism.* The Clarendon Press, Oxford, 1921.

KENNEDY, GEORGE ALEXANDER (1952a). 'Negatives in Classical Chinese', *Wennti* **1**, 1–16.

KENNEDY, GEORGE ALEXANDER (1952b). 'Tone in Archaic Chinese', *Wenti,* **2**, 17–32.

KENNEDY, GEORGE ALEXANDER (1964). *Selected Works.* Yale University, Far Eastern Publications, New Haven.

KIRCHER, ATHANASIUS (1652). *Oedipus aegyptiacus.* Mascardi, Rome.

KIRCHER, ATHANASIUS (1667). *China Monumentis, qua sacris qua profanis nec non variis naturae et artis spectaculis aliarumgue rerum argumentis illustrata.* Weyerstraet, Amsterdam.

KITAGAWA, H. (1960). 'A Note on the Methodology in the Study of Indian Logic', *Indogaku Bukkyōgaku Kenkyū,* **8**, 380–90.

KLAPROTH, JULIUS (1811). *Leichenstein auf dem Grabe der chinesischen Gelehrsamkeit des Herrn Joseph Hager, Doctors auf der Hohen Schule zu Pavia.* No place.

KLAPROTH, JULIUS (1812). *Abhandlung über die Sprache und Schrift der Uiguren. Nebst einem Wörterverzeichnisse und anderen Uigurischen Sprachproben, aus dem kaiserlichen Übersetzungshofe zu Peking.* Königliche Druckerei, Paris.

KNEALE, WILLIAM & KNEALE, MARTHA (1962). *The Development of Logic.* Oxford University Press, Oxford.

KNOBLOCK, J. H. (1988). *A Translation and Study of the Complete Works.* Stanford University Press, Stanford, 1988, 1990.

KOROTKOV, N. N. (1968). *Osnovnye osobennosti morfologicheskogo stroja kitajskogo jazyka* (Basic characteristics of the morphological structure of Chinese). Nauka, Moskow.

KÖSTER, H. (1967). (Tr.) *Hsün Tzu.* Stegler, Kaldenkirchen.

KOU, IGNACE PAO-KOH (1953). *Deux sophistes chinois: Huei Che et Kong-Souen Long.* Maisonneuve, Paris.

KRAFT, EVA (1976). 'Frühe chinesische Studien in Berlin', *Medizin-historisches Journal*, **11**, 92–128.

KRAMERS, R. P. (1950). *K'ung Tzŭ Chia Yü. The School Sayings of Confucius. Translation of Sections 1–10 with Critical Notes.* E. J. Brill, Leiden.

KRETZMANN, N. (1968). *William of Sherwood's Introduction to Logic*, tr. with an introduction and notes by N. Kretzmann. University of Minnesota Press, Minneapolis.

KRJUKOV, M. V. (1973). *The Language of the Shang Inscriptions.* Nauka, Moscow.

KROLL, J. L. (1987). 'Disputation in Ancient Chinese Culture', *Early China Journal*, 1985–7, **11–12**, 118–45.

KUANG, JU-SSU (1944). *Chinese Wit, Wisdom and Written Characters.* Pantheon, New York.

KUBLER, C. C. (1985). *A Study of Europeanized Grammar in Written Chinese.* Student Book Co., Taipei.

KÜHNER, R. & GERTH, B. (1890). *Ausführliche Grammatik der Griechischen Sprache.* 4 vols., Hahnsche Buchhandlung, Leipzig & Hannover, 1890–1904.

KÜHNERT, F. (1891). 'Die Partikel si in Lao Tsi's Tao-te-king', *Wiener Zeitschrift für die Kunde des Morgenlandes*, **5**, 327–42.

KUO, SHAO-YÜ (1938). 'The Flexibility of Chinese "words"', *Yenching Journal of Chinese Studies*, **24**, 238–77.

LABOV, W. (1972). 'The Logic of Nonstandard English'. In P. P. Giglioli (1972).

LACH, D. (1940). 'The Chinese Studies of Andreas Müller', *JAOS*, **60**, 565–75.

LAU, D. C. (1983a). *Confucius: The Analects.* Bilingual edition. Chinese University Press, Hong Kong, 1983.

LAU, D. C. (1983b). 'On Mencius' Use of the Method of Analogy in Argument', *AM*, 1963, **10**, part 2, 173–94; repr. in D. C. Lau (1983c).

LAU, D. C. (1983c). *Mencius,* bilingual edition, Chinese University Press, Hong Kong, 1983.

LAU, D. C. (1989). 'Taoist Metaphysics in the *Chieh Lao* and Plato's Theory of Forms'. In D. C. Lau (ed.), Wen Lin 2, Chinese University of Hong Kong, Chinese Language Research Centre, Hong Kong, pp. 101–22.

LEACH, EDMUND R. (1964). 'Animal Categories and Verbal Abuse'. In E.-H. Lenneberg (ed.), *New Directions in the Study of Language.* MIT Press, Boston.

LE BLANC, CHARLES I. (1978). 'The Idea of Resonance (kang-ying) in the Huai-nan Tzu. With a Translation and Analysis of Huai-nan Tzu Chapter Six'. Dissertation, University of Pennsylvania.

LE BLANC, CHARLES I. (1986). *Huai-nan Tzu. Philosophical Synthesis in Early Han Thought.* Chinese University Press, Hong Kong

LEE, CYRUS (1964). 'Mo-tzu on Time and Space', *Chinese Culture*, **6**, **1**, 68–78.

LEGGE, J. (1872). *The Chinese Classics, etc.* Vol. 5, Pts. 1 & 2, The '*Ch'un Ts'eu*' with the '*Tso Ch'uen*' ('Chhun Chhiu' and 'Tso Chuan'). Lane Crawford, Hong Kong. Trübner, London.

LEIBNIZ, G. W. (1735). 'Lettre sur la Philosophie Chinoise à Mons. de Remond.' In C. Kortholt (ed.), *Viri illustris Godefridi Guil. Leibnitii epistolae ad diversos.* 4 vols., Leipzig, 1734–42, vol. II, pp. 413–94.

LEMMON, E. J. (1965). *Beginning Logic.* Nelson, London.

LESLIE, DONALD (1964). *Argument by Contradiction in Pre-Buddhist Chinese Reasoning.* Australian National University Press, Canberra.

LESSING, F. (1925). 'Vergleich der wichtigsten Formwörter der chinesischen Umgangssprache und der Schriftsprache. Ein Versuch', *Mitteilungen des Seminars für orientalische Sprachen zu Berlin*, **28**, 58–138.

LEWICKI, MARIAN (1949). *La Langue Mongole des Transcriptions Chinoises du XIVe Siecle. Le Houa-yi yi-yu de 1389,* Academy Press, Wrocław. 2 vols., 1949 and 1959.

LEYEN, F. VON DER (1942). *Das Buch der deutschen Dichtung.* 3 vols., Dusel, Leipzig.

LIAO, W. K. (1939). *Han Fei Tzu, Works Translated from the Chinese.* Probsthain, London, vol. 1, 1939 & vol. 2, 1959.

LIDDELL, HENRY GEORGE & SCOTT, ROBERT (1869). *A Greek-English Lexicon.* Clarendon Press, Oxford.

LIEBENTHAL, W. (1952). 'The Immortality of the Soul in Chinese Thought', *MN*, **8**, 327–97.

LIGETI, L. (1966). 'Un vocabulaire sino-ouigur des Ming', *Acta Orientalia Hungarica*, **19**, 117–99 & 257–316.

LIGETI, L. (1968). 'Notes sur le lexique sino-tibétain de Touen-houang en écriture Tibétaine', *Acta Orientalia Hungarica*, **21**, 265–88.

LIGHT, T. (ed.) (1968). *Contributions to Historical Linguistics.* Brill, Leiden.

LINDQVIST, CECILIA (1991). *China, Empire of the Written Symbol:* tr. from the Swedish by Joan Tate; foreword by Michael Loewe. Harvill, Hammersmith, London.

LIOU, KIA-HWAY (1961). *L'esprit synthétique de la Chine.* Presses Universitaires, Paris.

LIU, ERIC S. (1965). 'Frequency Dictionary of Chinese Words'. PhD thesis, Stanford.

LIU, JAMES J. Y. (1975). *Chinese Theories of Literature.* University of Chicago Press, Chicago and London.

LIU, TS'UN-YAN (1964/65). 'A New Interpretation of the Canon of the Mohist', *New Asia Journal* (1964/65), **6** & **7**.

LLOYD, GEOFFREY E. R. (1966). *Polarity and Analogy.* Cambridge University Press, Cambridge.

LOEWE, MICHAEL (1979). *Ways to Paradise: The Chinese Quest for Immortality.* George Allen and Unwin, London.

LOOSEN, RENATE & VONESSEN, FRANZ (eds.) (1968). *Gottfried W. Leibnitz. Zwei Briefe über das binare Zahlensystem und die chinesische Philosophie. Aus dem Urtext neu editiert, übersetzt und kommentiert,* Fromann Verlag, Stuttgart.

LOUTON, JOHN MARSHALL (1981). 'The Lüshi Chunqiu: An Ancient Chinese Political Cosmology'. Dissertation University of Washington, Seattle.

ŁUKASIEWICZ, JAN (1957). *Aristotle's Syllogistic from the Standpoint of Modern Formal Logic*. 2nd enlarged edition. Oxford University Press, Oxford. (1st ed. 1951).

LUKES, STEVEN (1973). *Individualism*. Blackwell, Oxford.

LUNDBAEK, KNUD (1979). 'The First Translation from a Confucian Classic in Europe', *China Mission Studies (1550–1800) Bulletin*, **1**, 1–11.

LUNDBAEK, KNUD (1980). 'Une grammaire espagnole de la langue chinoise au xviie siècle'. In *Actes du IIe colloque international de sinologie*, Paris.

LUNDBAEK, KNUD (1983). 'Imaginary Chinese Characters', *China Mission Studies (1550–1800) Bulletin*, **5**.

LUNDBAEK, KNUD (1986). *T. S. Bayer (1694–1738), Pioneer Sinologist*. Curzon Press, London & Malmö.

MACINTOSH, A., SAMUELS, M. L. & BENSKIN, M. (1986). *A Linguistic Atlas of Late Medieval English*. Aberdeen University Press, Aberdeen.

MACKIE, J. L. (1964). 'Self-Refutation – a Formal Analysis', *Philosophical Quarterly*, **14**, 193–213.

MALMQVIST, N. G. D. (1960). 'Some Observations on a Grammar of Late Archaic Chinese', *TP*, **48**, 252–86.

MALMQVIST, N. G. D. (1964). *Problems and Methods in Chinese Linguistics*. Australian National University, Canberra.

MALMQVIST, N. G. D. (1968). 'Chou Tsu-mo on the Ch'ieh-yün', *BMFEA*, **40**, 33–78.

MALMQVIST, N. G. D. (1971). 'Studies on the Gongyang and Guuliang commentaries', *BMFEA*, 1971, **43**, 57–227; 1975, **47**, 19–69; 1977, **49**, 33–215.

MALORY, SIR THOMAS (1969). *Le Morte D'Arthur*. Ed. Janet Cowen. Penguin English Library, Harmondsworth.

MANUZIO, ALDO (1561). *Orthographiae ratio*. Aldus, Venice.

MARGOULIES, G. (1957). *La langue et l'écriture chinoises*. Payot, Paris.

MARSHMAN, J. (1814). *Elements of Chinese Grammar, Clavis Sinica*. Mission Press, Serampore.

MARTZLOFF, JEAN-CLAUDE (1988). *Histoire des Mathématiques Chinoises*. Masson, Paris.

MASPERO, HENRI (1920). *Le dialecte de Tch'ang-ngan sous les T'ang*. BEFEO, **20.2**, 1–24.

MASPERO, HENRI (1928). 'Notes sur la logique de Mo-tseu et de son école', *TP*, **25**, 1–64.

MASPERO, HENRI (1934). *La langue chinoise*. Conférence de l'Institut de Linguistique de l'Université de Paris, Paris.

MASPERO, HENRI (1935). 'Préfixes et dérivation en chinois archaïque', *Mémoires de la Société Linguistique de Paris*, **23**, 314–27.

MASPERO, HENRI (1952). 'Le Chinois'. In A. Meillet & M. Cohen 1952 eds., *Les Langues du monde*. E. Champion, Paris.

MASSON-OURSEL, P. (1918). 'Etudes de Logique Comparée', *Revue philosophique*, 1917, **84**, 57–76 & 1918, **85**, 148–66.

MATES, BENSON (1961). *Stoic Logic*. California University Press, Berkeley.

MATES, BENSON (1972). *Elementary Logic*. Oxford University Press, London.

MATHER, RICHARD (1976). (Tr.) *Shih-shuo Hsin-yü: A New Account of Tales of the World*. University of Minnesota Press, Minneapolis.

MATHIAS, J., ed. (1982). *Chinese Dictionaries. An Extensive Bibliography of Dictionaries in Chinese and Other Languages*. Greenwood Press, Westport, Connecticut.

MATHIEU, RÉMI (1985). *Guo Yu*. Collège de France, 1985.

MATILAL, BIMAL K. (1966). *The Navya Nyāya Doctrine of Negation*. Harvard Oriental Series, vol. 46, Cambridge, Mass.

MATILAL, BIMAL K. (1971). *Epistemology, Logic and Grammar in Indian Philosophical Analysis*. Janua Linguarum, Series Minor, vol. 111, Mouton, The Hague.

MATILAL, BIMAL K. (1977). 'Nyâya-Vaiśeṣika'. In J. Gonda (ed.), A *History of Indian Literature*, vol. 6, fasc. 2, Wiesbaden.

MATILAL, B. K. & SHAW, J. L. (1985) (eds.). *Analytical Philosophy in Comparative Perspective*, Reidel, Amsterdam.

McCOY, JOHN & LIGHT, T. (1986). *Contributions to Historical Linguistics*. Brill, Leiden.

McDERMOTT, A. & CHALENE, S. (1970). *An Eleventh-Century Buddhist Logic of 'Exists'*. Reidel, Dordrecht.

MEDHURST, W. H. (1942). *Notices on Chinese Grammar, Pt. 1*. Philo-Sinensis, Batavia.

MEI, TSU-LIN (1961). 'Chinese grammar and the linguistic movement in philosophy', *The Review of Metaphysics*, **14**, 135–75.

MEI, TSU-LIN (1982). 'Some Prosodic and Grammatic Criteria for Dating the Ballad "Southeast Fly the Peacocks"', *The Bulletin of the Institute of History and Philology, Academia Sinica*, **53**, Part II, Taipei, 227–49.

MEI, Y. P. (1929). *The Ethical and Political Works of Mo Tzu*. Probsthain, London.

MEI, Y. P. (1951). 'Hsün-tzu on Terminology', *Philosophy East and West*, 51–66.

MEI, Y. P. (1953). 'The Kung-sun Lung Tzu with a Translation into English', *HJAS*, **16**, 404–37.

MEILLET, A. & COHEN, M. (1952). (eds.) *Les Langues du monde*, E. Champion, Paris, 1952.

MEISTER, J. C. F. (1816). *Ganz neuer Versuch auch freien Denkern aus der Chinesischen Schriftsprache eine symbolische Ansicht zu eröffnen unter welcher das Gemüth empfänglicher wird für das Geheimnis der christlichen Dreieinigkeit*. (An entirely new attempt, on the basis of the written Chinese language, to open up a symbolic perspective under which the mind becomes more receptive to the mysteries of the Christian holy trinity). Leipzig.

DE MENDOZA, JUAN GONZALES (1588). *The Historie of the Great and Mightie Kingdome of China, and the Situation thereof: Togither with the Great Riches, Huge Citties, Politike Gouernement and Rare Inventions in the Same*, English tr. by R. Parke, London, I. Wolfe. First (Spanish) ed. Rome, 1585.

MENGE, H. (1960). *Repetitorium der lateinischen Syntax und Stilistik*. 13th ed., M. Huber, Munich.

MILL, JOHN STUART (1843). *A System of Logic*. John W. Parker, London.

MILLER, ROBERT PICKENS (1952). 'The Particles in the Dialogue of Yuan Drama; a Descriptive Analysis'. PhD thesis, Yale University.

MILLER, ROY ANDREW (1953). 'Problems in the Study of Shuo-wen chieh tzu'. PhD thesis, Columbia University, New York and in *Trends in Linguistics*, **13.2**. Also: Ann Artor Microfilms 1953.

MILLER, ROY ANDREW (1954). 'The Sino-Burmese Vocabulary of the *I shih chi-yü*', *Harvard Journal of Asiatic Studies*, **17**, 370–93.

MILLER, ROY ANDREW (1973). 'The Far East'. In Thomas A. Sebeok (ed.), *Current Trends in Linguistics*. Vol. 13.2, Mouton, The Hague.

MIRONOV, N. D. (1927). 'Dignāga's *Nyâyapraveśa* and Haribhadra's *tika* on it'. In *Festgabe für Richard von Garbe*. Erlangen, pp. 37–46.

MIRONOV, N. D. (1931). 'Nyâyapraveśa. I: Sanskrit text', *TP*, **1–2**, 1–24.

MISCH, J. (1935). 'Der Konditionalsatz im klassischen Chinesisch'. Dissertation, Berlin.

MISTELI, FRANZ (1887). 'Studien über die chinesische Sprache', *Internationale Zeitschrift für allgemeine Sprachwissenschaft*, **3**, 27–97.

MIYAZAKI, ICHISADA (1976). *China's Examination Hell: The Civil Service Examinations of Imperial China*. Tr. from the Japanese by Conrad Schirokauer. Weatherhill, New York and Tokyo.

MORITZ, RALF (1974). *Hui Shi und die Entwicklung des philosophischen Denkens im alten China*. Akademieverlag, Berlin.

MORRISON, ROBERT (1815). *A Grammar of the Chinese Language*. Serampore.

MOSER-RATH, E. (1964). *Predigtmärlein der Barockzeit. Exempel, Sage, Schwank und Fabel in geistlichen Ouellen des oberdeutschen Raumes*. De Gruyter, Berlin.

MOSSNER, F. (1964). 'Chinesische Schrift – Notation oder Begriff?', *Zeitschrift für Phonetik, Sprachwissenschaft und Kommunikationsforschung*, **17**, 33–49.

MULDER, J. W. F. (1959). 'On the Morphology of Negatives in Archaic Chinese', *TP*, **47**, 251–80.

MÜLLER, ANDREAS (1672). *Monumenti Sinici, quod anno Domini MDCXXV terris in ipsa China erutum . . . lectio seu phrasis, versio seu paraphrasis, translatio seu paraphrasis*. J. Volcker, Frankfurt/Oder.

MÜLLER, ANDREAS (1682). *Andreae Mülleri Besser Unterricht von der Sinenser Schrifft und Druck, als etwa in Hrn. D. Eliae Grebenitzen Unterricht von der Reformirten und Lutherischen Kirchen enthalten ist*. Ruugianus, Berlin.

MÜLLER, F. W. K. (1892). 'Vokabularien der Pa-i und Pah-po-Sprachen aus dem Hua-i i-yü', *TP*, **3**, 1–38.

MULLIE, J. L. M. (1942). 'Le mot-particule TCHE', *Tioung Pao*, **36**.

MULLIE, J. L. M. (1948). *Grondbeginselen van de Chinese letterkundige taal*. 3 vols., Dewallens, Louvain.

MUNAKATA, KIYOHITO (1983). 'Concepts of *lei* and *kan-lei* in Early Chinese Art Theory'. In (eds.), S. Bush & C. Murck *Theories of the Arts in China*. Princeton University Press, Princeton.

MUNGELLO, DAVID E. (1985). *Curious Land: Jesuit Accommodation and the Origin of Sinology*. Studia Leibnitiana, Supplementa 25, Steiner Verlag, Stuttgart.

MUNRO, DONALD (1939). *The Concept of Man in Early China*. Stanford University Press, Stanford.

MUNRO, DONALD (1985). *Individualism and Holism: Studies in Confucian and Taoist Values*. Michigan Monographs in Chinese Studies. No. 52. Center for Chinese Studies, University of Michigan, Ann Arbor.

NAESS, ARNE (1968). *Scepticism*. Routledge and Kegan Paul, London.

NAKAMURA, HAJIME (1958). 'Buddhist Logic Expounded by Means of Symbolic Logic'. *Indogaku Bukkyogaku Kenkyu*, **7**, 375–79.

NAKAMURA, HAJIME (1964). *Ways of Thinking of Eastern Peoples. India, China, Tibet, Japan*. Hawaii University Press, Honolulu.

NAKAMURA, HAJIME (1980). *Indian Buddhism: A Survey with Bibliographic Notes*. KUFS Publications, Osaka.

NAKAYAMA, SHIGERU (1973). 'Joseph Needham, Organic Philosopher'. In Nakayama and Sivin (1973), pp. 23–43.

NAKAYAMA, SHIGERU (1984). *Academic and Scientific Traditions in China, Japan, and the West*. University of Tokyo Press, Tokyo.

NAKAYAMA, SHIGERU & SIVIN, NATHAN (1973). *Chinese Science: Explorations of an Ancient Tradition*. M.I.T. Press Cambridge, Mass.

NEEDHAM, JOSEPH (1969). *The Grand Titration: Science and Society in East and West*. Allen & Unwin, London.

NIVISON, DAVID S. (1962). 'Aspects of Chinese Traditional Biography', *JAS*, **21**, 457–63.

NOCK, A. D. & FESTUGIERE, A. J. (1960). (eds.) *Corpus Hermeticum*. Les Belles Lettres, Paris.

NYLAN, MICHAEL (1982). 'Ying Shao's *Feng Su T'ung Yi*: An Exploration of Problems in Han Dynasty Political, Philosophical and Social Unity'. Dissertation, Princeton.

OCKHAM, WILLIAM (1957). *Philosophical Writings*. Ed. Philotheus Boehner. Nelson, London.

OSHANIN, V. M. (1952). *Kitajsko-russkij slovar'*. Nauka, Moskow.

PARTRIDGE, A. C. (1964). *Orthography in Shakespeare and Elizabethan Drama*. University of Nebraska Press, Lincoln.

PATZIG, GÜNTHER (1969). *Aristotle's Theory of the Syllogism*. Reidel, Dordrecht.

PATZIG, GÜNTHER (1982). 'Syllogistics', *Encyclopedia Britannica*, 15th ed., vol. 17, pp. 890–8.

PELLIOT, PAUL (1948). 'Le Hôja et le Sayyid Husain de l'Histoire des Ming.' *TP*, **38**, 81–292/207–90. Appendix III: Le Sseu-yi-houan et le Houei-t'ong-Kouan.

PERLEBERG, MAX (1952). *The Works of Kung-sun Lung Tzu*. Hyperion Press, Westport, Conn., 1973.

PERRETT, ROY W. (1984). 'Self-Refutation in Indian Philosophy', *Journal of Indian Philosophy*, **12**, **3**, 237–64.

PETERSON, WILLARD J. (1979). *Bitter Gourd: Fang I-chih and the Impetus for Intellectual Change*. Yale University Press, New Haven, Conn.

PETERSON, WILLARD J. (1982). 'Making Connections: Commentary on the Attached Verbalisations of the Book of Change'. *HJAS*, **42**, 1, 67–116.

PETROV, A. A. (1936). 'Wang Pi (+226 to +249)', *ANSSSR*, Moscow.

PETROV, N. A. (1961). 'Kistorii izuchenija kitajskogo jazyka v Rossii' (Towards a history of the study of Chinese in Russia). In *Dal'nyi Vostok, Sbornik statej*. Isdatel'stvo Vostochnoj Literaturj, Akademija Nauk, Moskow, 65–90.

PEVERELLI, PETER JAN (1986). 'The History of Modern Chinese Grammar Studies'. PhD dissertation, Leiden.

PEYRAUBE, ALAIN (1988). *Syntaxe diachronique du chinois: évolution des constructions datives du XIVe siècle av. J.C. au XVIIIe siècle*. Mémoires des Hautes Études Chinoises, vol. 29, Paris.

PFISTER, LOUIS (1932). *Notices biographiques et bibliographiques sur les Jésuites de l'ancienne mission de Chine, 1552–1773*. Variétés sinologiques no. 59, Shanghai.

PINBORG, JOHANNES & ROOS HENRICUS (1969). *Boethii Daci Opera*, vol. IV, Part 1, *Modi significandi sive quaestiones super priscianum mainorem*. Det Danske Sprog- og Litteraturselskab, Copenhagen.

PLATO (1907). *Platonis Opera*. Ed. J. Burnet, Oxford University Press, Oxford, 1900–07.

PLATO (1957). *Platon, Sämtliche Werke. In der Übersetzung von Friedrich Schleiermacher*, Rowohlts Klassiker, Hamburg.

PLOTINUS (1973). *Enneads*. Ed. Paul, Henry & Schwyzer, Hans-Rudolf *Plotini, Opera*, vol. 2, Desclée et Brouwer, Paris, 1951–73.

PLUTARCH (1914). *Plutarch's Lives*. Ed. and tr. B. Perrin. Loeb Classical Library Heinemann, London, 1914ff.

POHLENZ, M. (1939). *Die Begründung der abendländischen Sprachlehre durch die Stoa*. Nachrichten von der Gesellschaft der Wissenschaften zu Göttingen, Neue Folge 3.6. Göttingen.

POKORA, T. (1959). 'The Canon of Laws by Li K'uei – A double falsification? *ARO*, **27**, 96–121.

POPOV, P. P. (1888). *Kitajsko-Russkij slovar'*. 2 vols. Misssionary Press, Peking.

POTTER, K. H. (1977). *Encyclopedia of Indian Philosophy*, vol. I, part 1, Nyāya-Vaiśeṣika. Motilal, Delhi.

POUND, EZRA (1928). *Ta hio, The Great Learning, Newly Rendered into the American Language by Ezra Pound*. University of Washington Book Store, Seattle.

POUND, EZRA (1952a). *Confucius: The Unwobbling Pivot and the Great Digest, Translated by Ezra Pound with Notes and Commentary on the Text and the Ideograms, together with Chu Hsi's 'Preface' to the Chung Yung and Tseng's Commentary on the Testament*. No place.

POUND, EZRA (1952b). *Confucian Analects, Translated by Ezra Pound*. Kasper & Horton, New York.

POUND, EZRA (1954). *The Classic Anthology Defined by Confucius*. Faber & Faber, London.

PRANTL, C. (1870). *Geschichte der Logik im Abendlande*. 4 vols., Hirzel, Leipzig, 1855–70.

DE PRÉMARE, JOSEPH HENRI-MARIE (1831). *Notitia linguae sinicae*. Academia Anglo-Sinensis, Malacca. Reprint, Hong Kong, 1898.

PULLEYBLANK, E. G. (1959). '*Fei* 非, *Wei* 微 and Certain Related Words'. In E. Glahn & S. Egerod (eds.), *Studia Serica Bernhard Karlgren Dedicata*. Munksgaard, Copenhagen, pp. 179–89.

PULLEYBLANK, E. G. (1960). 'Studies in Early Chinese Grammar', *AM*, **8**, 36–67.

PULLEYBLANK, E. G. (1962). 'The Consonantal System of Old Chinese', *AM*, **9**, 58–114 & 206–65.

PULLEYBLANK, E. G. (1964). 'The Origins of the Chinese Tonal Systems'. In *Proceedings of the Ninth International Congress of Linguists*, The Hague.

PULLEYBLANK, E. G. (1972). 'Some Notes on Causative Constructions in Classical Chinese'. Paper presented to the Fifth International Conference of Sino-Tibetan Language and Linguistic Studies, October, 20–1.

PULLEYBLANK, E. G. (1978). 'Emphatic Negatives in Classical Chinese'. In David T. Roy & Tsuen-hsuin Tsien (eds.), *Ancient China: Studies in Early Civilization*. Chinese University Press, Hong Kong.

PULLEYBLANK, E. G. (1984). *Middle Chinese: A Study in Historical Phonology*. Vancouver University Press, Vancouver.

PULLEYBLANK, E. G. (1986). 'The Locative Particles *yü* 于 *yü* 於 and *hu* 乎 , *JAOS*, **106**, 1, 1–12.

PUTNAM, HILARY (1975). 'On Properties'. In H. Putnam, *Mind, Language and Reality*, vol. 1, Cambridge University Press, Cambridge.

QUINE, W. V. O. (1969). *Word and Object*. MIT Press, Cambridge, Mass. (1st ed. 1960).

QUINE, W. V. O. (1970). *Philosophy of Logic*. Prentice Hall, Englewood Cliff.

QUIRK, R. *et al.* (1985). *A Comprehensive Grammar of the English Language*. Longman, London.

RAMSEY, ROBERT (1989). *The Languages of China*. Princeton University Press, Princeton.

RANDLE, HERBERT N. (1930). *Indian Logic in the Early Schools*. Oxford University Press, Oxford.

RANKIN, B. K. (1965). 'On the Pictorial Structure of Chinese Characters'. PhD thesis, Philadelphia.

READ, B. E. (1976). *Chinese Materia Medica, Animals Drugs*. Southern Materials Center, Taipei.

REALLY, CONOR (1974). *Athanasius Kircher. Master of a Hundred Arts*. Edizioni del Mondo, Rome and Wiesbaden.

REDING, J. P. (1985). *Les fondements philosophiques de la rhétorique chez les sophistes grecs et chez les sophistes chinois*. Lang, Berne, Frankfurt & New York.

REDING, JEAN-PAUL (1986). 'Analogical Reasoning in Early Chinese Philosophy'. *Asiatische Studien*, **40**, 40–56.

REICHWEIN, A. (1923). *China und Europa. Geistige und künstlerische Beziehungen im 18. Jahrhundert*. Österheld, Berlin. (Eng. tr. Paul Trench, Trubner, London).

REIFLER, E. (1943). 'Théories sur l'origine et le développement des caractères chinois'. *Bulletin de l'Universite l'Aurore*, **3**, 684–707.

REINBOTHE, H. & NESSELRATH, H. G. (1979) (eds.). *Gottfried Wilhelm Leibniz: Novissima Sinica. Das Neueste von China*. Privately published, Deutsche China-Gesellschaft, Cologne.

RENAN, J. E. (1889). *De l'Origine du Langage*. 4th edn. Michel Levy, Paris, Shanghai. Eng. tr. in Watters (1889), p. 18.

RENOU, L. (1965) 'Recherches générales sur la phrase védique'. In *Symbolae linguisticae in honorem G. Kuryłovica*. Polish Academy of Sciences, Section for Linguistics, 1965, vol. 5. Polska Akademia Nauk, Warsaw.

RICCI, MATTEO (1615). *De Christiana expeditione apud Sinas*. Christoph Mangium, Augsburg.

RICHARDS, I. A. (1932). *Mencius on the Mind: Experiments in Multiple Definition*. Kegan Paul, London.

RICKETT, A. (1985). *Guanzi: Political, Economic & Philosophical Essays from Early China*, vol. 1, Princeton University Press, Princeton.

RIEMANN, F. (1980). 'Kung-Sun Lung, Designated Things and Logic'. *PEW*, **30**, 305–19.

RISCH, F. (1934). (Tr.) *Wilhelm von Rubruk. Reise zu den Mongolen 1253–1255*. Werner Scholl, Leipzig. For critical edition of the Latin see A. van den Wyngaert, *Sinica Franciscana I. Quaracchi*, Florence, 1929.

RITTER, J. & GRÜNDER, K. (1984). *Historisches Wörterbuch der Philosophie*, vol. VI, Schwab, Basel.

ROBIN, L. (1944). *Pyrrhôn et le Scepticisme Grecque*. Presses Universitaires, Paris.

ROBINS, R. H. (1967). *A Short History of Linguistics*. Longman, London.

ROBINSON, R. H. (1967). *Early Mādhyamika in India and China*. University of Wisconsin Press, Madison and London.

RODRIGUEZ, JUAN (1792). *A Grammar of the Chinese Language Expressed by the Letters that are Commonly Used in Europe. From the Latin of F. John-Anthony Rodriguez. With a Dedicatory Letter from the Translator, John Geddes, to the R. Honorable Mr. Dundas, dated May 29th 1792*. China Factory Records, vol. 20, India Office Records, Commonwealth Office, London, 55 pages, manuscript.

ROGET, PETER MARK (1852). *Thesaurus of English Words and Phrases*. Everyman's Library, Nos. 603, 631., Dent, London, 1930, 1st ed. 1912. (Original ed. 1852.)

ROSEMONT, HENRY (1974). 'On Representing Abstractions in Archaic Chinese', *Philosophy East and West*, **24**, 71–88.

ROSEMONT, HENRY (1978). 'Gathering Evidence for Linguistic Innateness', *Syntese*, **38**, 127–48.

VON ROSTHORN, A. (1898). 'Eine chinesische Darstellung der grammatischen Kategorien'. In *Xème Congrès International des Orientalistes*, Section V, 97–105.

VON ROSTHORN, A. (1941a). 'Zur Geschichte der chinesischen Schrift', *Wiener Zeitschrift für die Kunde des Morgenlandes*, **48**, 121–42.

VON ROSTHORN, A. (1941b). *Indischer Einfluss in der Lautlehre Chinas*. Sitzungsberichte der Akademie der Wissenschaften in Wien, Philosophisch-historische Klasse, Bd.219, 4. Abhandlung. Holder, Pichler, Tempsky, Wien.

VON ROSTHORN, A. (1942). 'Das Er-ya und andere Synonymiken', *Wiener Zeitschrift für die Kunde des Morgenlandes*, **49**, 126–44.

ROZHDESTVENSKIJ, Y. V. (1958). *Ponyatie formy slova v istorii grammatiki kitajskogo jazyka. Ocherki po istorii kitajevedenija*. (The concept of word form in the history of Chinese grammar. Studies on the history of Chinese grammer. Studies on the history of Chinese grammar). Nauka, Moskow.

RUBEN, WALTER (1929). (Tr.) *Die Nyâyasûtras: Text, Übersetzung, Erläuterung und Glossar*. Brockhaus, Leipzig.

RUBIN, V. (1965). 'Tzu-chhan and the City State of Ancient China'. *T'oung Pao*, 52.

RUGGIERI, MICHELE (1607). *Liber sinensium*. In Antonio Possevino, *Bibliotheca selecta qui agitur de ratione studiorum*. (1st ed. 1593) Gymnicum, Cologne.

RUMP, A. & CHAN, WING-TSIT (1979). *Commentary on the Lao Tzu by Wang Pi*. Hawaii University Press, Honolulu.

RUSSELL, BERTRAND & WHITEHEAD, ALFRED NORTH (1910). *Principia Mathematica*. Cambridge University Press, Cambridge.

RYGALOFF, A. (1958). 'A propos de l'antonomie: l'example du chinois', *Journal de Psychologie Normale et Pathologique*, **55**, 358–76.

SAILEY, JAY (1978). *The Master Who Embraces Simplicity. A Study of the Philosopher Ko Hung, AD 283–343*. Chinese Materials Center Inc., San Francisco.

SALMI, OLLI (1984). 'The Aspectual System of Soviet Dungan'. *Folia Fennistica et Linguistica*, 83–117.

SAMPSON, GEOFFREY (1985). *Writing Systems*. Stanford University Press, Stanford.

SANDYS, J. E. (1908). *A History of Classical Scholarship*. Cambridge University Press, Cambridge, 1903–8. (3rd ed., 1921).

SAWER, MICHAEL (1969). 'Studies in Middle Chinese Grammar – The Language of the Early Yeu Luh'. PhD thesis, Canberra.

SCHARFSTEIN, BEN-AMI, *et al.* (1978). *Philosophy East/Philosophy West. A Critical Comparison of Indian, Chinese, Islamic and European Philosophy.* Blackwell, Oxford.

SCHEFFLER, I. (1979). *Beyond the Letter. A Philosophical Inquiry into Ambiguity, Vagueness and Metaphor in Language.* Routledge, Kegan and Paul, London.

SCHELLING, FRIEDRICH WILHELM JOSEPH VON (1857). *Philosophie der Mythologie,* Cottascher Verlag, Stuttgart and Augsburg, 1857, as quoted in Adrian Hsia, *Deutsche Denker über China,* Insel Verlag, Frankfurt, 1985, p. 219.

SCHINDLER, B. & ERKES, E. (1918). 'Zur Geschichte der europäischen Sinologie', *Ostasiatische Zeitschrift,* 1916–18, **5/6**, 105–15.

SCHLEGEL, GUSTAVE (1895). *La loi du paralléllisme en style chinois. Demonstrée par la préface du Si-Yü Ki. La traduction de cette préface par Feu Stanislas Julien défendue contre la nouvelle traduction du Père A. Gueluy.* Brill, Leiden.

SCHOPENHAUER, ARTHUR (1960). *Sämtliche Werke.* Ed. Wolfgang Freiherr von Löhneisen. Cottascher Verlag, Stuttgart.

SCHRADER, M. & FÜRKÖTTER, A. (1956). *Die Echtheit des Schrifttums der Heiligen von Bingen.* Köln/Graz.

SCHUESSLER, AXEL (1976). *Affixes in Proto-Chinese.* Münchener Ostasiatische Studien vol. 18, Steiner, Wiesbaden.

SCHUESSLER, AXEL (1987). *A Dictionary of Early Zhou Chinese.* Hawaii University Press, Honolulu.

SCHULTZ, E. (1937). 'A Chinese grammar of the eighteenth century', *MS,* 1936–7, **2,** 423–5.

SCHWARTZ, BENJAMIN, I. (1964). *In Search of Wealth and Power: Yen Fu and the West. Western Thought in Chinese Perspective.* Harvard University Press, Cambridge, Mass.

SCHWARTZ, BENJAMIN, I. (1973). 'On the absence of reductionism in Chinese thought', *JCP,* i, **1,** 27–43.

SCHWARTZ, BENJAMIN, I. (1985). *The World of Thought in Ancient China.* Harvard University Press, Cambridge, Mass.

SCHWYZER, E. (1939). *Die Parenthese im engern und weitern Sinne,* Abhandlungen der Preussischen Akademie der Wissenschaften, Jahrgang 1939, phil. -hist. Klasse, Nr. 6, Berlin.

SEARLE, J. R. (1970). *Speech Acts.* Cambridge University Press, Cambridge.

SEBEOK, T. A. (1967) (ed.). *Current Trends in Linguistics II. Linguistics in East Asia and South East Asia.* Mouton, The Hague & Paris.

SEMEDO, ALVAREZ (1642). *Imperio de la China.* Madrid; reprint Oriente, Macao, 1956.

SENECA (1958). *Moral Essays,* vol. 2 *Ad Marciam de consolatione,* ed. John W. Basore. Loeb Classical Library, Cambridge, Mass.

SERJEANTSON, M. S. (1935). *A History of Foreign Words in English.* Routledge, K. Paul, London.

SERRUYS, P. L. M. (1943). 'Philologie et linguistique dans les études sinologiques', *MS,* **8,** 167–219.

SERRUYS, P. L. M. (1953). 'Une nouvelle grammaire du chinois littéraire', *HJAS,* **16,** 162–99.

SERRUYS, P. L. M. (1959). *The Chinese Dialects According to Han Time.* University of California Press, Berkeley, 1959.

SHERWOOD WILLIAN OF (1941). *Syncategoremata.* Ed. J. R. O'Donell, *Medieval Studies,* **3,** 46–93.

SHOPEN, TIMOTHY (1985). *Language Typology and Syntactic Description.* 3 vols., Cambridge University Press, Cambridge. (vol. 1: Clause structure; vol. 2: Complex constructions; vol. 3: Grammatical categories and the lexicon).

SHRYOCK, JOHN, K. (1937). *The Study of Human Abilities: the Jen wu chin of Liu Shao.* American Oriental Series, vol. 13. American Oriental Society, New Haven, Conn.

SIDNEY, SIR PHILIP (1868). *An Apologie for Poetrie.* E. Arbeir, London.

SIMON, WALTER (1924). 'Das erste etymologische Wörterbuch der chinesischen Sprache', *Deutsche Literaturzeitung,* Neue Serie, **1,** 1905–10.

SIMON, WALTER (1934). 'Die Bedeutung der Finalpartikel *i*', *Mitteilungen des Seminars für Orientalische Sprachen,* **37,** 143–68.

SIMON, WALTER (1937). 'Has the Chinese Language parts of speech?', *Transactions of the Philological Society,* London, 99–119.

SIMON, WALTER (1954). 'The functions and meaning of *erh*', *AM,* 1952, **2,** 179–202; 1953, **3,** 7–18 & 117–31; 1954, **4,** 20–35.

SIMPLIKIOS (SIMPLICIUS) (1882). *In Aristotelis Categorias commentarium.* In C. Kalbfleisch (ed.), *Commentarii in Aristotelem Graeci,* vol. 8. G. Reimer, Berlin 1882ff.

SIVIN, NATHAN (1966). 'Chinese conceptions of time', *Earlham Review,* **1,** 82–92.

SIVIN, NATHAN (1968). *Chinese Alchemy. Preliminary Studies.* Harvard University Press, Cambridge, Mass.

SIVIN, NATHAN (1969). *Cosmos and Computation in Early Chinese Mathematical Astronomy.* Brill, Leiden.

SIVIN, NATHAN (1982). 'Why the Scientific Revolution did not Take Place in China – or didn't it?', *Chinese Science,* **5,** 48.

SMITH, H. (1980). 'Western and Comparative Perspectives on Truth', *Philosophy East and West,* **30,** 425–37.

SOLNTSEVA, N. V. (1967). 'Chinese language study'. In USSR Academy of Science Institute of the Peoples of Asia, *Fifty Years of Soviet Oriental Studies,* Nauka, Moscow, pp. 3–28.

SOLNTSEVA, N. V. (1985). *Problemy tipologii isolirujushtchich jazykov* (Problems of the typology of isolating languages). Izdatel'stvo Nauka, Moscow.

SOLOMON, BERNARD S. (1969). 'The Assumptions of Hui-tzû', *MS,* **28,** 1–40.

SOMMERSTEIN, A. H. (1982). Aristophanes' *Clouds,* edited with translation and notes. Aris and Phillips, Warminster.

SOYMIÉ, MICHEL (1954). 'L'Entrevue de Confucius et de Hsiang T'o', *JA,* **242,** 311–91.

SPENCE, J. D. (1986). *The Memory Palace of Matteo Ricci.* Penguin, Harmondsworth.

SPILLETT, H. W. (1975). *A Catalogue of Scriptures in the Languages of China and the Republic of China*. British and Foreign Bible Society, London.

SPIZELIUS, THEOPHILUS (1660). *De Re litteraria Sinensium Commentarius*. Goebel, Augsburg.

STAAL, FRITS (1958). 'Means of Formalization of Indian and Western Thought'. In *Proceedings of the 12th International Congress of Philosophy*, Venice.

STAAL, FRITS (1962). 'Contraposition in Indian Logic'. In *Proceedings of the 1960 International Congress for Logic, Methodology and Philosophy of Science*, Stanford, pp. 634–49.

STAAL, FRITS (1965). 'Reification, Quotation and Nominalization'. In *Contributions in Logic and Methodology in Honor of I. M. Bochenski*. North Holland Publ. Company, Amsterdam, pp. 151–87.

STAAL, FRITS (1972). *A Reader on the Sanskrit Grammarians*. MIT Press, Cambridge, Mass.

STAAL, FRITS (1973). 'The Concept of Pakṣa in Indian Logic', *Journal of Indian Philosophy*, **2**, 2, 156–66.

STEINTHAL, H. (1860). *Charakteristik der hauptsächlichen Typen des menschlichen Sprachbaues*. F. Dummler, Berlin.

STEINTHAL, H. (1890). *Geschichte der Sprachwissenschaft bei den Griechen und Römern mit besonderer Rücksicht auf die Logik*. F. Dummler, 2nd ed., Berlin.

STRANDENAES, T. H. (1987). *Principles of Chinese Bible Translation*, Coniectanea Biblica, New Testament Series 19, Uppsala.

STRAWSON, P. F. (1952). *Introduction to Logical Theory*. Methuen, London.

SUGIURA, SADAJIRO & SINGER, EDGAR ARTHUR (1900). *Hindu Logic as Preserved in China and Japan*. Philadelphia.

SUNG, Z. D. (1969). *The Text of I Ching*. Paragon Reprint Corp., New York.

SZCZESNIAK, B. (1947). 'The Beginnings of Chinese Lexicography with Particular Reference to the Work of Michael Boym (1612–1659)', *JAOS*, **67**, 4.

TACHIKAWA, MUSASHI (1971). 'A Sixth-century Manual of Indian Logic', *Journal of Indian Philosophy*, **1**, 111–45.

TAI, T'UNG (1954). *The Six Scripts; Or The Principles of Chinese Writing*. Trans. by L. C. Hopkins. (1st pub. 1881). Cambridge University Press, Cambridge.

TAKEMURA, N. (1969). 'A Study of the Chinese Commentary on the Nyāyapraveśa by Wen Gui', *The Journal of the Ryukoku University*, **389** & **390**, 83–96.

T'ANG YUNG-T'UNG (1947). 'Wang Pi's New Interpretation of the *I Ching* and the *Lun Yü*', trans. by Walther Liebenthal, *HJAS*, **10**, 124–61.

TARSKI, A. (1935). 'Der Wahrheitsbegriff in den formalisierten Sprachen', *Studia Philosophica*, **1**, 261–405.

TCHANG, TCHENG-MING (1927). *L'écriture chinoise et le geste humain. Essai sur la formation de l'écriture chinoise*. T'ou-se-we, Shanghai.

TCHANG, TCHENG-MING (1937). *Le parallélisme dans les vers du Cheu King*. Editions Herakles, Paris & Shanghai.

TCHEN, TING-MING (1938). *Etude phonétique des particules de la langue chinoise*. Editions Herakles, Paris.

TENG SSU-YÜ (1968). *Family Instructions for the Yen clan. Yen-shih chia-hsün. Yen Chih-t'ai*. Brill, Leiden.

THERN, K. L. (1966). *Postface of the Shuo-wen Chieh-tzu. The First Comprehensive Chinese Dictionary*. Wisconsin China Series No. 1, Wisconsin.

THOMAS, F. W. & GILES, L. (1948). 'A Tibetan-Chinese Word-and-word Book', *BSOAS*, **12**, 753–68.

THRAX, DIONYSIOS. *Technē grammatikē*. (c. –100). Ed. Gustavus Uhlig, Olms, Hildesheim, 1979.

TJAN, TJOE SOM (1952). Tr. *Po Hu T'ung, the Comprehensive Discussions in the White Tiger Hall*. Brill, Leiden, 1949–52.

TŌDŌ, AKIYASU (1964). 'Development of Mandarin from the 14th Century to the 19th Century', *ACTAS*, **6**, 31–40.

TUBIANSKI, M. (1926). 'On the Authorship of the Nyāyapraveśa', *Bulletin de l'Académie des Sciences de l'URSS*.

TUCCI, GIUSEPPE (1928a). 'On the Fragments from Dinnāga', *JRAS*, 377–90.

TUCCI, GIUSEPPE (1928b). 'Is the Nyayapravesa by Dinnāga?' *JARS*, 7–15.

TUCCI, GIUSEPPE (1929a). *Pre-Dinnaga Buddhist Texts on Logic from Chinese Sources*, Oriental Institute, Baroda.

TUCCI, GIUSEPPE (1929b). 'Buddhist Logic Before Dinnāga', *JRAS*, 451–88.

TUCCI, GIUSEPPE (1930). *Nyāyamukha of Dignāga, After Chinese and Tibetan Materials*. Materialien zur Kunde des Buddhismus, Heft 15, Heidelberg.

TURNER, E. G. (1971). *Greek Manuscripts of the Ancient World*. Oxford University Press, Oxford.

TWITCHETT, DENIS, C. (1961). 'Chinese Biographical Writing.' In Beasley & Pulleyblank (1961), pp. 95–114.

TWITCHETT, DENIS, C. (1962). 'Problems of Chinese Biography'. In Wright & Twitchett (1962), pp. 24–42.

TYLER, STEPHEN (1978). *The Said and the Unsaid*. Academic Press, New York.

UHLE, F. M. (1880). *Die Partikel wei im Shu-King und Schi-King. Ein Beitrag zur Grammatik des vorklassischen Chinesisch*. Alexander Edelmann, Leipzig.

UHLENBECK, E. M. (1960). 'The Study of So-Called Exotic Languages and General Linguistics', *Lingua*, **9**, 417–30.

ULVING, TOR (1969). 'Indo-European Elements in Chinese?', *AN*, 1968/9, 943–59.

UNGER, U. (1957). 'Die Negationen im Shi-king – Ein Beitrag zur Erforschung des vorklassischen Chinesisch'. PhD thesis, Leipzig.

UNO, SEIICHI (1965). 'Some Observations on Ancient Chinese Logic', *Philosophical Studies of Japan*, **6**, 31–42.

VANDERMEERSCH, LÉON (1986). 'L'écriture partagée'. In L. Vandermeersch ed., *Le nouveau Monde sinisé*. Paris, pp. 127–51.

VANDERMEERSCH, LÉON (1988). 'Ecriture et divination en Chine'. In Anne-Marie Christin ed., *Espaces de la lecture. Actes du colloque de la Bibliothéque publique d'information et du Centre d'étude de l'écriture, université Paris VII*. Paris, pp. 66–73.

VARO, FRANCISCO (1703). *Arte de la lengua mandarina*. Mission Serafica de China, Canton.

VARRO, MARCUS TERENTIUS (1985). *De Lingua Latina*. Les belles Lettres, Paris.

VIDYABHUSANA, S. C. (1971). *A History of Indian Logic: Ancient, Medieval and Modern Schools*. Banarsidass, Delhi 1971. (1st ed. 1921).

VIDYABHUSANA, S. C. (1981). *The Nyāya Sūtras of Gotoma*. Motilal Banarsidass, Delhi (1st ed. 1921).

VOCHALA, J. (1964). 'A Contribution to the Problem of Delimiting Grammatical and Lexical Meanings of Elementary Linguistic Units in Chinese', *ARO*, **32**, 403–27.

VOGEL, CLAUS (1975). 'Indian Lexicography'. In *A History of Indian Literature*, ed. Jan Gonda, vol. 5, fasc. 4, pp. 303–401. Harassovitz, Wiesbaden.

WAGNER, RUDOLF, G. (1980). 'Interlocking Parallel Style: Laozi and Wang Bi', *Asiatische Studien*, **34**, no. 1, 18–59.

WALEY, ARTHUR (1938). *The Analects of Confucius*. Allen and Unwin, London.

WALEY, ARTHUR (1952). *The Real Tripitaka*. Allen and Unwin, London.

WATSON, B. (1958). *Ssu-ma Ch'ien, Grand Historian of China*. Columbia University Press, New York.

WATSON, B. (1963). *Hsün Tzu, Basic Writings*. Columbia University Press, New York.

WATSON, B. (1964). *Chuang Tzu, Basic Writings*. Columbia University Press, New York.

WATSON, B. (1967). (Tr.) *Basic Writings of Mo Tzu, Hsün Tzu and Han Fei Tzu*. Columbia University Press, New York & London.

WATTERS, T. (1889). *Essays on the Chinese Language*. Presbyterian Mission Press, Shanghai.

WEBB, JOHN (1669). *An Historical Essay Endeavouring a Probability That the Language of the Empire of China is the Primitive Language*. N. Brook, London. 2nd ed.: *The Antiquity of China, or an Historical Essay Endeavouring a Probability That the Language of the Empire of China is the Primitive Language Spoken Through the Whole World Before the Confusion of Babel*. R. Harford, London, 1678.

WEI, TAT (1973). *Ch'eng Wei Shih Lun. The Doctrine of Mere Consciousness by Tripitaka-Master Hsüan Tsang, English translation by Wei Tat*. Ch'eng Wei Shih Lun Publication Committee, Hong Kong.

WEIDMANN, H. (c. 1970). *Der Kausalsatz im klassischen Chinesisch, Untersuchungen zu Vorkommen und Funktion der grammatischen Formen zum Ausdruck kausaler Beziehungen in den Texten des Meng-tzu und Mo-Tzu. Ein Beitrag zur Grammatik des klassischen Chinesisch*. J. G. Bläschke Verlag, Darmstadt, no date.

WEINSTEIN, S. (1959). 'A Biographical Study of Tz'u-ên', *MN*, **15**, 119–49.

WELCH, HOLMES (1967). *The Practice of Chinese Buddhism, 1900–1950*. Harvard University Press, Cambridge, Mass.

WELCH, HOLMES (1968). *The Buddhist Revival in China*. Harvard University Press, Cambridge, Mass.

WELLER, F. (1930). 'Ein indisches Fremdwort im China des vierten vorchristlichen Jahrhunderts?', *AM*, **6**, 76–83.

WHITE, H. W. (1920). 'Chinese and Sumerian', *New China Review*, **2**, 37–43.

WHITEHEAD, A. N. (1929). *Process and Reality*. Cambridge University Press, Cambridge.

WIDMAIER, RITA (1983). *Die Rolle der chinesischen Schrift in Leibniz' Zeichentheorie*. Franz Steiner Verlag, Wiesbaden.

WIEGER, L. (1927). *Chinese Characters, their Origin, Etymology, History, Classification and Signification. A Thorough Study from Chinese Documents, translated into English by L. Davrout*. Dover Books, New York, 1961 (1st ed. Sienhsien, 1927).

WIERZBICKA, ANNA (1980). *Lingua Mentalis: The Semantics of Natural Language*. Academic Press, Sydney.

WIERZBICKA, ANNA (1985). *Lexicography and Conceptual Analysis*. Karoma, Ann Arbor.

WILD, N. (1946). 'Materials for the Study of the Ssu I Kuan', *BLSOAS*, 1943–46, **11**, 640.

WILHELM, H. (1930). 'Der Zusammenhang zwischen Schrift und Kultur in China', *Archiv für Schreib- und Buchwesen*, **4**, 161–64.

WILHELM, H. (1947). 'Schriften und Fragmente zur Entwicklung der staatsrechtlichen Theorie in der Chou-Zeit.' *MS*, **12**, 41ff.

WILHELM, R. (1928). *Frühling und Herbst des Lü Bv We* (the *Lü Shih Chhun Chhin*). Diederichs, Jena.

WILKINS, JOHN (1641). *Mercury, or the Secret and Swift Messenger*. Foundations of Semiotics vol. 6, J. Benjamin, Amsterdam.

WILKINS, JOHN (1668). *Essay towards a Real Character and a Philosophical Language*. S. Gellibrand, London.

WODEHOUSE, P. G. (1986). *Ukridge*. Penguin, Harmondsworth.

WRIGHT, A. F. (1947). Review of Petrov, '*Wang Pi, his place in the history of Chinese philosophy*', *HJAS*, **10**, 75–88.

WRIGHT, A. F. (1953). 'The Chinese Language and Foreign Ideas'. In A. F. Wright. *Studies in Chinese Thought*, ed. University of Chicago Press, Chicago, pp. 286ff.

WRIGHT, ARTHUR F. & TWITCHETT, DENIS C. (1962). (eds.) *Confucian Personalities*. Stanford University Press, Stanford, California.

WU, KUANG–MING (1987). 'Conterfactuals, Universals and Chinese Thinking', *PEW*, **37**, 84–94.

YAKHONTOV, S. E. (1965). *Drevnekitajskij jazyk*. (The Ancient Chinese Language). Nauka, Moscow.

YAKHONTOV, S. E. (1968). 'Ponjatie chastej rechi v obshchem i kitajskom jazykoznanii' (The concept of the parts of speech in general and Chinese linguistics). In V. M. Zhirmunskij & O. P. Sunik (eds.), *Voprosy teorii chastej rechi po materiale jazykov razlichnykh tipov*. Nauka, Moscow, pp. 70–9.

YANG, PAUL FU-MIEN (1960). Review of J. H. Prémare, *Notitia linguae Sinicae, Orbis*, **9**, 165–67.
YANG, PAUL FU-MIEN (1974). *Chinese Linguistics: A Selected and Classified Bibliography*. Chinese University Press, Hong Kong.
YANG, PAUL FU-MIEN (1985). *Chinese Lexicology and Lexicography, A Selected and Classified Bibliography*. Chinese University Press, Hong Kong.
YASKA (1967). (*c.* −5th century) *The Nighantu and the Nirukta*. Motilal Banarsidass, New Delhi.
YATES, FRANCES, A. (1974). *The Art of Memory*. University of Chicago Press, Chicago.
YEN, ISABELLA YI-YUN (1956). 'A Grammatical Analysis of Syau Jing'. PHD thesis, Cornell.
YEN, S. L. (1971). 'On Negation with *fei* in Classical Chinese', *JAOS*, 409–17.
YEN, S. L. (1978). 'On The Negative *wei* in Classical Chinese', *JAOS*, 469–81.
YULE, HENRY (1916). *China and the Way Thither: Being a Collection of Medieval Notices of China*. 1866, ed. and rev. by Henri Cordier. Hakluyt Society, London, 1913–16.

ZACH, ERWIN VON (1939). *Yang Hsiungs Fa-yen*, (*Worte strenger Ermahnung*): Druckerei Tong Ah, Batavia. Repr. San Fransaco: Chinese Materials Center, 1976.
ZACHARIAE, THEODOR (1897). *Die indischen Wörterbücher*. Trüner, Strassburg.
ZOGRAF, I. T. (1962). *Ocherk grammatiki srednekitajskogo yazyka*. Vostochnaya Literatura, Moscow.
ZOGRAF, I. T. (1972). Russian ('Outline of the Grammar of Medieval Chinese'). In *Byan'ven' o vozdayanii za milosti*. Vol. 2, Nauka, Moscow.
ZOGRAF, I. T. (1979). *Srednekitajskij yazyk, stanovlenie i tendencii razvitija*. (The Medieval Chinese Language, Origins and Tendencies of Development) Nauka, Moscow.
ZOGRAF, I. T. (1984). *Mongol'sko-kitajskaja interferencija. Jazyk Mongol'skoj kanceljarii v Kitae*. (Mutual Influence of Mongol and Chinese. The Language of the Mongol Chancellery in China.) Moscow: Izdatel'stvo Nauka.
ZÜRCHER, E. (1984). 'A New Look at the Earliest Chinese Buddhist Texts'. Cyclostyled.

GENERAL INDEX

Note: The alphabetical arrangement is letter by letter. In the arrangement of Chinese words, chh- and hs- follow normal sequence: *û* and *ü* are treated as *u*; *ë* and *ê* as *e*.

458

夏	HSIA kingdom (legendary?)			c. −2000 to c. −1520
商	SHANG (YIN) kingdom			c. −1520 to c. −1030
周	CHOU dynasty (Feudal Age)		⎧ Early Chou period	c. −1030 to −722
			⎨ Chhun Chhiu period 春秋	−722 to −480
			⎨ Warring States (Chan	−480 to −221
			⎩ Kuo) period 戰國	
First Unification	秦	CHHIN dynasty		−221 to −207
漢 HAN dynasty		⎧ Chhien Han (Earlier or Western)		−202 to +9
		⎨ Hsin interregnum		+9 to +23
		⎩ Hou Han (Later or Eastern)		+25 to +220
	三國	SAN KUO (Three Kingdoms period)		+221 to +265
First	蜀	SHU (HAN)		+221 to +264
Partition	魏	WEI		+220 to +265
	吳	WU		+222 to +280
Second	晉	CHIN dynasty: Western		+265 to +317
Unification		Eastern		+317 to +420
	劉宋	(Liu) SUNG dynasty		+420 to +479
Second		Northern and Southern Dynasties (Nan Pei chhao)		
Partition	齊	CHHI dynasty		+479 to +502
	梁	LIANG dynasty		+502 to +557
	陳	CHHEN dynasty		+557 to +587
	魏	⎧ Northern (Thopa) WEI dynasty		+386 to +535
		⎨ Western (Thopa) WEI dynasty		+535 to +554
		⎩ Eastern (Thopa) WEI dynasty		+534 to +543
	北齊	Northern CHHI dynasty		+550 to +577
	北周	Northern CHOU (Hsienpi) dynasty		+557 to +581
Third	隋	SUI dynasty		+581 to +618
Unification	唐	THANG dynasty		+618 to +906
Third	五代	WU TAI (Five Dynasty period) (Later Liang,		+907 to +960
Partition		Later Thang (Turkic), Later Chin (Turkic),		
		Later Han (Turkic) and Later Chou)		
	遼	LIAO (Chhitan Tartar) dynasty		+907 to +1125
		West LIAO dynasty (Qarā-Khiṭāi)		+1144 to +1211
	西夏	Hsi Hsia (Tangut Tibetan) state		+990 to +1227
Fourth	宋	Northern SUNG dynasty		+960 to +1126
Unification	宋	Southern SUNG dynasty		+1127 to +1279
	金	CHIN (Jurchen Tartar) dynasty		+1115 to +1234
	元	YUAN (Mongol) dynasty		+1260 to +1368
	明	MING dynasty		+1368 to +1644
	清	CHHING (Manchu) dynasty		+1644 to +1911
	民國	Republic		+1912

N.B. When no modifying term in brackets is given, the dynasty was purely Chinese. Where the overlapping of dynasties and independent states becomes particularly confused, the tables of Wieger (I) will be found useful. For such periods, especially the Second and Third Partitions, the best guide is Eberhard (9). During the Eastern Chin period there were no less than eighteen independent States (Hunnish, Tibetan, Hsienpi, Turkic, etc.) in the north. The term 'Liu Chhao' (Six Dynasties) is often used by historians of literature. It refers to the south and covers the period from the beginning of the +3rd to the end of the +6th centuries, including (San Kuo) Wu, Chin, (Liu) Sung, Chhi, Liang and Chhen. For all details of reigns and rulers see Moule & Yetts (I).